HANDBOOK

on the

IEE WIRING REGULATIONS
BS 7671

Revised Fifth Edition

A Handbook for Compliance

Trevor E. Marks

ARTCS, CEng, MIEE, MInstMC, FInstD

Copies can be obtained from:

William Ernest Publishing Limited
PO BOX 206
Liverpool
L69 4PZ
England

First published for the general public 1985
Reprinted: 1985, 1986, 1987
Revised Second Edition published 1988
Reprinted: 1988, 1989, 1990, 1991
Revised Third Edition published 1992
Reprinted 1992 (twice), 1993, 1994, 1995, 1996
Revised Fourth Edition published 1996
Revised Fifth Edition published 1999

William Ernest Publishing Limited

British Library Cataloguing in Publication Data

Marks, Trevor E. (Trevor Ernest), *1931–*
    Handbook on the IEE wiring regulations BS 7671. – Rev. 5th ed.
    Part I and Part II.
    621. 3 ' 1924

ISBN   0 - 9510156 - 6 - 4

DISCLAIMER
Although the contents have been checked, because of the large quantity of calculations involved no
responsibility for loss to any person, company or organisation for whatever reason will be accepted
by the author or publishers; this book is provided for guidance only.

Typset by T. E. Marks
Printed and bound in Great Britain by Phase Print Ltd, Nottingham

Book Contents

Preface to Fifth Edition

Two and a half years have past since the last update of this Handbook was published. Since then there have been numerous changes to British, European and international standards (BS, CENELEC and IEC) including a second amendment (AMD 9781) to BS 7671: 1992: Requirements for Electrical Installations (formerly the IEE Wiring Regulations). This fifth revision of the Handbook takes account of changes that have occurred over this period.

Part 2, which was completely rewritten and expanded for the fourth edition, continues to provide near-comprehensive guidance material for both teachers and students involved with the City and Guilds installation and BTEC courses for electrical installation work. Additionally, electrical installation designers will also find this Part invaluable in their endeavours to produce safe, cost-effective designs. As before, Part 2 embodies guidance on the selection of protective devices and gives advice for the determination of k factors for various conductor operating temperatures. Determination of fault current in parallel conductors is also addressed.

The chapter sequence remains in a logical manner which follows the chronological succession one might adopt when designing an electrical installation. The reference numbering of formulae continues to make identification and correlation easier to use when referring to texts and formulae. Although the many worked examples have been 'worked' to two or more decimal places to enable readers to follow the mathematical logic, in practice such resolution would not always to be of realistic application.

As before, Part 3 gives data for limits on earth fault loop impedance, Z_s, for circuit-breakers manufactured to meet the product specifications of BS 3871 and BS EN 60898. It should be noted however that BS 3871 has now been withdrawn, though the data relating to such overcurrent protective devices has been retained because of its usefulness, for checking existing installations. Part 4 remains unchanged providing as it does much data useful to the electrical installation designer.

As with every other revision, the opportunity has been taken to amend certain texts and figures to provide to even greater clarity in the guidance contained in this Handbook. Identification of the various chapters by colour has been retained for easy use by the student, teacher and designer alike. I sincerely hope that readers will find the additional information contained in this book useful.

T. E. M. – July 1998

Introduction

The original philosophy of the layout and function of the book has not changed and, apart from minor changes caused by changes to the regulations, this introduction is the same as in the first edition of the book.

This handbook has been produced in small format so that it can be carried in the pocket or toolbox. It is not intended as a substitute for the Wiring Regulations issued by the Institution of Electrical Engineers and the British Standards Institution, but to provide a pocket sized reference book. Used in this context, the handbook can be used by electricians, technicians, students, and engineers.

The handbook includes only those regulations which can be expected to be in use every day. Bearing this in mind, if the exact wording of a regulation is required, then the latest Edition of the Wiring Regulations should be consulted. To assist in this respect, the regulation numbers have been placed in the left hand margin below the text to which it refers.

The handbook is in four parts. Part One gives the regulations under subject headings, in an easily understandable language, and to obviate having to look in different sections for applicable regulations, all the appropriate regulations are included under the subject title. Certain regulations are therefore repeated under different subject headings, because they are associated with the subject concerned: for example, reference voltage drop, Regulation 525-01, will be found under each of the subject headings of Selection of equipment, Diversity, Selection of cables, Separated extra-low voltage, and Functional extra-low voltage.

Part Two gives explanations of the regulations, with numerous examples of the calculations required, and should therefore be of particular help to the electrician, teacher, lecturer, student and engineer.

Part Three comprises installation tables, with explanations and examples of how to use them, eliminating the need to carry out complicated calculations for small contracts, the calculations being confined to addition, subtraction, multiplication and division. The aim is to assist the contractor and electrician, although engineers may find them useful for checking final circuits, or seeing the effect grouping has on cable size, without having to do the calculations. They will also find the earth loop impedance tables to be useful when testing an installation.

Part Four contains tables and protective device characteristics giving information which is needed when designing an installation, and has been provided to assist the contractor and engineer.

The opinions and interpretations expressed in the handbook are the author's own, and do not purport to represent the official view of the Institution of Electrical Engineers.

T. E. M. July 1998

Preface to earlier editions

FIRST EDITION

The first copy of this handbook was produced to be used on the William Ernest Ltd. Fifteenth Edition Wiring Regulations training course for supervisors and electricians, and permission was given to Culham Laboratories to print 50 copies of the Handbook for use by their electricians.

They have subsequently been given permission to print 50 copies of this revised edition.

The amendments to the Fifteenth Edition published in May 1984 by the Institution of Electrical Engineers generally affected the text of the Handbook, and, in particular, the installation tables.

The Handbook has therefore been revised to take into consideration the May 1984 Amendments, and the whole of Part Three Installation Tables has been rewritten.

Revising the handbook has also enabled improvements to be made, one such improvement being to rearrange the table in Part Three, by placing tables of design data at the rear of the installation tables.

T. E. M - July 1984

SECOND EDITION

It is now some three years since the handbook was first published, during which time the regulations have been amended, the amendments of 1987 affecting the handbook most, particularly the tables giving the cable sizes for grouped cables and circuits.

It was for this reason, and the fact that the 16th Edition will not be published for at least two years, that it was decided to publish a revised edition, advantage being taken of the revision to include more information.

As before, each part of the book has been printed on coloured paper for easy identification. Part Four contains tables required when designing an installation, such as the CRG tables which give the current-carrying capacity required for conductors for all the various types of grouping for protective devices up to 250 amperes. This should prove useful to designers, since it enables them to see the effect of grouping cables without having to do any calculations. Part Four also contains a table of sines and cosines for those who want to take advantage of the calculations for determining the voltage drop in a conductor when the load has a power factor.

Part Three follows the same format as that used in the first edition, but has been completely rewritten. Again, the designer as well as the contractor and electrician may find the CSG tables helpful, since they give the actual size of cable required for various types of grouping; thus a quick comparison can be made on cable sizes with different grouping arrangements. As before, a chapter with worked examples is given to illustrate how the tables should be used, and where possible, instructions have been included with the tables.

Bearing in mind that the handbook is used by colleges and students as well as by contractors and engineers, Part Two has been completely rewritten, and gives more detail and worked examples, which show how to use the various formulae given in the regulations, as well as how to carry out the necessary calculations.

Part One is basically the same as the first edition by following the same format, but revised to bring it up to date with the amendments. It still lists the regulations in an understandable language under subject titles, but the subjects have now been put into alphabetical order, for although a comprehensive 'Contents - Index' has been provided, it was felt that the new arrangement would assist in locating a particular subject.

I have received many letters and phone calls from readers advising me of how useful they found the first edition; my only hope is that they find this edition even more useful.

T.E.M. - March 1988

THIRD EDITION

It is now some three years since the handbook was last revised. During this period there have been many changes to British and IEC Standards; in addition, the IEE Wiring Regulations have again been amended by the introduction of the 16th Edition. These changes have necessitated the publication of a new edition of the book.

Opportunity has been taken to include more information whilst revising the book which now contains 80 pages more than the first edition.

In many of the articles I wrote for the technical press, I included tables giving the maximum length of conductors to comply with the regulations; these have formed the basis of the 'INST' tables, which give the maximum length that cable from 1 mm^2 up to 16 mm^2 can be sized to comply with the regulations for different sized loads.

A further improvement to earlier editions is the re-arrangement of the tables that give the impedance of the armouring of armoured cables, so that they give the phase earth loop impedance at the design temperature, as well as the impedance of the cable armour.

Additionally, the table giving the resistance, reactance, and impedance of XLPE cables has been expanded to cover all sizes from 1 mm^2 up to 300 mm^2.

The table provided to save calculating the sine of the power factor angle (for voltage drop calculations) has been changed; originally it gave the conversion from cosines to sines; it now gives the conversion of power factor to sine, starting at 0.1 power factor, and finishing at 0.99.

A table of temperature conversion factors has also been included in part Four, and more MCB and MCCB characteristics have been included.

I have felt, for some time now, that the list of symbols and definitions were in the wrong part of the book; accordingly, they will now be found at the beginning of Part One.

The book has again been printed on coloured paper, for easy identification of each part. As before, the additional tables that are required by the engineer and contractor, when designing an installation, have been placed in Part Four.

Part Three has also been completely rewritten to bring it up-to-date. The tables giving the current-carrying capacity of cables have been rearranged, so that tables for similar types of cables are together: for instance, all armoured cables are grouped in one location. As before, tables giving the size of conductor required when grouped with other cables have been provided, since these assist the electrician and contractor, and enable the engineer and consultant to carry out spot checks when inspecting projects. Part Three also contains the tables giving the maximum length of a circuit with a given size of cable, to comply with the regulations. Tables are provided for 0.4 and 5 seconds disconnection times, as well as tables for socket outlet circuits, and voltage drop in lighting circuits.

Since the handbook is used by many colleges and teachers for the City and Guilds installation and BTEC electrical installation students, as well as by engineers and electricians, Part Two has been completely revised, and now includes chapters on the assessment of the installation, as well as selecting protective devices; as before, numerous examples are given of how to carry out the calculations.

Part One follows the same format as that used in previous editions, but has been completely revised in line with the 16th Edition of the IEE Wiring Regulations. As before, it still lists the regulations in an understandable language, under subject titles.

T.E.M. - December 1991

FOURTH EDITION

It is now over four years since the Handbook was last revised. Part One has been improved by the inclusion of more diagrams.

Part Two has been completely rewritten to give more information for both the students and teachers of City and Guilds installation and BTEC electrical installation courses as well as the installation designer. Additional material has been introduced concerning selecting protective devices, determination of the k factor for different conductor operating temperatures and fault current in parallel cables.

Part Three now gives the Z_s values up to 100A for all the BS 3871 and BS EN 60898 types of circuit breaker.

T.E.M. - July 1996

Acknowledgments

This handbook includes material reproduced by kind permission of the IEE from the IEE Wiring Regulations BS 7671. Such material may not be copied or reproduced without the permission in writing from the IEE. The IEE Regulations are amended and re-issued from time to time, and it is desirable that reference should always be made to the latest edition.

Extracts from IEC Standards have been used with the permission of the International Electrotechnical Commission, which retain the copyright.

Extracts from British Standards are quoted by kind permission of the British Standards Institution.

I would also like to thank the following companies for the information and help they have provided to enable me to construct the many tables and characteristics in this book:

GEC Alsthom Low Voltage Equipment Ltd.,

Wylex Ltd.,

BICC Pyrotenax Ltd.,

BICC Cables Ltd.,

Crabtree Electrical Industries Ltd.,

Delta Crompton Cables Ltd. (Formerly Aluminium Wire and Cable Co. Ltd.),

Walsall Conduits Ltd.,

City Electrical Factors Ltd.,

Merlin Gerin Ltd.,

the many other equipment and cable manufacturers for providing the information to enable me to construct the design tables in Part Four.

Finally, I would like to thank several people, even though they wish to remain anonymous, for reading the draft copy of the book, for the helpful suggestions they made, and for checking tables. They at least will know who I am referring to.

T. E. M. - July 1998

PART ONE

REGULATIONS UNDER SUBJECT TITLES

Contents -index

Chapter 1

Symbols

Introduction

Most of the following symbols will be found in the BS 7671. Others have been added so that there is less chance of making a mistake when carrying out calculations. For instance, I_p represents a three phase symmetrical fault, I_{pn} represents a phase to neutral fault, and I_{pp} a phase to phase fault. Using this notation keeps track of the type of fault being considered.

Conductors

PEN   Protective Earthed Neutral.

Cc   Circuit Protective Conductor.

S   Cross-sectional area of live conductor.

S_p Cross-sectional area of protective conductor.

t_f Final operating temperature of a conductor in °C.

t_p Maximum permitted operating temperature of a conductor in °C.

t_i Initial temperature of conductor °C.

t_L limit temperature of conductor's insulation °C.

Current

I_a Current causing automatic operation of a protective device in a specified time.

I_b Design or load current.

I_n Nominal rating or current setting of protective device.

I_z Current-carrying capacity of the circuit conductor.

I_2 Current causing the effective operation of an overload device (with overload current).

I_f Earth fault current = $U_{oc} + Z_S$.

I_p Prospective three-phase symmetrical short circuit current.

I_{pp} Prospective short-circuit current between two phases.

I_{pn} Prospective short-circuit current between a phase and neutral conductor.

I_t The minimum current-carrying capacity required for a conductor.

I_{tab} The actual tabulated current-carrying capacity for a conductor given in the tables.

$I_{\Delta n}$ Rated residual operating current of a residual current device (RCD).

mA   Milliamp (0.001 A).

Devices

gG   General purpose fuse having the full range breaking capacity and also suitable for motor protection.

gM   A full range breaking capacity fuse suitable only for motor protection.

MCB Miniature-circuit-breaker.

MCCB Moulded case circuit breaker.

RCD Residual current device.

RCCB Residual current circuit breaker.

Factors (correction)

G   Factor for grouping (C_g).

A   Factor for ambient temperature (C_a).

T   Factor for contact with thermal insulation (Ci).

S   Factor for protective device not having a fusing factor of 1.45 , (C4).

C_t Factor for correcting resistance for operating temperature of conductor.

Impedance and resistance

$R_1(Z_1)$ Resistance (impedance) of phase conductor from the origin to the end of the circuit.

$R_2(Z_2)$ Resistance (impedance) of the protective conductor from the origin to the end of the circuit.

R_A The sum of the resistance of the earth electrode and of the protective conductors connecting it to exposed-conductive-parts.

R_e The sum of resistances of earth electrodes.

Z_e Impedance of the phase/earth loop external to the installation, or of the system up to the point under consideration.

Z_{inst} Phase earth loop impedance in the installation (i.e., of the circuit being considered).

Z_s Phase earth loop impedance from the source of energy to the end of a final circuit.

Z_p Impedance (or resistance) of one phase.

Z_{pn} Impedance (or resistance) of phase and neutral .

Z_{pp} Impedance (or resistance) of two phases.

Voltage

U_{loc} Open circuit line voltage at distribution transformer.

U_{oc} Open circuit phase voltage at distribution transformer.

U_o Nominal voltage to earth or earthed neutral.

mV   millivolts (0.001 V).

Miscellaneous

t_a Ambient temperature °C.

t   Time (e.g. disconnection time of protective device).

k   Maximum thermal current capacity of a conductor.

IP   Index of Protection (see BS EN 60529).

Mathematical

<   Less than.

≤   Less than or equal to.

>   More than.

≥   More than or equal to.

≃   Approximately equal to.

Multiples and submultiples of units

10^{12}	tera	T	1 000 000 000 000	10^{-1}	deci	d	0.1
10^9	giga	G	1 000 000 000	10^{-2}	centi	c	0.01
10^6	mega	M	1 000 000	10^{-3}	milli	m	0.001
10^3	kilo	k	1 000	10^{-6}	micro	μ	0.000 001
10^2	hecto	h	1 00	10^{-9}	nano	n	0.000 000 001
10^1	deca	da	10	10^{-12}	pico	p	0.000 000 000 001
10^0			1	10^{-15}	femto	f	0.000 000 000 000 001

Chapter 2

Definitions

Before delving into the regulations in detail, it is wise to consider the terminology. Learning the more important definitions will enable a better understanding of the regulations. The following definitions are the basic ones which should be learnt, but also includes terminology such as 'Energy let-through', with which the plant engineer, electrician and electrical contractor, are less likely to be familiar.

AMBIENT TEMPERATURE.

The normal temperature of the air or other medium in the location where equipment is to be used.

ARM'S REACH.

The limit to which a person can reach without assistance, within a zone of accessibility, as outlined in the following diagram.

$$A = 1.25 \text{ m} \quad B = 2.5 \text{ m} \quad C = 0.75 \text{ m}$$

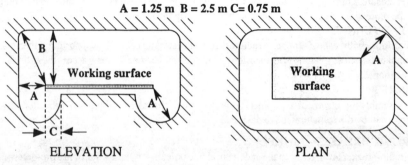

ELEVATION   PLAN

Fig 2.1 - Arm's reach

AUTOMATIC DISCONNECTION.

The protective device operates automatically when a fault occurs in the circuit it protects.

BANDS:

See "Voltage Bands"

BARRIER.

A component that protects against contact with live parts from any usual direction of access.

BONDING CONDUCTOR.

A conductor used for equipotential bonding

BUILDING VOID, ACCESSIBLE.

A space within the building structure, or its components, accessible at particular points, including partition spaces and those under floors and above ceilings, and types of door, architraves and window frames designed for the purpose of incorporating spaces.

BREAKING CAPACITY.

A value of fault current that a protective device is capable of breaking under defined conditions.

CHAPTER 2

CABLE CHANNEL
An above-ground enclosure, either ventilated or closed, which may or may not form part of the building structure, of dimensions which exclude entry by persons but which permits access to conduits, cables and the like throughout the length of the channel, both during and after installation.

CABLE DUCTING
A metal or non-metallic enclosure which is for the purpose of enclosing cables installed after the ducting has been erected.

CABLE LADDER
Cable support system made up of main longitudinal members with a number of transverse supports permanently fixed to main members.

CABLE TRAY
Uncovered continuous-base (which may be perforated) cable support system with turned-up edges.

CABLE TRUNKING
An enclosure, normally of rectangular section, with one side arranged to be removable or hinged, for the protection of cables or to house electrical equipment.

CABLE TUNNEL
A corridor, with cable support structures and similar facilities for supporting other wiring components such as joints, with dimensions which permit persons to freely access the entire length of the tunnel.

CIRCUIT
An assembly of electrical equipment supplied from the same origin and protected against overcurrent by the same protective device(s).

CIRCUIT-BREAKER (LINKED)
A circuit-breaker arranged to make and break all the poles at the same time, or in a pre-determined sequence.

CIRCUIT CONDUCTOR
Any conductor in a system which is intended to carry electric current in normal conditions, or to be energised in normal conditions and includes a combined neutral and earth conductor,* but does not include a conductor solely to perform a protective function by connection to earth or other reference point as shown in Figure 2.2. *See Protective Multiple Earthing and TN-C-S System.

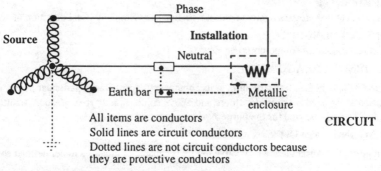

Figure 2.2 - **Illustrating conductors and circuit conductors**

CIRCUIT PROTECTIVE CONDUCTOR
A conductor connecting exposed-conductive-parts to the main earthing terminal .

CONDUCTOR
Any material in its solid, liquid or gaseous state which will conduct electricity.

CLASS I EQUIPMENT
Equipment having only basic insulation, and required to be earthed via a protective conductor.

CLASS II EQUIPMENT
Equipment that does not rely solely on the basic insulation for protection against electric shock i.e., the equipment is either all-insulated or double-insulated. The equipment does not provide for, or requires, the connection of a protective conductor. The installation may still require a protective conductor at accessories such as double insulated ceiling roses and Class II luminaries.

CLASS III EQUIPMENT
Equipment where protection against electric shock relies upon the voltage not exceeding 50 V a.c. or 120 V ripple free d.c., the electrical supply complying with the requirements for SELV.

CURRENT-CARRYING CAPACITY
The maximum current which equipment or cable can carry under specified conditions without exceeding the designed operating temperature of the equipment or cable, with a given ambient temperature.

DANGER
Means risk of injury. Some Electricity at Work Regulations mention preventing danger; to prevent danger therefore means to prevent the risk of injury.

DIRECT CONTACT
Contact by persons or live stock with normally energised phase or neutral parts that may result in electric shock.

DISCONNECTOR
Another name for an isolator.

DISTRIBUTION CIRCUIT
A band II circuit connecting an item of switchgear, controlgear or a distribution board, to which at least one final circuit is connected, with the origin.

Such a circuit may connect remote buildings and separate installations.

EARTHING
Connecting exposed-conductive-parts of an installation to the main earthing terminal of that installation.

EARTH FAULT CURRENT
Current caused by a fault of negligible impedance between a live conductor and earth, or a live conductor and an earthed exposed-conductive-part.

CHAPTER 2

EARTH FAULT LOOP IMPEDANCE

The impedance of the phase conductor and earth return from the source of energy to the point of a phase to earth fault, or the point of measuring the phase earth loop impedance including the impedance of the source. The earth return is either the actual earth (ground) in a TT and IT system, or protective conductors in a TN system.

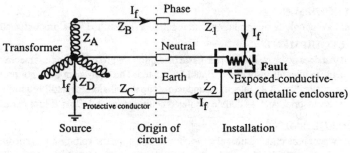

Phase earth loop $Z_S = Z_A + Z_B + Z_1 + Z_2 + Z_C + Z_D$

Figure 2.3 - Earth fault loop impedance in TN Systems

EARTHED EQUIPOTENTIAL ZONE

An area in which the exposed-and-extraneous-conductive-parts are bonded together, so that the magnitude and duration of the voltage appearing between them when an electrical fault occurs external to the installation will not lead to an electric shock as defined.

ELECTRICAL EQUIPMENT

Abbreviated to equipment in BS7671, and includes all items used in an electrical installation, including wiring systems, accessories and appliances.

ENERGY LET-THROUGH

The total energy let-through experienced by a protective device before the protective device finally interrupts the fault current flowing into the circuit concerned.

EQUIPOTENTIAL BONDING

An electrical connection between exposed-conductive-parts and extraneous-conductive-parts which puts them at approximately the same potential. See Figure 2.4.

EXPOSED-CONDUCTIVE-PART

Metalwork of electrical equipment which can be touched and which is not a live part but may become live under fault conditions.

EXTRANEOUS-CONDUCTIVE-PART

Metalwork that is not part of the electrical installation, and which is liable to introduce a potential, generally earth potential.

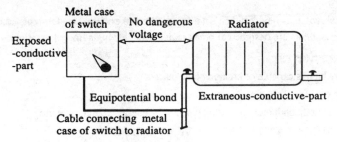

Figure 2.4- Equipotential bonding

EXTRA-LOW VOLTAGE
A voltage not exceeding 50 V a.c., or 120 V ripple free d.c between conductors, or between any conductor and earth. See definition of ripple free.

FAULT CURRENT
A current caused by a fault of negligible impedance between either live conductors or a phase conductor and earth. A fault between live conductors is referred to as a short-circuit current and between a phase conductor and earth as an earth fault.

FUNCTIONAL EXTRA-LOW VOLTAGE (FELV)
A system whose voltage does not exceed extra-low voltage but does not comply with all the regulations for SELV and PELV

FUNCTIONAL SWITCH
A device for switching 'on' or 'off' or varying the supply of energy in the course of the normal operation of an installation.

GAS INSTALLATION PIPE
Any pipe on the consumer's side of the mains service gas meter that carries gas to a gas appliance or within a gas appliance including all accessories used.

HAZARDOUS-LIVE-PART
A live part, which under certain conditions of external influence, can give an electric shock.

HIGHWAY DISTRIBUTION CIRCUIT
A band II circuit connecting street furniture or a remote highway distribution board with the origin of the installation and such a circuit connecting the highway distribution with street furniture.

INDIRECT CONTACT
Contact with exposed-conductive-parts (metalwork surrounding live parts)

ISOLATION
The cutting off of the electrical supply from all, or a section of, a circuit or installation, so that the circuit or installation is separated from every source of electrical energy to enable work to be carried out in safety.

CHAPTER 2

ISOLATOR
A mechanical switching device which is not designed to make or break load current, although certain high voltage isolators are designed to break fault current having a limited duration.

ISOLATING SWITCH
An isolator which is capable of breaking load current.

LIVE PART
A phase or neutral conductor, or part intended to be energised in normal use.

LV SWITCHGEAR AND CONTROLGEAR ASSEMBLY
A type-tested or partially type-tested assembly of LV electromechancial or electronic switching devices for the function of control, signalling, protection, measuring and regulating under the responsibility of the manufacturer complete with all mechanical, electrical and structural parts (see BS EN 60439-1)

MAKING CAPACITY
The peak fault current that a protective device is capable of making on to under defined conditions.

OVERCURRENT
A current exceeding the nominal current rating of a circuit. The nominal current rating of conductors is their current-carrying capacity.

OVERLOAD CURRENT
A current exceeding the nominal current rating of an electrically sound circuit or equipment.

PELV (PROTECTIVE EXTRA-LOW VOLTAGE)
An extra-low voltage system which, with the exception of it being connected to earth at one point only, satisfies the requirements for SELV circuits.

PES (PUBLIC ELECTRICITY SUPPLIER)
A person or company who supplies electrical energy, including the person or company owning the transmission lines and cables used to deliver the electrical energy.

PROSPECTIVE FAULT CURRENT
The current that could result from a fault of negligible impedance between live conductors having a difference in potential between them, or between a live conductor and an exposed-conductive-part, or earth.

PROTECTIVE CONDUCTOR
A conductor used as a means of protection against electric shock by connecting extraneous-conductive-parts together, or to exposed-conductive-parts, or connecting any of them to the main earth terminal.

PROTECTIVE DEVICE
In the text of this Handbook, it is a device installed in a circuit to protect the circuit by disconnecting the supply to it if there is an overload, short-circuit, or a fault to earth.

PROTECTIVE MULTIPLE EARTHING (PME)

That part of the system where the neutral conductor is also used as the earthing conductor for the system neutral thus becoming a protective earthed neutral (PEN) conductor. See Figure 2.5

REDUCED LOWVOLTAGE SYSTEM

An electrical supply in which the voltage between phases is 110 volts, and the phase to earth voltage does not exceed 63.5 volts for three-phase, or 55 volts for single-phase supplies.

RESIDUAL CURRENT

The vector sum of the instantaneous current flowing in all live conductors of a circuit at a point in the installation. If the circuit is balanced, the residual current is zero.

RESIDUAL CURRENT DEVICE

A current-operateeearth-leakage operated circuit breaker. Abbreviated to RCD.

RIPPLE FREE

An extra-low voltage d.c supply, in which the nominal voltage does not exceed 120 V and the maximum peak value of a rectified a.c. voltage does not exceed 128.5 V for a sinusoidal ripple or 140 V for a non-sinusoidal ripple.

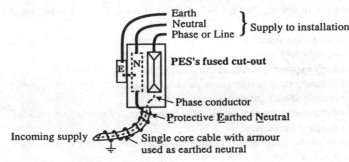

Figure 2.5 - PEN conductor

SEPARATED EXTRA LOW VOLTAGE (SELV)

A system not connected with earth or other systems and not exceeding extra-low voltage, so that a first fault will not give rise to electric shock.

CHAPTER 2

SHORT-CIRCUIT CURRENT
A current caused by a fault of negligible impedance between either phase conductors, or phase and neutral conductors.

SIMULTANEOUSLY ACCESSIBLE
Equipment or parts which can be touched at the same time by a person, or where applicable, by livestock.

SWITCH
A mechanical switching device capable of making, breaking and carrying its rated current. Where specified, it may also be capable of operating under overload conditions. It may also be capable of making, but not breaking, short-circuit and earth fault currents.

SWITCH (LINKED)
A switch arranged to make and break all the poles at the same time or in a predetermined sequence.

SYSTEM
An electrical system in which all the electrical equipment is, or may be, electrically connected to a common source of electrical energy, and includes such source and such equipment. A system comprises a single source of electrical energy and an installation.

There are five types of system, TN-S, TN-C, TN-C-S, TT and IT, these being explained and illustrated as follows:

T   The neutral of the source of energy directly connected with earth, or the installation exposed-conductive-parts are connected directly to earth.

N   Installation's exposed-conductive-parts connected to the source earth.

C   Protective conductor connecting the installation's exposed-conductive-parts to the source earth is combined with the neutral conductor.

S   Protective conductor connecting the installation's exposed-conductive-parts to the source earth is a separate conductor.

I   The source of energy is either not connected with earth, or is connected to earth through a high impedance.

TN-S.   **Separate neutral and protective conductor system.**

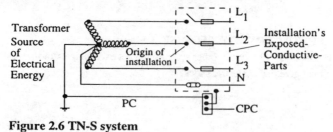

Figure 2.6 TN-S system

TN-C.   **Combined neutral and protective conductor system.**

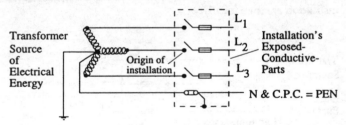

Figure 2.7 TN-C system

TN-C-S   Neutral and protective conductor combined from the source to the origin of the installation, and then separate within the installation.

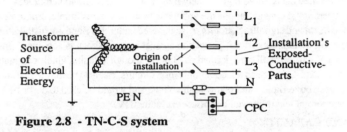

Figure 2.8 - TN-C-S system

TT.   **System earthed at source, but installation earthed locally**

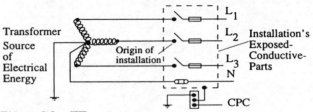

Figure 2.9 - TT system

IT.   Source either isolated from earth or earthed through a high impedance;
       insulation earthed locally.

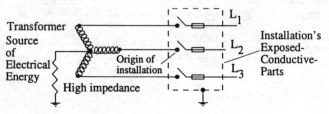

Figure 2.10 - IT system

VOLTAGE BANDS

There are two defined bands of voltage, namely Band I and Band II. Band I embraces installations in which for operational considerations the voltage is limited, such as in telecommunications and bell systems, control and signalling circuits. Also covered under this Band are installations that, by virtue of applying certain conditions, protection against electric shock is achieved. Installations operating at ELV can normally be defined as Band I, though circuits which originally exhibit voltages higher than ELV, such as telecommunication circuits, can also be defined. Band II covers all other voltages above Band I and below high voltage and would be used to describe voltages to domestic, commercial and industrial premises (excluding HV).

WITHSTAND CAPACITY

The maximum current flowing through a device that the device can withstand without being damaged, or causing a danger to arise.

Chapter 3

Assessment of the installation

Before an alteration, an extension or a new installation is started, an assessment has to be made of the characteristics of the installation. This involves determining the purpose for which the installation is required, the maximum demand of the installation, the characteristics of the supply, the distribution and circuit arrangements, external influences, compatibility with other equipment and services, and the frequency and quality of the maintenance it will receive.

Purpose of the installation
The first assessment is of the purpose for which the installation is intended. This will include such items as: the type of building structure, whether the building will be used for handicapped people; are there any hazardous areas; are corrosive substances being used?
300-01

Maximum demand
The first requirement is to work out the total electrical load that will be connected to the installation. Diversity should be taken into account when assessing the maximum demand of the installation. When assessing the maximum demand, starting currents or in-rush currents of equipment can in some instances be ignored.
311-01

Supply characteristics
The next step is to determine the number and types of live conductors. This to some extent will depend upon the purpose for which the installation is intended, and the electricity supplier.

The next requirement is to determine the type of system and earthing arrangements of which the installation will form part. Will it be TN-C, TN-S , TN-C-S, TT or IT?
312-01, 312-02, 312-03

The supply company must also be asked for the supply characteristics, to enable the protection within the installation to be designed. These characteristics are still required, even if the supply is from a private source such as a generator. The characteristics required can be listed as follows:

1. Nominal voltage and voltage range.
2. Nature of current and frequency - (a.c. 50 Hz or 60 Hz etc., or d.c).
3. The single-phase and three-phase prospective short -circuit current at the origin of the installation.
4. Type and rating of the overcurrent protective device at the origin of the installation.
5. The earth fault loop impedance (Z_e) external to the origin of the installation.

313-01

Standby or safety services
Where the installation concerns the provision of standby or safety services, the characteristics of the source of any standby power supplies shall be determined.

Additionally the capacity, reliability, and rating of such sources shall be determined, along with the time required to change-over in the event of an interruption in the normal supply source.

Where the installation receives its normal supply from a supplier, then arrangements shall be made with the supplier concerning the switching arrangements, such that both sources of power are either prevented from operating in parallel, or are arranged to operate in parallel.

CHAPTER 3

Structure of the installation

Separate circuits have to be provided for those parts of the installation that need to be separately controlled to avoid danger and minimize inconvenience. This is to ensure that other circuits remain energised when there is a fault on one circuit. This means making certain that there is discrimination between protective devices. Consideration has to be given to what would happen when a protective device operates. For instance, what would happen if all the lights went out due to a fault in a power circuit? Could there be an accident due to the lights going out?

300-01, 314-01, 531-02-09, 533-01-06

Arrangement of circuits

Every installation has to be broken down into circuits to avoid danger and minimize inconvenience in the event of a fault, so that the installation can be operated and maintained safely. This also allows inspection and testing to be carried out safely. Each final circuit in an installation shall be connected to a separate way in a distribution board , making it electrically separate from every other circuit. This means that where single-phase circuits are derived from a three-phase supply, a separate neutral shall be installed for each single-phase circuit.

The number of circuits required shall be planned, so that the circuits and installation will comply with the requirements for current-carrying capacity, overcurrent protection, and isolation .

314-01

Environmental conditions

The environmental conditions applicable to the installation have to be taken into account when selecting equipment and wiring materials. Consideration shall be given to the ambient temperature, mechanical stresses, corrosive substances, presence of water and moisture, and to any other conditions which can foreseeably be seen to affect the installation now and in the future.

Chapter 32, 522-01

Compatibility

Consideration has to be given to what effects the installation is likely to have on other equipment and services. For instance, installing mains cables in close proximity to telemetry cables can affect the signals carried by those cables. Starting currents of large motors can cause a dip in voltage which will affect electronic control circuits. The supplier of the electricity will also need to be informed of any equipment which is likely to affect their supplies, as this would affect other customers connected to the same service. The designer has to consider all the following characteristics:

    transient overvoltages;
    rapidly fluctuating loads;
    starting currents;
    harmonic currents;
    d.c.feedback;
    high-frequency oscillations;
    earth leakage currents;
    the necessity for additional connections to earth;

331-01, EMC Regulations

Maintainability

When selecting the protective devices for the installation, consideration has to be given to the frequency and quality of maintenance that the installation will receive throughout its intended life. This is required so that the protective measures taken to provide safety and the reliability of the equipment selected remain effective throughout the intended life of the installation. In addition, consideration has to be given to any periodic inspection, testing, maintenance and repairs that may be necessary on the installation during its intended life.

341-01

Chapter 4

Agricultural and horticultural installations

Where requirements are applicable

With the exception of dwellings used solely for human habitation, the requirements concerning agricultural and horticultural installations apply to the fixed installation of the whole site, including both outdoor and indoor installations. For example, hay lofts, storage areas and piggeries etc are included in the scope.
605-01

Protection by separated extra-low voltage (SELV)

No matter how low the voltage used for the SELV circuit is, protection against direct contact shall be provided by either barriers or enclosures giving a minimum protection against the standard finger 80 mm long and 12 mm in diameter (IP2X), or by complete protection against a finger entering the enclosure (IPXXB), or by insulation that will withstand 500 V a.c. rms for 1 minute.
605-02-02

Socket-outlets

Except for SELV circuits, all socket-outlet circuits shall be protected by an RCD manufactured to British Standards. The operating current shall not exceed 30 mA, and it will have a maximum operating time of 40 ms with a residual current of 150 mA. Since all socket-outlet circuits are RCD protected Regulations 471-08-06 and 471-16-01 are not applicable.
605-03 , 605-09-03

Protection against indirect contact

Where protection against indirect contact is provided by earthed equipotential bonding and automatic disconnection of the supply for areas in which it is intended to keep livestock , then Regulations 605-05 to 605-09 shall apply supplementing the general requirements.
605-04-01

TN system

The limiting values of earth fault loop impedance given in Table ZS 9 and Regulation 605-05 for installations forming part of a TN system are only applicable where the exposed-and-extraneous-conductive-parts are within the equipotential zone created by the main equipotential bonding.
605-09-01

Circuit protective conductors shall connect all exposed-conductive-parts in the installation to the main earth terminal which in turn shall be connected to the earthed point of the supply.
413-02-06

Protection against indirect contact shall be by either or both of the following devices:
   1) overcurrent protective device.
   2) residual current device.
Either device gives protection by automatic disconnection.
413-02-07

The characteristic of each overcurrent protective device, and the earth fault loop impedance of each circuit, has to be such that : $Z_S \times I_a = Uo$ for TN Systems

Where Z_S is the earth fault loop impedance of the circuit, U_0 is the nominal voltage to earth, and I_a is the current causing the protective device to disconnect within a specified time. As explained in Part 2, for design purposes U_0 is replaced by U_{oc}, the open circuit voltage to earth.

413-02-08

The specified disconnection time in locations in which livestock is kept is:

1. In accordance with Table 605A for final circuits feeding:
   (a) portable equipment intended for manual movement during use or,
   (b) hand-held Class I equipment.
2. 5 seconds for:
   (a) a distribution circuit or,
   (b) a final circuit supplying only stationary equipment or,
   (c) a final circuit not used for hand-held Class 1 equipment or,
   (d) a circuit feeding equipment which is not going to be moved manually.
3. 5 seconds for circuits complying with the reduced low voltage requirements.

Maximum disconnection times for TN systems (605A)	
U_0	t (seconds)
120	0.35
220 to 277	0.2
400, 480	0.05
580	0.02

605-05-01, 605-05-02, 605A, 605-05-06, 471-15-06

Where the nominal voltage to earth U_0 is 230 V, and an overcurrent protective device is used to satisfy the disconnection times in Table 605A, the maximum value of Z_S for each type of protective device is given in Table ZS 11 in Part 3.

605-05-03, 605-05-04, 605-05-07

For reduced low voltage installations, where the voltage between phases is 110 V, the maximum values of earth loop impedance are given in Table ZS 10.

471-15-06

Where the nominal voltage to earth U_0 is 230 V, and the disconnection time allowed is 5 seconds, the maximum values of Z_S allowed are given in ZS 11 in Part 3.

605-05-06

The limitation of impedance of the circuit protective conductor by application of Table 41C or PCZ8 is not allowed.

605-05-05

Where the disconnection time allowed for a final circuit is 5 seconds, and the actual disconnection time of the circuit exceeds that given in Table 605A, whilst another final circuit which requires disconnecting within the time specified in Table 605A is connected to the same distribution board or distribution circuit, then one of the following conditions shall be fulfilled:

1. The protective conductor impedance from the distribution board to the main earth bar for equipotential bonding shall not exceed $25 \text{ V} \times Z_S + U_{oc}$, where Z_S is the earth loop impedance for 5 seconds disconnection time. Where the nominal voltage to earth U_0 is 230 V, the formula becomes $0.104 \times Z_S$.

2. Equipotential bonding shall be carried out at the distribution board to the same types of extraneous-conductive-parts as those to which the main equipotential bonding is connected; the size of the bonding conductors shall be the same as that of the main equipotential bonding.

Where the disconnection times detailed above cannot be met by using an overcurrent protective device, then one of the following remedies shall be provided:

1. Local supplementary equipotential bonding shall be installed, connecting together the exposed-conductive-parts of the circuit concerned (including the earthing terminal of socket-outlets) and extraneous-conductive-parts. The resistance of the supplementary bonding conductor shall be less than or equal to $25 V \div I_a$, where I_a is the minimum current needed by an overcurrent device to disconnect the circuit in 5 seconds, or is the residual operating current of an RCD.

2. Protection shall be provided by an RCD complying with $Z_S \times I_{\Delta n} \leq 25$ V.

605-05-08, 605-05-09, 413-02-15, 413-02-16

Where exposed-conductive-parts are simultaneously accessible, they shall be connected to the same earthing system. They can be connected individually, collectively, or in groups.

413-02-03

Use of residual current devices (RCDs)

If the overcurrent protective devices cannot disconnect the low voltage circuits within the time specified in Table 605A, or disconnect fixed equipment circuits and reduced low voltage circuits within 5 seconds, then protection shall be provided by a RC D, such that the operating current of the RCD $I_{\Delta n} \times Z_S \leq 25$ V.

605-05-08

Where RCDs are used for automatic disconnection, exposed-conductive-parts need not be connected to the TN system's protective conductors, providing they are connected to an earth electrode such that $R_A \times I_a \leq 25$ V where R_A is the resistance of the protective conductor from the exposed-conductive-part to, and including the earth electrode. The value I_a is the current that will cause disconnection of the overcurrent protective device within 5 seconds, or is the residual operating current $I_{\Delta n}$ of an RCD. The installation is treated as a TT installation.

413-02-17, 413-02-20

Equipment outside the equipotential zone

Where a circuit which feeds fixed equipment with exposed-conductive-parts is outside the earthed equipotential zone, and where it can be touched by a person directly in contact with the ground, the disconnection time allowed for the circuit must comply with Table 605A.

605-09-02

Where the system is TT

Where exposed-conductive-parts are protected by a single protective device, they shall be connected via a protective conductor to the main earthing terminal, and then to a common earth electrode. Where several protective devices are in series, the exposed-conductive-parts may be connected to a separate earth electrode associated with each protective device.

The protective device can be either an overcurrent device, or an RCD, the preferred protection

being a residual current device.
413-02-18, 413-02-19

Each circuit must comply with the formula $R_A \times I_a \leq 25$ V.
605-06

Where the system is TT, and protection is by equipotential bonding and automatic disconnection, every socket-outlet circuit shall be protected by an RCD complying with $R_A \times I_n \leq 25$ V.
471-08-06, 605-05-09, 413-02-16

Where the system is IT
Where the system is IT, Regulations 413-02-21 to 413-02-26 are applicable.
605-07-01

Regulation 413-02-26 is applicable, except that Table 41E is changed to 605E, and I_a is the current which either disconnects the circuit within the time specified in Table 605E, or within 5 seconds, when this disconnection time is permitted.

Maximum disconnection times for IT systems (605E)

Voltage between phase and neutral U_0 Voltage between phases U volts	Neutral not distributed t (seconds)	Neutral distributed t (seconds)
120 to 240	0.4	1
220/380 to 277/480	0.2	0.5
400/690	0.06	0.2
580/1000	0.02	0.08

605-07-02, 605E

Supplementary equipotential bonding
The minimum size of supplementary bonding conductors shall be the larger of either:
1. The size determined by the formula $I_a \times R \leq 25$ V, or
2. The size in accordance with Section 547-03.

605-08-01

In areas where livestock is kept, supplementary equipotential bonding shall be installed. The supplementary equipotential bonding shall connect together all exposed- and -extraneous-conductive -parts (including non-insulating floors) which can be touched by livestock.

Where a metallic grid is laid in non-insulating floors for the purpose of supplementary bonding, it shall be connected to the protective conductors of the installation.
605-08-02, 605-08-03

Cable installation
Cables shall be sized in accordance with the overload, short-circuit, isolation and protective conductor regulations.
600-01, 600-02

Calculations
The temperature rise of an overcurrent conductor due to fault current shall be taken into account

CHAPTER 4

when checking that the circuit is protected against indirect contact. Where a protective device complies with the characteristics in Appendix 3 and the Z_s values given in the tables are not exceeded, this requirement is deemed to be complied with, unless manufacturers' characteristics are used.
413-02-05

Protection against fire
With the exception of equipment essential to the welfare of livestock, an RCD having a maximum residual operating current of 500 mA shall be installed for the supply to equipment.
605-10-01

To minimise the risk of burns to livestock and of fire, heating appliances shall be installed so as to maintain a suitable distance from combustible material and livestock. Where the heaters are radiant heaters, the suitable clearance distance shall be as recommended by the manufacturer, but in any event not less than 500 mm.
605-10-02

Installation and materials
Electrical equipment shall at least be protected to IP44 when installed for normal use. It shall also be suitable for the environmental conditions applicable to the installation, such as water, dust, corrosion, flora, fauna, solar radiation, wind, vermin, and mechanical stresses.

Devices installed for emergency switching or emergency stopping shall be installed so that they are inaccessible to livestock. The operation of such safety devices must not be impeded by livestock, due consideration being given to the conditions that could arise if the livestock panicked.
605-11 , 605-13

Electric fence controllers
Electric fence controllers shall comply with the following British Standards:

BS EN 61011-1 Battery-operated electric fence controllers suitable for connection to the supply mains.

BS EN 61011-2 Battery-operated electric fence controllers suitable for connecting to the mains.
The effects of induced voltages shall be taken into account when installed near overhead lines.
605-14-01

The installation of mains operated fence controllers, so far as is reasonably practicable, shall be free from risk of mechanical damage or unauthorised interference; they shall not be installed on a pole carrying an overhead power or telephone line, unless the overhead power line is insulated, and is the supply for the fence controller.
605-14-02, 605-14-03

An earth electrode that is connected to a fence controller shall be kept completely separate from the earthing system of any other circuit, and its resistance area must not overlap the resistance area of any electrode used for protective earthing.
605-14-04

The electric fence controller, the electric fence or conductor shall be installed so that it will not be liable to come into contact with any other equipment or conductor; not more than one controller shall be connected to each electric fence or similar system of conductors.
605-14-05, 605-14-06

Chapter 5

Bath and shower rooms

Socket-outlets

Socket-outlets or any provision for connecting portable equipment are not allowed in a shower room or a room containing a fixed bath unless they comply with the following requirements.

The circuit is derived from a SELV safety source with a nominal voltage not exceeding 12 V r.m.s. a.c., or d.c. The safety source is out of reach of a person in a bath or shower. Irrespective of the SELV voltage all cables and equipment shall give protection against direct contact by insulation that will withstand a test voltage of 500 V a.c.rms. for one minute.

Where a socket-outlet is installed from a 12 V SELV source, the socket-outlet must not have any accessible metal parts, and shall be insulated or protected with barriers or enclosures against direct contact.

Where a shower is installed in a room that is not a bathroom, sockets shall be at least 2.5 metres from the shower cubicle.

A shaver-socket-outlet must have been manufactured to BS 3535 and the protective conductor of the final circuit feeding the shaver-socket shall be connected to the earth terminal of the shaver-socket. The shaver-socket can be independent or part of a luminaire.

601-03, 601-09, 601-10

Lampholders

If a lampholder is within 2.5 metres of a fixed bath or shower it shall be constructed of, or shrouded in insulating material, and have an insulating shroud complying with BS 5042 or BS EN 61184. Alternatively, luminaires that are totally enclosed may be used.

601-11

Switches

Switches and control devices shall be mounted so that they are inaccessible to a person using a fixed bath or shower.

The only switches or controls that are allowed to be accessible in a room with a fixed bath or shower are: shaver-sockets manufactured in accordance with BS 3535, the switches or controls of water heaters and shower pumps manufactured to the appropriate British Standard, insulating cords of cord-operated switches which comply with BS 3676, switches which form part of a circuit derived from a SELV safety source with a nominal voltage not exceeding 12 V r.m.s. a.c., or d.c., and switches or controls operated by mechanical actuators with insulating linkages.

The safety source for the SELV circuit shall be out of reach of a person using the bath or shower, the switches must not have any accessible metal parts, and shall be insulated or protected with barriers or enclosures against direct contact.

601-08, 601-03

Equipment or wiring not allowed

No electrical equipment shall be installed in the interior of a bath tub or shower basin.

Installations carried out on the surface shall not use metal conduit, metal trunking, a cable with an exposed metallic sheath , or an exposed earthing or bonding conductor.

601-02, 601-07

CHAPTER 5

Protective measures not allowed

Protection against electric shock, by means of obstacles, by placing out of reach, by non-conducting location, or earth free local equipotential bonding, is not allowed.
601-05, 601-06

Electrical equipment

Stationary appliances which have heating elements, including silica glass sheathed elements that can be touched, must not be installed within reach of a person using a fixed bath or shower.

Where electrical equipment is installed in the area below a bath, access to that area must only be available by using a key or tool, and supplementary bonding shall be installed between exposed-conductive-parts and between exposed-and extraneous-conductive-parts, unless the equipment is supplied from a SELV source.
601-12-01, 601-04-03

Electric heating

Heating units embedded in the floor and intended for heating the room may be installed, providing they are either covered by a metallic grid, or by an earthed metallic sheath, which is connected to the local equipotential bonding conductors.
601-12-02

Supplementary bonding

Except for equipment supplied from a SELV circuit, supplementary equipotential bonding shall be provided between all conductive parts that are simultaneously accessible in a room containing a fixed bath or shower. This includes all exposed- and extraneous-conductive parts, such as the metalwork of electrical equipment, metal waste pipes, metal baths or shower-trays, metal water pipes, central heating pipes, or exposed structural metalwork. It includes electrical equipment that is installed in the space below a bath, even though it is only accessible by use of a key or tool, but equipment supplied from a SELV circuit is excluded.
601-04-02, 601-04-03

Size of supplementary bonds

The minimum size of supplementary bonding conductors shall be:

1. For exposed-conductive-parts bonded together:
If sheathed or otherwise mechanically protected, not less than the conductance of the smallest c.p.c. connected to the exposed- conductive-parts. If not mechanically protected, the minimum size shall be 4 mm^2.

2. For exposed-conductive-parts bonded to extraneous conductive parts:
If sheathed or otherwise mechanically protected, not less than half the conductance of the c.p.c. connected to the exposed-conductive-part. If not mechanically protected, the minimum size shall be 4 mm^2.

3. For extraneous -conductive-parts bonded together
If sheathed or otherwise mechanically protected, not less than 2.5 mm^2. If not mechanically protected, 4 mm^2. If any of the extraneous-conductive-parts are bonded to an exposed-conductive-part, the conductor size shall be determined in accordance with paragraph 2.
547-03-01, 547-03-02
547-03-03, 547-03-04

Exposed-conductive-part to exposed-conductive-part

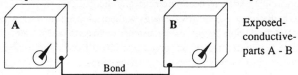

Exposed-
conductive-
parts A - B

Same conductance as the smallest cpc feeding A or B, providing that
it is mechanically protected: otherwise minimum size is 4 mm².

Exposed-conductive-part C to extraneous- conductive-part

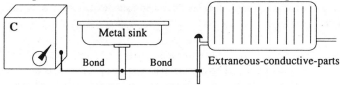

Half conductance of the cpc in 'C' if sheathed or mechanically
protected: otherwise minimum size required is 4 mm².

Extraneous-conductive-part to extraneous-conductive-part

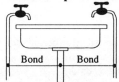

Minimum size 2.5 mm² if sheathed or otherwise mechanically
protected: 4 mm² if not mechanically protected.

Figure 5.1 - Minimum size of supplementary bonding conductors

Supplementary bonding a fixed appliance

Where a fixed appliance has to be included in the supplementary bonding, and that appliance is
supplied from a flex outlet type accessory through a short length of flexible cable containing a c.p.c.,
then the supplementary bonding connection can be made at the earth terminal in the flex outlet
accessory, thereby using the cpc of the fixed appliance also as the supplementary bonding conductor.
547-03-05

Disconnection time

Except for SELV circuits, where electrical equipment is simultaneously accessible to the metalwork
of other electrical equipment, or to extraneous-conductive-parts, the electrical equipment shall be
disconnected by the protective device in 0.4 seconds in the event of an earth fault in the equipment.
601-04-01

CHAPTER 5

Earthing

Where exposed-conductive-parts may become electrically charged, they shall be connected to earth in such a manner as to discharge the electrical energy without danger. Either overcurrent or RCD's shall be used to protect the circuits against the persistence of an earth fault current or earth leakage current. Where exposed-conductive-parts are connected with earth and are accessible to extraneous-conductive-parts, the extraneous-conductive-parts shall be connected to the main earthing terminal of the installation.
130-04

Instantaneous showers

Every instantaneous shower heater must contain an automatic device for preventing a dangerous rise in temperature.
554-04

The metal water pipe through which the water supply to the heater is provided shall be solidly and mechanically connected to all metal parts (other than live parts) of the heater or boiler. There must also be an effective electrical connection between the metal water pipe and earth independent of the c.p.c.
554-05-02

Where the heater is installed in a room containing a fixed bath or shower, the switch shall be out of reach of a person using the bath or shower.
601-08-01

A switch or control device is not considered accessible to a person using a bath or shower, if it cannot be touched and is operated by an insulating cord, or by insulated mechanical actuators, or if the controls comply with the appropriate British Standard.
601-08-01

Where a shower tray is made from a metallic material, or where it is made from concrete, it shall be bonded to the metal water pipe in the shower heater. In the case of a concrete shower tray, the reinforcing can be used for the bonding connection.
INF

Where regulations are applicable

The regulations for bath and shower rooms apply to bath and shower tubs, including their surroundings, where the risk of electric shock is increased due to a reduction in body resistance.
601-01

Cable and equipment installation

Except where modified by the above requirements, the installation shall comply with all the other requirements of the BS7671 such as; isolation, overload protection, fault current protection, protection against electric shock and thermal capacity of the protective conductor.
600-01, 600-02

Chapter 6

Caravan electrical installations

General requirements

Except where modified by the following requirements the general requirements of the British Standard are applicable to caravans and motor caravans and the absence of any reference to a Regulation, Section or Chapter means that the general requirements are applicable.

The particular requirements of Section 608, (Division One) of BS7671 do not apply to mobile homes, fixed recreational vehicles, transportable sheds and the like, nor to temporary premises.

600-01, 600-02

The requirements for caravans and motor caravans only apply where the nominal voltage does not exceed 250/440 V. Neither do they apply to extra-low voltage installations carried out to BS 5765 Part 3, or electrical circuits and equipment embraced by the Road Vehicles Lighting Regulations 1989.

608-01

Protection against direct contact

The only methods allowed for protection against direct contact are:

    a) protection by insulation of live parts,

    b) protection by barrier or enclosure.

Where supplementary protection is provided by an RCD it shall have an operating current $I_{\Delta n}$ not exceeding 30 mA and shall operate within 40 ms with a residual current of 150 mA.

608-02, 412-01, 412-06

Protection against indirect contact

The only methods of protection against indirect contact allowed are:

    a) earthed equipotential bonding and automatic disconnection of the supply,

    b) Class II equipment or equivalent insulation.

Where method (a) is used, a double pole 30 mA RCD shall be provided which complies with the appropriate British Standard and which operates within 40 ms with a 150 mA residual current flowing.

608-03-01, 608-03-02

Protective conductors

Where protection is provided by equipotential bonding and automatic disconnection of the supply a circuit protective conductor shall be provided throughout the installation, which shall be connected to: the earth terminal of the incoming supply, the exposed-conductive-parts within the caravan and the earth terminal of socket-outlets. All protective conductors must be insulated. Where the protective conductor is installed separately, i.e., not enclosed in the cable, or enclosed in conduit or trunking, it shall have a minimum cross-sectional area of 4 mm^2.

608-03-02, 608-03-03, 608-06-03

Equipotential bonding

Unless the caravan or motor caravan is manufactured from insulating material, such that any extraneous-

conductive-parts are unlikely to become live in the event of a fault, all extraneous -conductive -parts shall be connected to the circuit protective conductor with a 4 mm^2 minimum cross-sectional area conductor. Such bonding shall be carried out for each isolated extraneous-conductive-part.
608-03-04

Overcurrent protection

The overcurrent protective device shall disconnect all phase and neutral conductors in a final circuit. The other requirements for overcurrent protection shall be as stated in the BS7671.
608-04-01

Size of conductor

The minimum cross-sectional area of every conductor shall be 1.5 mm^2. The following types of wiring complying with the appropriate British Standard shall be used:

a) flexible single core cable, such as 2491X BS 6004 PVC insulated cable, enclosed in plastics conduit, or

b) conduit wiring cables having a minimum of seven strands, such as 6491X BS 6004 cables enclosed in plastics conduit, or

c) sheathed flexible cables, such as 3183Y BS 6500 cables.

Any non-metallic conduit can be used providing it is not pliable polyethylene conduit.
608-06-02, 608-06-01

Cable installation

LV and ELV circuits must be installed separately so that there is no risk of contact between the two systems.

Flexible conduit, and cables not installed in rigid conduit, must be supported at least every 250 mm when installed horizontally and 400 mm when installed vertically.

All wiring must be protected against mechanical damage due to vibration by additional protection on sharp edges or abrasive parts, or by locating the wiring where it cannot be damaged.

Where a cable is to be installed through metalwork the hole in the metalwork should have a bush or grommet securely fixed in position to protect the cable.
608-06-04, 608-06-05, 608-06-07

Mains intake

Unless the demand for the motor caravan or caravan exceeds 16 A, the appliance inlet power point to the caravan shall be: a two pole and earthing contact with key position 6h plug complying with BS EN 60309-2. The inlet shall be installed up to a maximum of 1.8 m above the ground in a readily accessible position and installed in a suitable enclosure, with cover, on the outside of the caravan or motor caravan.
608-07-01, 608-07-02

The incoming electrical supply shall be taken to a main isolating switch which disconnects both the phase and neutral conductors. This switch is to be mounted in an accessible position within the caravan or motor caravan. The switch can be replaced by an overcurrent protective device, which disconnects all phase and neutral conductors, if only one final circuit is installed.
608-04-01, 608-07-04

Where there is more than one independent electrical installation each installation shall be supplied

from an independent inlet device, LV and ELVcircuits being separated.
608-05-01, 608-04-01

The connection from the caravan/motor caravan to the caravan site socket-outlet shall be by means of either a flexible cord to BS 6500 (Harmonised code H05VV-F3) or BS 6007 (Harmonised code H07RN-F3) e.g. 3183Y flexible cord or a 6383P flexible cord. The length of the flexible cord shall be between 23 m and 27 m long and shall have a plug on one end and a connector, suitable for the inlet plug installed on the caravan, at the other end. Both plug and connector complying with BS EN 60309-2.

The cross-sectional area of the connecting cable for the caravan's rated current shall be: 2.5 mm^2 for 16 A, 4.0 mm^2 for 25 A, 6.0 mm^2 for 32 A, 16 mm^2 for 63 A and 35 mm^2 for 100 A.
608-08-08

Notices
A notice in durable material shall be permanently fixed at the electrical inlet point of the caravan/ motor caravan giving the nominal voltage, frequency and rated current of the installation.

A notice in durable material shall also be permanently fixed adjacent to the main isolating switch in the caravan giving the instructions in 608-07-05 (see Identification and Notices).

The characters on all notices shall be indelible and easily legible.
608-07-03, 608-07-05

Socket-outlets
Except for low voltage socket-outlets supplied from an individual winding of an isolating transformer, all socket-outlets shall be provided with an earthing contact.

Low voltage socket-outlets shall not be interchangeable with socket-outlets supplied with extra-low voltage. Additionally, all extra-low voltage socket-outlets shall have their voltage clearly marked on them.

No socket-outlet shall have accessible conductive parts and where a socket-outlet is exposed to moisture or the effects of moisture it shall comply with IP55.
608-08-01, 608-08-02, 608-08-03

Other accessories
No accessory shall have accessible conductive parts and where it is installed so that it is exposed to the effects of moisture it shall comply with IP55.

Every item of equipment that is not connected to the supply through a plug and socket shall either; be controlled by a switch on the appliance, or a switch adjacent to the appliance.
608-08-01, 608-08-04, 608-08-05

Lighting fittings
Luminaires should, preferably, be fixed to the structure of the caravan/motor caravan. Where pendant luminaires are installed, provision should be made for securing the pendant so as to prevent damage when the caravan/motor caravan is moved. Additionally, the lampholder, flex and ceiling rose should be suitable for any weight they have to support.
608-08-06

Luminaires manufactured for dual voltage supplies shall have a separate lampholder for each

voltage which will also prevent the interchange of lamps of different voltages. The lamp wattage and voltage shall be displayed adjacent to each lampholder. The fitting must also be suitable for both lamps operating at the same time without any damage occurring and have adequate separation between the different voltage supplies, such as low voltage and extra-low voltage.
608-08-07

Extra-low voltage supplies
The standard voltages of 12 V, 24 V, and 48 V can be used for d.c. extra-low voltage power sources, and when a.c. extra-low voltages are used the following r.m.s. voltages are allowed: 12 V, 24 V, 42 V, and 48 V.

If extra-low voltage is used in a caravan it shall comply with Section 411-02 detailed as follows:
1) extra-low voltage shall not be exceeded in the circuit concerned,
2) the supply shall be from one of the following safety sources:
   a) safety isolating transformer to BS 3535 with output winding fully isolated, or
   b) a supply from a motor generator having windings that give electrical separation equal to that of a BS 3535 transformer, or
   c) a battery or engine driven generator, or
   d) special electronic devices.
3) where the extra-low voltage system is supplied from a higher voltage source which does not have electrical separation equivalent to a BS 3535 transformer it cannot be considered as a SELV supply.
4) where a mobile source is being used for the SELV supply it shall be selected and erected in accordance with the protection by Class II regulations.
5) It must also comply with the remaining SELV regulations of Section 411-02 (see the Chapter on SELV).
608-08-09

Prevention of explosion risk
The electrical installation, including electrical equipment and accessories, shall not be installed in compartments intended for the storage of gas cylinders.
608-06-06

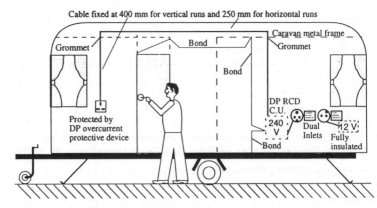

Figure 6.1 - Caravan installation

Chapter 7

Caravan site installations

General requirements
Except where modified by the following requirements the general requirements are applicable to caravans and motor caravans and the absence of any reference to a Regulation, Section or Chapter means that the general requirements of the BS7671 are applicable.
600-01, 600-02

These regulations are applicable to that part of the site that provides electricity for caravans, motor caravans, leisure accommodation vehicles and tents, where the nominal voltage does not exceed 250/440 V. They are not applicable to the electrical installations in caravans, mobile homes, recreational vehicles, transportable sheds etc., or for temporary premises, temporary structures or motor caravans.
608-09-01

Protection against direct contact
The only methods allowed for protection against direct contact are:
   a) protection by insulation of live parts,
   b) protection by barrier or enclosure.
Where supplementary protection is provided by an RCD it shall have an operating current $I_{\Delta n}$ not exceeding 30 mA and shall operate within 40 ms with a residual current of 150 mA.
608-10-01, 412-01, 412-06

Protection against indirect contact
The only methods of protection against indirect contact allowed are:
   a) earthed equipotential bonding and automatic disconnection of the supply,
   b) Class II equipment or equivalent insulation.
608-11-01

Electrical supply to site pitch
Equipment intended to be used for the pitch electrical supply shall be mounted adjacent to the pitch and in any event shall not be more than 20 m from any point it is intended to serve on the pitch . *As a guide, the maximum distance from the centre of the pitch to comply with this requirement, expressed as the radius of a circle is:*

   a) *for a rectangular pitch., supply must be within a radius of* $20\ m - \dfrac{\sqrt{L^2 + B^2}}{2}$

   where L is the length of the pitch and B is the breadth of the pitch.
   b) *for a circular pitch the radius from the centre of the circle is 20 m minus the radius of the pitch.*
608-13-01

It is preferable for the cable supply to be underground and, unless provided with additional mechanical protection, it should be installed outside the area of the pitch or any area in which items such as, tent pegs or equipment securing spikes may be used.
608-12-01, 608-12-02

CHAPTER 7

Where the supply to the pitch is overhead the cables must be of a suitable construction and completely covered with insulation capable of withstanding any mechanical, electrical, thermal or chemical stresses to which it may be subjected; additionally, the insulation must only be removable by destruction.

The overhead conductors must not be installed within 2 m of the outside boundary of any pitch and any poles, straining wires or cable supports must be installed in positions where they are unlikely to be damaged by vehicle movement. The minimum mounting height for overhead lines is 6 m where there is vehicle access and 3.5 m elsewhere.

Each pitch must be provided with at least one socket-outlet.
608-12-03, 608-13-02

Socket-outlets

Socket-outlets at the pitch supply point should comply with BS EN 60309-2, have an IP code rating of IPX4 and have a minimum current rating of 16 A. The mounting height to the bottom of the socket-outlet should be between 800 mm and 1500 mm from the ground.
608-13-02

An overcurrent protective device shall be provided for each individual socket-outlet. Additionally, an RCD complying with BS 4293, which will operate within 40 ms with a residual current of 150 mA, shall protect each socket-outlet individually or in groups not exceeding three socket-outlets. The RCD earth terminal must not be connected to a PME supply.

Where the supply is PME the circuit protective conductor for each socket-outlet circuit shall be bonded to an earth electrode such that: $R_A I_{\Delta n} \leq 50$ V, where R_A is the sum of the c.p.c. resistance and the earth electrode resistance and $I_{\Delta n}$ is the operating current of the RCD. Each pitch must be provided with at least one socket-outlet.
608-13-02, 608-13-04, 608-13-05, 413-02-18 to 20

Where sockets are grouped they must be connected to the same phase.
608-13-06

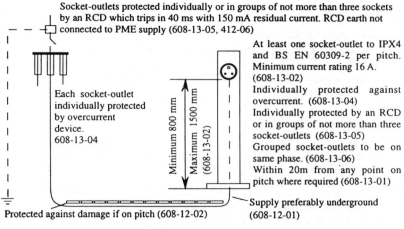

Socket-outlets protected individually or in groups of not more than three sockets by an RCD which trips in 40 ms with 150 mA residual current. RCD earth not connected to PME supply (608-13-05, 412-06)

At least one socket-outlet to IPX4 and BS EN 60309-2 per pitch. Minimum current rating 16 A. (608-13-02)

Each socket-outlet individually protected by overcurrent device. 608-13-04

Minimum 800 mm / Maximum 1500 mm (608-13-02)

Individually protected against overcurrent. (608-13-04)
Individually protected by an RCD or in groups of not more than three socket-outlets (608-13-05)
Grouped socket-outlets to be on same phase. (608-13-06)
Within 20m from any point on pitch where required (608-13-01)

Protected against damage if on pitch (608-12-02)

Supply preferably underground (608-12-01)

Figure 7.1 - Outline of the rules

42

Chapter 8

Circuit protective conductors

General requirements

Where a voltage may appear on exposed-conductive-parts, then the exposed-conductive-parts shall be earthed so that the electrical energy is discharged without danger.
130-04-01

Where exposed-conductive-parts are earthed, the circuits shall be protected by either overcurrent protective devices, or RCD's, to prevent the persistence of dangerous earth leakage currents.

Where the phase earth loop impedance is too high to allow the prompt disconnection of the circuit by overcurrent protective devices because the prospective earth fault current is insufficient, then RCD, or other effective device, shall be used for circuit protection.
130-04-02, 130-04-03

Where exposed-conductive-parts are earthed, and are simultaneously accessible with extraneous - conductive-parts of other services, the extraneous-conductive-parts shall be connected to the main earthing terminal of the installation.
130-04-04

The only protective device allowed in an earthed neutral is a linked circuit-breaker, or switch, where the related phase conductors are arranged so as to be substantially disconnected at the same time.
130-05-02

Protective conductors shall be protected against mechanical and chemical deterioration, and electrodynamic effects.
543-03-01

Where a protective conductor is installed through several installations, each one of which has its own earthing arrangement, then either of the following requirements shall be complied with:

a)   the protective conductor shall be earthed within one installation only, and insulated from the earthing arrangement in the other installations, or

b)   be capable of carrying the maximum fault current likely to flow through it from any of the installations.

Where the protective conductor is part of a cable, it shall only be earthed in the installation containing the protective device for the circuit.
542-01-09

Combined protective and functional purposes

Protective conductors can be used jointly or separately for protective or functional purposes to suit the requirements of the installation, providing the functional purpose does not adversely affect the protective measures of the conductor.
542-01-06, 546-01-01

Where the neutral and protective conductor is combined (PEN conductor) the requirements of Section 546 can only be applied if:

a)   authorisation for using a PEN conductor has been obtained and the conditions of the authorisation are applied, or

b) where the supply is from a privately owned transformer or convertor so that there is, other than an earth connection, no connection with the public supply, or the supply is obtained from a private generating plant.

546-02-01

Providing the particular installation is not supplied through an RCD the following cables can be used as a PEN conductor :

1. A conductor with a cross-sectional area of at least 10 mm^2 if it is copper or 16 mm^2 if it is aluminium, providing the cable is used for a fixed installation and not subject to flexing.

2. A concentric cable complying with British Standards which has an outer conductor with a cross-sectional area of at least 4 mm^2 and providing the outer conductor is not used for more than one final circuit. It must also comply with Regulations 546-02-03 to 08.

546-02-02, 546-02-03

At 20 °C the conductance of the outer conductor of a concentric cable shall be the same as the internal conductor of a single core cable, or if the cable is a multicore cable feeding several points in a final circuit, or it has internal conductors connected in parallel, then the outer conductor shall have a conductance not less than the conductors connected in parallel. The exception is a multicore cable in a multiphase or multipole circuit complying with BS 5593, when the conductance of the outer conductor must be not less than one internal conductor.

546-02-04

An additional conductor must be installed at every joint in the outer conductor and at every termination of a concentric cable to ensure continuity irrespective of how good the joints or terminations are made. The additional conductor's conductance shall be determined by Regulation 546-02-04. Additionally, the outer conductor of a concentric cable must not be broken by inserting an isolator or a switch in it.

546-02-05, 546-02-06

If the PEN conductor is divided at any point into a neutral and protective conductor, the neutral and protective conductor must not be joined again beyond that point. Where the PEN conductor is separated into a neutral and protective conductor separate terminals or bars have to be provided for each conductor. The conductance of the terminal link shall be not less than that required by Regulation 546-02-04. Additionally, the PEN conductor shall be insulated or have an insulated covering, unless it is manufactured to BS 6207 and installed in accordance with the manufacturer's instructions; such insulation or covering must cater for the highest voltage to which it may be exposed.

546-02-08, 546-02-07

Marking of protective conductors

Protective conductors up to 6 mm^2 shall be covered throughout, the same as conduit wiring cables to BS 6004 or BS 7211, unless they form part of a multicore cable or cable enclosure, such as conduit or trunking.

543-03-02

The uninsulated protective conductor in a sheathed cable shall be protected with green/yellow insulating sleeving complying with BS 2848, where the sheath is removed at joints and terminations, if the size of the protective conductor is less than 6 mm^2.

543-03-02, 514-03-01

When a bare conductor or busbar is used as a protective conductor, it shall be marked by equal green and yellow stripes each stripe being between 15 and 100 mm wide. The stripes shall be placed close together, and be positioned either throughout the length of the conductor, or in each compartment and at each accessible position. Where adhesive tape is used for marking, it shall be coloured green/yellow.

514-03-01

Ring circuits

Every ring final circuit protective conductor, unless formed by conduit, trunking, cable armour or the metal sheath of a cable, shall also be installed in the form of a ring, and both ends shall be connected to the same earth terminal at the origin of the circuit.

543-02-09

When conduit, trunking, ducting, cable armour or the metal sheath of a cable is used as the protective conductor for socket-outlet circuits, the earthing terminal of each socket-outlet shall be connected by a separate insulated protective conductor to an earth terminal in the associated socket-outlet box or enclosure.

543-02-07

Radial circuits

When conduit, trunking, ducting, cable armour or the metal sheath of a cable is used as the protective conductor for radial circuits, the earthing terminal of each outlet accessory shall be connected by a separate insulated protective conductor to an earth terminal in the associated outlet box or enclosure.

543-02-07

Size of conductor

The size of a circuit protective conductor shall either be calculated in accordance with 543-01-03, or determined from Table 54G

543-01-01

Where the phase conductor has been sized for protection against short-circuit currents, and the earth fault current is likely to be less than the short-circuit current, then the size of the protective conductor shall be determined by calculation.

543-01-01

Where the circuit protective conductor is not conduit, trunking, ducting, cable armour or the sheath of a cable, or enclosed in a cable, or contained in an enclosure formed by a wiring system, then the minimum size of conductor must be equivalent to 2.5 mm^2 copper if protection against mechanical damage is provided and 4 mm^2 if not. Protective conductors shall be protected against mechanical, chemical deterioration, and electrodynamic effects.

Where a protective conductor is buried in the ground then the minimum cross-sectional area specified in Table 54A is applicable.

543-01-01, 543-03-01

When one protective conductor is used for several circuits, its cross-sectional area shall be either:

1. Selected in accordance with Table 54G by using the largest phase conductor of the group of circuits concerned, or
2. By selected by determining the fault current and disconnection time for each circuit, and

CHAPTER 8

then using the values of the circuit with the most unfavourable value of fault current and disconnection time, to size the conductor, by using the formula of Regulation 543-01-03.
543-01-02
Where the size of the protective conductor is determined by calculation, the following formula shall be used.

$$S\ mm^2 \geq \frac{\sqrt{I^2 t}}{k}$$

The value of 'k' is obtained from tables 54B to 54F in the BS7671, or from Table K1A & K1B in Part 4. The most important item to watch when selecting 'k' are the initial and final temperatures of the conductor. S mm² is the minimum size of protective conductor required, so in most cases, the next largest standard conductor size will have to be chosen. 'I' is the phase to earth fault current in amperes, and 't' is the actual time taken in seconds for the protective device to disconnect the circuit.
543-01-03
Where the disconnection time 't' is less than 0.1 seconds, then $I^2 t$ from the manufacturers' $I^2 t$ characteristics shall be less than $k^2 S^2$ for the protective conductor.
434-03-03
Table 54G can be used instead of calculating the minimum cross-sectional area of the protective conductor. A modified table 54G based on 'k' factors, giving multiplying factors for the c.p.c. for copper, aluminium and steel is contained in PCA 1 in Part 4, as M54G.

SIZE OF MAIN EARTHING CONDUCTOR REQUIRED		
Area of phase conductor in mm² S	Area of protective conductor in mm² when it is the same material as the phase conductor	Area of protective conductor in mm² when it is a different material to phase conductor
S not exceeding 16	S	$S \times \dfrac{k_1}{k_2}$
S from 16 to 35	16	$16 \times \dfrac{k_1}{k_2}$
S exceeding 35	$\dfrac{S}{2}$	$S \times \dfrac{k_1}{2\,k_2}$

where S is the cross-sectional area of the phase conductor, k_1 is the 'k' value for the phase conductor, and k_2 is the 'k' value for the c.p.c. material.
Table 54G, 543-01-04
Table 54G gives the minimum sizes of protective conductor required, so in most cases, the nearest larger standard size of cable will have to be installed.
543-01-04

Items which can be used as protective conductors

A circuit protective conductor can be one of the following:

1. A single core cable.
2. An insulated or bare conductor contained in a cable.
3. The metal sheath, screen or armour of a cable.
4. Metal conduit, trunking, ducting or any other type of metallic enclosure.
5. An insulated or bare conductor in a common enclosure with insulated live conductors.
6. An extraneous-conductive-part which complies with Regulation 543-02-06

The protective conductor shall be copper where its size does not exceed 10 mm^2.
543-02-02, 543-02-03

With the exception of gas pipes, oil pipes and flexible or pliable conduit, an extraneous-conductive-part may be used as a protective conductor, providing:

1. Its electrical continuity can be assured in such a way that it is protected against mechanical, chemical or electrochemical deterioration.
2. Its cross-sectional area complies with Regulation 543-01-01.
3. Precautions against its removal, unless alternative measures are provided, are taken and consideration has been given to such use, and if necessary, it has been suitably adapted.

However, protective conductors must be incorporated in or in close proximity to phase and neutral conductors when overcurrent protective devices are used for protection against electric shock.
543-02-06, 544-01

Where the sheath of an MICC cable, conduit, trunking, ducting or the metal sheath of a cable is used as a protective conductor it shall:

1. Have its electrical continuity ensured by its construction or by its connection.
2. Be protected against mechanical, chemical , and electrochemical deterioration.
3. Have a cross-sectional area not less than that given by the formula of 543-01-03, or Table 54G, or by type-testing in accordance with BS EN 60439-1

543-02-05

Factory built equipment

Where the case, frame, or enclosure of busbar trunking, switchgear, control gear and distribution boards etc., is used as a protective conductor, it shall:

1. Have its electrical continuity ensured, by its construction or by its connection.
2. Be protected against mechanical, chemical and electrochemical deterioration.
3. Have a cross-sectional area not less than that given by the formula of 543-01-03, or Table 54G, or by testing in accordance with BS EN 60439-1, and
4. Allow the connection of other protective conductors at every pre-determined tap-off point.

543-02-04

Accessibility of joints

With the exception of cables buried underground, conduit, trunking or ducting, all joints in protective conductors shall be accessible for inspection, unless the joint is made by welding, soldering, brazing or compression tool.
543-03-03, 526-04-01

Joints that can be disconnected for test purposes are allowed in protective conductors.
543-03-04

Items not allowed

In general, protective conductors must not have a switch or isolator installed in them, but a multipole switching device, or plug in device is allowed, providing the protective conductor circuit is opened last, and closed no later than the associated phase and neutral conductors.

There is the special case, where an installation is supplied from more than one source of energy which requires independent earthing, and it is necessary to make sure that not more than one means of earthing is connected at any one time. In this case, a switch may be installed between the neutral point and the means of earthing, providing the switch is linked so as to switch the phase and neutral conductors at the same time as the earthing conductor.
543-03-04, 460-01-05

The plain colours of yellow or green shall not be used in flexible cables or cords. Neither shall any bi-colour, other than green/yellow, be used.
514-07-02

In fixed wiring, the single colour green shall not be used.
514-06-02

Plain slip or pin grip sockets shall not be used for conduit joints when they are used as a protective conductor.

Joints in conduit shall be screwed or made with substantial mechanical clamps, so that they are electrically continuous.
543-03-06

Flexible or pliable conduit is not allowed to be used as a protective conductor. Neither is a gas or oil pipe allowed to be used as a protective conductor.
543-02-01

Except for switchboards, busbar trunking, or the metal sheath of a cable, the exposed metalwork of equipment cannot be used as part of the circuit protective conductor of other equipment, except as given in Regulations 543-02-02, 543-02-04 and 543-02-05.
543-02-08

Residual current devices

Where an RCD is used for protection, only the phase and neutral conductors are to pass through the magnetic circuit, the protective conductor being outside that circuit.
531-02-02

Protective conductor colour

The colour combination green/yellow shall only be used for protective conductors. One of the colours shall cover between 30% and 70% of the surface, with the other colour the remainder.

The marking of bare conductors shall be either with multicoloured adhesive tape or other means by equal green and yellow stripes close together. The width of which shall be between 15 mm and 100 mm wide and installed either throughout the conductor's length or in each compartment and at each accessible position.
514-03-01

Caravan sites

Where the installation for mobile touring caravans forms part of a TN-C-S system where the public supply is PME. the protective conductor of each socket-outlet shall be connected to an earth electrode. The socket-outlet circuit shall be protected by a residual current device, and $R_A I_a$ must not exceed 50 volts, where R_A is the total resistance of the earth electrode and protective conductors up to the socket-outlet, and I_a is the residual operating current $I_{\Delta n}$.
608-13-05, 413-02-18, 413-02-19, 413-02-20

Where an RCD is used to protect a mobile touring caravan socket-outlet, or a group of three socket-outlets, the earthing terminal of the socket-outlets must not be connected to the PME terminal.
608-13-05

Data processing equipment complying with BS 7002

Where the earth leakage current from equipment complying with BS EN 60950 does not exceed 3.5 mA, no special precautions are generally necessary unless the cumulative leakage current of a number of items of equipment is of concern.
607-02-02

The earthing arrangements for installations supplying equipment such as industrial control equipment, micro-electronics or computers, having an earth leakage current in excess of 3.5 mA, shall comply with the following regulations:
607-01

Where an installation is supplied through an RCD, and there is more than one item of equipment having an earth leakage current in excess of 3.5 mA, the total leakage current must not exceed 25% of the operating current of the RCD.
607-02-03

Where the total leakage current exceeds 25% of the tripping current of the RCD, the equipment shall be supplied through a double wound transformer or other device, in which the input and output circuits are electrically separate. The protective conductor shall connect the exposed- conductive-part of the equipment to the secondary winding of the transformer. The protective conductor shall comply with one of the arrangements outlined in Regulation 607-02-07.
607-02-03

Where an item of stationary equipment has an earth leakage current between 3.5 mA and 10 mA, it shall be permanently connected to the fixed wiring, or connected by means of a BS EN 60309-2 socket-outlet.
607-02-04

Where an item of stationary equipment has an earth leakage current exceeding 10 mA in normal use, it shall preferably be permanently connected to the fixed wiring of the installation. Alternatively, the equipment can either be:
   a)   connected through a BS EN 60309-2 socket-outlet, providing the protective conductor in the flexible connecting cable is supplemented by a separate additional 4 mm^2 conductor, and an additional contact in the plug and socket-outlet, or the cable complies with 607-02-07 (iii), or
   b)   an earth monitoring system is installed complying with BS 4444 which will automatically disconnect the supply in the event of a loss of continuity of the protective conductor.
607-02-05

Where a final circuit feeds a number of socket-outlets, in an area where it can be expected that the total earth leakage current from the equipment connected to the socket-outlets will exceed 10 mA, the circuit shall be provided with a high integrity protective connection complying with items (i) to (vi) of Regulation 607-02-07.

Alternatively, a ring final circuit can be used, providing there are no spurs taken from that circuit, the minimum size of protective conductor used is 1.5 mm^2, and the ends of the protective conductor are connected into separate earth terminals at the source of the circuit.

607-02-06

Where the fixed wiring of every final circuit is installed in a location intended to accommodate several items of stationary equipment in which the total earth leakage current exceeds 10 mA, the circuit shall be provided with a high integrity protective connection complying with one or more of the following:

1. A single protective conductor with a minimum cross-sectional area of 10 mm^2.

2. A separate duplicated protective conductor, having independent mechanically and electrically sound terminations, protected from mechanical damage and vibration, and made by compression-sockets, soldering, brazing, welding, or mechanical clamps, each protective conductor having a minimum cross-sectional area of 4 mm^2.

3. Duplicate protective conductors with live conductors in a multicore cable, providing that the total cross-sectional area of all of the conductors in the cable is not less than 10 mm^2. One of the protective conductors can be formed by the metallic armour, sheath or braid of the cable, providing its electrical continuity can be assured, and providing it has a minimum cross-sectional area calculated in accordance with Regulation 543-01.

4. An earth monitored protective conductor, the circuit being disconnected in the event of a failure in that protective conductor; the circuit shall comply with the automatic disconnection and equipotential bonding regulations.

5. Duplicate protective conductors formed by trunking and conduit, sized in accordance with Regulation 543-01-03, and having at least a 2.5 mm^2 conductor installed in the same enclosure and connected in parallel with it.

6. Equipment being connected to the supply through a double wound transformer having electrically separate primary and secondary circuits. The circuit protective conductor shall be connected to the exposed- conductive-parts of the equipment, and to a point on the secondary side of the transformer. The protective conductors for the circuit shall be installed in accordance with one of the arrangements (1) to (5) above.

Items 1 to 4 above shall comply with Regulation 607-02-06 unless the protective conductor complies with Regulation 607-02-05.

607-02-07

Where the installation forms part of a TT system and the stationary items of equipment have an earth leakage current exceeding 3.5 mA, the product of the total earth leakage current and twice the resistance of the installation earth electrodes shall not exceed 50. Where this arrangement cannot be achieved, the equipment shall be supplied through a double wound transformer described in (6) of Regulation 607-02-07.

607-03

Where the installation forms part of an IT system, equipment having a high earth leakage current shall not be directly connected.
607-04

Each exposed-conductive-part of data processing equipment, together with the metallic enclosures of Class II and Class III equipment, and any FELV circuit earthed for functional reasons, shall be connected to the main earthing terminal.
A protective conductor which is only being used for a functional purpose, and not as a protective conductor, does not need to comply with Section 543 of the BS7671.
607-05

The manufacturer of the equipment shall be consulted when a 'clean" (low noise) earth is specified, in order to confirm that the arrangements made are satisfactory and suitable for the functional purposes.
545-01

The requirements of protective measures shall take precedence over function purposes when protective and functional earthing purposes are combined.
546-01

Protective conductors compulsory
With the exception of lampholders having no exposed-conductive-parts, a c.p.c. shall be installed to every point in the wiring, and to each accessory in an installation protected by automatic disconnection of the supply.
471-08-08

Chapter 9

Conduit

Material

Conduit and conduit fittings used should have been manufactured in accordance with British Standards:

BS 31 Specification steel conduit and fittings for electric wiring.

BS 731 Flexible steel conduit and adaptors.

BS 4568 Part 1 Steel conduit bends and couplers metric version of BS 31.

BS 4568 Part 2 Plain and threaded fittings for use with conduit in Part 1.

BS 4607 Insulating conduits, fittings and components.

BS EN 60423 Outside diameters of conduits and threads for conduits and fittings.

BS EN 50086 Specification of general requirements for conduit.

BS 6099 Part 2 Section 2.2 Specification of rigid plain conduits of insulating material
511-01-01, 521-04

Corrosion and moisture

Protection shall be provided against corrosion that may occur in damp situations, where metallic conduit is installed in contact with building materials containing magnesium chloride, corrosive salts, lime, and acidic woods, *such as oak.*

Protection includes coating the conduit, or preventing contact by separating with plastics or other suitable material.

Special care shall be taken when installing aluminium conduit in damp situations, particularly with fixing clips. Generally, contact between dissimilar metals shall be avoided, especially where brass with a high copper content is involved, since dissimilar metals can set up electrolytic action in damp situations. (Galvanised steel would however be suitable for fixing clips).

Metal conduit and its fixings shall be corrosion-resistant when exposed to the weather, or installed in damp situations, and must not be in contact with dissimilar metals likely to set up electrolytic action, mutual or individual deterioration.

INF, 522-05-01, 522-03-02, 522-05-03

Drainage outlets are to be provided in conduit systems where moisture may collect e.g., through condensation, and protection against mechanical damage shall be provided where it is subject to waves.

522-03-02, 523-03-03

Conduit bends

The radius of every conduit bend shall be not less than 2.5 times the overall diameter of the conduit. Where cables other than conduit wiring cables are installed in conduit, the radius of conduit bends shall be increased from the minimum stated above, as determined by the cable radius required from the manufacturers (See Table 3 under Selection of Cables).

522-08-03

Solid elbows or tees are only allowed immediately behind an accessible outlet box, inspection fitting or luminaire. Where a conduit run does not exceed 10 metres between outlet points, and the sum of all bends between the outlets is equivalent to only one 90 degree bend or less, then a solid elbow may be installed, up to 500 mm from an accessible outlet box.

INF

Terminations and junctions

Conduit ends shall be free from burrs. Terminations in equipment not fitted with spout entries shall be treated to prevent damage to the cables. *(A female bush on the end of the conduit is a suitable means of achieving this.)*
522-08-01

Where conduit is buried in the structure of the building the conduit shall be completely erected before any cable is drawn in. Adequate means of access for drawing cables in and out of a conduit system shall be provided. Where cables are connected to a conduit system, substantial boxes with ample capacity shall be installed.
522-08-02

Every outlet for cables from a conduit system to a duct, or ducting system, shall be so made that it is mechanically strong, and cables must not be damaged whilst being installed.
522-08-01, 522-08-02

The termination of conduit, non-sheathed cables, or sheathed cables that have had the outer sheath removed, shall be enclosed in fire resistant material that complies with British Standards, unless they are terminated in a box, accessory or luminaire complying with British Standards.

The building structure may form part of the enclosure referred to above.
526-03-03, 526-03-02

Ambient temperature

Conduits shall be suitable for the maximum and minimum values of ambient temperature to which they will be exposed in normal service.

Where non-metallic or plastics boxes are in contact with a luminaire, they shall be suitable for the suspended load, and the temperature to which they will be subjected.
522-01-01

Long straight lengths of rigid plastics conduit must allow for expansion and contraction; *using slip joints in plastics conduit is a means of achieving this.*
522-12-01

Colour code

Where conduit is to be distinguished from other services, it shall be coloured orange in compliance with BS 1710
514-02

Prefabricated systems

Where conduit systems are prefabricated and not wired in situ, allowance shall be made for tolerances in the buildings dimensions. Precautions against damage to such systems shall be taken, so that the conduits or cable ends are not damaged during erection, or during building operations.
522-06-01, 522-08, 522-12-01.

Fire barriers

Where conduits pass through floors, walls, partitions or roofs, the opening through such walls or floors etc., shall be sealed to the same degree of fire resistance as that of the material through which it is passing.
527-02-01

Where conduits pass through walls and floors etc., that have a specified degree of fire resistance, the hole round the conduit in the wall or floor etc., shall be sealed to the same degree of fire resistance. Additionally, the conduit shall be internally sealed to maintain the degree of fire resistance of the material through which the conduit is passing.

Where the conduit system has an internal cross-sectional area not exceeding 710 mm^2, and is a non-flame propagating system, it need not be internally sealed. *(With the exception of 38 mm conduit, all conduits have an internal area less than 710 mm^2.)*
527-02-02

Conduit fixings

The conduit system has to be selected and erected to minimise any damage occurring to the sheath and insulation of the conductors during the installation, use and maintenance, of the system.
522-08-01

The conduit system has to be so supported that the conductors in the conduit are not exposed to undue mechanical strain, and so that there is no undue mechanical strain on the terminations of the conductors. Consideration shall be given to the conductors' own weight.
522-08-05

Cables that are installed in a vertical conduit shall, where necessary, be provided with additional support, so that the cables are not damaged by their own weight.
522-08-04

Distance between fixings for conduit

Conduit size in mm	Maximum distance between fixings in metres					
	Metal		Plastics		Pliable	
	Horizontal	Vertical	Horizontal	Vertical	Horizontal	Vertical
16	0.75	1.0	0.75	1.0	0.3	0.5
20	1.75	2.0	1.5	1.75	0.4	0.6
25	1.75	2.0	1.5	1.75	0.4	0.6
32 & 40	2.00	2.25	1.75	2.0	0.6	0.8
Over 40	2.25	2.5	2.0	2.0	0.8	1.0

INF

Overhead wiring

Cables sheathed with PVC, or having an oil-resisting, flame-retardant, or h.o.f.r sheath for overhead wiring between separate buildings, may be installed in heavy gauge steel conduit providing the area is inaccessible to vehicular traffic.

Where conduit is used to span between buildings above ground, the maximum length of span allowed is 3 metres. No joints are allowed in its span, and the minimum size of conduit shall be 20 mm, and the minimum height allowed above ground 3 metres.
INF

Underground wiring

PVC insulated cables can be installed in conduit underground providing they are buried at sufficient depth to avoid damage by foreseeable ground disturbance.
522-06-03

Conduit used as protective conductor

Plain slip or pin grip sockets shall not be used for conduit joints. Joints shall be screwed or made with substantial mechanical clamps.
543-03-06

Where metal conduit is used as a protective conductor, its electrical continuity shall be assured and its cross-sectional area shall be determined either by application of the formula of 543-01-03, or Table 54G.

Table M54G (Part)

Cross-sectional area of copper conductor	Area required for steel conduit
$S \leq 16$	$2.63 \times S$
$S > 35$	$1.31 \times S$

Area of steel conduit is given in PCA 1 Part 4. The area required for the conduit given above is based on the worst factor given in the tables. *In practice, conduit will always be large enough to act as a protective conductor.*
543-02-05

Where a conduit is used as a protective conductor, its electrical continuity has to be assured, and has to be protected against mechanical, chemical or electrochemical deterioration.
543-02-05, 543-02-04

Where a conduit is used as the circuit protective conductor for more than one circuit, it shall be sized either by:
  a) calculation, using the formula 543-01-03 and the most onerous value of fault current and disconnection time encountered in each of the various circuits, or
  b) selected by using Table 54G, with the cross-sectional area of the largest cable enclosed in the conduit.
543-01-02

When conduit is used as a protective conductor, it shall be protected against mechanical, chemical and electrodynamic effects.
543-03-01

The joints in conduit when used as a circuit protective conductor do not need to be accessible.
543-03-03, 526-04

Conduit shall not be used as a PEN conductor.
546-02-02

Where a protective conductor is used for both protective and functional purposes, the requirements for the protective measures shall prevail.
546-01

Flexible conduit

Flexible or pliable conduit shall not be used as a protective conductor.
543-02-01

CHAPTER 9

Conduit wiring

When conduit is used as a protective conductor, the earthing terminal of each accessory and socket-outlet shall be connected by a protective conductor to an earthing terminal fixed in the associated box or other enclosure.
543-02-07

Conduit installed underground must provide sufficient mechanical protection for the cables installed through it.
522-06-03, 522-08-01

Earthing of equipment with high earth leakage current

Where a final circuit supplies an item of stationary equipment having an earth leakage current exceeding 10 mA in normal service, or where a final circuit supplies a number of socket-outlets in an area where it is expected or known that several items of equipment installed in that area will have a total earth leakage current exceeding 10 mA in normal service, a high integrity protective connection shall be made by duplicate protective conductors. This can comprise a metal conduit sized in accordance with Regulation 543-01, and a conductor having a cross-sectional area not less than 2.5 mm^2 installed in the same conduit and connected in parallel with the conduit. The electrical continuity of the conduit shall be assured by checking that it will not deteriorate due to mechanical, chemical or electrochemical damage.
607-02-07, 607-02-06, 543-02-04

Band I and Band II Circuits.

Band I circuits must not be installed in the same conduit as Band II circuits, unless the Band II circuits are insulated to the highest Band I voltage present.
528-01-01

Fire detection and alarm circuits and emergency lighting circuits must be segragated from all other circuits according to BS 5839 and BS 5266.
528-01-04

Where controls or accessories are mounted in or on a common box, switch plate or block, Band I and Band II cables and terminations shall be segregated from each other by a partition. If the partition is metal, it shall be earthed.
528-01-07

Conduit used for mechanical protection

Short lengths of metal conduit not exceeding 150 mm in length which are inaccessible, do not require protection against indirect contact, *i.e., for earthing purposes.*
471-13-04 (v)

Conduit can be used for protecting cables to be installed outdoors on walls, providing it is suitably protected against corrosion.
522-05-01

Damage during installation

The number of cables drawn into a conduit shall be such that no damage occurs to the cable or conduit during the installation of the cables.
522-08-01

Wiring capacity of conduit

See Table CCC 1 (Cable Capacity Conduit) in Part 3 .

Chapter 10

Construction site installations

Extent of the requirements for construction sites

Except where modified by the following requirements, the general requirements in BS7671 are applicable to construction sites. Thus the requirements for protection against direct contact, overload protection, short-circuit protection, and the sizing of circuit protective conductors, isolation etc., are still applicable.
600-01, 600-02

Types of construction site

Construction sites fall into different categories depending upon the type of work being carried out; the following are the names of different types of site which can be considered construction sites:

1. New building construction, which would include housing sites, factory sites etc.
2. Extensions to existing buildings and the demolition of existing buildings.
3. Alterations and repairs to existing buildings.
4. Works of engineering construction , such as the Thames barrier, Channel tunnel.
5. Earthworks, such as excavating for roads, reservoirs, railway cuttings.
6. Work of a similar nature to the above where electrical, mechanical, civil or building work is being carried out.

All these types of works require compliance with the special requirements for construction sites.
604-01

Where not applicable

The special requirements for construction sites are not applicable to the management buildings on a site. Management buildings can comprise site offices, cloakrooms, site meeting rooms, canteens, restaurants, dormitories and toilets. These buildings and areas are considered to be fixed installations to which the normal requirements apply.

The fixed installations on a construction site are limited to those mentioned above, and to the main switchgear for the site in which the principal protective devices are contained. Any installation other than the management buildings, on the load side of the main switchgear, are considered to be moveable.
604-01

Voltage of equipment

Equipment shall be identified for the particular supply from which it is energised, and shall be suitable for the supply voltage and frequency available. The equipment shall only contain voltages derived from the same source of supply, unless they are control or signalling circuits, or from a standby supply. On large sites, high voltage distribution is allowed.

The maximum voltage that can be used in various locations is as follows:

1. 25 V single-phase SELV for use with portable hand lamps in damp and confined locations.
2. 50 V single-phase, centre tapped to earth, for portable hand lamps in damp and confined locations.
3. 110 V single-phase reduced low voltage, centre tapped to earth for portable hand lamps in general use, portable hand-held tools, and local lighting up to 2 kW.

4. 110 V three-phase reduced low voltage, centre tapped to earth, for single-phase portable hand lamps in general use, portable hand-held tools, and local lighting up to 2 kW, as well as small single-phase and three-phase mobile plant up to 3.75 kW.

5. 230 V single-phase for fixed floodlighting.

6. 400 V three-phase for fixed and movable equipment larger than 3.75 kW.

Where safety and standby supplies are connected, devices shall be provided to prevent the interconnection of different supplies.

604-02-01, 604-02-02, 604-11-06

Protection against indirect contact

Protection against indirect contact shall be by either or both of the following methods:

1. Overcurrent protective device.
2. Residual current device.

Either method gives protection by automatic disconnection. Where an RCD is used for protection the protective conductor must not be combined with the neutral on the load side of the RCD.

413-02-07

Where protection is by means of automatic disconnection of the supply, Regulations 604-04 to 604-08 shall apply such that:

$$Z_s \leq \frac{U_o}{I_a}$$

where Z_S is the earth fault loop impedance of each circuit, U_o nominal phase voltage to earth, and I_a is the current causing the protective device for the circuit to disconnect the circuit within the following times for TN systems:

Voltage to earth U_O volts	Disconnection time 't' seconds
120 V	0.35
From 220 to 277 V	0.2
400 to 480 V	0.05
580 V	0.02

As explained in Part 2, for design U_o is replaced by $U_o c$, the open circuit voltage to earth.

604-04-01, Table 604A, 413-02-08

The above maximum disconnection times are applicable to all movable installations, (i.e. all equipment downstream of the main switchgear and principal protective devices) whether they are directly connected to the supply, or connected through a plug and socket. The exception, however, is reduced low voltage systems installed in accordance with the reduced low voltage requirements and a fixed installation, where the disconnection time allowed is 5 seconds.

604-04-01, 604-04-02, 604-04-06

Where the voltage to earth U_o is 230 V, the earth loop impedances corresponding to a disconnection time of 0.2 seconds for all types of protective devices are given in Table ZS 11 Part 3. For reduced low voltage installations, where the voltage between phases is 110 V, the maximum values of earth loop impedance are given in Table ZS 8. RCDs are to be used if disconnection times cannot be met.

604-04-03, 604-04-04, 604-04-07

Where exposed- conductive-parts are simultaneously accessible, they shall be connected to the same earthing system. They can be connected individually, collectively, or in groups.
413-02-03

Every exposed-conductive-part of the installation shall be connected by protective conductors to the main earthing terminal of the source of the supply.
413-02-06

Use of residual current devices (RCDs)

If the overcurrent protective devices cannot disconnect the low voltage circuits within the time specified in Table 604A, or the fixed installation, or the reduced low voltage circuits within 5 seconds, then protection shall be provided by RCDs, such that $Z_s \times I_{\Delta n} \le 25V$.
604-04-08

Where RCDs are used for automatic disconnection, exposed conductive parts need not be connected to the TN system protective conductors, providing they are connected to an earth electrode, such that $R_A \times I_a \le 25$ V, where R_A is the resistance of the c.p.c and earth electrode.
413-02-17

Where the system is TT

Where exposed-conductive-parts are protected by a single protective device, they shall be connected via a protective conductor to the main earthing terminal, and then to a common earth electrode. Where several protective devices are in series, the exposed-conductive-parts may be connected to a separate earth electrode associated with each protective device. The protective device can either be an overcurrent device or a residual current device, the preferred protection being a residual current device.
413-02-18, 413-02-19

Each circuit must comply with the formula $R_A \times I_a \le 25$ V, where R_A is the resistance of the protective conductor from the exposed conductive part to the earth electrode, plus the resistance of the earth electrode. The value I_a is the current that will cause disconnection of the overcurrent protective device within 5 s, or is the residual operating current $I_{\Delta n}$ of an RCD.
604-05

Where the system is IT

An IT system should not be used unless there is no other system to provide the supply. If an IT system is used, then permanent earth fault monitoring shall be provided.
604-03-01

Where the system is IT, Regulations 413-02-21 to 413-02-25 are applicable, except that the voltage in Regulation 413-02-23 and 413-02-28 is changed from 50 V to 25 V.
604-06-01, 604-07-01

CHAPTER 10

Regulation 413-02-26 is applicable, except that table 41E is changed to the following:

Voltage between phase and neutral U_o and voltage between phases U (volts)	Neutral not distributed t (seconds)	Neutral distributed t (seconds)
120 to 240 volts	0.4	1
220/380 to 277/480 volts	0.2	0.5
400/690 volts	0.06	0.2
580/1000 volts	0.02	0.08

604-06-02

Requirements that are not applicable to construction sites
The group of requirements concerned with installations within the equipotential zone, or circuits feeding fixed equipment or socket-outlets outside the equipotential zone, are not applicable to a construction site. This means that the following requirements are not applicable: 471-08-02 to -05.
604-08-01, 604-08-02, 604-08-03

Socket-outlets
Every socket-outlet shall be protected by one or more of the following:
1. Reduced low voltage system using automatic disconnection.
2. A 30 mA RCD, manufactured to British Standards, having an operating time of 40 ms with a residual current of 150 mA.
3. A SELV supply complying with the SELV requirements.
4. Electrical separation complying with the requirements for electrical separation, each socket-outlet being supplied from a separate transformer.

604-08-03, 604-08-04

Where the system is TT, and where protection is by automatic disconnection, every socket-outlet circuit shall be protected by an RCD complying with $Z_s \times I_{\Delta n} \leq 25$ V.
471-08-06, 604-04-08

Every socket-outlet shall comply with BS EN 60309-2, and be incorporated into a unit manufactured in accordance with BS 4363 and BS EN 60439-4.
604-12-01, 604-12-02

Cable couplers must comply with BS EN 60309-2 and luminaire supporting couplers are not allowed.
604-12-03, 604-13

Distribution and distribution equipment
An assembly comprising the main control gear and principal protective devices shall be installed at the origin of each installation. Each supply and distribution assembly shall incorporate a means of isolating and switching the incoming supply, the isolator being suitable for locking in the off position.
Emergency switching shall be provided on the supply to equipment where it may be necessary to disconnect all phase and neutral conductors in order to remove a hazard.
604-11-01, 604-11-02, 604-11-03, 604-11-04

The general requirements concerning isolation are applicable to construction sites.
600-01, 600-02

Every circuit supplying equipment shall be fed from a distribution assembly complying with BS 4363 and BS EN 60439-4. Each distribution assembly shall comprise:

1. Overcurrent protective devices.
2. Devices affording protection against indirect contact.

If required, socket-outlets can be incorporated in the distribution unit.
604-09-01, 604-11-05

Except for assemblies manufactured to BS 4363 and BS EN 60439-4, equipment shall be protected to at least IP44.
604-09-02

Cable installation
Adequate protection shall be given to cables installed across site roads and walkways, and cables shall be so installed that no strain is placed on the terminations of conductors, unless the termination has been designed to withstand such strain.
604-10-01, 604-10-02

Calculations
Cables shall be sized in accordance with the overload, short-circuit, and protective conductor requirements taking into account the necessary correction factors given in the general requirements. Protection against indirect contact detailed in the requirements for construction sites shall be complied with.
600-01, 600-02

Tables 41C in the BS7671, or PCZ 10 in Part 3 are not applicable to construction sites, since these tables limit the touch voltage by limiting the impedance of a c.p.c.
604-04-05

Testing
The earth loop impedance Z_s should not exceed the value given in the tables when the conductors are working at their normal operating temperature. The Z_s values given in the tables have to be adjusted if the conductors are not at their normal operating temperature when tested.
604-04-04

Chapter 11

Data processing equipment

Extent of requirements

Except where modified by the following requirements, the general requirements in BS 7671 are applicable and the absence of any reference to a Regulation, Section or Chapter means that the general wiring requirements should be referred to.
600-01, 600-02

Manufacturer to be consulted

The manufacturer of the equipment shall be consulted when a low noise earth is specified for a particular item of equipment, to confirm that the arrangements detailed in this Chapter will be suitable for functional purposes.
607-02-01

Leakage current limited

Where the earth leakage current from equipment complying with BS EN 60950 does not exceed 3.5 mA, no special precautions are necessary.
607-02-02

Earthing arrangements

The earthing arrangements for installations supplying equipment such as industrial control equipment, microelectronics, or computers, having an earth leakage current in excess of 3.5 mA, shall comply with the following requirements.
607-01

Where an installation is supplied through an RCD, and where there is more than one item of equipment having an earth leakage current in excess of 3.5 mA, the total leakage current must not exceed 25% of the rated residual operating current of the RCD.
607-02-03, 531-02-04

Where the total leakage current exceeds 25% of the rated residual operating current of the RCD, the equipment shall be supplied through a double wound transformer, or other device, in which the input and output circuits are electrically separate. The protective conductor shall connect the exposed-conductive-part of the equipment to the secondary winding of the transformer. The protective conductor shall comply with one of the arrangements outlined in Regulation 607-02-07.
607-02-03

Where an item of stationary equipment has an earth leakage current between 3.5 mA and 10 mA, it shall be permanently connected to the fixed wiring, or connected by means of a BS 4343 socket-outlet.
607-02-04

Where an item of stationary equipment has an earth leakage current exceeding 10 mA in normal use, it shall preferably be permanently connected to the fixed wiring of the installation with a flexible cable. The equipment can be connected through a BS EN 60309-2 socket-outlet, providing the protective conductor in the flexible connecting cable is supplemented by a separate additional 4 mm^2 conductor, and an additional contact in the plug and socket-outlet. Alternatively, a monitoring

system, complying with BS 4444, which automatically disconnects the supply in accordance with the requirements for earthed equipotential bonding and automatic disconnection of the supply for TN systems, TT systems and IT systems Regulations 413-02-06 to 17, 413-02-18 to 20 and 413-02-21 to 26 respectively, can be used.
607-02-05

Socket-outlets
Where a final circuit feeds a number of socket-outlets, in an area where it can be expected that the total earth leakage current from the equipment connected to the socket-outlets will exceed 10 mA, the circuit shall be provided with a high integrity protective connection complying with items 1 to 6 of Regulation 607-02-07.

Alternatively, a ring final circuit can be used, providing there are no spurs taken from that circuit, the minimum size of protective conductor used is 1.5 mm², and the ends of the protective conductor are connected into separate earth terminals at the source of the ring circuit.
607-02-06

Circuits feeding several items of stationary equipment
Where the fixed wiring of every final circuit is installed in a location intended to accommodate several items of stationary equipment, and where the total earth leakage current exceeds 10 mA, the circuit shall be provided with a high integrity protective connection complying with one or more of the following:

1. A single protective conductor with a minimum cross-sectional area of 10 mm².
2. A separate duplicated protective conductor, having independent mechanically and electrically sound terminations, protected from mechanical damage and vibration and made by compression-sockets, soldering, brazing, welding, or mechanical clamps, each protective conductor having a minimum cross-sectional area of 4 mm².
3. Duplicate protective conductors with live conductors in a multicore cable, providing that the total cross-sectional area of all of the conductors in the cable is not less than 10 mm². One of the protective conductors can be formed by the metallic armour, sheath or braid of the cable, providing its electrical continuity can be assured, and providing it has a minimum cross-sectional area calculated in accordance with Regulation 543-01.
4. Duplicate protective conductors formed by trunking and conduit sized in accordance with Regulation 543-01-03, and having at least a 2.5 mm² conductor installed in the same enclosure and connected in parallel with it.
5. An earth monitored protective conductor, the circuit being disconnected in the event of a failure in the protective conductor. The circuit shall comply with the automatic disconnection and equipotential bonding requirements.
6. Equipment being connected to the supply through a double wound transformer having electrically separate primary and secondary circuits. The cpc shall be connected to the exposed-conductive-parts of the equipment, and to a point on the secondary side of the transformer. The protective conductors for the circuit shall be installed in accordance with one of the arrangements 1 to 5 above.

The protective conductors in 1 to 4 above shall comply with Regulation 607-02-06; the only exception is where Regulation 607-02-05 applies.

607-02-07

TT Systems

Where the installation forms part of a TT system, and the stationary items of equipment have an earth leakage current exceeding 3.5 mA, the product of the total earth leakage current and twice the resistance of the installation earth electrodes shall not exceed 50. Where this arrangement cannot be achieved, the equipment shall be supplied through a double wound transformer described in (6) of Regulation 607-02-07.

607-03

IT systems

Where the installation forms part of an IT system, equipment having a high earth leakage current shall not be directly connected.

607-04

Each exposed-conductive-part of data processing equipment, together with the metallic enclosures of Class II and Class III equipment and any FELV circuit earthed for functional reasons, shall be connected to the main earthing terminal.

A protective conductor which is only being used for a functional purpose and not as a protective conductor, does not need to comply with Section 543 of BS 7671.

607-05

Low noise earthing: safety

The manufacturer of the equipment shall be consulted when a 'clean' (low noise) earth is specified, to confirm that the arrangements made are satisfactory, and suitable for the functional purposes.

545-01

The requirements of protective measures shall take precedence over functional purposes when protective and functional earthing purposes are combined.

546-01

The metallic enclosures of Class II and Class III equipment and the exposed-conductive-part of data processing equipment shall be connected to the main earthing terminal.

When a FELV circuit is earthed for functional reasons, it, too, shall be connected to the main earthing terminal.

Compliance with Section 543 is not required when an earth conductor is used only for a functional purpose.

607-05

Chapter 12

Diversity

Maximum demand
The maximum demand of the installation has to be calculated in amperes.
311-01

Diversity
When calculating the maximum demand current for an installation or circuit, diversity may be taken into account.
311-01

Where the standard circuit arrangements are used for socket-outlets, no allowance for diversity is to be used, because diversity has already been taken into account.
INF

Final circuits
The full-load current of each piece of equipment connected to the circuit is taken, after allowing for any appropriate diversity. The total of all the individual loads on the circuit are then added together to obtain the total load on the circuit. Table 1 can be used to work out the load of each piece of equipment.

Typical final circuit loads

TABLE 1

2 A socket-outlets	Minimum 0.5 A.
Socket-outlets exceeding 2 A.	Rated current I_n
Tungsten lighting	Connected load with minimum allowance of 100 Watts per lampholder
Household cooker	10 A plus 30% of the remaining cooker load plus 5 A if socket is in cooker unit.
Equipment not exceeding 5 VA (Clocks etc.)	Ignore.
Discharge lighting	Rated lamp watts × 1.8 divided by the voltage, providing the circuit is power factor corrected to 0.85 lagging
All other stationary equipment	Full load current (F.l.c.)

INF

TABLE 2 DIVERSITY

ALLOWANCES FOR DIVERSITY AT A DISTRIBUTION BOARD FOR SIZING FEEDER CABLE

Type of circuit connected at distribution board	Diversity to be applied depending upon the type of premises		
	Individual household dwelling & individual dwellings in a large complex	Small business premises Small shops and stores. Very small factories.	Small commercial premises including hotels, boarding houses & guest houses
Heating excluding thermal storage heaters & cookers	Full load up to 10A plus 50% of connected load in excess of 10A	100% of full load of largest appliance plus 75% full load of remaining appliances	100% full load of largest appliance +80% full load 2nd largest appliance +60% full load remaining appliances
Lighting	66% of connected load	90% of connected load	75% of connected load
Cooking appliances	10A + 30% of the remaining cooker load + 5A if socket outlet in cooker control	100% full load of largest appliance plus 80% of the full load of the second largest appliance plus 60% of the full load of any remaining appliances.	
Final circuits inst- alled in accordance with Appendix 5	100% full load of largest circuit plus 40% full load of every other circuit	100% of full load current of largest circuit plus 50% of full load current of every other circuit	
Instantaneous water heaters	100% full load of largest appliance plus 100% full load of second largest appliance plus 25% full load of any remaining appliances		

This table is provided as a guide for those occasions when it is not possible to work out diversity from experience.

TABLE 2 DIVERSITY continued

ALLOWANCES FOR DIVERSITY AT A DISTRIBUTION BOARD FOR SIZING FEEDER CABLE

Type of circuit connected at distribution board	Diversity to be applied depending upon the type of premises		
	Individual household dwelling & individual dwellings in a large complex	Small business premises Small shops and stores. Very small factories.	Small commercial premises including hotels, boarding houses & guest houses
Socket-outlets and stationary equipment not installed in accordance with Appendix 5	100% full load current of the largest point of utilisation + 40% of full load current of every other point of utilisation.	100% full load current of the largest point of utilisation + 75% of full load current of every other point of utilisation current of every other point of utilisation.	100% full load current largest point of utilisation + 75% full load current for main rooms in constant use + 40 full load
Motors excluding lift motors		100% full load current of the largest motor + 80% full load current of 2nd largest motor + 60% full load current of remaining motors	100% full load current largest motor plus 50% full load current of the remaining motors.
Thermostatically controlled water heaters Off peak heating		DIVERSITY NOT ALLOWED	

This table is provided as a guide for those occasions when it is not possible to work out diversity from experience.

CHAPTER 13

The earthing of the installation shall be:
a) continuously effective,
b) able to carry earth fault currents and earth leakage currents without danger, particularly from thermal, thermomechanical, and electromechanical stresses,
d) adequately robust, or have additional suitable mechanical protection,
e) arranged so that the risk of damage to other metallic parts through electrolysis is avoided.

542-01-07, 542-01-08, EW Regulation 8

Where a protective conductor is installed through more than one installation, each one of which has its own earthing arrangement, then either of the following requirements shall be complied with:
a) the protective conductor shall be capable of carrying the maximum fault current likely to flow through it from any of the installations, or
b) the protective conductor shall be earthed within one installation only, and insulated from the earthing arrangements of the other installations; in this case, where the protective conductor is part of a cable, it shall only be earthed in the installation containing the protective device for the cable circuit.

542-01-09

Where the earthing system of an installation is subdivided, then each subdivision must comply with the earthing requirements.

541-01-02

Lightning protection installed

Where lightning protection is installed, the bonding of the lightning protection system to the main earth bar of the electrical installation, shall comply with BS 6651.

541-01-03

Main earthing terminal

The consumer's main earthing terminal shall be connected to the earthed point of the source of electrical energy for a TN-S system, by the supply undertaking to the neutral of the source of electrical energy for a TN-C-S system, and connected by an earthing conductor to an earth electrode for a TT or IT system.

542-01-01, 542-01-02, 542-01-03, 542-01-04

A main earthing terminal or bar shall be provided for every installation, to connect the circuit protective conductors, main equipotential bonding conductors, any functional earthing conductors, and the lightning bonding conductor to the main earthing conductor.

542-04-01

The main earthing terminal shall be accessible, to enable the earthing conductor to be disconnected for test purposes, The joint shall be mechanically strong, reliably maintain electrical continuity, and be capable of disconnection only by means of a tool.

542-04-02

Provision shall be made for the connection of PEN conductors to the main earthing terminal, when the system is TN-C.

542-01-05

Earth electrodes

The following types of earth electrodes may be used:

Chapter 13

Earthing

Fundamental requirements

Where a voltage may appear on exposed-conductive-parts, due to the insulation of live conductors becoming defective, or due to a fault in equipment, then the exposed-conductive-parts shall be earthed so that the electrical energy is discharged without danger.
130-04-01

The arrangement of circuits shall be such that the persistence of dangerous earth leakage currents is prevented.
130-04-02

Where exposed-conductive-parts are earthed, the circuits shall be protected either by overcurrent protective devices, or RCDs, to prevent the persistence of dangerous earth leakage or earth fault currents. Where the phase earth loop impedance is too large to cause the prompt disconnection of the circuit by overcurrent protective devices, because the prospective earth fault current is insufficient, then RCDs shall be used for the circuit's protection.
130-04-03

Where exposed-conductive-parts are earthed, and are simultaneously accessible to extraneous conductive parts of other services, the extraneous-conductive-parts shall be connected to the main earthing terminal of the installation.
130-04-04

The only protective device allowed in an earthed neutral is a linked circuit-breaker, where the related phase conductors are arranged to be disconnected at the same time.
130-05-02

Earthing general requirements

An assessment shall be made to make certain:
1. Any earth leakage currents do not have a harmful effect upon other circuits or equipment.
2. Whether additional connections with earth are required independent of the installation's main earth, to prevent interference with the correct operation of any equipment.
331-01

The supplier (of electricity) shall not connect an installation unless they are satisfied that it complies with the Electricity Supply Regulations 1988, as amended.
Electricity Supply Regulations 1988, amended

Before commencing an installation the type of earthing arrangement shall be determined i.e. the type of system which the installation and supply will comprise.
312-03

Depending upon the requirements of the installation, joint or separate earthing arrangements may be used both for protection or functional purposes, such as equipment that requires an earth for its operation, (quick start fluorescent fittings or computers). Any functional use must not affect the conductor's protective qualities.
542-01-06, 546-01-01, 607

CHAPTER 12

Distribution

The full-load current (without diversity) of each final circuit connected to the distribution board is taken, and the diversities given in Table 2 are then applied to obtain the total current demand on the distribution board.

Table 2 must not be applied to circuits whose load has been reduced by applying Table 1.

The designer of the installation is allowed to apply diversity to a group of circuits that have already had a diversity factor applied to them, but the designer is not allowed to use Table 2 to determine the current demand of that group of circuits.

A suitably qualified electrical engineer may use other methods of determining maximum demand.

Distribution boards shall be capable of carrying the connected load, and not just the current after diversity has been taken into account.

INF

Voltage drop

When calculating voltage drop in a circuit, or in a distribution circuit cable to a group of circuits, the current demand can be used after diversity has been taken into account.

INF

1. Earth rods or pipes, (providing the pipes are not gas or water services).
2. Earth tapes or wires.
3. Earth plates.
4. Underground structural metalwork embedded in foundations.
5. Except for pre-stressed concrete, welded metal reinforcing of concrete embedded in the ground.
6. Metallic sheaths or other coverings of cables, providing:
   a) they are not liable to deterioration through excessive corrosion, and
   b) the cable sheath is in effective contact with earth, and the consent of the cable owner has been obtained to use the cable as an earth electrode, as long as arrangements are made with the cable owner to notify the consumer of any change that is made to the cable, which will affect its suitability as an earth electrode.

542-02-01, 542-02-04, 542-02-05

Earth electrode material

The design used, and the material from which earth electrodes are made, shall be able to withstand damage due to corrosion. The earthing arrangements shall also take into account any increase in earth resistance caused by corrosion. The earth electrode should be buried at a depth such that the earth electrode resistance is not increased by the soil drying or freezing.

542-02-02, 542-02-03

Compliance with other requirements

Every earthing conductor shall also comply with the regulations for protective conductors and where buried in the ground shall comply with Table 54 A.

542-03-01, 541-01-01

Earth electrode connections

Connections to earth electrodes shall be soundly made, so that they are electrically and mechanically satisfactory; they shall have a label to BS 951 permanently attached in a visible position, durably marked in legible type 4.75 mm high, with the words 'Safety electrical connection - do not remove'. The connection must also be suitably protected against corrosion.

542-03-03

Size of earthing conductor

The minimum size of earthing conductors shall be determined either by calculation or by using Table 54G.

By calculation means using the formula:

$$ S \geq \frac{\sqrt{I_f^2 \, t}}{k} \ \text{mm}^2 $$

where S is the cross-sectional area required for the earthing conductor, I is the fault current, t is the disconnection time of the protective device and k is a factor from Table K1A or K1B in Part 4.

Using Table 54G: the requirements being outlined in the table on the following page.

The thickness of any tape or strip conductor used should withstand mechanical damage and corrosion. The calculated minimum sizes are further limited in minimum size when they are buried in the ground. (BS 7430 gives further information)

CHAPTER 13

TABLE 54 G

SIZE OF MAIN EARTHING CONDUCTOR REQUIRED		
Area of phase conductor in mm² S	Area of earthing conductor in mm² when it is the same material as the phase conductor	Area of earthing conductor in mm² when it is a different material to phase conductor
S not exceeding 16	S	$S \times \dfrac{k_1}{k_2}$
S from 16 to 35	16	$16 \times \dfrac{k_1}{k_2}$
S exceeding 35	$\dfrac{S}{2}$	$S \times \dfrac{k_1}{2\,k_2}$

where k_1 is the value of 'k' for the phase conductor and 'k_2' is the value of k for the earthing conductor.
543-01-03, 543-01-04, 542-03-01

Buried earthing conductors

Where cables are buried in the ground they shall be subject to the following minimum cross-sectional areas:

MINIMUM CROSS-SECTIONAL AREAS OF BURIED EARTHING CONDUCTOR		
	Protected against mechanical damage	Not protected against mechanical damage
Protected against corrosion by a sheath	543-01-03 formula or Table 54G	16 mm² copper 16 mm² coated steel
Not protected against corrosion	25 mm² copper 50 mm² steel	25 mm² copper 50 mm² steel

542-03-01

Installation with more than one supply

Where there is more than one source of supply to an installation, and where one of the sources of electricity requires its earth to be independent, a switch may be inserted between the neutral point and earth, providing the switch is linked, so that it connects and disconnects the live conductors at the same time as the earth connection.
460-01-05

Chapter 14

Emergency switching

Emergency switching

Emergency switching is used to remove any hazard to persons, livestock, or property, by rapidly cutting off the electrical supply.

One device can be used to perform all the following functions:

1. Emergency switching.
2. Isolation.
3. Switching off for mechanical maintenance.
4. Functional switching.

Where one device is used for one or more of the above functions, it shall be verified that the device complies with the requirements for each function it performs.

476-01-01

Where required

Emergency switching shall be provided for every part of an installation where it is necessary to prevent or remove a hazard by disconnecting the supply rapidly.

463-01-01

Every machine driven by electricity, and which may give rise to danger, shall be provided with an emergency stopping device.

476-03-02

A means of interrupting the supply on load shall be provided for every appliance not connected to the supply by a plug and socket-outlet designed for the most onerous use intended. The operating device shall be so placed as not to put the operator in danger.

The operating device can be either: part of the equipment or in a readily accessible position external to the equipment.

One means of interrupting the supply can be provided where two or more appliances are installed in the same room.

476-03-04

How installed

Where possible, the emergency switch should act directly on the supply conductors, and be so arranged that a single initiative action will cut off the supply.

463-01-02

An emergency switch shall be installed in a readily accessible position where the hazard might occur, and if appropriate, at any other position from which the device for emergency switching may need to be operated. Additionally, further devices shall be provided where additional emergency switching may be required.

537-04-04

Accessibility

Where emergency switching is provided for machines driven by electricity, and which may give rise to danger, it shall be readily accessible and easily operated by the person in charge of the machine.

Where there is more than one device for manually stopping the machine, the restarting of which could cause danger, provision shall be made to prevent the unintentional or inadvertent restarting of the machine.
476-03-02

The intended use of the premises shall be taken into account, so that access to the emergency switch is not impeded when a foreseeable emergency condition arises.
476-03-01

Emergency switches shall be readily accessible, and suitably durably labelled.
463-01-04

Operational safety

The arrangement of emergency switches must not interfere with the complete operation necessary to remove a hazard, or create another hazard when operated.
463-01-03

Emergency switches shall only be operated by skilled or instructed persons, where the inappropriate operation of an emergency switch could give rise to a greater danger.
476-03-03

The release of an emergency switching device shall not re-energise the equipment concerned.
537-04-05

Emergency stopping

Where movements caused by electrical means may give rise to danger, emergency stopping shall be provided.
463-01-05

Emergency stopping may include energising equipment to effect an emergency stop, *e.g. d.c., injection braking.*
537-04-02(ii)

Emergency stopping may include the supply remaining on for electrical breaking facilities.
537-04-02

Fireman's emergency switch

Exterior electrical installations and internal discharge lighting installations, operating at a voltage exceeding 1000 V a.c., or 1500 V d.c., shall be provided with a fireman's emergency switch in the low voltage circuit. When the voltage exceeds extra low voltage, a fireman's emergency switch shall also be provided if :

   a)   the installation is installed in a shopping mall,
   b)   the installation is installed in an arcade,
   c)   the installation is installed in a covered market.

A temporary installation in a permanent building used for exhibitions is not an exterior installation.
476-03-05

Exclusions

Portable discharge lighting luminaires or signs, not exceeding 100 watts, fed from an accessible local socket-outlet do not need a fireman's switch.
476-03-05

Requirements for fireman's switches

Fireman's switches shall comply with the requirements of the local fire authority, and the following:

1. The switch shall be mounted exterior to the building and adjacent to the equipment when the installation is exterior to the building. Alternatively, a notice is to be mounted adjacent to the equipment, giving the position of the fireman's switch, with a notice mounted at the switch, making it clearly distinguishable,
2. The switch shall be mounted in the main entrance for interior installations, or at a position agreed with the local fire authority,
3. The switch shall be mounted not more than 2.75 metres from the ground, in a conspicuous position, accessible to firemen, unless agreed differently with the local fire authority,
4. Where more than one fireman's switch is installed on any one building, each switch shall be clearly marked to indicate the part of the installation it controls,
5. Where practicable, one master fireman's switch shall be mounted on the building to control all the exterior installations operating at a voltage higher than 1000 V a.c., or 1500 V d.c. A separate master fireman's switch shall be installed in the building to control all the interior discharge lighting.

476-03-06, 476-03-07

Colour, labelling and position of switch handle

Every fireman's emergency switch shall:

1. Be coloured red.
2. Have a 150 mm x 100 mm minimum size nameplate mounted adjacent to it, with the words 'FIREMAN'S SWITCH' in 12.7 mm high minimum size characters marked on it. Where necessary, the size of the nameplate and lettering shall be increased in size, so that it is clearly legible from a distance appropriate to the site conditions,
3. Have the OFF position at the top, and clearly indicate the ON and OFF positions in lettering that is clearly legible by a person on the ground,
4. Be prevented from being inadvertently returned to the ON position, and arranged to facilitate operation by a fireman.

537-04-06

Requirements for emergency switches

The handles or push buttons of emergency switching devices shall be clearly identifiable, and preferably coloured red.

537-04-04

Where practicable, emergency switches shall be manually operated and directly interrupt the main circuit.

537-04-03

Circuit breakers or contactors operated by remote emergency switches shall open on de-energisation of the coils, or be opened by another suitable method.

537-04-03

The emergency device used for interrupting the supply shall be capable of cutting off the full load current of the relevant parts of the installation, taking into account stalled motor conditions.

537-04-01

CHAPTER 14

An emergency switching device shall be capable of being restrained in the OFF or STOP position, or have a latching mechanism keeping the device in the OFF position.
537-04-05

The release of an emergency switching device shall not re-energise the equipment concerned.
537-04-05

An emergency switching device which automatically resets is permitted where both the operation of the emergency device, and the means of re-energising the equipment, are under the control of one person.
537-04-05

Emergency switching shall comprise either:
1. A single device that directly cuts off the incoming supply, or
2. A group of devices, operated by a single initiation, which will remove the hazard by cutting off the appropriate supply.

Devices not allowed

Plugs and socket-outlets shall not be selected as devices for emergency switching. *It is, of course, impossible to stop the client using plugs and socket-outlets as a means of emergency switching, but the installation design should allow for correct emergency switching devices to be installed at the appropriate positions, thus making the use of plugs and socket-outlets unnecessary.*
537-04-02

Chapter 15

Equipotential bonding

Basic measure of protection

One of the basic measures of protection against indirect contact is equipotential bonding and automatic disconnection of the supply. Its application shall be based on the requirements of the system of which the installation forms a part.
413-01-01(i), 413-02-01

Object of bonding

Connecting the extraneous- conductive-parts within the installation to the main earthing terminal and bonding the exposed-conductive-parts to the same terminal creates a zone in which the magnitude of touch voltages appearing between accessible (exposed-and-extraneous-) conductive-parts is reduced, as can be seen from Figure 15.1. *Supplementary bonding minimises touch voltages.*
471-08-01

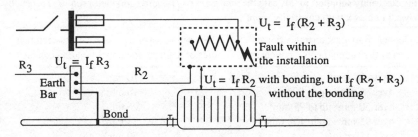

Figure 15.1 - Shock voltage with bonding

Main equipotential bonding

Extraneous-conductive-parts that extend throughout the installation shall be connected to the main earth terminal at the origin of an installation by main equipotential bonding conductors, which may include the following items:

1. Main water pipes.

2. Gas installation pipes.

3. Main service pipes and ducting.

4. Central heating and air conditioning systems.

5. Exposed metallic parts of the building structure.

6. The lightning conductor system.

7. Metallic sheath of telecommunications cable, subject to the owner's or operator's consent.

8. Any extraneous-conductive-part which is in direct contact with earth.
413-02-02, Electricity Supply Regulations

CHAPTER 15

Supplies to other buildings

Where an installation also feeds other buildings, main equipotential bonding shall be carried out in each building.
413-02-02

Conductors not allowed

Conductors or terminations which are liable to initiate electrolytic action, or cause deterioration or hazardous degradation between them, shall not be placed in contact with each other.
522-05-02, 522-05-03

Size of main equipotential bonding conductors

The cross-sectional area of main equipotential bonding conductors shall be at least half the cross-sectional area of the main earthing conductor, the minimum size allowed being 6 mm^2 copper.

The cross-sectional area of the bonding conductor need not exceed 25 mm^2, if made from copper. A material other than copper can be used, providing it has the same conductance as the one made from copper.

If the supply to the installation is PME, the size of the main bonding conductor is determined by the electricity supplier; in this case the minimum size bonding conductor required is usually as follows, but check with electricity supplier:

Copper equivalent cross-sectional area of the supply neutral conductor	Minimum copper equivalent cross-sectional area of main equipotential bonding conductor
Up to 35 mm^2	10 mm^2
over 35 mm^2 up to 50 mm^2	16 mm^2
over 50 mm^2 up to 95 mm^2	25 mm^2
over 95 mm^2 up to 150 mm^2	35 mm^2
over 150 mm^2	50 mm^2

547-02-01, Electricity Supply Regulations

Where bonding connections are required

The main equipotential bonding connection to water and gas services shall be on the consumer's side of any insulating insert in the pipework, and as near as possible to the point at which the service enters the premises. The connection shall be made on the consumer's side of any meter, and between the meter connection and any branch pipework. Where possible the bonding connection shall be made within 600 mm of the meter, as illustrated in Figure 15.2
547-02-02

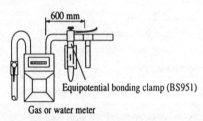

Gas or water meter

Figure 15.2 - Bonding with meters

Supplementary bonding exposed- to extraneous-conductive-parts

With the exception of special situations, such as bathrooms, additional protection is only required if the automatic disconnection times specified for the circuit cannot be met.

Where the automatic disconnection times cannot be fulfilled, then either local supplementary equipotential bonding, or protection by an RCD, has to be used.

413-02-04

Where local supplementary equipotential bonding is to be installed, it shall connect together the exposed- conductive-parts of the circuit concerned, including the earth terminal in socket-outlets, and the extraneous-conductive-parts.

The resistance of the supplementary bonding conductor shall not exceed 50 V divided by the minimum current that will disconnect the circuit to the exposed-conductive-part within five seconds, and providing the bonding conductor is sheathed or otherwise mechanically protected, its conductance shall not be less than half that of the cpc connected to the exposed- conductive-part. Where the bonding conductor is not mechanically protected, its minimum cross-sectional area shall be 4 mm^2.

547-03-02, 413-02-27, 413-02-28

Where an RCD is used instead of equipotential bonding, then $Z_S \times I_{\Delta n} \leq 50$ V. The resistance of the bonding conductor, subject to the minimum sizes specified in the preceding paragraph, shall not exceed

$$R_{bond} = \frac{50 \text{ V}}{I_{\Delta n}}$$

413-02-16, 413-02-28

Bonding two extraneous-conductive-parts together

A supplementary bonding conductor bonding two extraneous-conductive-parts together shall be subject to the following minimum sizes:

1. Not less than 2.5 mm^2, if sheathed or otherwise mechanically protected,
2. A minimum size of 4 mm^2, if not mechanically protected.

547-03-03

Bonding two exposed-conductive-parts together

Where two exposed-conductive-parts are to be bonded together, the supplementary bonding conductor, subject to the above minimum sizes, shall have a conductance not less than that of the smallest cpc installed to the exposed-conductive-parts. See Figure 15.3.

547-03-01

Bonding two extraneous-conductive-parts to an exposed-conductive-part

Where an extraneous-conductive-part is to be bonded to another extraneous-conductive-part, which in turn, is bonded to an exposed-conductive-part, then, subject to the minimum sizes given above, the supplementary bonding conductor shall have a conductance equal to half that of the cpc connected to the exposed-conductive-part and the bonding conductor whose resistance R shall not exceed 50 V + I_a, where I_a is the minimum current needed to disconnect the circuit to the exposed-conductive-part within 5 s. See Figure 15.3.

547-03-03, 413-02-28

CHAPTER 15

Bonding exposed-to extraneous-conductive-parts

Where an exposed-conductive-part is to be bonded to an extraneous-conductive-part, the supplementary bonding conductor, subject to the above minimum sizes, shall be half the size of the cpc connected to the exposed-conductive-part, and the bonding conductor resistance R shall not exceed $50V + I_a$, where I_a is the minimum current needed to disconnect the circuit to the exposed-conductive-part within 5 s. See Figure 15.3.

547-03-02, 413-02-28

Exposed-conductive-part to exposed-conductive-part

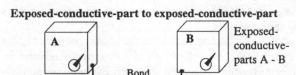

Exposed-conductive-parts A - B

Same conductance as the smallest cpc feeding A or B, providing that it is mechanically protected: otherwise minimum size is 4 mm².

Exposed-conductive-part C to extraneous-conductive-part

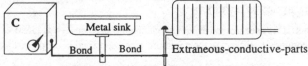

Half conductance of the cpc in 'C' if sheathed or mechanically protected: otherwise minimum size required is 4 mm².

Extraneous-conductive-part to extraneous-conductive-part

Minimum size 2.5 mm² if sheathed or otherwise mechanically protected: 4 mm² if not mechanically protected.

Figure 15.3 - Minimum size of supplementary bonding conductors

General

Supplementary bonding shall be provided by permanent and reliable conductive parts, or by supplementary conductors, or by both. Where a fixed appliance is to be connected to a supplementary bonding conductor, and the appliance is connected to an accessory by a short length of flexible cable, the protective conductor in the flexible cable can also serve as the bonding conductor from the appliance to the earth terminal in the accessory.

547-03-04, 547-03-05

Extraneous-conductive-parts used for bonding

With the exception of gas pipes, oil pipes and flexible or pliable conduit, structural metalwork and extraneous-conductive-parts can be used as protective conductors, providing it can be assured that the construction and connection will maintain the electrical continuity, and protects against mechanical, chemical or electrochemical deterioration, and the cross-sectional area is at least equal

to that calculated from 543-01-03, or Table 54G. The part to be used as a protective conductor must have been suitably adapted and, unless an alternative suitable item providing protection is provided, precautions taken against its removal. However, protective conductors must be incorporated in or in close proximity to phase conductors when overcurrent protective devices are used for protection against electric shock.
543-02-06, 544-01

Bath and shower rooms

With the exception of equipment fed from a SELV circuit, local supplementary bonding shall be carried out in a room with a fixed bath or shower, even when the following conditions apply:

1. When there are no exposed-conductive-parts in the bath or shower room,
2. When the extraneous-conductive-parts are connected back to the main equipotential bonding conductors with metal to metal joints of negligible impedance.

Local supplementary bonding shall be carried out between simultaneously accessible exposed-and extraneous-conductive-parts, and between simultaneously accessible extraneous-conductive-parts.
601-04-02

Electrical equipment installed in the space below the bath shall only be accessible by using a key or tool, and shall be included in the supplementary equipotential bonding in the bathroom.
601-04-03

Caravans and motor caravans

The minimum cross-sectional area of equipotential bonding conductors shall be 4.0 mm^2. These shall be used to bond extraneous-conductive-parts to the protective conductor of the caravan installation. Furthermore, bonding shall be applied at more than one place if the construction of the caravan does not provide continuity. If the caravan is made of insulating material, then isolated metal parts which cannot become live in the event of a fault, need not be bonded. Where part of the structure of the caravan is made from metal sheets, these need not be considered as extraneous-conductive-parts.
608-03-04

When bonding is not required

The following list gives those situations where bonding need not be carried out:

1. Overhead line insulator wall brackets, or any metal connected to them, providing they are out of arm's reach.
2. Inaccessible steel reinforcement in reinforced concrete poles.
3. Exposed-conductive-parts that cannot be gripped or contacted by a major surface of the human body , providing a protective conductor connection cannot be readily made or be reliably maintained; this exception includes isolated metal bolts, rivets, up to 50 mm x 50 mm nameplates, and cable clips.
4. Fixing screws of non-metallic parts, providing there is no risk of them contacting live parts.
5. Short lengths of metal conduit or similar items which are inaccessible, and do not exceed 150 mm in length.
6. Metal enclosures used mechanically to protect equipment complying with the regulations for Class II (double insulated) equipment.

7. Inaccessible, unearthed street lighting (furniture) supplied from an overhead line.
471-13-04

Instantaneous showers having immersed and uninsulated heating elements

The metal water pipe through which the water supply to the heater is provided shall be solidly and mechanically connected to all metal parts (other than live parts) of the heater or boiler. There must also be an effective electrical connection between the metal water pipe and the main earthing terminal, independently of the circuit protective conductor.
554-05-02

The supply to the heater shall be through a double pole linked switch. The switch shall be separate from and within easy reach of the heater. Alternatively the switch can be incorporated in the heater. The electrical connection from the boiler or heater shall be made directly to the switch, and not through a plug and socket.

Where the heater is installed in a room containing a fixed bath or shower, the switch shall be out of reach of a person using the bath or shower; an insulated cord operated pull switch would be acceptable.

The installer of the boiler or heater shall confirm that the neutral cannot be independently opened by any single pole switch etc., from the origin of the installation up to the boiler, before the boiler or heater is connected.
554-05-03, 554-05-04

Where a shower tray is made from a metallic material, or where it is made from concrete, it shall be bonded to the metal water pipe in the shower heater; *in the case of a concrete shower tray, the reinforcing can be used for the bonding connection.*
601-04-02, INF

Chapter 16

FELV and PELV systems

When all the SELV requirements cannot be met with extra-low voltage systems, then FELV systems shall comply with Regulations 471-14-03 to 06 and PELV systems shall comply with Regulation 471-14-02 to give protection against direct and indirect contact.
471-14-01

Extra-low voltage(Band I)
Extra-low voltage does not exceed 50 V r.m.s., a.c., or 120 V ripple free d.c., between conductors, or between conductors and earth.
Definition.

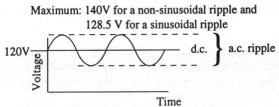

Maximum: 140V for a non-sinusoidal ripple and
128.5 V for a sinusoidal ripple

Figure 16.1 - Ripple created by rectified a.c.

Voltage drop
The voltage at the terminals of fixed equipment must not be less than the minimum value specified in the British Standard relevant to the fixed equipment. Where no British Standard is available for the fixed equipment, the voltage drop within the installation must not exceed a value appropriate to the safe functioning of equipment.

Where the supply is given in accordance with the tolerance allowed by the Electricity Supply Regulations 1988 as amended, the previous requirements are satisfied if the voltage drop from the origin of the installation up to the terminals of the fixed equipment or the socket-outlet terminals does not exceed 4% of the nominal voltage of the supply.
525-01

Functional extra-low voltage (FELV)
FELV is a system which does not comply with all the protective measures required for SELV or PELV, the voltage of which must not exceed 50 V r.m.s., a.c., or 120 V ripple-free d.c.
471-14, Definition

If for functional reasons the requirements for SELV are not complied with, by for instance, the insulation of contactors, relays, switches or transformers not being as required by the SELV requirements, then Regulations 471-14-03 to 06 shall be used to provide protection against direct and indirect contact.
471-14-01, 411-03, HD384.4.47 S2

Extra-low voltage of a FELV system shall not be used as the only means of protection against electric shock.
411-01-02

CHAPTER 16

Protection against direct contact

Where the extra low-voltage system does not comply with the regulations for SELV, other than by live or exposed-conductive-parts being in contact with earth or protective conductors of other systems, then protection against direct contact shall be provided by one or more of the following:

1. Barriers or enclosures that give protection against the insertion of a solid body that exceeds 12 mm in diameter, and up to 80 mm long (IP2X), or contact with a finger (IPXXB).

2. Insulation corresponding to the minimum test voltage required for the primary circuit.

Additionally, Regulation 471-14-04 or 471-14-05 shall be used to ensure protection against indirect contract.

Where the extra-low voltage circuit supplies equipment whose insulation does not comply with the minimum test voltage required for the primary circuit, then any insulation which is accessible shall be reinforced during erection to withstand a test voltage of 1500 V r.m.s., a.c., for one minute.
471-14-03

Protection against indirect contact

Where the primary circuit of the source of extra-low voltage is protected by automatic disconnection of the supply, then exposed-conductive-parts of the extra-low voltage system shall be connected to the primary circuit protective conductor.
471-14-04

Where the primary circuit of the source of supply for the FELV is protected by electrical separation, then the exposed-conductive-parts of the FELV circuit shall be connected to the primary circuit's non-earthed protective conductor.
471-14-05

Barriers or enclosures

Live parts shall be in enclosures or behind barriers giving protection against direct contact, giving a minimum protection against the insertion of a solid body exceeding 12 mm in diameter and up to 80 mm in length. (IP2X) or against a finger touching live parts (IPXXB). Where an opening larger than this is necessary for the replacement of parts, or to avoid interference with the functioning of the equipment, then both the following requirements shall be met:

a) unintentional contact with live parts shall be prevented, and

b) provision shall be made so that a person is sure to be aware that live parts can be touched through the opening, and that they should not be touched.

A larger opening than IP2X or IPXXB is only allowed for equipment or accessories complying with British Standards, where it is impracticable to contain the opening to IP2X or IPXXB, due to the function of the equipment, e.g. a lampholder. Where such a deviation is used, the opening shall be as small as possible.
412-03-01, 471-05-02

Where horizontal top surfaces of barriers or enclosures are readily accessible, then it must not be possible to make contact with live parts with small tools, wires or other thin objects which are more than 1 mm thick (IP4X).
412-03-02

Barriers or enclosures shall be suitable for the normal working conditions to which they will be subjected in service, and be firmly secured in place, to provide stability and durability in order to maintain the degree of separation and protection required.
412-03-03

Where it is necessary to remove a barrier or to open an enclosure, then one of the following precautions has to be taken:

1. Opening an enclosure or removing a barrier must only be possible by using a key or tool, or
2. It must only be possible to open an enclosure or remove a barrier after the supply is disconnected, and it must only be possible to restore the supply after the barrier has been replaced, or the unit reclosed, or
3. An intermediate barrier giving a degree of protection of at least IP2X or IPXXB is installed to prevent contact with live parts.

The above does not apply to BS 67 ceiling roses, BS 3676 cord operated switches, BS 5042 or BS EN 61184 B.C. lampholders or to BS EN 60238 E.S. lampholders.
412-03-04

Socket-outlets
Socket-outlets and luminaire supporting couplers used for functional extra-low voltage shall not admit plugs for different systems (voltages or SELV) used in the same premises.
471-14-06

Protective extra-low voltage (PELV)
Where an extra-low voltage system complies with the Regulations for SELV (separated extra-low voltage) except that it is electrically connected to earth at one point only, it is known as a PELV system.
Definition

Protection against direct contact has to be provided by either:

a) barriers or enclosures which provide protection against hazardous parts being touched with a finger (IPXXB) or against items larger than 12.5 mm diameter and 80 mm long (IP2X), or

b) protection is given by insulation that will withstand a voltage of 500 V d.c. for 1 minute, when one point of the extra-low voltage circuit is connected with earth, and all the other requirements for SELV circuits having been complied with.
471-14-02

Where the equipment is within a building where main equipotential bonding is carried out, protection against direct contact is not required providing the equipment is used in a dry location and contact by large parts of the body are not expected to occur and the voltage does not exceed 25 V a.c. r.m.s. or 60 V ripple free d.c.

For all other situations within the equipotential zone the voltage must not exceed 6 V a.c. r.m.s. or 15 V ripple free d.c.
471-14-02

Chapter 17

Generators

Generators - Scope
Amendment No 2 to BS 7671 introduced some new particular requirements for LV and ELV installations incorporating generating equipment. Requirements are separately identified for generating sets supplying installations which are not supplied by the public supply, for generators employed as an alternative source to that of the public supply and for those supplying installations in parallel with the public supply, or for any appropriate combination of the above.

Regulation Group 551-01 sets the scope as applying to LV and ELV installations incorporating generating sets. It does not refer to combined self-contained ELV source and load for which a specific product exists that includes electrical safety requirements.

Before a generator set is installed in an installation connected to the public supply the electricity supplier must be informed and the supplier's specific requirements must be ascertained.

The scope also identifies the types of power sources which are considered, namely, turbines, combustion engines, electrochemical sources, photovoltaic cells, electric motors or other suitable sources. The types of generators, considered in terms of their electrical characteristics, are mains excited and separately-excited synchronous generators, mains-excited and self-excited synchronous generators, and mains-commuted and self-commuted static inverters with or without bypass facilities. The purpose for which generators are to be employed are given as supplies to permanent installations, supplies to temporary installations, and supplies to portable equipment which is not connected to a permanent fixed installation.
551-01

Generator characteristics
The means of excitation and commutation of the generator must be appropriate for the intended use so as to not impair the safe functioning of other sources. The prospective fault current must be assessed for every independent source (or combination of sources), and the short-circuit capacity of protective devices must be selected so as to be suitable for all the sources likely to be used to supply the installation.

Where a generating set supplies an installation, either a sole source or as a standby, the capacity and operating characteristics must be such as not to be a danger or cause damage on connection or disconnection of the load. Means of load-shedding must be provided where loading is totally beyond the generator capacity.
551-02

Extra-low voltage systems supplied from more than one source
More than one SELV or PELV sources may be used to supply equipment, but either all sources must be SELV or all sources to be PELV. SELV source requirements must be met for the sources. Where one of

the sources does not comply with requirements for SELV sources the requirements relating to FELV must be applied.

Where it is necessary to maintain a supply to an ELV system when one or more parallel sources fail, it is essential that the remaining source is capable of supplying the load on its own. Additionally, provision must be made so that, on the loss of the LV supply to the ELV system, no damage is caused or danger can result to other ELV systems.
411-02, 413-03-01, 551-03

Protection against indirect contact including automatic disconnection
Protection against indirect contact must be provided in respect of each source which may be run independently. For example, this means that an installation fed by standby generator must have suitable protective devices for automatic disconnection of supply, bearing in mind the likely reduced fault level. The protective measure against indirect contact must be EEBAD, except as modified by Regulation 551-04-04, 551-04-05 or 551-06.

Where installations incorporate a generating set which provides a switched alternative supply to the public supply (standby systems), the generators for standby systems must not rely on the public supply means of earthing. A suitable independent earth electrode must be provided.

For protection against indirect contact which relies on automatic closure of a bypass switch and the protective devices on the supply side does not disconnect within the time limits, supplementary bonding must be carried out on the load side of the inverter and the following equation must be met:

$$R \leq \frac{50}{I_a}$$

Where: R is the resistance of the supplementary bonding conductor,
I_a is the maximum fault current which can be supplied by the static inverter alone, for a period up to 5 seconds.

Precautions must be taken so that the correct operation of protective devices is not impaired by dc currents generated by a static inverter, or by associated filters. Alternatively, selection of devices must take account of such dc currents.

There are additional requirements for protection by automatic disconnection where the installation and generating sets are not permanently fixed. Generating sets which are not fixed, portable or those intended to be moved, protective conductors between equipment must be selected to Table 54G and must be incorporated in suitable cord or cable. Additionally, irrespective of the system type, an RCD (30mA or less) must be employed for protection against indirect contact.
551-04, 413-02-27 and 413-02-28

Protection against overcurrent
The means of detecting overcurrent in a generating set, as opposed to the conductors, must be located as near as practicable to the generator terminals.

For a generating set operating in parallel, either with the public supply or with another generating set, the circulating harmonic currents must be limited to the thermal rating of the conductors. There

arc five options for limiting the circulating harmonic currents: generating sets equipped with compensation windings, provision of suitable impedance in connection of the generator star points, provision of interlocking switches to interrupt the circulatory circuit, but which do not impair protection against indirect contact, provision of filters, or other suitable means.
551-05

Standby systems
Additional requirements apply for installations where the generator serves as a switched alternative to the main supply. Precautions must be taken to prevent the generating set operating in parallel with the public supply by one or more of the available options, which include, mechanical interlock with one key, electrical-controlled changeover device, three-position break-before-make changeover switch, or other means providing equivalent security.
For TN systems, where the neutral is not distributed, any fitted RCD must be positioned so that malfunction due to any parallel neutral earth path is avoided.
551-06

Generating sets operating in parallel with the public supply
When a generating set is to run in parallel to the public supply, care must be taken to avoid adverse effects to the public supply and other installations, in consideration of: power factor, voltage changes, harmonic distortion, unbalance, starting, synchronising, and voltage variation. The use of automatic synchronising systems which consider frequency, phase and voltage is preferred.

In the event of loss of public supply, deviation in voltage or frequency of supply, protection must be provided to disconnect the generating set from the supply. Such protection must be agreed with the Public Electricity Supplier.

Means must be provided to prevent connection of the generating set to the public supply if the voltage or frequency is outside the normal limits of the public supply.
Means of isolation, accessible to the Public Electricity Supplier at all times, must be provided for the generating set. If the generator set is also to serve as standby set, the requirements of Regulation Group 551-06 also apply.
551-07

Chapter 18

Identification and notices

Identification of equipment

Unless there is no possibility of confusion, switchgear and controlgear shall be labelled, and where danger could arise because it is remote from the operator, a suitable indicator complying with BS EN 60073 shall be placed at a position visible to the operator.

514-01

Protective devices for circuits, including those in a distribution board, shall be identified so that the circuits protected may be easily recognised.

514-08

The nominal current rating appropriate to the circuit shall be on, or adjacent to, every fuse or circuit breaker. The nominal current rating for rewirable fuses to BS 3036 should be the manufacturer's recommendation. Where these are not available then the table in the overcurrent chapter can be used.

533-01-01, 533-01-02

Where it is practicable, cables should be arranged or marked so as to make it easily identifiable for carrying out: inspections, tests, repairs or alterations to an installation.

514-01-02

Diagrams and charts

Diagrams, charts, or tables, or their equivalent, have to be provided, showing the size and type of conductors, and the points served by a circuit, together with the location of the circuit's protective devices, isolators, and switches, and the information necessary to identify these items, any symbols used complying with BS EN 60617.

A description has to be given of the method chosen for protection against indirect contact (from 413-01-01), together with the installation's characteristics that will give that protection.

514-09

Voltage exceeding 250 V or different voltages

Where it is not expected that equipment or an enclosure contains a voltage exceeding 250 volts, the equipment shall be labelled with the maximum voltage present, the label being clearly visible before access to the equipment is possible.

Where a voltage in excess of 250 volts exists between two simultaneously accessible live parts in separate enclosures, a label shall be provided, warning of the maximum voltage present, which label shall be clearly visible before access to live parts can be made.

Where access may be available to live parts, a notice of the voltages present shall be made where different nominal voltages exist between different parts of equipment or switchgear,

514-10

Earthing or bonding label

At the point of connection of every earthing conductor to an earth electrode, and a bonding conductor to an extraneous-conductive-part, and a main earthing terminal not part of the main switchgear, a permanent label shall be fixed. The label shall be durably marked, complying with BS 951, containing the words 'Safety Electrical Connection - Do Not Remove'.

514-13-01

CHAPTER 18

Periodic inspection and testing notice

On completion of an installation, a notice having a minimum character size of 14 point, and made from durable material so that the notice will remain readable throughout the life of the installation, shall be fixed as close to the origin of the installation as possible. *(Usually the incoming main switch.)* The notice stating that the installation shall be periodically inspected and tested, giving the dates of the last inspection, and the recommended date of the next inspection, is illustrated below.

> IMPORTANT
>
> This installation should be periodically inspected and tested, and a report on its condition obtained, as prescribed in BS 7671 (formerly the IEE Wiring Regulations for Electrical Installations) published by the Institution of Electrical Engineers.
> Date of last inspection ....
> Recommended date of next inspection ....

This notice need not be installed for highway power supply cables, where the installation is subject to a programmed Inspection and Testing procedure.
514-12-01, 611-04-04

Where earth-free local equipotential bonding or protection by electrical separation is used, a warning notice shall be installed in a prominent position adjacent to every access point to the location concerned. The warning notice shall be durably marked with 14 point characters, and contain the following warning:

> The equipotential protective bonding conductors associated with the electrical installation in this location
> MUST NOT BE CONNECTED TO EARTH
> Equipment having exposed-conductive-parts connected to earth must not be brought into this location.

514-13-02

Colour identification

Where electrical services such as conduit are required to be distinguished from other services or pipelines, the colour used for identification of the electrical services shall be orange in compliance with BS 1710.
514-02

The colours reserved exclusively for identification of protective conductors are green and yellow, and this combination must not be used for any other purpose. One of the two colours shall cover between 30% and 70% of the surface, the other colour covering the remaining surface.

Where a bare conductor or busbar is used as a protective conductor, it shall be identified with green and yellow stripes, each being between 15 mm and 100 mm wide, and placed close together. The stripes shall either be installed throughout the length of the conductor, or placed in each compartment, and at each accessible position. Where adhesive tape is used for this purpose it shall be bicoloured green and yellow.
514-03

The single colour green shall not be used.
514-06-02

Every single core and multicore non-flexible cable shall be identified throughout its length, red, yellow, or blue indicating the phase conductors, and black the neutral conductors. Paper-insulated cables, or PVC insulated armoured auxiliary cables, can have cores numbered in accordance with British Standards. Generally 1, 2 and 3 are used for phase conductors, and 0 for the neutral.

MICC cables can be identified at their terminations with tapes, sleeves, or discs of the appropriate colour.

Any colour used for identification of a switchboard busbar shall comply with the colours given for cables.

A bare conductor shall be identified where necessary by tape, disc, or sleeve of the same colour as that used to identify cable cores.

514-06-01, 514-06-04, 514-06-03

Red, yellow, and blue may be used as phase colours for cables up to final circuit distribution boards, but single-phase circuits from a final distribution board shall be coloured red and black.

The colour cream shall be used for the identification of functional earthing conductors.

Table 51A

Every core of a flexible cable or cord shall be identified throughout its length, Brown for phase, Blue for neutral, Green/yellow for the protective conductor.

514-07-02

Fireman's emergency switch

Where a fireman's emergency switch is not located adjacent to the equipment, a notice shall be placed adjacent to the equipment, indicating where the switch is located. A further notice shall be placed at the switch, so that it is clear which equipment that switch operates.

476-03-07

Where more than one switch is installed on any one building, each switch shall be clearly marked to indicate which installation it controls.

476-03-07

Every fireman's switch shall be labelled with a nameplate 150 mm x 100 mm, containing the letters 'FIREMAN'S SWITCH'. The size of these words shall be appropriate to the circumstances, but in any event, not less than 36 point characters shall be used.

537-04-06

Isolation

Where an installation is supplied from more than one source, a durable warning notice shall be permanently fixed in such a position that any person wanting to operate any of the main switches is warned that all switches shall be operated to isolate the complete installation.

460-01-02

Where a single device does not isolate all the live parts in an enclosure or equipment, a durable warning notice shall be permanently fixed warning that additional isolators require to be operated before access is made to live parts.

514-11

All devices used for isolators shall be clearly identifiable, by either; durable marking or its position to indicate the installation or circuits which they isolate.

461-01-05

CHAPTER 18

Mechanical maintenance
Devices used for switching off for mechanical maintenance shall be readily identifiable by durable labelling where necessary.
462-01-02

Emergency switches
Devices used for emergency switching shall be durably marked.
463-01-04

Residual current devices
A notice using indelible characters, not smaller than 14 point, shall be fixed in a prominent position at or near the origin of the installation when the installation incorporates a residual current device. The notice shall read:

> 'This installation, or part of it, is protected by a device which automatically switches off the supply if an earth fault develops. Test quarterly, by pressing the button marked "T" or "Test". The device should switch off the supply, and should then be switched on to restore the supply. If the device does not switch off the supply when the button is pressed, seek expert advice.'

514-12-02

Areas for skilled and instructed persons
Warning signs shall be provided clearly and visibly to indicate areas reserved for skilled and instructed persons.
471-13-03

Street lighting and highway supplies
Except where the method of installation prohibits the marking of underground cable, it shall be marked with cable covers or marking tape.

For the purposes of identification, ducting, marker tape, or cable tiles used with highway power supply cable, shall either be marked or colour coded, so that they are distinct from other services.
611-04-02, 611-04-03, 522-06-03

Caravans
A notice that will permanently be legible and easily readable throughout the life of the installation, shall be installed at the caravan or motor caravan electrical intake point. This notice shall give the nominal voltage and frequency for which the caravan's, or motor caravan's, installation has been designed, and the rated current of the caravan or motor caravan installation.
608-07-03

All extra-low voltage socket-outlets shall have their nominal voltage clearly marked on the socket-outlet.
608-08-03

A label manufactured from durable material, with easily legible characters, shall be provided next to the main switch in every caravan, advising the user what to do before connecting the electrical supply to the caravan, and what to do before disconnecting the supply from the caravan on leaving the caravan site. This label is illustrated as follows:

INSTRUCTIONS FOR ELECTRICITY SUPPLY

To connect

1. Before connecting the caravan installation to the mains supply, check that -
   a) the supply available at the caravan pitch supply point is suitable for the caravan electrical installation and its appliances, and
   b) the caravan main switch is in the OFF position.
2. Open the cover to the appliance inlet provided at the caravan supply point, and insert the connector of the supply flexible cable.
3. Raise the cover of the electricity outlet provided on the pitch supply point, and insert the plug of the supply cable.
   THE CARAVAN SUPPLY FLEXIBLE CABLE MUST BE FULLY UNCOILED TO AVOID DAMAGE BY OVERHEATING.
4. Switch on the caravan main switch.
5. Check the operation of residual current devices, if any, fitted in the caravan by depressing the test buttons.
   IN CASE OF DOUBT, OR IF AFTER CARRYING OUT THE ABOVE PROCEDURE THE SUPPLY DOES NOT BECOME AVAILABLE, OR IF THE SUPPLY FAILS, CONSULT THE CARAVAN PARK OPERATOR OR THE OPERATOR'S AGENT, OR A QUALIFIED ELECTRICIAN.

To disconnect

6. Switch off at the caravan or motor caravan and unplug both ends of the cable.

Periodical Inspection

Preferably not less than once every three years, and more frequently if the vehicle is used for more than normal average mileage for such vehicles, the caravan electrical installation and supply cable shall be inspected and tested, and a report on their condition obtained, as prescribed in BS 7671 (formerly the Regulations for Electrical Installations) published by the Institution of Electrical Engineers.

608-07-05

Chapter 19

Index of protection

Regulation 6 of the Electricity at Work Regulations 1989, requires the correct selection of equipment for the environmental conditions that can foreseeably occur.

The IP (Index of Protection) codes can be of assistance in determining the type of enclosure required to comply with Regulation 6. They give a means of specifying your requirements to manufacturers, or determining whether manufacturers' equipment is suitable for the conditions envisaged.

This chapter gives an outline of the IP codes, and how the numbering system works. Full information on the IP codes and the tests to be carried out for the various degrees of protection are given in BS EN 60529, which follows IEC 529, by providing the optional extension of the IP codes by an additional letter A, B, C, or D. These letters indicating that the actual protection of persons against access to dangerous parts is higher than indicated by the first characteristic numeral.

The letters can be used on their own to show the protection given, however in this instance the letter is placed after the second characteristic number, for example IPXXB. Additionally, a supplementary letter can be used after the second characteristic numeral giving supplementary information.

The basic code only needs to contain two characteristic numbers, the letters IP being followed by two numerals. In France, a third number is used to indicate the degree of protection against mechanical impact (UTE C 20-010 French Standards) which would have been well worth incorporating into the European Standard. For completeness details of the third characteristic numeral are given.

Figure 18.1 shows how the characters are laid out in the code, except that in the UK the third characteristic numeral is not shown.

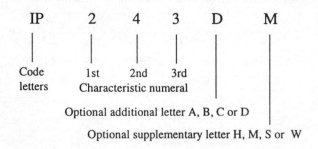

Figure 18.1 - Outline of the code

Details of each characteristic number and letter will be given with a brief explanation of how to use the IP codes. For additional information BS EN 60529 should be consulted.

First characteristic numeral

The first characteristic numeral defines two types of protection. First it defines what degree of protection the enclosure gives against persons gaining access to dangerous parts. Secondly it defines what degree of protection the enclosure gives against the ingress of solid foreign bodies. To comply with the protection against access to dangerous parts adequate clearance must be provided between the test probe and the dangerous parts.

0   No protection of equipment against the ingress of solid objects, and no protection against contact with live or moving parts.

1   No protection against deliberate access, but protection against solid objects exceeding 50 mm diameter.

2   Protection against contact with live and moving parts inside an enclosure by fingers and solid objects exceeding 12 mm. diameter and up to 80 mm. in length.

3   Protection against contact by tools and wires, or other solid objects exceeding 2.5 mm. diameter.

4   Protection against objects of a thickness greater than 1.0 mm.

5   Complete protection against contact with live or moving parts within the enclosure, and protection against the ingress of dust sufficient in quantity to interfere with the satisfactory operation of the enclosed equipment.

6   Complete protection against contact with live or moving parts, and against the ingress of dust.

Second characteristic numeral

The second characteristic numeral defines the degree of protection against the ingress of water. Compliance with any characteristic numeral up to 6 also means compliance with any numeral below 6. However, an enclosure which has the characteristic numeral 7 or 8 is not suitable for categories 5 or 6 (water jets), unless these categories are also specified. This is usually done by specifying both degrees, e.g., IPX5/IPX7.

0   No protection against the ingress of water.

1   Protection against drops of condensed water which will have no harmful effect on the enclosure.

2   Protection against drops of liquid which will have no harmful effect when the enclosure is tilted at any angle up to 15 degrees from its normal position.

3   Protection against rain or spray at an angle up to 60 degrees from the vertical position.

4   Protection against liquid splashed from any direction.

5   Protection against water projected from a nozzle against the enclosure from any direction.

6   Protection from water from heavy seas or water projected in powerful jets.

7   Protection against water entering the enclosure in harmful quantities, when the enclosure is immersed in water under defined conditions of time and pressure.

8   Protection against water entering the enclosure, when the enclosure is immersed for an indefinite period in water under specified pressure.

Third characteristic numeral

The third characteristic numeral is not part of BS EN 60529, but is derived from the French standard mentioned earlier. It is useful since it specifies the degree of protection against mechanical impact that equipment will withstand.

0   No protection.

1   Protection against an impact energy of 0.225 joule, i.e. impact of 150 g from a height of 150 mm.

2   Protection against an impact energy of 0.375 joule, i.e. impact of 250 g from a height of 150 mm.

3   Protection against an impact energy of 0.500 joules, i.e. impact of 250 g from a height of /200 mm.

5   Protection against an impact energy of 2.00 joules, i.e. impact of 500 g from a height of 400 mm.

7   Protection against an impact energy of 6.00 joules, i.e. impact of 1.5 kg from a height of 400 mm.

9   Protection against an impact energy of 20.00 joules, i.e. impact of 5 kg from a height of 400 mm.

Optional additional letter

An additional letter A, B, C or D indicates that the protection offered to persons against contact with dangerous parts in the enclosure is higher than that indicated by the first characteristic numeral. The letter can also be used on its own to indicate that it is the only protection against touching dangerous parts. In theses circumstances the first characteristic numeral is replaced with an X with the letter following it. It also implies that the protection against the ingress of foreign bodies is not specified.

A   Protected, by there being adequate clearance from dangerous parts, against the back of the hand touching dangerous parts.

B   Protected, by adequate clearance from dangerous parts, against a 12 mm diameter by 80 mm long finger touching dangerous parts.

C   Protected, by adequate clearance from dangerous parts, against access with a 2.5 mm diameter by 100 mm long tool.

D   Protected, by adequate clearance from dangerous parts, against access with a 1.0 mm diameter by 100 mm long wire.

Supplementary letter

Supplementary letters have been introduced to give more information on the degree of protection. These supplementary letters can be introduced into product standards, but must conform to the basic safety standards of BS EN 60529. So far, four supplementary letters have been agreed.

It is important to note that if the IP code does not contain the letters S or M then the protection is all embracing and is available whether or not the equipment is stationary or in motion.

H   High voltage apparatus.

M   Tested against the ingress of water having harmful effects when the moving parts of the equipment are in motion.

S   Tested against the ingress of water having harmful effects when the moving parts of the equipment are stationary.

W   Suitable with specified weather conditions and provided with additional protection for such conditions.

Using the IP code

The letters IP can be followed by either two or three numerals, with or without additional letters and/ or supplementary letters. Each numeral indicating the degree of protection either required or provided, any additional letter indicating the additional protection provided for persons, with supplementary letters giving additional information.

Where a degree of protection is not specified or provided, the numeral is replaced by an "X", this means that there may be some protection, but the degree of protection is not specified, e.g. IP2X indicates that the degree of protection against the ingress of liquid is not specified. The "X" is used instead of "0" since "0" would indicate that no protection was given.

Where a degree of protection is given, then protection is afforded against all the lower degrees, e.g. equipment specified as having IP66 protection would indicate that it was also protected against 5, 4, 3, 2, 1 and 0, for both the first and second numerals. The exception is for IPX7 and IPX8, which do not indicate protection against the lower degrees of protection.

When the degree of protection against impact is agreed, and BS EN 60529 is amended, then the IP code will be expressed with three numerals, e.g. IP 243 which indicates:

IP 2	First characteristic numeral.
IP§ 4	Second characteristic numeral.
IP§ § 3	Third characteristic numeral (not given in BS EN 60529).
IP§§§D	Additional letter.
IP§§§§M	Supplementary letter.
IP243DM	The above in code form.

In the above examples the numerals and letters convey the following information:

2. Persons are protected against access to dangerous parts with their fingers. Solid foreign bodies having a diameter of 12.5 mm or larger are stopped from entering the enclosure.
4. The enclosure will not suffer any harmful effects by water being splashed at it from any direction.
3. The equipment will withstand an impact energy of 0.5 joules, which is 250 gram from a height of 200 mm.
D. Protects persons handling wires with a diameter of 1.0 mm or larger, and a maximum length of 100 mm, from touching dangerous parts even if the wire penetrates to its maximum length.
M. No harmful effects due to the ingress of water when rotating parts of the machine are moving. This implies that protection is not available when the equipment's moving parts are stationary. If it did the letter M would be omitted from the code.

Chapter 20

Inspection and testing

Introduction

A complete chapter explaining Inspection and Testing is given in Part 2, and the following is only a reminder of the requirements relating to inspection and testing.

Requirement for testing

On completion of a new installation or of an extension or alteration to an existing installation, it has to be verified, by inspection and testing, that the requirements of the Fundamental Requirements for Safety in the BS 7671 have been met. Periodic inspection recommendations shall be given to the person ordering the work by the person carrying out the inspection and test or by a person acting on their behalf.
130-10

Inspection and test have to be made to verify that the regulations have been complied with. The inspection and test can be carried out during the erection of the installation, or on completion of the installation, but shall be carried out before the installation is put into service.
711-01-01, 721-01-02

General requirements

Precautions have to be taken to ensure that any tests carried out do not cause danger to persons, livestock or property, and must not damage installed equipment. Inspection must always precede testing.
711-01-01, 712-01-01

The information on which the installation design was based has to be made available to the person carrying out the inspection and testing, along with the drawings or schedules for the installation.

The inspection and test has to verify that the requirements of the BS 7671 have been complied with; this includes the basic design requirements, such as the characteristics of the supply:

1. Nominal voltage,
2. Nature and frequency of current,
3. Prospective short-circuit current at the origin of the installation,
4. Type and rating of protective device at the origin of the installation,
5. Suitability of the requirements for the installation, including maximum demand,
6. The external phase earth loop impedance.

Other design requirements:

1. Protection against direct and indirect contact,
2. Protection against short circuit currents,
3. The requirements for overload protection,
4. Correct size of earthing and protective conductors,
5. Discrimination between protective devices.

711-01-01, 711-01-02, 311-01, 312-01, 313-01

Drawings or schedules shall be given to the person carrying out the inspection and testing showing:

1. The position and type of isolators, switches, etc,

2. The position and type of outlet points,
3. The position, type, and size of protective devices,
4. The size and type of conductors, together with the method of wiring,
5. Any equipment likely to be damaged by a particular test.

Where symbols are used to show equipment etc. on drawings or in schedules, they are to comply with BS 3939.

Additionally, the person carrying out the inspection and test has to be given a statement advising the method used for protection against electric shock from both direct and indirect contact. **711-01-02, 514-09**

Visual inspection

An inspection has to be carried out, to check that the materials are correctly erected, comply with British Standards, are the same as stated on the design drawings, and have not been damaged during erection, and that they have been installed in safe zones or protected against mechanical damage. The inspection will include making certain that all connections are tight, conductors are correctly identified, correct connections have been made at accessories and equipment and that equipment is properly secured in place. Confirming compliance with the requirements for protection against indirect contact will include checking protective conductors, main equipotential bonding conductors, and supplementary bonding conductors, as well as checking that earthing conductors have been properly installed and connected.

Visual inspection includes checking that fire barriers, isolators, switches, and protective devices have been installed, and that such items have been correctly labelled where necessary.

Protection against direct contact has to be verified by checking insulation of live parts, protection by barriers or enclosures, obstacles or placing out of reach.

Inspection includes checking that Band I and Band II circuits have not been incorrectly mixed, that all protective devices are of the correct rating and properly labelled and that equipment and accessories are of the correct type for the environmental conditions.

Whether sufficient space has been allowed for access and maintenance of switchgear and equipment has to be checked. **712-01-03**

Sequence of tests

The following tests have to be carried out in the following sequence:
1. Continuity of protective conductors, including main and supplementary bonding conductors.
2. Continuity of ring final circuit conductors.
3. Insulation resistance.
4. Insulation applied on site.
5. Protection by electrical separation.
6. Protection by barriers or enclosures provided during erection.
7. Insulation of non-conducting floors and walls.
8. Polarity.
9. Earth electrode resistance.
10. Earth fault loop impedance.
11. Operation of RCDs.
12. Functional testing of switchgear and equipment.

If an installation fails any of the above tests, then any previous test already carried out, and affected by the fault indicated, shall be repeated after remedial work has been carried out.
713-01

Continuity of protective conductors

Every protective conductor has to be tested to make sure that it is electrically sound and properly connected. The test can be made with a d.c. ohmmeter having a voltage range 4 V to 24 V and a test current of at least 200 mA. Where ferrous protective conductors are used such as conduit, trunking or the armour of cables, and the inspecting engineer is in doubt concerning the protective conductor with the test just described, the test can be made with a phase earth loop impedance tester passing a test current of about 20 to 25 A for 40 ms. (*This is the better test for ferrous protective conductors since the reactance of the circuit is included in the test.*) If the inspecting engineer is still in doubt as to the validity of the result a further test can be made at a voltage not exceeding 50 volt a.c. or d.c. at 1.5 times the design current of the circuit up to a maximum test current of 25 A. (*If the earth loop impedance test does not give a satisfactory result a rigorous inspection of all joints should be carried out before continuing with the last test, since something is obviously wrong with the installation.*) Details of the tests required are given in Part 2 'Inspection and Testing.'
713-02

Continuity of ring circuit conductors

A test has to be made to make sure that a proper ring circuit has been installed. Additionally, the continuity of all the ring circuit conductors, including the protective conductor, has to be checked.
713-03

Insulation resistance

All insulation tests are to be carried out with direct current and the insulation test equipment shall be capable of producing the test voltage with a load of 1 mA.

The insulation resistance between live conductors has to be measured before the installation is energised. Any circuit containing electronic devices shall be tested between the protective earthing conductor and the phase and neutral conductors connected together, additional precautions may be necessary so that the electronic devices are not damaged by the test voltage.

The insulation resistance between each live conductor and earth has also to be measured, the PEN conductor in a TN-C system being considered as part of the earth.

Phase and neutral conductors may be connected together for testing purposes where it is appropriate for the tests concerned.

The insulation resistance between live conductors, and live conductors and earth, shall be considered satisfactory if the main switchboard, and each distribution circuit when tested separately with all final circuits connected, but current-using equipment disconnected, has at least the insulation values as follows:

Extra-low voltage circuits

SELV and PELV circuits supplied from an isolating transformer to BS 3535:
Test voltage 250 V d.c. - minimum insulation resistance 0.25 megohm.

Low-voltage up to 500 V

Test voltage 500 V d.c. - minimum insulation resistance 0.5 megohm

Low-voltage over 500 V

Test voltage 1000 V d.c. - minimum insulation resistance 1 megohm.

713-04-01, 713-04-02, 713-04-03, 713-04-04

Insulation applied on site

Where insulation is applied on site for protection against direct contact, it shall be verified that it will withstand the tests laid down in the British Standard for similar type tested equipment.

713-05-01

Where supplementary insulation is applied on site to protect against indirect contact, tests shall be made to verify that the degree of protection of the insulating enclosure is not less than IP2X or IPXXB, and that it will withstand an applied voltage test laid down in British Standards for similar types of equipment.

713-05-02

Protection by electrical separation

Where protection by electrical separation provides protection against electric shock, the electrical separation of the separated circuit shall be verified as follows:

SELV (Separated Extra-Low Voltage)

The extra-low voltage circuits shall be checked to make sure that the nominal voltage does not exceed extra-low voltage, and that the source of supply is from one of the following:

1. Safety isolating transformer manufactured and tested to BS 3535,
2. Motor generator whose windings give the same degree of electrical separation as a BS 3535 transformer,
3. Battery or motor driven generator,
4. Electronic devices restricted to make sure that an internal fault cannot result in a voltage at the output terminals being greater than extra-low voltage, or if it so results in a greater voltage, it is immediately reduced on contact, to extra-low voltage.

The circuits and equipment shall also be checked that they comply with the SELV requirements. The separation of live parts from other circuits and from earth shall be checked, by measuring the insulation resistance, the insulation resistance being at least 0.25 MΩ with a 250 V d.c. test voltage.

PELV (Protective Extra-low Voltage)

A check should be made that the circuit or circuits are earthed at one point only, and that protection against direct contact is provided where the circuits are outside the equipotential zone, or the voltage exceeds 25 V a.c. r.m.s or 60 V ripple free d.c. for dry conditions or 6 V a.c. r.m.s. or 15 V ripple free d.c. for all other conditions, within the equipotential zone.

Additionally, the separation of live parts from other circuits has to be checked by measuring the insulation resistance, which should be at least 0.25 MΩ with a test voltage of 250 V d.c.

Electrical separation

The source of supply has to be checked to determine it is either an isolating transformer to BS 3535 or a motor-generator having the same electrical separation between windings as an isolating transformer. That no live part of the separated circuit is connected at any point to another circuit or to Earth and that the insulation is sufficient to avoid the risk of a fault to Earth. That flexible cables or cords are visible throughout their length if they are liable to mechanical damage. That all the other requirements of Regulation 413-06 are complied with.

An insulation resistance test has to be made of the separated circuit in accordance with the requirements given under insulation resistance.
713-06, 411-02, 471-14, 413-06

Functional Extra-low Voltage

The insulation resistance of functional extra-low voltage circuits shall comply with the requirements of low voltage circuits.
713-06-01, 411-02-01, 411-02-02

Protection by barriers or enclosures against direct contact

Where protection against direct contact is made using barriers or enclosures, then they shall be inspected to make sure that the barriers and enclosures are firmly secured in place, and will maintain the degree of protection required. They must also be checked to make sure that they offer at least IP2X or IPXXB protection, and where there are horizontal top surfaces which are readily accessible, the minimum protection is increased to IP4X. The opening of the barrier or enclosure should only be by means of a tool, and should not be possible until the supply to live parts has been disconnected.
713-07, 412-03

Non-conducting floors and walls

Where a non-conducting location is used for protection against indirect contact, a check shall be made to make sure that the installation complies with the requirements, and with the specification for the installation, since this method of protection is not recognised by the requirements for general use.

Where the location has an insulating floor and walls, one or more of the following arrangements applies:

1. The distance between exposed-conductive-parts, and between exposed-and-extraneous-conductive-parts is not less than 2 m, or where out of arm's reach, 1.25 m,
2. Obstacles, preferably insulating, are installed between exposed-conductive-parts and extraneous-conductive-parts. These are considered effective if they extend the distance between exposed- and extraneous-conductive-parts to 2 m, or 1.25 m, if out of reach. If any obstacle is non-insulating, then it must not be in contact with earth, or with exposed-and extraneous-conductive-parts,
3. Extraneous-conductive-parts are insulated with material having adequate electrical and mechanical strength.

Where (3) is used for protection then the insulation resistance shall not be less than 0.5 MΩ when tested at 500 V d.c. and shall be capable of withstanding a test voltage of at least 2 kV r.m.s. a.c. and in normal use it shall not pass a leakage current exceeding 1 mA.

A resistance measurement shall be made between three points on each wall and floor and the main protective conductor. One measurement shall be not less than 1 m or more than 1.2 m from any extraneous-conductive-part in the location, all the others being at a greater distance. Where the voltage to earth does not exceed 500 V the minimum insulation resistance shall be 50 kΩ and where the voltage is between 500 V and 1000 V the insulation resistance shall be 100 kΩ.
713-08-01, 713-08-02, 413-04-04, 413-04-07

Polarity

It shall be verified that all protective devices, single pole switches, control switches, and the centre

contact of bayonet or Edison screw lampholders, are connected only in the phase conductor, and that socket-outlets and similar accessories have been correctly connected.
713-09

Earth electrode resistance
Where protection involves the use of an earth electrode, the earth electrode resistance shall be measured.
713-10

Earth fault loop impedance
Where protection against indirect contact uses automatic disconnection involving the phase and circuit protective conductors, the relevant impedances shall be measured or calculated. These must then be corrected for temperature and checked against the appropriate Z_s value for the protective device used.

Where protection is provided only by limiting the impedance of the cpc, this impedance shall be measured and checked against the appropriate table 41C or PCZ10 in Part 3.
713-11

Functional testing (including RCDs)
The effectiveness of RCDs has to be verified by a test simulating a fault condition in the circuit. All loads normally supplied through the circuit breaker are disconnected, and a test is then made on the load side of the circuit breaker between the phase conductor and the cpc so that a suitable residual current flows. This test is independent of the test button on the RCD.

Switchgear, controlgear, control circuits and interlocks etc, are to be tested to ensure that they are functioning properly and correctly installed.
713-12

Alterations and additions to an installation
The rating and condition of the existing installation and its earthing arrangements shall be checked, to make sure that it is capable of carrying the additional load. The check shall be made before any alteration or addition, either temporary or permanent, is made to the existing installation.
130-09

The inspection and testing of any alteration or addition to an installation shall be the same as a new installation. Additionally, it shall be verified that every alteration or addition does not impair the safety of the existing installation.
721-01-01, 721-01-02

An Electrical Installation Certificate signed by a competent person has to be handed to the person ordering the work after all defects found during the inspection and testing have been rectified. Any defects found in related parts of the existing installation shall be reported in writing by the person or contractor responsible for the new work to the person ordering the work.
743-01

Periodic inspection and testing
The inspection and testing shall be carried out with care to make sure that the installation still complies with the relevant requirements, and will not cause danger to persons or livestock, or damage to property. This may involve partial dismantling.
731-01-01, 731-01-02, 732-01-03

CHAPTER 20

The extent of the inspection and test and results shall be recorded in a Periodic Inspection Report, and signed by the person carrying out the inspection (or a person authorised to act on his behalf). The report shall then be sent to the person ordering the inspection and test.
744-01-01, 744-01-02

Where it is found that a dangerous condition exists it shall be recorded on the Periodic Inspection and Testing Report along with any limitations to the inspection and testing.
744-01-02

The report shall contain a record of any damage, defects, deterioration and dangerous conditions arising from non-compliance with the BS 7671. Any limitations of the inspection and testing shall also be recorded in the report.
744-01-02

Frequency of testing

The frequency at which inspection and tests are carried out shall be determined by the type of installation, the way it is used and operated, external influences, and the frequency of maintenance.

Danger shall not be caused to persons or livestock, or damage caused to property and equipment whilst the inspection and testing is carried out.
732-01-01, 732-01-02

RECOMMENDED PERIODS BETWEEN INSPECTIONS

Type of premises or installation	Maximum time between inspections - in years
Domestic	10
Highway power supplies	6
Commercial, educational and hospitals	5
Industrial	3
Agricultural, horticultural, caravans and motor caravans	3
Emergency lighting	3
Churches under five years old	2
Cinemas, leisure complexes, restaurants	1
Hotels, places of public entertainment	1
Theatres, churches over five years old	1
Caravan sites, fire alarms, launderettes	1
Petrol filling stations	1
Temporary electrical installations	3 months

Electrical Installation Certificate

An Electrical Installation Certificate complete with schedules of test results carried out to comply with the BS 7671 has to be handed to the person ordering the work, together with recommendations for periodic inspection and testing after all defects found during the inspection and testing have been rectified. The certificates will be of the form shown in Appendix 6 and signed or otherwise authenticated by a competent person.
741-01, 742-02, 130-10

Any defects found in the installation shall be made good before the Electrical Installation Certificate is issued
742-01-01

Testing notice

A notice in durable material which will remain easily legible throughout the life of the installation, shall be fixed in a prominent position as close to the main incoming switch as possible, and will contain the date of the last inspection and the recommended date of the next inspection.

IMPORTANT

This installation should be periodically inspected and tested and a report on its condition obtained, as prescribed in BS 7671 (formerly the IEE Wiring Regulations for Electrical Installations) published by the Institution of Electrical Engineers.

Date of last inspection ....

Recommended date of next inspection ....

The size of the characters on the label should not be less than 14 point. This notice need not be installed for highway power supply cables where the installation is subject to a programmed Inspection and Testing procedure.
514-12-01, 611-04-04

Chapter 21

Isolation

Object of isolation

Isolation is used to make live parts of an installation securely dead, so that work can then be carried out on those parts in safety by electrically-skilled persons.

What is an isolator?

A low voltage isolator is basically an off load mechanical switching device, but an isolating switch may be capable of switching a load at infrequent intervals. A basic isolator is not designed to make or break load current, nor to make onto or break fault currents. It is often referred to as a disconnector.

What is isolation?

Cutting off the electrical supply from every source to an installation, a circuit, or an item of equipment for safety purposes.

One device can be used to perform the functions of:

1. Isolation,
2. Switching off for mechanical maintenance,
3. Emergency switching,
4. Functional switching.

Where a device is used for one or more of the above functions, it shall be verified that the device complies with the requirements for each function it performs.
476-01-01, 530-01-03, 537-01

General requirements

Every installation shall be provided with a means of isolation and switching.
476-01-01

Non-automatic isolation and switching shall be provided to prevent danger associated with the electrical installation or electrically powered equipment and machines.
460-01-01

An isolator or a switch shall not be installed in a protective conductor or a PEN conductor. The exception to this is where an installation is supplied from more than one source of electrical energy, one of which requires earthing independently of the other sources, and where it is necessary to make sure only one means of earthing is used at any one time. A switch is then allowed between the neutral point and the means of earthing, providing it is linked to disconnect and connect live conductors at the same time.
460-01-03, 460-01-04, 543-03-04

Where BS 7671 requires all the live conductors to be disconnected, it must not be possible to disconnect the neutral before the phase conductors are disconnected. It is recommended that the neutral be re-connected at the same time as the phase conductors, or before the phase conductors are connected.
530-01-01

Arrangements have to be made for disconnecting the neutral in the isolator or switch.
460-01-06

No switch or circuit breaker shall be inserted in a neutral conductor in a TN or TT system, unless it is linked with the phase conductors.
530-01-02

Circuits to be isolated

To prevent or remove danger, isolators shall be installed to cut off all voltage from every installation, from every circuit, and from all equipment. The device used shall be suitably placed for ready operation.
130-06-01

Every circuit shall be provided with a means of isolation. Where the circuit is part of a TN-S or TN-C-S system, all phase conductors shall be isolated. Where it can be reliably regarded that the neutral conductor is at earth potential, the neutral does not need to be isolated or switched. However, if it is intended that unskilled persons are to operate the main switch of a single phase supply, then both the phase and neutral conductor must be switched. Where the circuit is part of a TT or IT system, provision shall be made to isolate all live conductors.

One device may be used to isolate a group of circuits, e.g. an isolator feeding a distribution board.
461-01-01, 461-01-02, 461-01-03, 537-02-01

All poles of the supply have to be isolated in a d.c. system.
460-01-02

Prevention of re-energisation

Precautions shall be taken to prevent any equipment being unintentionally or inadvertently energised.
461-01-02

Remote isolators

Where isolators are remote from the circuits or equipment they are intended to isolate, then both the following requirements shall be met:

1. The isolator shall be secured against inadvertent reclosure whilst being used to isolate the equipment or circuits it controls.
2. Where an isolator is secured against inadvertent reclosure by a lock or removable handle, the key or handle shall not be interchangeable with any others used for a similar purpose within the installation.

476-02-02

Motors

An efficient means of disconnection, which is readily accessible, easily operated, and so placed as to prevent danger, shall be provided for every motor.
130-06-02

Every motor circuit shall be provided with an isolator (disconnector). The isolator shall disconnect the motor and its associated equipment, including any automatic circuit breaker.
476-02-03

Discharge lighting

Circuits or luminaires used for discharge lighting, where the open circuit voltage exceeds low voltage, shall use one or more of the following methods for isolation.

1. Self contained luminaires shall have, in addition to the switch normally used for controlling the circuit, an interlock which automatically disconnects the supply before access can be made to live parts.
2. In addition to the switch normally used to control the circuit, an effective local means of isolating the circuit from the supply shall be provided.
3. A lockable distribution board, lockable switch, or removable handle shall be provided. The locks and removable handle shall be unique within the same installation, and so placed or guarded that they can be operated only by skilled persons, and cannot inadvertently be returned to the ON position.

476-02-04

Isolation of discharge lighting circuits shall comply with the relevant requirements for isolation and erected and connected in accordance with BS 559.

554-02, BS 559 1991

Isolation at the origin

Isolation shall be provided as near as possible to the origin of the installation, by a linked switch or linked circuit breaker capable of interrupting the supply on load.

This means that there should not be any other equipment upon which work may be required to be carried out, between the origin and the means of isolation provided. The means of isolation should comply with all of the relevant regulations for isolation.

A main switch or circuit breaker at the origin of the installation shall switch the following conductors of the incoming supply:

1. Both live conductors when the supply is single-phase a.c. and its operation is by unskilled persons.
2. All poles of a d.c. supply.
3. All phase conductors in a T.P. or T.P. & N. supply in TN-S or TN-C-S systems, where the neutral is reliably at earth potential.
4. All live conductors in a T.P. or T.P.& N. supply in TT or IT systems.

One switching device is allowed to isolate a group of circuits.

460-01-02, 461-01-01,
476-01-03, 537-02-01

Where an isolator is used in conjunction with a circuit breaker to enable maintenance of switchgear to be carried out, it shall either be interlocked with the circuit breaker, or be so placed and guarded that it can be operated only by skilled persons.

476-02-01

Where the supplier of electricity provides switchgear complying with BS 7671 at the origin of the installation, and agrees that it can be used to isolate the installation from the origin up to the main distribution at which isolation is provided, this agreement satisfies the requirement for having a main switch at the origin.

476-01-01

Arrangements have to be made in the isolator or switch for disconnecting the neutral conductor. Any joint used for this purpose has to be accessible and only disconnected by means of a tool. The joint must be mechanically strong and reliably maintain electrical continuity.

460-01-06

Isolating switchgear

A switch or isolator shall not be placed in a protective conductor or PEN conductor. The exception to this is where an installation is supplied from more than one source of electrical energy, one of which requires earthing independently of the other sources, and where it is necessary to make sure only one means of earthing is used at anyone time.

A switch is then allowed between the neutral point and the means of earthing, providing it is linked so that it disconnects or reconnects live conductors at the same time as the earthing conductor.
460-01-03, 460-01-04

Where an installation is supplied from more than one source of electricity, each source shall be controlled by its own main switch. A permanent durable warning notice must, however, be installed, warning that all the switches need to be switched off if the whole installation is to be isolated.
460-01-02

A means of interrupting the supply on load shall be provided for every circuit and final circuit.
476-01-02

Identification

All isolators shall be identified by position or durably marked to identify the circuits or installat:on they are controlling.

A notice, stating that it must only be operated off-load, shall be placed on the front of each isolator.
461-01-05, BS EN 60947-3, 537-02-09

Where a single device is not capable of isolating the live parts of equipment or those contained within an enclosure, then either of the following precautions shall be taken:

1. A durable warning notice shall be placed so that any person gaining access to live parts will be warned to use the appropriate isolating devices, or
2. An interlocking arrangement is provided so that all the circuits concerned are isolated before access can be made.

461-01-03

Capacitors & inductors

Adequate means shall be provided to discharge capacitors or inductors to prevent danger. Similarly, capacitive circuits or inductive circuits shall be discharged to prevent danger.
461-01-04

Requirements for isolators

The isolating distance between contacts when in the open position shall not be less than that specified for disconnectors in BS EN 60 947-3. (Note: not yet specified, but can be taken to be at least 3 mm for 230 volts r.m.s. to earth installations.).
537-02-02

Semiconductor devices shall not be used as isolators.
537-02-03

The isolated position shall only be indicated when the specified isolating distance has been achieved in each phase or pole. The position of the isolator contacts or other means of isolation shall either be externally visible or clearly and reliably indicated.
537-02-04

Isolators shall be selected and installed so that unintentional reclosure by persons or mechanical shock or vibration is prevented. Provision shall also be made to secure against the unauthorised and inadvertent operation of off-load devices used for isolation.
537-02-06, 537-02-07

Isolation shall be achieved by the use of a single multipole device cutting off all the appropriate poles of the supply. All the appropriate poles of the supply can be isolated by single pole devices, providing they are mounted adjacent to each other.
537-02-08

Where a link is installed in the neutral conductor, it shall only be removable by the use of a tool, or it shall be accessible only to skilled persons.
537-02-05

Accessibility

Isolators and switches shall be installed so that adequate and safe means of access and working space are provided to enable operation, maintenance, inspection and testing to be carried out. These facilities shall not be impaired by equipment being mounted in enclosures or compartments.
130-07, 513-01, 529-01-02

Where a joint is used for disconnecting the neutral conductor, it shall be in an accessible position.
460-01-06

Chapter 22

Mechanical maintenance

Switching off for mechanical maintenance
Switching off for mechanical maintenance is to stop mechanical movement of parts by electrical means, such as by a motor, in order to enable mechanical maintenance or cleaning of non-electrical parts of equipment, plant or machinery. It is also used for the replacement or cleaning of lamps. One device can be used to perform the functions of:
1. Switching off for mechanical maintenance.
2. Isolation.
3. Emergency switching.
4. Functional switching.

Where one device is used for one or more of the above functions, it shall be verified that the device complies with the requirements for each function it performs.
476-01-01, 530-01-03, 537-01

Provisions for safety
Where mechanical maintenance may involve a risk of physical injury, provision shall be made for switching off for mechanical maintenance.
462-01-01

Precautions shall be taken in the selection and erection of devices, so that any equipment switched off is not unintentionally or inadvertently re-energised. *(For instance, by mechanical shock or vibration.)*
537-03-03

Except where the switch is continuously under the control of the person carrying out the mechanical maintenance, provision shall be made to enable precautions to be taken to stop any equipment or machines being unintentionally or inadvertently started whilst maintenance is being carried out.
462-01-03

Identification
The equipment used shall be durably labelled and mounted in a position which is convenient for its intended use. i.e. local to, and within easy reach of, the equipment it controls.
462-01-02

Position of devices
Devices shall be inserted in the main supply circuit, but if additional precautions which provide the same degree of safety as the main supply being interrupted are taken, the device can be inserted in the control circuit.
537-03-01

Requirements for devices
Devices used for switching off for mechanical maintenance, including control switches for such devices, shall have an externally visible contact gap, or clearly and reliably indicate the OPEN or OFF position only when the OFF or OPEN position has been achieved on each pole.
537-03-02

Devices used for switching off for mechanical maintenance shall be manually operated.
537-03-02

Switches used for switching off for mechanical maintenance shall be capable of switching off the full load current of that part of the circuit the switch controls.
537-03-04

Socket-outlets used for 'switching off for mechanical maintenance' shall not have a rating exceeding 16 A.
537-03-05

Provision of devices
Every fixed motor shall be provided, for mechanical maintenance purposes, with a means of switching off: this device shall be readily accessible, easily operated, and so placed as to prevent danger.
130-06-02

Where the rating of a plug and socket-outlet does not exceed 16 A it may be used for switching off for mechanical maintenance.
537-03-05

Accessibility
Safe access and working space shall be provided for equipment that requires operation or maintenance to be carried out.
130-07

Access to equipment shall be arranged so as to give access to each connection. Additionally, access shall be arranged so as to enable its operation, maintenance and inspection to be safely carried out. Access shall not be impaired by equipment mounted in enclosures or compartments.
513-01, 529-01-02

Chapter 23

Overcurrent protection - 'General requirements'

General requirements
All electrical conductors shall be of sufficient size and current-carrying capacity.
130-02-03

Single pole fuses and circuit breakers shall only be placed in phase conductors. Additionally, no fuse or circuit breaker (unless linked) shall be installed in an earthed neutral. Where a linked circuit breaker is used, it must disconnect all the related conductors.
130-05-01, 130-05-02

Every installation and every circuit shall be protected against overcurrent by devices which will operate automatically, at values of current and time for the circuit, so that no danger is caused. The devices must have adequate breaking capacity, and, where necessary, making capacity. They shall be located so that there is no danger from overheating, arcing, or the scattering of hot particles, and they shall be able to restore the supply without danger.
130-03

Where switchgear or fusegear is supplied by the supplier of electricity at the origin of the installation, and where they give permission for it to be used for isolation from the origin to the main installation switch, it will not be necessary to protect the tails from the origin to the incoming main switch of the installation with additional overcurrent devices.
476-01-01

Protection of live conductors shall be given by one or more devices which will interrupt the supply automatically in the event of an overload or fault current.

Where the source is incapable of supplying a current exceeding the current-carrying capacity of the conductors, the conductors are considered to be protected against both overload and fault currents, so overcurrent protective devices are not generally required.
431-01-01, 436-01

Where a common device is used for both overload and fault current protection, it shall be capable of breaking (and, for circuit-breakers, making) any overcurrent at the point at which the device is installed.
432-02

Assessment of general characteristics
The prospective short circuit current at the origin of the installation shall be calculated, ascertained or determined.
313-01-01(iii)

The type and rating of the overcurrent protective device at the origin of the installation shall be ascertained.
313-01-01(vi).

Automatic re-closure
Where danger is likely to be caused by a protective device reclosing, the reclosure shall not be automatic.
451-01-06

Now find out what the law requires you to do

ELECTRICITY AT WORK & RELATED REGULATIONS
Second EDITION
Soon to be revised
Trevor E Marks

A Handbook for compliance

- Details the inspection and tests for fixed and portable equipment
- Gives periods between inspection and testing and test results required
- Inspection and testing computers and business equipment
- Details who is allowed to carry out inspection and tests
- Regulations given in italics for quick and easy reference
- Gives basic rules for isolation and cutting off the supply
- Explains the extent of employers and employees' responsibilities
- Guidance in drawing up your own "Code of Safe Working Practice"
- Explains the principles of protection against electric shock and what is meant in the regulations by "strength and capability"
- Qualifies: competency of personnel and environmental conditions.

ISBN 0-9510156-4-8   WILLIAM ERNEST PUBLISHING

Overcurrent detection

A device that will detect and disconnect an overcurrent in each phase conductor shall be provided.

The device is not required to disconnect other live conductors, but precautions have to be taken where the disconnection of one phase conductor could cause danger, e.g. three-phase motor single phasing.
473-03-01

There is no need to install overcurrent detection in a neutral conductor that is part of a TN or TT system, if the neutral has the same cross-sectional area as the phase conductor(s).
473-03-03

When the neutral does not have the same cross-sectional area as the phase conductors in a TN or TT system, overcurrent detection shall be provided in the neutral, to disconnect the phase conductors, unless the neutral is protected against fault currents by the protective device in the associated phase conductors, and the current carried by the neutral conductor under normal circuit operation is much less than the current-carrying capacity of the neutral conductor.
473-03-04

Where the neutral is not distributed in a TT system and the load is therefore between two phases, an overcurrent protective device need not be installed in one of the phases, providing a differential protection relay is installed in the same circuit or on the supply side, and the neutral is not derived from an artificial neutral point on the load side of the differential protection provided.
473-03-02

Discrimination

Where it is necessary to prevent danger, protective devices shall be sized so that the smaller rated device operates before with the larger rated device with which it is in series, in the event of a fault or overload, so that the intended discrimination is achieved.
533-01-06

Coordination of overload and fault current protection

Coordination is required between overload devices and fault current protective devices, such that the energy let-through of the fault current protective device is within the withstand capacity of the overload device. Manufacturers' advice shall be obtained for motor starters.
435-01, 432-03

Replacing fuses that are energised

When fuse links are likely to be removed or replaced whilst the circuit they protect is energised, they shall be of a type that can be removed or replaced without danger. *(Live parts shrouded.)*
533-01-03

Identification

Every overcurrent protective device shall have its nominal current rating either marked on it, or placed adjacent to it. The current rating to be indicated for semi-enclosed (rewirable) fuses shall be in accordance with the table given below[1]. Unless a fuse link is going to be replaced by a skilled or instructed person its type should preferably prohibit replacement with one of a higher current value.
533-01-01

---

[1] It is preferable to check with the manufacturer of the fuse holder, otherwise excessive burning may occur when a fault occurs.

To enable easy recognition of which protective device protects a circuit, the protective device shall be identified.
514-08

The size of fusewire used in semi-enclosed (rewirable) fuses shall be in accordance with the manufacturers' instructions. Where no instructions are available, the fusewire shall be selected from the following table. Cartridge type fuses are preferred.
533-01-04

Size of fuse wire for semi-enclosed fuses		
Rating of fusewire in amperes	Nominal diameter of fusewire in mm	Approximate S. W. G.
3	0.15	38
5	0.20	36
10	0.35	29
15	0.50	25
20	0.60	23
25	0.75	22
30	0.85	21
45	1.25	18
60	1.53	17
80	1.80	15
100	2.00	14

Fuse replacement by unskilled persons
Where fuse links are likely to be replaced by unskilled persons, the fuses should be of a type which cannot be replaced by a fuse having a higher fusing factor of the same nominal rating, or there is marked, either on the fuse or adjacent to it, the type of fuse link intended to be used.
533-01-02

Circuit-breakers operated by unskilled persons
Where circuit-breakers can be operated by unskilled persons, it should not be possible for the setting or calibration of their overcurrent releases to be modified without the use of a key or tool, which results in a visible indication of its setting.
533-01-05

Disconnection times
Where overcurrent protective devices are also used for protection against electric shock in TN and TT systems, they shall be selected so that their disconnection time is such that the permissible final temperature of the phase conductors and circuit protective conductors is not exceeded, when there is a fault of negligible impedance between a phase conductor and an exposed conductive part. The disconnection time must also comply with the shock protection requirements for the type of circuit and environmental conditions applicable, which, under normal conditions is 0.4 seconds for socket-outlet circuits, and 5 seconds for certain fixed equipment circuits, where the declared voltage to earth is 230 V.
531-01-01

Lampholders

The size of device protecting a lampholder and its wiring, which is not contained in earthed metal or insulating material having the ignitability characteristic 'P', as specified in BS 476 Part 5, or protected by its own protective device, shall be limited by the type of lampholder used as follows.

Type of lampholder	Cap type		Maximum amperes of circuit protective device
BS 5042 or BS EN 61184	B15	SBC	6
Bayonet type	B22	BC	16
BS EN 60238	E14	SES	6
Edison Screw	E27	ES	16
	E40	GES	16

553-03-01

IT systems

If an overcurrent protective device is to be used for protection against electric shock in the event of a second fault and the exposed-conductive-parts are connected together, then the overcurrent protective device shall be selected so that its disconnection time is such that the permissible final temperature of the phase conductors and any bonding/protective conductors is not exceeded, when there is a fault of negligible impedance between a phase conductor and an exposed-conductive-part.

The disconnection time must also comply with the shock protection requirements for the type of circuit and environmental conditions applicable, which, under normal conditions is 0.4 seconds for socket circuits, and 5 seconds for fixed equipment circuits, where the declared voltage to earth is 230 V.

531-01-02, 531-01-01

Every circuit's neutral shall be provided with overcurrent detection where the neutral is distributed, which shall disconnect all phase and neutral conductors of the associated circuit. The exception to this requirement is where the neutral is protected against fault current on the supply side by a device having a breaking capacity that is not less than the prospective short-circuit current or earth fault current, or where the circuit is protected by an RCD having a residual operating current not more than 0.15 times the current-carrying capacity of the circuit's neutral conductor, the RCD disconnecting both phase and neutral conductors.

473-03-05, 434-03-01

Chapter 24

Overcurrent protection - 'Fault currents'

Conductor protection

One or more protective devices that will operate automatically shall be used to protect phase and neutral conductors against prospective fault currents, unless the current-carrying capacity of the conductors is greater than the current that can be supplied from the source of energy.
431-01-01, 436-01

Withstand capacity

The characteristics of the fault current protective device have to be co-ordinated with the overload device, so that the energy let-through of the fault current protective device is within the withstand capacity of the overload device. The manufacturer should be consulted where a motor starter is used.
431-01-02, 435-01

Where a common device is used for both overload and fault current protection, it shall be capable of breaking (and for circuit-breakers, making) any overload or fault current at the point the device is installed, unless 434-03-01 is applicable.
432-02

Devices installed to protect against fault current shall be capable of breaking (and for circuit-breakers, making) the prospective fault current available, unless 434-03-01 is applicable.
432-04

Limitation of mechanical and thermal effects

Fault current shall be interrupted by a protective device before such current can cause damage or danger due to thermal or mechanical effects in conductors or connections.

The nominal current rating of a protective device used for fault current protection may be greater than the current-carrying capacity of the conductor which the device is protecting.
434-01

Position of fault current protective devices

Fault current protective devices shall be placed at the point where a reduction in the conductor's current-carrying capacity occurs. Reduction in current-carrying capacity can be caused by:
1. Reduction in conductor cross-sectional area,
2. Change in ambient temperature,
3. Contact with thermal insulation,
4. Change in the number of cables grouped together,
5. Change in the method of installation,
6. Change in the type of cable conductor.
473-02-01

A fault current protective device can be placed at any point on the load side of a reduction in current-carrying capacity of a conductor, providing:
1. The distance from the point of reduction in current-carrying capacity to the protective device does not exceed 3 metres, and
2. The conductors are so erected that the risk of short-circuit, earth fault, the risk of fire, or the risk of danger to persons is reduced to a minimum, and

3. The protective device is not being used to give protection against indirect contact, because the earth fault loop impedance is too high.
473-02-02

Except for where there is a high risk of fire or explosion or a special situation exists, a fault current protective device may be placed at a point other than where the current-carrying capacity of a conductor is reduced, provided that there is a protective device on the supply side of the change in current-carrying capacity, which will protect the conductors against fault current in accordance with **434-03-03, 473-02-03**

Omission of fault current protection

Fault current protective devices need not be provided on the supply side of conductors connecting generators, transformers, rectifiers, or batteries with their control panels in which fault current protective devices are installed, provided the conductors are erected in such a manner that the risk of fault current, fire, or danger to persons is reduced to a minimum.

Certain measuring circuits, or circuits where disconnection would cause a greater danger than the fault current need not be protected with short-circuit protective devices, provided that the conductors are erected in such a manner that the risk of short-circuit, fire, or danger to persons is reduced to a minimum.

Where the supplier of electricity provides fault current protective devices at the origin of the installation , and agrees that these may be used to protect the installation up to the main distribution point, there is no need to duplicate the protective devices at the origin of the installation. The first fault current protective device shall be installed at the main distribution point.
473-02-04

Prospective fault current

The prospective short-circuit and earth fault current shall be ascertained by measurement, calculation at every point of change in conductor current-carrying capacity. The calculations shall take into account both maximum and minimum fault current conditions.
434-02, 533-03

Fault current protective devices

The breaking capacity of protective devices installed must not be less than the prospective fault current at the point the device is installed. *(For three-phase circuits, the symmetrical short-circuit current is taken at the point at which the device is installed.)*
434-03-01

Protective devices with a lower breaking capacity than the prospective fault current available at the point of installation can be installed, provided that:

1. There is a protective device, with the necessary breaking capacity, on the supply side of the lower rated protective device, and
2. The characteristics of both protective devices are co-ordinated, so that the energy let-through of the supply side device will not damage the lower rated device, or the cable conductors protected by both devices.

434-03-01

Fault current capacity of conductors

Where a protective device is used for both overload and fault current protection, and has a breaking capacity at least equal to the prospective fault current at the point it is installed, and the conductors on the load side of the device have been sized in accordance with the overload requirements, it can be assumed that the conductors are protected against fault current. This assumption shall be checked where conductors are installed in parallel and it may not be valid for non-current limiting types of circuit-breaker. In this case it shall be checked that the conductors are protected against fault current, by using 434-03-03.

434·03-02

Where a protective device is only providing protection for short-circuit currents and earth fault currents, a check shall be made to ensure that the rise in temperature of the circuit conductors, caused by a fault at any point in the circuit, does not exceed the limit temperature for the conductor or conductor's insulation.

The time t, in which a given fault current will raise the conductor temperature from its normal working temperature to its limit temperature, can be approximately determined from the following formula:

$$t = \frac{k^2 S^2}{I^2}$$

where S is the cross-sectional area of the conductor, I is the prospective fault current and k is the thermal capacity factor based on the conductor material and the conductor's insulation.

Values of k for live conductors:

1. For PVC insulated copper conductors up to 300 mm^2 115.
2. For PVC insulated copper conductors over 300 mm^2 103.
3. For PVC insulated aluminium conductors up to 300 mm^2 76.
4. For PVC insulated aluminium conductors over 300 mm^2 68.
5. For MICC copper conductors, PVC covered, or exposed to touch   115.
6. For impregnated paper insulated copper conductors   108.
7. For tin soldered joints in copper conductors   100.

Other values for k are given in Table K1A and K1B Part 4.

When the disconnection time t is 0.1 seconds or less, the energy let-through of the protective device, as given by the manufacturer, must not exceed the thermal capacity of the conductor: i.e., $I^2 t$ shall be less than $k^2 S^2$.

434-03-03

Discrimination

Where necessary to prevent danger, protective devices which are in series, shall be sized so that the smaller rated device will operate before the larger rated device, so that intended discrimination is achieved.

533-01-06

Conductors in parallel

Where two or more conductors are connected in parallel and are protected by a single device, the characteristics of the device and the parallel conductors shall be coordinated. The cables shall be checked by calculation for being protected against fault currents, account being taken of what would happen if the fault did not occur in all the conductors. (*Conductors should be of the same type, construction, cross-sectional area, length and arrangement and installed to carry equal currents. There should be no branch circuits throughout their length.*)

434-04

Chapter 25

Overcurrents - 'Overloads'

Conductor protection
One or more protective devices that will operate automatically shall be used to protect phase and neutral conductors against possible overload currents, unless the current-carrying capacity of the conductors is greater than the current that can be supplied from the source of energy.
431-01-01

Withstand capacity
The characteristics of the overload devices have to be coordinated with the fault current protective devices, so that the energy let-through of the fault current protective device is within the withstand capacity of the overload device. The manufacturer should be consulted where a motor starter is used.
431-01-02, 435-01

Overload protective devices may have a breaking capacity less than the prospective fault current at the point at which they are installed. *(But see previous paragraph.)*
432-03

Where a common device is used for both overload and fault current protection, it shall be capable of breaking (and for circuit-breakers, making) any overload or fault current at the point at which the device is installed, unless 434-03-01 is applicable.
432-02

Limitation of temperature rise
Overload current shall be interrupted by a protective device before the overload current causes a temperature rise which could damage the conductor's insulation, joints, terminations or the surroundings of the conductor.
433-01

Duration of overload
Small overloads of long duration should not be allowed to occur, and circuits shall be designed so that such overloads are unlikely to occur. *(Large overloads of short duration are permissible, as in the case of starting current, or inrush current, when equipment is switched on.)*
433-01

Device size
The nominal rating of the protective device I_n must not be less than the full-load current (design current) of the circuit I_b.
433-02-01(i)

Conductor size
The effective current-carrying capacity I_z of the smallest conductor in the circuit must not be less than the nominal rating of the protective device I_n. (See Derating semi-enclosed fuses.) For ring final circuits, I_z must not be less than 0.67 times the rated current or the current setting, I_n, of the

overcurrent protective device.
433-02-01(ii), 433-02-04

Device fusing current

The tripping (operating current on overload) current I_2 for the protective device must not exceed 1.45 x I_z for the smallest conductor in the circuit. (See Derating for semi-enclosed fuses.)
433-02-01(iii)

Devices that comply with requirements for fusing current

Providing the smallest conductor in the circuit has a current-carrying capacity not less than the nominal rating of the protective device, and its rating is not less than the full-load current (design current) of the circuit, the regulations for overload protection are satisfied, when general purpose type 'gG' BS88 Part 2 HRC fuses, BS 88 Part 6 HRC fuses, BS 1361 Cartridge fuses, BS 1362 cartridge fuses and BS 3871 Part 1 or BS EN 60898 circuit breakers are used as the overload protective device.
433-02-02

Derating semi-enclosed fuses

Semi-enclosed fuses are often referred to as BS 3036 fuses, or rewirable fuses. Due to the requirement that the current causing the effective operation of the protective device (I_2) must not exceed 1.45 times the current-carrying capacity of the smallest conductor in the circuit (I_z), and due also to the fact that a semi-enclosed fuse has a fusing factor twice its normal rating, the minimum current-carrying capacity of the smallest conductor in the circuit, in order to guarantee overload protection shall be $\dfrac{I_n}{0.725}$ when used with semi-enclosed fuses, to comply with 433-02-01(iii).
433-02-01(iii), 433-02-03

Conductors in parallel

With the exception of a ring final circuit, and providing conductors in parallel are of the same type, have the same cross-sectional area, are the same length, and are installed together following the same route, and there are no branch circuits throughout their length, they can be protected by the same protective device. The minimum required current-carrying capacity of each conductor will be:-

$$I_t = \frac{I_n}{\text{Number of conductors in parallel}}$$

433-03

Position of overload devices

Overload protective devices shall be placed at the point where a reduction in the conductor's current-carrying capacity occurs. Reduction in current-carrying capacity can be caused by:

1. Reduction in conductor cross-sectional area,
2. Change in ambient temperature,
3. Contact with thermal insulation.
4. Change in the number of cables grouped together,
5. Change in the type of overload protective device.

473-01-01

Exception to normal overload position

With the exception of installations in locations of increased fire risk, explosions or where special requirements apply, overload protective devices can be installed along the run of the conductors, providing there are no branch circuits or current-using equipment connected between the point the conductor size is reduced and the position of the overload device.
473-01-02

Omission of overload devices

No overload protective devices need be installed where the unexpected opening of the circuit could cause danger. Overload protective devices should not therefore be placed in the secondary circuit of a current transformer, circuits to lifting magnets, fire extinguishing equipment or exciter circuits of rotating machines.
473-01-03

An overload protective device is not required at the point where a reduction in the current-carrying capacity of a conductor occurs, if the protective device on the supply side of the reduction in conductor current-carrying capacity protects the conductor with the reduced current-carrying capacity.
473-01-04(i)

Overload protective devices are not required for conductors if the characteristics of the load are not likely to cause an overload in the conductors. *(Such as resistive loads.)*
473-01-04(ii)

Overload protective devices are not required at the origin of an installation if the supplier provides equipment at the origin, and agrees that his equipment can be used for overload protection between the origin and the main distribution point containing the next set of overload protective devices.
473-01-04(iv)

Discrimination

Where necessary to prevent danger, any intended discrimination in the operation between protective devices shall be achieved.
533-01-06

To avoid unintentional operation, the peak value of current may have to be taken instead of I_b when sizing conductors. For cyclic loads, the thermally equivalent constant load shall be worked out, and I_n, I_b, I_z, and I_2 shall be chosen accordingly. *(Find r.m.s. value of peaks.)*
533-02

Overload replacement

Where non-skilled persons are likely to replace fuses, the fuses should preferably be of a type that cannot be replaced with one of a higher fuse rating. Additionally, it should have marked on it or adjacent to it, the type of fuse that shall be used, or be of a type that cannot inadvertently be replaced by fuse having a higher fusing factor than the one being replaced. Where a circuit-breaker can be operated by non-skilled personnel, it shall not be possible to alter the setting or calibration of the breaker without using a key or tool. Any alteration must result in the display of a visible indication of its setting.
533-01-01, 533-01-02, 533-01-05

CHAPTER 25

If a fuse is of the type that is likely to be removed or replaced whilst the supply is connected, that fuse shall be capable of being removed and replaced without danger. *(Shrouded.)*
533-01-03

Preferably, fuses shall be of the cartridge type. (BS 88, BS 1361, BS 1362.) The manufacturers, instructions shall be used for determining the size of fuse element for BS 3036 (rewirable fuses), but in the absence of any instructions the table below can be used.
533-01-04

Ring final circuits

A ring final circuit can be deemed to comply with 433-02-01 for overload providing:

a) it is protected by a 30 A or 32 A protective device, *or lower rating.*
b) the protective device complies with BS 88, BS 1361, BS 3036, BS 3871 or BS EN 60898,
c) the accessories supplied by the circuit comply with BS 1363,
d) copper conductors with a minimum cross-sectional area of 2.5 mm^2 or 1.5mm^2 two core mineral-insulated cables to BS 6207 are used,
e) the current-carrying capacity of the conductors I_z is not less than 0.67 times the overload protective device rating I_n.

433-02-04

Identification of overload devices

The intended nominal current rating of every overload protective device shall be provided on or near it.
The nominal current shall be related to the circuit being protected.

The nominal current to be indicated for BS 3036 rewirable fuses should be that recommended by the manufacturer. Where no recommendations are available the following table can be used.
533-01-01

Size of fuse wire for semi-enclosed fuses		
Rating of fusewire in amperes	Nominal diameter of fusewire in mm	Approximate S. W. G.
3	0.15	38
5	0.20	36
10	0.35	29
15	0.50	25
20	0.60	23
25	0.75	22
30	0.85	21
45	1.25	18
60	1.53	17
80	1.80	15
100	2.00	14

Chapter 26

Protection against heat and fire

Location of equipment
The installation and location of fixed equipment shall make sure that the heat thereby produced does not cause a fire hazard to fixed materials installed adjacent to the equipment, or to materials which may foreseeably be placed in the same vicinity of the equipment. Any installation instructions provided by the manufacturer shall be met.
422-01-01

Hazardous areas
Equipment shall be so constructed or protected, and special precautions shall be taken, when equipment is installed in an area susceptible to the risk of fire or explosion.
130-08-02

High temperature equipment
Where electrical equipment has a surface operating temperature that could cause a risk of fire, be harmful to adjacent materials, or produce harmful effects, one of the following methods of installation shall be utilised:

a) The equipment shall be mounted on a support, or placed in an enclosure which is capable of withstanding the temperature generated by the equipment without being set on fire, or creating a harmful effect. Any support used should limit the transmission of heat from the equipment by having a low thermal conductance, or

b) the equipment is screened by material which will withstand, and not be affected by, the heat generated by the equipment, or

c) the equipment is fixed so as to allow the safe dissipation of heat and at such a distance from, other materials as to avoid fire and harmful effects.

422-01-02

Protection against arcing
Equipment that is likely to emit an arc or high temperature particles shall be installed by one of the following methods:

a) the equipment is totally enclosed in arc resistant material,

b) materials that could be affected are screened by arc-resistant screening material,

c) the equipment is so mounted that any emissions are safely extinguished at a safe distance from and before they reach materials which would be affected by the emissions.

Any materials used to give protection must be arc-resistant and non-ignitable, they must be of sufficient thickness to make them mechanically strong and have a low thermal conductivity.
422-01-03

Heat barriers in vertical ducts etc.
Where vertical channels, ducts or trunking contain cables, they shall be equipped with internal barriers to prevent the air at the top becoming excessively hot. *The distance between barriers shall be, either the distance between floors, or 5 metres, whichever distance is the lesser*

INF, 522-01-01

CHAPTER 26

Terminations

Live conductor terminations or joints made between live conductors shall be made in accordance with one or more of the following:
a) an electrical accessory or enclosure complying with British Standards,
b) an enclosure manufactured from material complying with the 'glow wire test requirements' of BS 6458 Section 2.1,
c) an enclosure using building materials which are considered to be incombustible when tested to BS 476 Part 4,
d) an enclosure which is partly or completely formed by the building structure, which has a BS 476 Part 5 ignitability characteristic 'P'.
422-01-04

Protection against burns

Table 42A(reproduced below) specifies the maximum temperature which an accessible part of equipment that is within arm's reach must not exceed, unless a British Standard specifies the maximum temperature. Where any part of the fixed installation is likely, for short periods of time, to exceed the temperature limits specified in Table 42A under normal load conditions, it shall be guarded so that accidental contact with it is prevented.

Maximum temperature of equipment within arm's reach at full load		
Part of equipment	Construction of accessible surface	Temperature allowed °C
Operation by hand or hand-held	Metallic	55
	Non-metallic	65
A touchable part, but not hand-held	Metallic	70
	Non-metallic	80
A part that does not need to be touched in normal use	Metallic	80
	Non-metallic	90

423-01

For safety reasons, equipment having electrically heated surfaces that can be touched, shall be provided with a switch.
476-01-02

Any person who is adjacent to electrical equipment shall be protected against the harmful effects of heat developed, of burns, or thermal radiation, developed by the electrical equipment. Similarly, any fixed equipment or materials that are adjacent to electrical equipment shall also be protected against the same hazards.

Protection must be given against any combustion, ignition or degradation of materials or any reduction in the safe functioning of equipment and electrical equipment must not present a fire hazard to adjacent materials.
421-01

Flammable liquid

Where electrical equipment, such as switchgear or transformers, in one location, contain more than 25 litres of a flammable liquid, such as oil, precautions shall be taken to prevent any burning liquid, smoke, and toxic gases spreading to other parts of the building.

422-01-05

> *Note:* *One location does not mean each individual switch in a switchboard: it could comprise a complete switchboard in the corner of a factory, or a room containing several separate items of switchgear etc, the total content for all items exceeding 25 litres.*

Prevention of overheating

Precautions shall be taken to prevent danger from overheating, arcing, or the scattering of hot particles during the operation of switchgear.

130-03

One or more of the following methods shall be used to protect the wiring system from the effects of heat arising from external sources, or from solar gain:

  a) the installation shall be shielded,
  b) the installation shall be installed out of range of the heat,
  c) wiring materials that are suitable for the additional heat shall be selected,
  d) the current-carrying capacity of the wiring is reduced, *i.e., install larger cables.*
  e) the insulation is reinforced, or changed in the area affected by heat.

Cables or flexible cords installed within an accessory, appliance, or luminaire shall be suitable for the temperature in that equipment. *Covering each cable core individually with glass sleeving is one way of satisfying this requirement.*

522-02-01, 522-02-02

With the exception of central-storage heaters, the electrical heating elements of forced air heating systems shall be interlocked so that they cannot be switched on until the air flow is correct and will be switched off if the air flow is stopped or reduced below the prescribed amount. Additionally, two independent temperature limit thermostats shall switch off the heating elements in the event of the temperature in the air ducts rising above the permissible temperature.

The heating elements shall be enclosed in a non-ignitable frame and enclosure.

424-01

All electrically heated water appliances shall be protected against overheating in all service conditions. Where the appliance does not comply with an appropriate British Standard, protection shall be provided by a non-self-resetting device which functions independently of the appliance thermostat. If the appliance does not have a water outlet to the atmosphere, it shall also be fitted with a device that limits water pressure.

424-02

Reducing the risk of fire

The risk of the spread of fire shall be by selection of the appropriate wiring materials, and installing them so that the integrity of the building structure is not reduced. Where materials manufactured to a British Standard are used, and the standard does not specify testing for the propagation of flame, the wiring system shall be enclosed in non-combustible building material having the characteristic P, to BS 476 Part 5.

527-01

CHAPTER 26

Where fixed equipment is of a type that can cause a focusing or concentration of heat, *such as tungsten halogen lighting*, then the distance it must be mounted from any other materials must be such that it does not raise the temperature in other materials to a level that can be dangerous.
422-01-06

Materials should comply with the resistance to heat and fire of the appropriate BS product standard when used for the construction of enclosures, or where no standard exists enclosures must be suitable for the highest temperature the electrical equipment can produce in normal use.
422-01-07

Fire barriers
Where cables, conduits, ducts, trunking, or other items of a wiring system, pass through walls, floors, roofs or ceilings of a building, any part of the hole that is left round the electrical material shall be made good to the same degree of fire resistance as that required for the element being passed through. Additionally, internal barriers that give the same degree of fire resistance shall be installed in conduits, trunking, ducting, busbars and busbar trunking, where the walls, floors, ceilings and roofs have a specified fire resistance. Where the wiring system is non-flame propagating, and has a maximum internal cross-sectional area of 710 mm², it need not be internally sealed.

If the fire resistance is for less than one hour, this requirement is satisfied if the sealing of the wiring system has been type tested in accordance with BS 476 Part 23.
527-02-01, 527-02-02

Any sealing arrangement used must comply with all the following:
    a) there shall be no incompatibility between the wiring system and the sealing material,
    b) the wiring shall be permitted to expand and contract due to heat, without damage,
    c) it shall be possible to remove the sealer without damage to the cables,
    d) it shall withstand the same environmental conditions as the wiring system.
527-02-stet

During erection or during an alteration, temporary sealing arrangements shall be made. Any sealing that is disturbed shall be reinstated as soon as possible. Each sealing arrangement shall be inspected during the installation to make sure that it conforms to the manufacturers' erection instructions. The inspection shall be recorded.
527-03-01, 527-03-02, 527-04

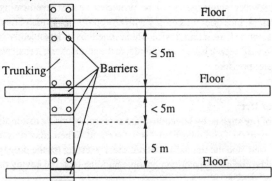

Figure 25.1 - Provision of barriers in trunking

Cables meeting the requirements of BS 4066: Part 1(flame propagation) may be installed without special precaution, unless there is a high risk of fire. Cables not meeting these requirements should only be used for short-length connections to appliances and which must not transgress between fire compartments.

Conduit and trunking with fire propagation properties of BS EN 50085 and BS EN 50086, and other products having similar flame propagation properties, may be used without further precautions being taken.
527-01-04, 527-01-05

Chapter 27

Protection against direct contact

Methods allowed

The following methods are illustrated in Figure 1:

1) Protection by insulation of live parts.
2) Protection by barriers or enclosures.
3) Protection by obstacles.
4) Protection by placing out of reach.

412-01.

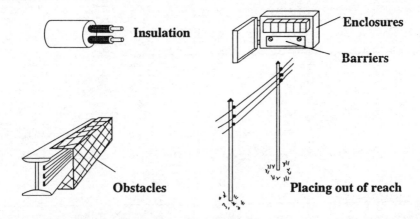

Figure 26.1 - Protection against direct contact

Method 1 – Insulation

Insulation shall be able to withstand any mechanical, electrical, thermal or chemical stresses to which it may be subjected; additionally, the insulation must only be removable by destruction.

Insulation is to prevent contact with live parts when used in conjunction with protection against indirect contact. For equipment, other than factory-built equipment, insulation must be capable of durably withstanding mechanical, electrical, chemical and thermal influences which may prevail. Varnishes, paint and similar applications, such as lacquer, do not generally provide adequate insulation for the purpose of protection against direct contact. When applied on site it shall be tested in accordance with British Standards.

412-02, 471-04

Method 2 – Barriers or enclosures

This method is intended to prevent or deter contact with live parts when used in conjunction with protection against indirect contact. Barriers or enclosures shall be secured firmly in place, and must maintain the required degree of protection during normal service. Live parts shall be behind a barrier or in an enclosure, any opening in which must give protection against objects larger than 12 mm diameter, and up to 80 mm long (IP2X) or against being touched by a finger (IPXXB). Where the opening is in a top surface of the enclosure, the opening must not exceed 1 mm in diameter (IP4X).

412-03-01, 412-03-02, 412-03-03, 471-05-01

Unintentional contact with live parts shall be prevented. Where openings in enclosures are required for replacement of parts, precautions shall be taken to prevent unintentional contact with live parts, and people shall be made aware that live parts accessible through the opening must not be touched.
412-03-01

With the exception of ceiling roses manufactured to BS 67, ceiling switches manufactured to BS 3676 and bayonet lampholders complying with BS 5042, or ES lampholder complying with BS EN 60238, one or more of the following requirements shall be satisfied where it is necessary to open an enclosure or remove a barrier:

1. It is possible to remove a barrier or open an enclosure only by a key or tool,
2. The supply is disconnected before access to an enclosure or the removal of a barrier is possible, and the supply can only be restored after barriers are replaced or the equipment is re-closed,
3. Contact with live parts is prevented by an additional barrier which can be removed only by a key or tool, and which prevents objects 12 mm diameter and up to 80 mm long coming into contact with live parts (IP2X) or the human finger (IPXXB).

412-03-04, 471-05-02

Method 3 – Protection by obstacles

This method does not stop intentional contact with live parts. Its application is limited to areas accessible only to skilled persons or instructed persons under skilled supervision. Obstacles do not need to be removable with a key or tool, but shall be secured to prevent unintentional removal, and to stop unintentional bodily approach to live parts, or the unintentional contact with live parts, when operating live equipment.

Where only skilled or supervised instructed persons are allowed into an area, it is sufficient to provide against unintentional contact with live parts by using obstacles, or by placing out of reach in accordance with the relevant requirements.
412-04-01, 412-04-02, 471-06, 471-13-01

Method 4 – Placing out of reach

This method is intended to prevent unintentional contact with live parts. The application of 412-05-02 to 04 is limited to areas accessible only to skilled or instructed persons. Overhead lines between buildings shall be installed in accordance with the Electricity Supply Regulations.
412-05-01, 471-07-01

Where bare live parts are out of arm's reach, but may still be accessible, they shall not be within 2.5 m of exposed-or extraneous-conductive-parts, or bare live parts of other circuits.
412-05-02.

Where body movement is restricted in the horizontal plane by an obstacle which would allow access of objects larger than 12 mm diameter and 80 mm long, then arm's reach shall extend from that obstacle.

Where long or bulky objects are handled, the limit of arm's reach shall be increased by the length or size of the object being handled.
412-05-03, 412-05-04.

Protection by residual current device

A RCD cannot be used as the sole means of protection against direct contact.

CHAPTER 27

Where protection by insulation, by barriers or enclosures, by obstacles, or by placing out of reach is used, and the RCD has a rated residual operating current of 30 mA and will operate in 40 ms with a residual current of 150 mA in accordance with British Standards, the RCD is recognised as reducing the risk of electric shock.
412-06-01, 412-06-02

Chapter 28

Protection against indirect contact

General requirements

Where exposed-conductive-parts are earthed, the circuits shall be protected either by overcurrent protective devices, or by residual current devices, to prevent the persistence of dangerous earth leakage currents.

Where the prospective earth fault current is too low to give prompt disconnection of the circuit because the phase earth loop impedance is too great, then a residual current protective device shall be used for circuit protection.

130-04-03

Basic measure of protection

One of the basic measures of protection against indirect contact is equipotential bonding and automatic disconnection of the supply.

413-01

Basic rule for protection

The characteristics of protective devices have to be coordinated with the installation earthing arrangements and circuit impedances, so that the magnitude and duration of earth fault voltages appearing between simultaneously accessible exposed-and extraneous-conductive-parts anywhere in the installation, are limited so as not to cause danger.

413-02-04

Protection when the voltage to earth is 230 V up to 277 V in a TN system

The basic rule for protection against indirect contact, inside the equipotential zone, can be considered satisfied, if the circuits comply with the following rules:

   a) when a phase to earth fault occurs, every socket-outlet and every circuit feeding portable equipment intended for manual movement, or hand-held Class I equipment, is disconnected within 0.4 s, or

   b) it is permissible to increase the disconnection time to 5 s when the impedance of the protective conductor, from the end of the final circuit back to the earth bar where the equipotential bonding is carried out, complies with Table PCZ 10 in Part 3, or if not given in the table, is limited to a value of $50 \times Z_S \div U_{oc}$ Ω, where Z_S is the earth loop impedance for a disconnection time of 5 s. (Obtained from Tables ZS 1 to ZS 9 Part 3.)

   c) where 110 volt reduced low voltage circuits that comply with the reduced low voltage requirements are used, a phase to earth fault on a socket-outlet, or on fixed equipment, is disconnected within 5 s,

   d) at every point of utilisation of fixed or stationary equipment, or distribution circuit, a phase earth fault is disconnected within 5 s,

   e) where fixed equipment is connected by means of a plug and socket-outlet, and where it can be guaranteed that the socket-outlet cannot be used for supplying hand-held equipment, a phase earth fault is disconnected within 5 s,

f) where circuits requiring disconnection within 0.4 s are connected to the same distribution board as circuits requiring disconnection within 5 s, then one of the following protective measures is taken:

1. From the main earth bar at which the main equipotential bonding conductors are connected, to the distribution board to which the mixed circuits are connected, the impedance of the cpc does not exceed the values given in Table PCZ 10 in Part 3, or where the device is not listed in the table it does not exceed $50 \times Z_S \div U_{oc}$ Ω. (Where Z_S is the earth loop impedance for 5 s disconnection time.)

2. Equipotential bonding shall be carried out with the same size of bonding conductors as the main equipotential bonding conductors, from the distribution board with the mixed circuits, to the same types of extraneous-conductive-parts as the main equipotential bonding.

413-02-08, 413-02-09,413-02-12, 413-02-13

Where the voltage to earth is 230 V, then for fuses, the Z_S values in Table 41B1 can be used for 0.4 s, and Table 41D for 5 s disconnection times. Table 41B2 can be used for circuit-breakers for 0.4 s and 5 s disconnection times. For both 0.4 s and 5 s disconnection times, Tables ZS 1 to ZS 9 in Part 3 can be used.

413-02-10, 413-02-11, 413-02-14

Protection at different voltages to earth in a TN system

The basic rule for protection against indirect contact, within the equipotential zone, can be considered satisfied, if the protective device for the circuit, and the phase earth loop impedance ensure that automatic disconnection of the supply occurs within a specified time, when a phase to earth fault occurs anywhere in the installation.

Requirements of 413-02-04 can be considered to be satisfied if :

$$Z_s \leq \frac{U_{oc}}{I_a}$$

where Z_S is the phase earth loop impedance, U_{oc} is the open circuit voltage to earth, and I_a is the current causing disconnection within the time specified in Table 41A of socket-outlet circuits, or circuits for portable equipment which can be manually moved, or is hand-held Class I equipment.

TABLE 41A

Maximum disconnection times allowed	
Nominal Voltage to earth , U_o volts	Disconnection time 't' s
120	0.8
230	0.4
277	0.4
400	0.2
Over 400	0.1

For intermediate values of voltage the next higher voltage should be used and where the voltage is within the tolerance band the disconnection time for the nominal voltage applies.

413-02-08, 413-02-09, 413-02-12, 413-02-13, 41A

Resistance to be used in calculations

The resistance of conductors should be taken at the average of the maximum permitted operating temperature of the conductor and the limit temperature for the conductor's insulation, for fault current calculations.

However, where the protective device complies with the characteristics given in Appendix 3 of the British Standard and the values of Z_s for each circuit comply with the values given in Tables ZS 1 to ZS 9, they are deemed to comply with this requirement. *If manufacturers' characteristics are used or there is no value for Z_s in the tables, then the first paragraph applies. (Safer to use first paragraph.)*

Automatic disconnection when using RCDs in a TN system

Where the disconnection times for overcurrent devices cannot be met, where residual current devices are used to comply with the disconnection times, and where the installation is part of a TN system, the residual operating current $I_{\Delta n} \times Z_s$ must not exceed 50 V.
413-02-04, 413-02-16

Where a circuit extends outside the earthed equipotential zone, and automatic disconnection of the circuit is provided by a residual current device, exposed-conductive-parts need not be connected to the cpc's of the TN system, providing they are connected to an earth electrode whose resistance is appropriate to the operating current of the RCD. In this case the circuit forms part of a TT system.

In this case, every circuit shall comply with $R_a \times I_{\Delta n} \leq 50$ V, where R_a is the resistance of the earth electrode added to the resistance of the cpc's connected to the exposed-conductive-part of the circuit. Generally, unless there are known reasons explaining high electrode resistances, values exceeding 100ohm should be considered to indicate instability (see BS 7430).
413-02-17, 413-02-20

RCDs shall not be used for automatic disconnection with circuits incorporating a PEN conductor.
471-08-07

Where several residual current devices are connected in series, the exposed-conductive-parts may be connected to separate earth electrodes for each RCD.
413-02-18

The risk of electric shock (direct contact) is reduced where earthed equipotential bonding and RCD protection is used. The operating current of the RCD shall be 30 mA, and it must disconnect a circuit within 40 ms when a residual current of 150 mA is applied, in accordance with BS 4293, BS 7071, BS 7288, BS EN 61008-1 or BS EN 61009-1.
412-06-02

Z_S for different voltages

Where the open circuit voltage to earth U_{OC} is not 240 volts, the value of Z_S in Tables ZS 1 to ZS 9 shall be modified as follows: Revised Design $Z_S = (Table\ Z_S \times U_{OC}) \div 240\ V.$
INF

Equipotential zone

The values of Z_S given in Tables ZS 1 to ZS 9 (41 B1, 41 B2, and 41D) are only applicable inside the equipotential zone created by the main equipotential bonding conductors. Furthermore, the limiting values of earth fault loop impedance and cpc impedance given by the formulae:
$Z_S \times I_a \leq U_o$ and $Z_{cpc} = 50 \times Z_S \div U_o$, and $Z_S \times I_{\Delta n} \leq 50$ V, are only applicable inside the equipotential

zone. Where the specified disconnection time cannot be met, 413-02-04 shall apply.
471-08-02, 413-02-04

Equipment outside equipotential zone

Where equipment is outside the equipotential zone, and that equipment can be touched by a person in direct contact with earth, and where the supply to the equipment is obtained from within the equipotential zone, disconnection shall occur within 0.4 s when the voltage to earth is 230 V, and an earth fault occurs. For other voltages to earth , the phase earth loop impedance shall ensure that the circuit is disconnected within the times specified in Table 41A.
471-08-03

Where a socket-outlet rated up to 32 A is fitted in an installation, and where it can reasonably be expected that it may be used to supply portable equipment for use outdoors, or where a flexible cable, having a current-rating of 32 A or less, not connected through a socket outlet supplies portable equipment outdoors, it shall be protected by an RCD having a residual operating current not exceeding 30 mA and an operating time not exceeding 40 ms, when subjected to a residual current of 150 mA.
471-16-01,412-06-02, 471-16-02

Compliance with Tables 41A to 41 D (TN system)

The intention of automatic disconnection is to limit the magnitude and duration of voltages appearing between simultaneously accessible conductive parts, and includes all methods of earthing exposed-conductive-parts.
The limiting values of Z_s given in tables 41A to 41 D shall be applied within the equipotential zone where normal body resistance applies. In situations where a lower body resistance can be expected, lower values of Z_s shall be achieved, or an alternative method of protection shall be used.
471-08-03, 471-08-02

Normal body resistance

Normal body resistance can be expected where the person is clothed, the floor is carpeted, the skin is dry, or moist with perspiration, and involves contact with one hand and both feet. Low body resistance occurs where the hands or feet are wet, the floor is uninsulated, or a large portion of the body is in contact with conductive parts, or where the shock current path is not through the extremities of the body.
INF

Protection in TN systems

Protective conductors shall connect all exposed-conductive-parts of the installation to the installation main earthing terminal, and that terminal shall be connected to the means of earthing provided for the source of supply in accordance to the system used.
413-02-06

Protective devices allowed in TN system

Protective devices shall be either:
   a) overcurrent protective devices or
   b) residual current devices
RCDs must not be used with circuits using combined protective earthed neutral conductors i.e. PEN conductors. Where the system is TN-C-S the PEN conductor shall not be used on the load side of

the RCD. Connection of the circuit protective conductor shall be made to the PEN conductor on the supply side of the RCD.
413-02-07, 471-08-07

Where the specified disconnection times cannot be achieved by the overcurrent protective devices, then either local supplementary equipotential bonding shall be used, or protection shall be given by an RCD, providing $Z_S \times I_{\Delta n} \leq 50$ V.
413-02-O4, 413-02-16

Protection in TT systems

Exposed-conductive-parts protected by a single protective device are to be connected to a common earth electrode(s). Where several protective devices are in series, each may have its own earth electrode, and the exposed-conductive-parts of each circuit may be connected to that protective device's earth electrode. Protective conductors can be installed individually, in groups, or collectively.

The addition of the cpc resistance and the electrode resistance (R_A) multiplied by the current required to disconnect the circuit within 5 s (I_a) must not exceed 50 V.
413-02-18, 413-02-20

Protective devices allowed in TT systems

Protective devices shall be either rcd's or overcurrent protective devices. However, the former are preferred.
413-02-19

Bathrooms

Where equipment installed in a room containing a fixed bath or shower is not in a SELV circuit, and where that equipment is simultaneously accessible with either other exposed-conductive-parts or extraneous-conductive-parts, the characteristics of the protective device for the circuit supplying the equipment shall be co-ordinated with the earthing arrangements, so that disconnection occurs within 0.4 s.
601-04-02

No electrical equipment shall be installed in the interior of a bath tub or shower basin, even if the supply is from a SELV source.
601-02

Protection by non-conducting location, or protection by means of earth-free local equipotential bonding shall not be used in a bath or shower room.
601-06

All means of electrical control, including switches and a SELV source of supply shall be placed out of reach of a person using a bath or shower, and all SELV circuits and equipment shall be insulated.
601-04-01, 601-08-01

Reduced low voltage systems

Reduced low voltage systems using 110 volts r.m.s., between phases, shall use overcurrent low protecive devices in each phase conductor, or residual current devices to give protection by automatic disconnection against indirect contact, and all exposed-conductive-parts of the reduced low voltage system shall be earthed. The value of Z_S at each point of utilisation shall ensure

that disconnection occurs in 5 s. Where an RCD is used, $I_{\Delta n} \times Z_S$ must not exceed 50. Where circuit-breakers or fuses are used. *Values of Z_S allowed are given in Table ZS 10 Part 3.*
471-15-06

Agricultural installations

In agricultural buildings or in situations accessible to livestock, electrical equipment shall have at least IP44 protection, which must be increased to suit the environmental conditions encountered.
605-11-01

Where automatic disconnection is used for protection against indirect contact, the disconnection time for final circuits supplying portable equipment intended for being moved during its use, or hand-held equipment, is reduced to 0.2 s, when the voltage to earth is between 220 V and 277 V. *Table ZS 11 Part 3 gives the maximum design value of Z_S for 0.2 disconnection time.*
605-05-01

For distribution circuits, and final circuits for stationary equipment, the disconnection time is 5 s, so Tables ZS 1 to ZS 9 Part 3 are applicable.

Where circuits requiring disconnection in 0.2 s and 5 s are connected to the same distribution board, one of the following conditions shall be applied:

a) the impedance of the protective conductor from the main earth bar, to which the main equipotential bonding conductors are connected and the distribution board, shall not exceed $25 \times Z_S \div U_o$, where Z_S is the earth fault loop impedance for a disconnection time of 5 s,

b) equipotential bonding shall be carried out at the distribution board to the same extraneous-conductive-parts as connected to the main equipotential bonding conductors.

Protection against indirect contact by limiting the impedance of the protective conductor is not allowed. Therefore, Table 41 C in the British Standard, and PCZ 10 Part 3 cannot be used.
605-05-06, 605-05-05

The following formulae are changed so that the voltage is 25 V.

$Z_S \times I_{\Delta n} \leq 25$ V,   413-02-16
$R_A \times I_a \leq 25$ V,   413-02-20
$R_A \times I_d \leq 25$ V,   413-02-23
605-05-09, 605-05-06, 605-07-01

Caravans and motor caravans

Each mobile touring caravan or motor caravan shall be supplied from a 16 A BS EN 60309-2 socket-outlet which is protected by a 30 mA RCD, with an operating time not exceeding 40 ms when subjected to a residual current of 150 mA.
608-13-02, 608-13-05

Where protection by automatic disconnection of the supply is used for the internal installation of every caravan and motor caravan, protection shall be by a 30 mA double pole RCD with an operating time not exceeding 40 ms when subjected to a residual current of 150 mA.
608-03-02

Socket-outlets shall incorporate an earthing contact, and a protective conductor shall be installed throughout each circuit of the caravan installation from the protective contact of the inlet.
608-03-02

Exceptions for protection against indirect contact

The following list gives those situations where bonding need not be carried out.

1. Overhead line insulator wall brackets, or any metal connected to them, providing they are out of arm's reach.
2. Inaccessible steel reinforcement in reinforced concrete poles.
3. Exposed-conductive-parts that cannot be gripped or contacted by a major surface of the human body, providing a protective conductor connection cannot be readily made, or be reliably maintained. This provision includes isolated metal bolts, rivets, nameplates up to 50 mm x 50 mm, and cable clips.
4. Fixing screws of non-metallic parts, providing there is no risk of them contacting live parts.
5. Short lengths of metal conduit or similar items which are inaccessible, and do not exceed 150 mm in length.
6. Metal enclosures used to mechanically protect equipment complying with the requirements for Class II (double insulated) equipment.
7. Inaccessible, unearthed street lighting (furniture) supplied from an overhead line

471-13-04

Protective conductors

A bare conductor or single core cable can be used as a protective conductor, and where complying with Section 543, it need not be enclosed in conduit, trunking or ducting. Similarly the metal sheath, screen or armour of a cable can be used as a protective conductor, as can rigid metallic conduit and trunking.

543-02-02, 521-07-03

Earthing conductors can be used for protective and functional purposes, but the protective qualities of the conductor must not be affected by the functional purpose.

546-01, 542-01-06

TN system disconnection times

Equipotential zone	Equipment type or location	Disconnection time allowed	Protective device allowed
Inside	Fixed	5 seconds	Overcurrent or R.C.D
	Socket-outlets or portable equipment	0.4 seconds	
	Bathrooms	0.4 seconds	
	110V supplies	5 seconds	
Equipment outside fed from inside the equipotential zone	Fixed	0.4 seconds	Overcurrent or R.C.D
	110V supplies	5 seconds	
	Socket-outlets for fixed equipment * See note	0.4 seconds	
	Socket-outlets	0.4 seconds	R.C.D only

413-02-08, 413-02-09, 413-02-13,
471-08-03, 471-15-06, 471-16-01
*Note for previous table: an overcurrent protective device can be used for socket-outlets feeding fixed equipment, providing the socket-outlet cannot be used for hand-held equipment.
413-02-09

Detrimental influences
The installation shall be checked to make sure that no detrimental influence can occur between the various protective measures in the same installation.
470-01-02

Supplementary bonding
When the overcurrent protective device cannot disconnect the circuit within the specified time in table 41A, then either:
a) local supplementary bonding shall be installed, or
b) protection shall be provided by a 30 mA RCD having an operating time not exceeding 40 ms when subjected to a residual current of 150 mA.
413-02-04, 413-02-16

Maintainability
Where a protective measure has to be removed for maintenance, reinstatement of the protection shall be made without reducing the original degree of protection. Adequate access to all parts of the wiring system shall be possible for maintenance to be carried out safely.
529-01

Local protection by earth- free equipotential bonding

This method of protection should only be used in special circumstances, such as electronic test rooms.
413-05-01

Protection by electrical separation

The use of this type of protection for more than one circuit from a single source is only recognised where the installation is under the effective supervision of a qualified electrical engineer.
471-12-01

The source of supply must be either an isolating transformer to BS 3535 where the secondary winding is isolated from earth, a motor generator set with separation equivalent to BS 3535, a mobile or fixed source complying with the Class II equipment requirements and the voltage must not exceed 500 V.

Where this type of protection is given to a single circuit the live parts have to be kept separate from any other circuit and earth, with sufficient insulation to ensure there is not a fault to earth.

Where mechanical damage can occur to any flexible cable or cord used, the cable must be visible throughout its length.

None of the exposed-conductive-parts of the separated circuit are to be connected to any other exposed-conductive-part or to the protective conductor of the source of supply.
413-06-01, 413-06-02,
413-06-03, 413-06-04

Chapter 29

Reduced low voltage systems

Reduced low voltage system

Where an extra-low voltage system cannot be used, and a SELV system is not needed, then a reduced low-voltage system can be used.
471-15-01

Maximum voltage

Reduced low-voltage circuits shall not exceed 110 V between phases, giving 63.5 V to earthed neutral for a three phase supply, and 55 V to the mid-point for a single phase supply.
471-15-02

Source of supply

The source of supply shall be:
  a) a double wound isolating transformer to BS 3535 Part 2, or
  b) a motor generator whose windings provide the equivalent isolation of an isolating transformer, or
  c) a diesel generator, or other independent source of supply.
471-15-03

The mid-point, or star point of the secondary winding of a transformer or generator shall be connected to earth.
471-15-04

Protection against direct contact

Protection against direct contact shall be the same as for low voltage installations.
471-15-05

Protection against indirect contact

Overcurrent devices in each phase conductor, or residual current devices, shall be used to give protection against indirect contact. All exposed-conductive-parts of the reduced voltage system shall be earthed. The value of Z_S at each point of utilisation (including socket-outlets) shall ensure disconnection occurs within 5 s. Where an RCD is used, $I_{\Delta n}$ x Z_S must not exceed 50. *Maximum values of Z_S for various types of protective device are given in Table ZS 10 Part 3.*
471-15-06

For protective devices that are not given in Table ZS 10, the value of Z_S is determined from the formula: $Z_S = U_{oc} + I_a$, where I_a is the current required to disconnect the protective device within 5 s, and U_{oc} is the open circuit voltage to earth, or to the neutral point.
471-15-06, 413-02-08

Plugs and socket-outlets

Plugs, socket-outlets, and cable couplers shall have a protective conductor contact, and not be interchangeable with plugs, socket-outlets, and cable couplers of different voltages or frequencies in the same installation.
471-15-07

Residual current devices

Fundamental requirement for safety

Where exposed-conductive-parts of electrical equipment may become charged with electricity due to the insulation of a conductor becoming defective, or due to a fault in the equipment, the exposed-conductive-parts shall be earthed by a method which will discharge the electrical energy without danger, or an equally effective measure that will prevent danger shall be used, e.g. Class II insulation.
130-04-01

Protective devices that will prevent the persistence of dangerous earth leakage currents shall be provided for every circuit.
130-04-02

Overcurrent protective devices shall be provided to stop the persistence of an earth fault current, and where an earth fault current is insufficient to cause the rapid operation of an overcurrent protective device, a residual current device shall be used to protect against dangerous earth fault currents.
130-04-03

Maximum phase earth loop impedance allowed

Where RCDs are used in TN systems to comply with the disconnection times required for the voltage to earth that is available, then $Z_s \times I_{\Delta n}$ must not exceed 50 V.

In a TT system, $R_A \times I_{\Delta n} \leq 50$ V, where R_A is the resistance of the protective conductor from the exposed-conductive-part to the earth rod, and includes the earth electrode resistance.
413-02-16, 413-02-20

PEN conductors

A PEN conductor shall not be used on the load side of an RCD. When the system is TN-C-S, circuit protective conductors shall be connected to the PEN conductor on the supply side of the RCD.
413-02-07

TT system installations

Residual current devices are preferred for protection against indirect contact. If a single RCD is used for protection, it shall be placed at the origin, unless the installation from the origin up to the RCD complies with the requirements for protection by Class II equipment.
413-02-19, 531-04

Socket-outlets in TT system

Where automatic disconnection is used as protection against indirect contact in an installation forming part of a TT system, all socket-outlets are to be protected by a residual current device, and the product of $I_{\Delta n} \times Z_s \leq 50$ and $R_A \times I_{\Delta n} \leq 50$ V.
471-08-06, 413-02-20

IT system installations

Where protection against indirect contact in an IT system uses RCDs, each final circuit shall be separately protected.
413-02-25

Electrical equipment outside the equipotential zone

Where fixed equipment is outside the equipotential zone, and that equipment can be touched by a person in direct contact with earth, and where the supply to the equipment is obtained from within the equipotential zone, disconnection shall occur within 0.4 s when an earth fault occurs, if the voltage to earth is 230 to 277 V. For other voltages, the disconnection times shall be: 120 V disconnection 0.8 s, for 400 V disconnection 0.2 s, and over 400 V disconnection 0.1 s.
471-08-03

Where any socket-outlets rated up to 32 A are fitted in an installation, and where it is possible that they may be used to supply portable equipment for use outdoors, or where a flexible cable, having a current-rating of 32 A or less, not connected through a socket-outlet, supplies portable equipment outdoors, they shall be protected by a 30 mA RCD having a residual operating time not exceeding 40 ms, when subjected to a residual current of 150 mA, in order to comply with BS 4293, BS 7071, BS 7288, BS EN 61008-1 or BS EN 61009-1.
471-08-04, 471-16-01, 471-16-02

Indirect contact

Where the disconnection times for protection against indirect contact cannot be achieved using overcurrent protective devices, protection shall be made by either local supplementary bonding or by an RCD. Where protection is provided by an RCD, $Z_s \times I_{\Delta n} \leq 50$ V (25 V in special locations).
413-02-04, 413-02-16

Direct contact

An RCD cannot be used as the only means of protection against direct contact, but it can be used to reduce the risk of direct contact, providing:
  a) it has a residual operating current not greater than 30 mA, and has a residual operating time not exceeding 40 ms, when subjected to a residual current of 150 mA, and
  b) protection against direct contact is provided by insulation, barriers, enclosures, obstacles, or placing out of reach.
412-06-02

Reduced low voltage systems

Automatic disconnection shall be used for protection against indirect contact by:
  a) an overcurrent protective device in each phase conductor, or by
  b) a residual current device, and
  c) all exposed-conductive-parts shall be connected to earth.
The Z_s at every point of utilisation (including socket-outlets) shall ensure a disconnection time of 5 s, where RCDs are used, then: $Z_s \times I_{\Delta n} \leq 50$ V.
471-15-06

Caravan site installation

Each caravan shall be supplied from a socket protected by a 30 mA RCD, which must not be the sole means of protection against direct contact. (i.e. protection against direct contact shall be by insulation, barriers or enclosures.) The RCD must disconnect within 40 ms, when subjected to a residual current of 150 mA. Not more than 3 socket-outlets shall be protected by one RCD. Where the supply is PME, the protective conductor of each socket-outlet shall be connected to an earth electrode, and the installation treated as a TT system.
608-13-05

CHAPTER 30

Selection and erection of residual current devices

All phase and neutral conductors shall pass through the coil of the magnetic circuit. Protective conductors shall be outside the magnetic circuit, *i.e., does not pass through the coil of the magnetic circuit.* Additionally, the RCD shall be capable of disconnecting all the phase conductors at substantially the same time.

531-02-01, 531-02-02

The residual operating current $I_{\Delta n}$ shall not exceed $50 \text{ V} \div Z_s$

The loads and circuits protected by an RCD shall be so arranged, and the selection of the RCD shall be so made, that any leakage currents which may be expected to occur during normal use will not cause nuisance tripping of the RCD.

531-02-03, 531-02-04

Where separate auxiliary supplies are relied upon for the operation of an RCD and the RCD does not automatically trip if the auxiliary source of supply fails, one of the following conditions shall be complied with:

1. Protection against indirect contact shall be maintained when the auxiliary supply fails, or
2. The device shall be regularly supervised, maintained and tested by skilled or instructed persons.

531-02-06

RCDs shall be located outside the magnetic fields of other equipment, unless it can be ascertained that the RCDs' operation will not be impaired.

531-02-07

RCDs used with, but separately from, short-circuit current protective devices, shall be able to withstand the short-circuit current that would flow from a short-circuit at the outgoing terminals of the RCD.

531-02-08

Where equipment in part of a TN system cannot comply with the disconnection times of Table 41A, then protection can be given by an RCD for that equipment. The exposed-conductive-parts for that part of the installation shall be connected to the TN earthing system cpc, or to an earth electrode. Where an earth electrode is used the system becomes TT and the earth electrode impedance shall be appropriate to the operating current of the RCD.

531-03

Discrimination

Where two or more residual current devices are installed in series, and are being used to protect against indirect contact, and where discrimination in their operation is required to prevent danger, then their characteristics shall be such that discrimination is achieved. *(One way of achieving discrimination is to select devices with different delay times.)*

531-02-09

Operation by ordinary persons

Where an RCD can be operated by a non-electrically skilled or instructed person, the device must be designed such that it is not possible to alter the settings or calibration of $I_{\Delta n}$ or time delay without a deliberate act involving a tool or key, and with setting visibly indicated.

531-02-10

Protection not provided

Protection against indirect contact is not provided if there is no cpc connected to the circuit's exposed-conductive-parts, even if the RCD has an operating current of 30 mA. *(That is, the human body is not to be used to complete the phase earth loop circuit.)*
531-02-05

Chapter 31

Road lighting and highway power supplies

Extent of requirements

Except where modified by the following requirements, the general requirements are applicable and in the absence of any reference to a Regulation, Section or Chapter means that reference should be made to the general wiring regulations.
600-01, 600-02

The regulations concerning highway power supplies and street furniture are not applicable to electric lines, supports and apparatus of, or under, the control of an electricity supplier, but apply to all other highway distribution circuits or any fixed equipment which is not directly associated with the use of the highway, but is located on it (street furniture). The requirements are also applicable to any other area used by the public but not called a highway or building. *They can also be used for areas between factory buildings or commercial premises which are not generally open to the general public.*
611-01-01, 611-01-02

Protection against direct contact

Protection against direct contact shall be provided by either:
  a) insulation of live parts capable of withstanding any mechanical, electrical, thermal or chemical stresses to which it may be subjected, the insulation being capable of removal only by destruction, or
  b) protection by barriers or enclosures such that the live parts are protected to at least IP2X or IPXXB, except where access is required for replacement of parts. Where access is required to live parts both the following requirements shall be complied with:
      i)   protection shall be provided to prevent the unintentional touching of live parts, and
      ii) the person gaining access will be made aware that live parts can be touched and should not be touched, or
  c) placing out of reach, providing it only applies to low voltage overhead lines constructed in accordance with the Electricity Supply Regulations 1988 amended, and maintenance of equipment is only carried out by skilled persons specially trained for carrying out maintenance under such conditions if the overhead line is within 2.5 m of the street lighting equipment. Where overhead lines are within 1.5 m of the lighting equipment and the maintenance staff have not been specially trained, placing out of reach cannot be used for protection against direct contact.

Protection by obstacles cannot be used for protection against direct contact.
611-02-01

A door in switchgear, control cubicles or lighting columns does not qualify as a barrier or enclosure for protection against direct contact. For a barrier or enclosure to qualify as protection against direct contact it must comply with one or more of the following requirements:
  a) it must only be possible to open the enclosure or remove the barrier by using a key or tool,
  b) it must only be possible to open the enclosure or remove the barrier after the electrical supply to the equipment has been disconnected and it must only be possible to restore the supply after the unit has been re-closed or the barriers replaced.

An intermediate barrier which is only removable by using a key or tool and giving a degree of protection IP2X or IPXXB shall be provided to prevent contact with live parts.
611-02-02

Protection against indirect contact

Protection against indirect contact can only be achieved by equipotential bonding and automatic disconnection of the supply or by protection by Class II equipment.
611-02-03

The maximum disconnection time allowed for fixed circuits is 5 s and where earthed equipotential bonding and automatic disconnection of the supply is used for protection against indirect contact there shall be no equipotential bonding from the main earth bar to any metallic structures that are not connected to or part of the fixed equipment for the highway power supplies. Metallic structures that are not connected to, or part of, the highway power system shall not be treated as extraneous-conductive-parts.
611-02-04, 611-02-05

Isolation and switching

Providing precautions are taken to prevent any equipment from being inadvertently or unintentionally energised and isolation and switching is carried out only by skilled or instructed persons, then isolation can be carried out by using a suitably rated fuse carrier in a TN system. Providing the approval of the electricity supplier is obtained, the supplier's cut out can be used as the means of isolation.
611-03

Identification and notices

Where the highway power supply is subject to a regular inspection and test, the periodic inspection and testing notice that has to be placed at the origin of each installation can be omitted.
611-04-04

Unless impractical, armoured or metal sheathed cables buried underground shall be marked by cable covers or marking tape, which is colour coded to distinguish it from other services; underground conduit or ducts shall also be suitably identified.

The colour orange shall be used as the identification colour for conduit.
611-04-02, 611-04-03, 514-02-01

Detailed records shall be provided of power supplies, distribution circuits etc., in accordance with Identification and Notices (Section 514) with the Electrical Installation Certificate.
611-04-01

Temporary supplies

Any temporary supplies taken from the fixed equipment shall not affect the safety of the permanent installation. A durable label giving the maximum continuous current that can be taken from each temporary supply unit shall be externally mounted on the equipment.
611-06

Chapter 32

SELV (Separated extra-low voltage)

Extra-low voltage
Extra-low voltage does not normally exceed 50 V r.m.s. a.c. or 120 V ripple free d.c. between conductors or between conductors and earth.
Definition.

Ripple-free
A d.c supply from a rectified a.c. source is considered to be ripple free if the ripple content does not exceed 10% r.m.s. The maximum voltage allowed for a sinusoidal ripple voltage for a 120 V ripple free d.c. supply is 140 V and 70 V for a 60 V ripple free d.c supply.
411-02-09

Protection against direct and indirect contact
Using SELV is a means of providing protection against direct and indirect contact. SELV must not exceed 50 V r.m.s. a.c. or 120 V ripple-free d.c.
411-02-01

SELV sources
The extra-low voltage shall be obtained from one of the following sources.

1. A safety isolating transformer having no connection between the secondary winding and the core of the transformer, or the primary earthing circuit. The transformer must also comply with BS 3535, with no connection to earth on the secondary side.
2. A motor generator whose windings etc. provide the same degree of safety as an isolating transformer made to BS 3535.
3. A battery or other source which is independent of higher voltages.
4. Electronic devices manufactured and tested to British standards, the output terminal voltage of which cannot exceed extra-low voltage, even with an internal fault condition. A voltage higher than extra-low voltage is permissible at the output terminals with these electronic devices, providing that, where direct or indirect contact does occur, the voltage at the output terminals is immediately reduced to extra-low voltage. Compliance with this requirement can be verified by testing the output with a voltmeter having an internal resistance of at least 3000 Ω, the voltmeter indicating a voltage not exceeding extra-low voltage.

411-02-02

Where extra-low voltage is obtained from a higher voltage system, the device reducing the higher voltage must have the necessary electrical separation between the higher voltage and the extra-low voltage. (As between the primary and secondary windings of a BS 3535 transformer.)

Where extra-low voltage is provided from a higher voltage system through an auto transformer, potentiometer, semiconductor device etc., it is not a SELV system.
411-02-03

The requirements for Class II equipment, or for equivalent insulation shall be used for the selection and erection of a mobile source for SELV.
411-02-04

If there is likely to be intentional or fortuitous contact between the exposed-conductive-parts of an extra-low voltage system and exposed- conductive-parts of any other system, the former system shall be treated as a PELV system if the circuits are earthed at one point only, all other instances being treated as a FELV system.
411-02-08, 471-14

Segregation of live and exposed parts
Live parts of SELV circuits must not be connected to earth, protective conductors, or live parts of other circuits.
411-02-05

Exposed-conductive-parts of SELV circuits must not be connected to earth, to an exposed-conductive-part, or to the protective conductors or extraneous- conductive-parts of another system.
411-02-07

Where equipment has to be connected to extraneous-conductive-parts, any voltage that may appear on the extraneous-conductive-part must not exceed extra-low voltage.
411-02-07

Live parts of SELV equipment shall be electrically separate from higher voltages. The separation required is the equivalent of that obtained between the primary and secondary windings of a safety isolating transformer. Example: A SELV relay in a control panel using higher voltages. Additionally, a live part of a SELV system must not be connected to a live part, or to a protective conductor of another system, or to earth.
411-02-05

Segregation of conductors
SELV circuit conductors shall be kept physically separate from conductors of other circuits.
Where conductors of SELV circuits cannot be kept separate from other circuits, then one of the following requirements shall be satisfied:
1. The conductors shall be insulated to the highest voltage present.
2. They shall be enclosed in an insulating sheath in addition to basic insulation.
3. They shall be separated by an earthed metallic screen or sheath.
4. When grouped with other cables, or enclosed in a multicore cable, they shall either be insulated individually or collectively for the highest voltage present.

In (2) and (3) the basic conductor insulation beneath the sheath or screen need only be sufficient for the voltage of the SELV circuit.

For SELV circuit components, such as relays, switches, contactors and the like, the electrical separation between the SELV circuit and all other circuits must be maintained.
411-02-06

Plugs and socket-outlets
Plugs used for SELV circuits must not be capable of insertion into socket-outlets for higher voltages, or for non-SELV circuits.
Socket-outlets must not be capable of having plugs for other voltages used on the same premises inserted into them.

The socket-outlets shall not have a protective conductor connection.
411-02-10

Where a luminaire supporting coupler has a protective conductor contact, it shall not be installed in a SELV system.
411-02-11

Protection against touching live parts (direct contact)

Where the SELV exceeds 25 V a.c. or 60 V ripple-free d.c., one or more of the following has to be adopted for protection against direct contact:

1. Protection by barriers or enclosures, giving protection against direct contact by the insertion of a solid body more than 12 mm in diameter, and up to 80 mm long, (IP2X) or the human finger (IPXXB).
2. The insulation shall be capable of withstanding 500 V d.c. for one minute.

411-02-09

If the nominal voltage does not exceed 25 V r.m.s. a.c. or 60 V ripple-free d.c. there is no need to insulate the cables, or to protect against direct contact, unless the normal body resistance is reduced, in which case the voltages of 25 V a.c. and 60 V d.c. must also be reduced. *(Body resistance can be reduced by a large portion of the body being in contact with earth, or in wet conditions etc.)*
411-02-09

Voltage drop

The voltage at the terminals of fixed equipment must not be less than the minimum value specified in the British Standard relevant to the fixed equipment. Where no British Standard is available for the fixed equipment, the voltage drop within the installation must not exceed a value appropriate to the safe functioning of equipment.

Where the supply is given in accordance with the Electricity Supply Regulations 1988 amended, the previous requirements are satisfied if the voltage drop from the origin of the installation up to the terminals of the fixed equipment or a socket-outlet does not exceed 4% of the nominal voltage of the supply.

A greater voltage drop is permissible for short periods with equipment having an inrush current when started, such as motors. The amount of voltage drop during start-up in these circumstances should not exceed the values given in the appropriate British Standard or recommended by the manufacturer of the equipment.
525-01

Other extra-low voltage systems

Where extra-low voltage is used, but not all the requirements for SELV are complied with, additional measures shall be taken to protect against both direct and indirect contact, as laid down in the requirements for PELV and FELV.
411-03

Chapter 33

Sauna heaters

Extent of requirements

Except where modified by the following requirements, the general requirements are applicable and in the absence of any reference to a Regulation, Section or Chapter means that reference should be made to the general wiring requirements.
600-01, 600-02

Type of appliance

Where electric hot air sauna heaters are used, they must comply with BS EN 60335-2-53, which is the standard for Safety of Household and Similar Electrical Appliances Part 2: Particular requirements, Electric sauna heating appliances, which has also to be read in conjunction with BS 3456 Part 201.
603-01

Temperature zones

The temperature zones illustrated in Figure 32.1 shall be taken into account when assessing the general characteristics of the installation.
603-02

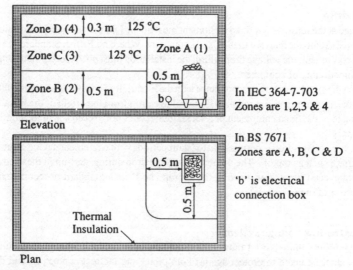

Figure 32.1 - Zones of ambient temperature

Electrical equipment

The only items that are allowed in a sauna are items of switchgear built into the sauna, thermostats, thermal cut-outs, and luminaires being so mounted in the sauna as to prevent overheating. All other equipment shall be mounted outside the sauna.
603-08 , 603-09

The minimum degree of protection for all equipment shall be IP24.
603-06-01

Wiring
Only flexible cords having 180 °C insulation and complying with BS 6141 shall be used, and shall be mechanically protected by double insulation, i.e. no metallic sheaths or metallic conduits are allowed for mechanical protection.
603-07-01

Installation
No equipment is allowed in zone A other than the sauna heater and equipment directly associated with it. Equipment in zone B can have a standard temperature rating, there being no special requirements for the heat resistance of equipment. Any equipment installed in zone C shall be suitable for an ambient temperature of 125 °C. In zone D, only luminaires and the control devices for the sauna shall be installed. These, together with the cables, shall be suitable for an ambient temperature of 125 °C.
603-06-02

Where SELV is used, protection against direct contact has to be provided by either: insulation that will withstand a type-test voltage of 500 V a.c. rms for one minute, or barriers, or enclosures that will give protection to at least IP24 to BS EN 60529.
603-03

Protection by means of obstacles, or by placing out of reach, is not allowed as a protection against direct contact, and protection by non-conducting location, or protection by earth free equipotential bonding, is not allowed for protection against indirect contact.

All other requirements of the British Standard are applicable to the installation.
603-04 , 603-05

Chapter 34

Selection and erection of cables

Materials and workmanship

Proper materials shall be used, and all electrical conductors shall be of sufficient size and current-carrying capacity: to prevent danger, they shall be placed in such a manner as to be safeguarded, or insulated and where necessary, effectively protected. All joints and connections shall be properly constructed with regard to insulation, conductance, mechanical strength, and protection.
130-01, 130-02-03, 130-02-04, 130-02-05

Electrical equipment

This term is abbreviated to 'equipment' in the British Standard, and includes wiring materials (such as cables), accessories, appliances, wiring systems and luminaires.
Definitions

Selection of cables

Low voltage non-flexible cables shall be one of the following types:
1) non-armoured PVC-insulated to BS 6004, or BS 6346, or BS 6231,
2) armoured PVC-insulated to BS 6346,
3) split-concentric copper-conductor PVC insulated to BS 4553,
4) rubber insulated to BS 6007 or BS 6883,
5) impregnated-paper-insulated, lead sheathed to BS 6480,
6) armoured with thermosetting insulation to BS 5467,
7) mineral-insulated to BS 6207,
8) consac cables to BS 5593.

The cable shall comply with either a British or Harmonized Standard.

Any of the listed cables sheathed with PVC, lead, h.o.f.r sheath, or an oil-resisting and flame retarding sheath may incorporate a catenary wire or include hard drawn copper conductors.
521-01-01

The phase conductors in a.c. circuits and live conductors in d.c. circuits shall have the following minimum sizes:
1) power and lighting circuits, 1.0 mm^2 copper or 16 mm^2 aluminium,
2) signalling and control circuit, 0.5 mm^2 copper, except for multicore cables having 7 or more cores, used for electronic equipment, when 0.1 mm^2 conductors are allowed,
3) bare conductors for power circuits, 10 mm^2 copper (16 mm^2 aluminium) and 4 mm^2 copper for control and signalling circuits,
4) insulated flexible connections and conductors, 0.5 mm^2 for any application except for electronic circuits outlined and complying with note 2,
5) insulated flexible connections and conductors for extra-low voltage circuits for special applications, 0.5 mm^2.

Connectors used for terminations for aluminium conductors, shall be suitable for this use.
524-01

Where a conductor other than a cable is used for overhead lines at low voltage, it shall be selected from one of the following types:

THIS IS THE LAW

A Handbook for Compliance

William Ernest Publishing Limited

P O Box 206, Liverpool L69 4PZ

Table 1

Cable selection

Cable type	Application	Comments
Fixed Wiring		
Single core pvc or rubber insulated unsheathed cable to BS 6004.	Used in conduit, trunking or ducting.	Not to be used in conduit underground.
Single core and flat pvc insulated and sheathed cable to BS 6004.	General indoor use. On exterior walls. Overhead between buildings. Underground in conduit.	Protect against severe mechanical stresses.
PVC insulated, armoured and pvc sheathed to BS 6346.	General indoor and outdoor applications.	Take precautions against damage when installed underground.
Mineral-insulated cable to BS 6207.	General indoor and outdoor applications.	Requires pvc covering when exposed to weather or corrosion, or if installed in underground or in concrete ducts.
Flexible cords		
E.P. rubber insulated CSP (HOFR) sheathed to BS 6500 Ref: 3183TQ.	For use with heating appliances, such as immersion heaters.	Not to be used on portable domestic appliances.
V.R. insulated and braided unkinkable cord to BS 6500 Ref: 2213.	Indoors in domestic or commercial premises.	For portable appliances: only suitable for low mechanical stress.
PVC insulated, pvc sheathed, cords to BS 6500. Ref: 3183Y.	Indoor and outdoor use on domestic or commercial premises, including damp situations.	Not allowed outdoors for agricultural or industrial applications.

1) hard-drawn copper or cadmium-copper conductors to BS 125,
2) hard-drawn aluminium and steel reinforced aluminium conductors to BS 215,
3) aluminium-alloy conductors to BS 3242,
4) conductors covered with PVC for overhead power lines to BS 6485 type 8.

521-01-03

Flexible cables used in an installation shall be selected from one of the following types:

1) insulated flexible cords to BS 6500,
2) rubber-insulated flexible cable to BS 6007,
3) PVC-insulated non-armoured flexible cable to BS 6004,
4) braided travelling cables for lifts to BS 6977,
5) rubber-insulated flexible trailing cables for quarries and miscellaneous mines to BS 6708.
6) single core flexible copper, aluminium or aluminium alloy welding cables to BS 638

521-01-01

Standard of cables

All cables shall comply with the current edition of the applicable British Standard. Where the British Standard takes account of a CENELEC Harmonisation Document (H.D. for short) then cables made to a foreign standard based on the same H.D. can be used, providing they are just as safe as if manufactured to the British Standard. This also applies to cables manufactured to an IEC Standard or a foreign standard based on an IEC Standard.

511-01-01, Preface.

If no standard exists, or if the cables are used in a way different from that which allowance is made in the standard, then the designer or the person responsible for specifying that the cables are used in that way, must certify that the degree of safety will not be less than that achieved by compliance with the British Standard.

511-01-02

Current-carrying capacity

The current-carrying capacity of every cable shall be not less than the normal full-load current flowing through it, so that its operating temperature does not exceed the value given in the current rating tables for the type of conductor and insulation concerned.

Where a conductor operates at a temperature higher than 70 °C, it shall be verified that the equipment to which it is connected is suitable for that temperature.

523-01, 512-02

Thermal insulation

Cables should not be installed in areas where thermal insulation is likely to be installed, but if they have to be installed in such areas, they should be installed so that they are not covered by the insulation.

If cables have to be installed so that they are completely covered by insulation, then they must have their current-carrying capacities decreased. Where the cables are in contact with thermal insulation on one side only, the other side being in contact with a thermally conductive surface, the current-carrying capacities can be obtained from the tables in Part 3.

The derating factors for single cables totally enclosed in thermal insulation are given in the table on page 149.

Table 2
Spacing of supports for cables in accessible positions

Maximum spacing of cable supports in millimetres

Overall diameter of cable in mm	Non-armoured PVC, or lead sheathed cables				Armoured multicore PVC or XLPE cable				Mineral-insulated copper sheathed or aluminium sheathed cables	
	Generally		In caravans		Horizontal		Vertical			
	Horizontal	Vertical	Horizontal	Vertical	Copper	Aluminium	Copper	Aluminium	Horizontal	Vertical
Up to 9	250	400							600	800
10 to 15	300	400	150	250	350		450		900	1200
16 to 20	350	450	for all	for all	400	1200	550	550	1500	2000
21 to 40	400	550	sizes	sizes	450	2000	600	600		
41 to 60					700	3000	900	900		
over 60					1100	4000	1300	1300		

Note: The major axis is taken for flat cables. 'Horizontal' includes angles up to a maximum of 60 degrees from the horizontal

Length in mm of conductors totally enclosed in thermal insulation	Derating factor
50	0.89
100	0.81
200	0.68
400	0.55
500 or over	0.5

523-04

Temperature
The current-carrying capacity of every cable shall take into account the maximum operating temperature likely to occur in normal service, including any heat transferred from accessories or appliances.
522-01-01

When determining the type cable to install, account shall be taken of the minimum and maximum ambient temperature that will be encountered, and installation should only be carried out within the temperature range of the cable to avoid mechanical damage occurring to the cable.
522-01, 522-08-01

The current-carrying capacity of a cable shall be reduced when installed in an ambient temperature higher than 30 °C.
App 4

Where cables in the same enclosure have different permitted operating temperature ratings, they shall be sized and treated as though they all have the lowest permitted operating temperature rating. *The following table gives correction factors.*

Cable with the lower operating temperature in the group	Correction factor to be applied to the I_{tab} of the higher operating temperature conductor in the group		
	70 °C	85 °C	90 °C
70 °C	1	-	-
85 °C	0.852	1	-
90 °C	0.816	0.957	1

Penetration of load-bearing structual elements by wiring systems, including cables, shall only be carried out where the integrity of the load-bearing element can be assured after such penetration.
521-07-03

522-12-03

Grouping cables or circuits
Where cables or circuits are grouped together the current-carrying capacity of the conductor has to be derated to take into account the reduction in heat dissipated from the conductor. *(See Part 2 for the various calculations for grouping and Part 4 for the grouping factors.)*
App 4

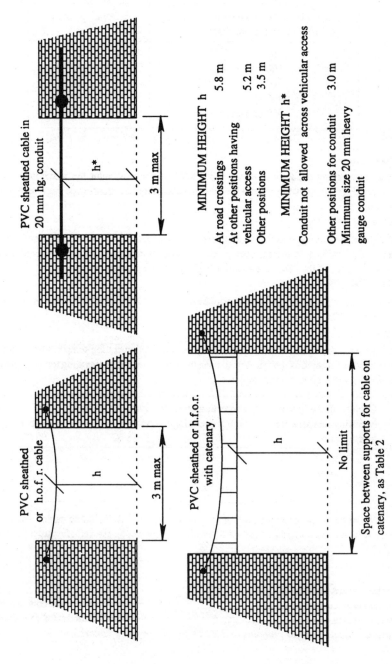

Figure 33.1 - The three most common methods of overhead wiring between buildings

Overload protection
Where a overcurrent protective device is providing overload protection for a circuit, the current-carrying capacity of the circuit conductors must not be less than the nominal rating of the protective device.
Additionally, where the fusing factor of the protective device exceeds 1.45 times its nominal rating, conductors having a higher current-carrying capacity shall be used, as determined by the formula:

The tabulated current required I_t = $\dfrac{I_n \times \text{fusing factor}}{1.45}$

433-02-01, 433-02-02, 433-02-03

Ring circuits
A ring final circuit is deemed to comply with the overload requirements if:
   a) the circuit is protected by a 30 A or 32 A protective device complying with BS 88, BS 1361, BS 3036, fuses or BS 3871 or BS EN 60898 MCBs, and
   b) the accessories of the ring circuit comply with BS 1363, and
   c) copper conductors with a minimum cross-sectional area of 2.5 mm^2 are used, or 1.5 mm^2 if mineral -insulated, and
   d) the calculated current-carrying capacity (I_t) of the cable is not less than:

$$I_t = \frac{I_n \times 0.67}{G \times A \times T}$$

where I_n is the rating of the overcurrent protective device, G is the grouping factor, A is the ambient temperature factor and T is the factor for thermal insulation. The circuit does not have to be derated if a BS 3036 semi-enclosed fuse is used as the protective device.
433-02-04

Neutral conductors
The neutral conductor shall have a cross-sectional area not less than the phase conductor.
524-02-01

Where the load of a three-phase and neutral circuit is not balanced due to the load, or where an imbalance occurs due to harmonic currents *(3rd harmonics from fluorescent fittings)*, power factor or discharge lighting circuit, the neutral conductor shall be capable of carrying the current that is likely to flow through it. *(Usually, it is safer to make the neutral the same size as the phase conductor.)*
Where an imbalance, such as from inequality of phase-loading, harmonic currents and/or inequality in power factor, is unlikely to occur in a three-phase and neutral circuit in normal use, multicore cables using a reduced size neutral can be used provided conductor temperature remains within limits, but the neutral with single core cables should be sized to the current expected to flow through it. Three-phase circuits supplying discharge lighting must have a neutral csa not less than that of the associated phase-conductor.
542-02-02, 524-02-03

Cables in parallel
Cables connected in parallel shall be of the same type, construction, size, length and disposition, have no branch circuits and shall be arranged to carry equal currents. The current-carrying capacity I_z of the group being the sum of all the cables. *But don't forget cables in parallel are grouped.*
Where 50 mm^2 copper or 70 mm^2 aluminium single core cables are installed in either trefoil or flat

Table 3

Minimum internal bending radius for fixed wiring cables

Conductor material	Conductor insulation	Cable finish	Overall diameter	Internal radius of bend given by diameter of cable times factor from below	
				For single core cables installed in conduit, trunking or ducting	Other installation methods
Copper or aluminium	Rubber or P.V.C. circular or	Non-armoured	less than 10 mm	2	3
			11 mm to 25 mm	3	4
			Over 25 mm	6	6
		Armoured	Any		6
Solid aluminium or shaped copper	P.V.C.	Armoured or non-armoured	Any		8
Copper or aluminium	Impregnated paper	Lead sheath	Any		12
Copper or aluminium	Mineral	All types MICC cable sheaths	Any		6

Note: Flat cables have the factor applied to the major axis, i.e., width of cable, not thickness.

formation, the phases should be arranged so as to equalise the load currents. *(Transpose the phase conductors.)*
523-02, 433-03
Every cable shall be sized so that it can withstand the thermal and mechanical stresses imposed on it due to any current, or fault current, it may have to carry.
521-03

Eddy currents
Where a circuit is wired in single core non-magnetic metal sheathed or armoured cables the sheaths or armour shall be bonded together at both ends of the circuit. It is permitted to bond the sheath or armour at one point only where:

    i) the cables have an insulating sheath, and
    ii) the size of the conductor exceeds 50 mm^2, and
    iii) the unbonded ends of the circuit are insulated, and
    iv) when the circuit is carrying full load current the following conditions are complied with:
        a) the voltages from sheath or armour to earth is limited to 25 V, and
        b) corrosion is not caused by circulating currents in the sheath or armour, and
        c) no danger or damage to property will arise when carrying fault currents.
523-05-01
Magnetic sheaths or armour shall not be used for single core cables used on an a.c. supply and all live conductors installed in magnetic enclosures should be contained in the same enclosure along with any protective conductors. Where single core a.c. conductors have to enter or leave a magnetic (ferrous) enclosure they should not be individually surrounded by ferrous material so as to stop circulating eddy currents. See Figure 33.2
521-02-01

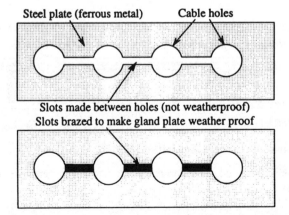

Figure 33.2 - Avoiding eddy currents in ferrous plate

Rotating machines
Cables carrying the starting, accelerating, and load currents of a motor shall have a current carrying-

capacity at least equal to the full load current rating of the motor. Where a motor is used intermittently, and for frequent starting and stopping, account shall be taken of the cumulative effect of the starting periods upon the temperature rise of the cables.
552-01-01

Care is needed when sizing cables feeding rotors of slip-ring or commutator induction motors since the rotor currents often exceed the full load current rating of the motor.
INF

Where a motor is frequently started and stopped, account shall be taken of the rise in temperature caused by the frequent starting and stopping upon the equipment and circuit conductors.
552-01-01

Mechanical damage
All conductors and cables shall either be constructed to withstand, or protected against, any risk of mechanical damage to which they may be subjected in normal use.
522-06-01

Cables shall be protected against abrasion when passing through holes in metalwork (*i.e., cables protected by bush or grommet*).
522-06-01

When assessing the risk of mechanical damage to cables, allowance shall be made for the strains imposed during installation.
522-08-01

Where cables are installed in conduit or ducting which is buried in the structure of the building each circuit shall be completely erected before cables are drawn in. Where the wiring of the installation is to be withdrawable, sufficient access shall be provided for drawing the cables in and out. See Figure 33.3
522-08-02

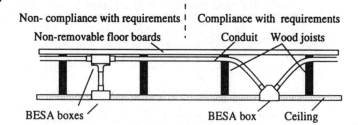

Non- compliance with requirements | Compliance with requirements
Non-removable floor boards · Conduit Wood joists

BESA boxes BESA box Ceiling

Figure 33.3 Compliance with 522-08-02

The number of cables drawn into an enclosure shall be such that no damage occurs to the cables during installation.
522-08-02

Conduit shall be completely erected before any cable is drawn in, unless a prefabricated system is being used.
522-08-02

Where cables are installed through timber joists in the ceiling or floor, they shall be not less than 50 mm from the top or bottom of the joist, unless mechanical protection that will stop the cable being

penetrated by nails or screws etc. is provided. (150 mm long lengths of conduit need not be earthed 471-13-04). Alternatively, cables not protected by an earthed armour metallic sheath can be enclosed in an earthed steel conduit which is securely supported. Cables of insulated concentric construction are also suitable for this purpose.
522-06-05

Cables and conductors used for fixed wiring shall be supported so that there is no mechanical strain, including the strain caused by the weight of the cable on them, or their terminations. (See table 2.)
522-08-01, 522-08-05

Supports for conductors shall be installed at intervals, if the conductor is not continuously supported, to stop the conductor being damaged by its own weight.
522-08-04

Where a wiring system is installed on a structure subject to vibration, the cable and its connections shall be suitable for the conditions expected.
522-07

Cable buried directly in the ground shall be of circular construction, incorporate an earthed metal sheath, or armour, or both, marked either by cable covers or marking tape. The cable shall be buried at a depth sufficient to avoid any damage by any ground disturbance likely to occur. Sufficient protection shall be given to any wiring system buried in a floor so that it will not be damaged by the normal use of the floor. Alternatively, an insulated concentric cable may be used.
522-06-03, 522-06-04

Cables installed in vertical conduits shall be supported at 5 metre intervals to avoid mechanical stress and compression of the cables. Any cable installed in a vertical trunking or duct shall be supported at 5 metre intervals.

Unarmoured cables installed in vertical runs, where they are inaccessible and not likely to be disturbed, shall be supported at 5 metre intervals if sheathed in rubber or PVC and at 2 metre intervals if lead covered.

Armoured or sheathed cables installed horizontally in inaccessible positions, where they are unlikely to be disturbed, need not be fixed providing the surface on which they are installed is smooth.
INF

Cable installation

Temperature
Where cables are installed in a heated part of a building, or in a heated floor, the maximum operating temperature of the heated floor, or that part of the building in which the cables are installed, shall be taken as the ambient temperature for the cables.
522-01-01

The insulation of conductors connected to busbars or bare conductors shall be suitable for the operating temperature of the busbar.
523-03

Where cables enter accessories, appliances, or luminaires, they shall be suitable for the temperatures likely to be encountered, or shall have each conductor sleeved with insulation suitable for the temperature encountered *(e.g. glass sleeving)*, so that there is no risk of short circuit between conductors or of an earth fault.

522-02-02

Internal barriers shall be provided in every vertical channel, duct, ducting or trunking, at the ceiling of each floor, or at 5 metre intervals, whichever is the less, to limit the temperature rise.
INF

Fire barriers

Where cables pass through a wall or floor that is a fire barrier, the hole round the cable shall be sealed to the same degree of fire resistance as the wall or floor.

Where conduits, ducts, channels or trunking pass through a fire barrier, the interior of the conduit, duct, channel or trunking, shall also be sealed with suitable fire resistant material. Steel conduit, trunking or ducting having an internal area of 710 mm² need not be internally sealed.

Where a fire barrier is installed in a trunking, there is no need to fix an additional barrier at that point to stop the air becoming excessively hot.
527-02-01, 527-02-02

Bending radii

The internal radius of every bend in a non-flexible cable shall be such that conductors and cables are not damaged.
522-08-03

Lift shafts

Only cables that form part of a lift installation are allowed to be installed in a lift or hoist shaft.
528-02-06

Cables installed in walls, floors and ceilings

Where cables are installed through timber joists in the ceiling or floor, they shall not be less than 50 mm from the top or bottom of the joist, unless mechanical protection that will stop the cable being penetrated by nails or screws etc., is provided (150 mm long lengths of conduit need not be bonded, 471-13-04). Alternatively, cables not protected by an earthed metallic sheath can be enclosed in an earthed steel conduit which is securely supported. Insulated concentric cables may also be used without further mechanical protection.
522-06-05

Where a cable is to be installed in a wall or partition at a depth of less than 50 mm, it shall be placed in an area within 150 mm from the top of the wall or partition, or within 150 mm of an angle formed by two walls or partitions, as illustrated by the shaded area in Figure 33.4
522-06-06

Where the cable feeds a point, switchgear or accessory on the wall outside the above zones, the cable shall be installed in straight vertical or horizontal runs to the point or accessory, as illustrated by 'A' and 'C' in figure 33.4. *For house wiring, check the NHBC Rules in case there is a difference.*
522-06-06

Where the above requirements are not practicable, the cable shall be enclosed in earthed metallic conduit, trunking, ducting, or shall incorporate an earthed metallic covering, all complying with the requirements for protective conductors, as illustrated by 'B' in figure 33.4. Insulated concentric cables may also be used without further mechanical protection.
522-06-07

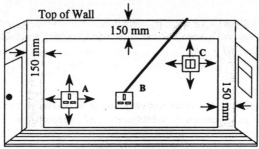

Figure 33.4- Installation Zones

Other than protective conductors, non-sheathed cables shall be enclosed in conduit, trunking or ducting.
521-07-03

Cables shall be sheathed and/or armoured when installed in underground ducts, conduit, or pipes.
522-06-01

Cables shall be sheathed and/or armoured, or in conduit or other enclosure, giving suitable protection when installed on walls outdoors.
522-06-02

Cables buried directly in the ground shall have a metallic sheath and/or armour or be of insulated concentric construction, be marked with cable covers or marking tape, and be buried at sufficient depth to avoid damage to the cable.
522-06-03

Sufficient protection shall be given to a wiring system buried in a floor, so that it is not damaged.
522-06-04

Conductors installed overhead between buildings must comply with the Electricity Supply Regulations 1988 as amended and the appropriate British Standard.
412-05-01, 521-01-03

Low voltage cables installed between buildings shall be placed beyond risk of mechanical damage.
522-06-01, 522-08-05

In areas inaccessible to vehicular traffic, cables may be installed in heavy gauge conduit (minimum size 20 mm without joint in span) , or in other enclosure, giving suitable protection (Figure 33.1). Cables and conductors shall be supported throughout their length, so that they and their terminations are not subjected to undue mechanical strain. *This includes the weight of the cable or conductor.*
522-08-05

Fauna, vermin and sunlight

Where cables are liable to attack by animals, they shall be of a suitable type, or suitably protected. Cables exposed to direct sunlight or other sources of UV radiation shall be resistant to damage by ultra violet light, or shielded from the solar radiation. *Insulated and sheathed PVC cables are all right since the inner cores are protected against direct sunlight, as are single core cables installed in conduit.*
522-10 , 522-11

Plants and mould growth

Special protective measures have to be taken if the wiring system is not selected to cater for the hazard of flora or of mould growth.

522-09-01

Installation design

The installation design shall take into account any structural movement that is likely to occur in the building structure, *e.g., expansion or subsidence* cable supports and mechanical protection allowing some movement to occur so that cables are not damaged. Flexible wiring shall be used where the structure is either flexible or likely to be unstable.

522-12

Excessive tensile and torsional stresses shall be avoided in the conductors and connections where a flexible wiring system is used.

522-08-06

Where wiring systems, including cables, penetrate a load-bearing building element, the integrity of the element shall not be impaired by such a penetration.

522-12-03

Band I and Band II circuits

Band I and Band II circuits shall not be contained in the same wiring system as high voltage conductors, unless every cable is insulated to the highest voltage present, and unless one or more of the following methods is used:

a) each conductor is insulated to the highest voltage present in a multicore cable, or contained in an earthed metallic screen, which has a current-carrying capacity equal to the largest conductor contained in the screen, or

b) The cables are installed in a separate compartment of cable trunking or ducting, and insulated to their normal system voltage, or have an earthed metallic screen.

Band I circuits shall be separated from Band II circuits.

528-01-01, 528-01-02

Band I covers extra-low voltage circuits and Band II embraces low voltage circuits.

Definitions

Fire alarm and emergency lighting circuits must be segregated from other circuits and from each other in accordance with BS 5266, BS 5839 and BS 6701 where appropriate.

528-01-04

Controls or outlets mounted in a common box, block or switchplate enclosing Band I and Band II circuits must have the different bands segregated from each other by an effective partition which, if metal, must be earthed.

528-01-07

Joints and terminations

Joints shall be accessible for inspection unless they are compound filled, encapsulated, or made by welding, brazing, or compression lugs, and contained in suitable enclosures. Apart from the exceptions listed above, and the cold tail connection made between a cold tail and a heating element, joints shall be accessible for inspection.

526-04

Every cable termination or joint shall be:
a) electrically and mechanically sound,
b) protected against mechanical damage and vibration,
c) shall not cause harmful mechanical damage to the conductor, or place a strain on the fixings of the connection.
Mechanical clamps and compression lugs must retain all the wires of the conductor.
526-01

Terminations shall be of the correct size and type for the conductor to be used, and suitably insulated for the nominal voltage present. They must take into consideration the temperature attained by the conductor connections and its effect on conductor insulation. Where connections or joints are soldered, they must take into consideration mechanical stress, creep, and temperature rise under fault conditions. Where vibration or thermal cycling occurs adequate locking facilities shall be provided.
526-02-01

Note: Where soldered connections are used, then the 'k' factor used in fault calculations shall be based on the soldered joint, if this is lower than the 'k' factor of the conductor.
INF

All terminations and joints in conductors shall be made in the appropriate enclosure manufactured to British Standards. Any enclosure used shall provide adequate mechanical protection against external influences.
526-03-02, 526-03-01

Where the sheath has been removed from sheathed cables the cores shall be enclosed, within at conduit, trunking, ducting etc., and single core unsheathed cables shall be in an enclosure.
526-03-03

Contact with non-electrical services
Protection shall be provided against indirect contact, and against the hazards that are likely to arise when a wiring system is located in close proximity to non-electrical services.
528-02-01

Unless a wiring system is suitably protected, and derated, where appropriate, it shall not be installed in the vicinity of a service which produces heat, smoke, fumes, or condensation likely to be detrimental to the wiring.
528-02-02, 528-02-03

Where non-electrical services and electrical services are installed close to each other, it shall be so arranged that work being carried out on either service will not cause damage to the other.
528-02-04

Moisture and corrosion
All cables likely to be exposed to weather, corrosive atmospheres, or other adverse conditions, are to be constructed or protected to prevent danger.
130-08-01

Wiring systems shall not be exposed to rain, water, or steam unless designed to withstand these external influences. When exposed to the weather, or in damp situations, all materials used shall be corrosion-resistant, enclosures shall be damp-proof, and every cable and joint shall be protected

against the effects of moisture.
522-03-01, 528-02-03

Drainage points shall be provided in a wiring system where moisture may collect due to condensation. Drainage points shall be suitably located to allow the harmless escape of the condensation.
522-03-02

Protection against mechanical damage shall be provided if the wiring is subjected to waves.
522-03-03

Materials shall not be placed in contact with each other, where they are liable to initiate electrolytic action. *For example, precautions shall be taken to maintain electrical continuity when aluminium conductors are connected to brass or to other metals with a high copper content.*
522-05-02, 522-05-03

Wiring systems shall be suitably protected where they are subjected to corrosion and deterioration from corrosive or polluting substances. *For example, aluminium conductors connected to earth electrodes, or to water pipes subject to condensation.*
522-05-01

Hazardous areas

All cables in surroundings susceptible to risk of fire or explosion shall be constructed or protected to prevent danger.
130-08-02

Where cables are installed in potentially explosive atmospheres, the additional precautions detailed in BS 5345 shall be taken.
110-01-01(xi)

Voltage drop

The voltage drop in the installation shall be such that the voltage at the terminals of fixed equipment is not less than that value required by the British Standard for such equipment.

Where no British Standard exists for the equipment, then the voltage drop within the installation must not exceed a value appropriate to the safe functioning of the equipment.

Where the supply to the premises is provided in accordance with the Electricity Supply Regulations 1988 as amended, the above requirements are deemed to be complied with if the voltage drops between the origin of the installation and the fixed equipment or a socket-outlet does not exceed 4% of the nominal supply voltage. With a 400/230 V supply, this is 9.2 V single-phase, and 16.0 V three-phase. With a 415 V / 240 V supply this is 16.6 V and 9.6 V respectively.
525-01-01, 525-01-02

When calculating voltage drop, starting currents or inrush currents can be ignored.
525-01-02

Cable colours

The colour combination green/yellow shall be used only for protective conductors. The single colour green or yellow alone or any bi-colour other than the combination green/yellow shall not be used in flexible cables or cords.

514-03, 514-07-02

Protective conductors shall be coloured green and yellow, but a functional earth *(clean earth with computer equipment)* shall be coloured cream. The phase conductors of multi-phase a.c., circuits shall be red, yellow and blue, with the neutral coloured black. The phase conductor of single-phase a.c., circuits shall be red and the neutral black.

For d.c., supplies, the positive of a two wire circuit shall be red and the negative black. The positive of a d.c., three wire circuit shall be red, the middle wire black and the negative blue. The functional earth for telecommunications shall be cream.

Table 51A

Every single core and multicore non-flexible cable conductor shall be identified at its terminations, and preferably throughout its length, red, yellow or blue indicating the phase conductors and black the neutral conductor. Paper-insulated cables, PVC-insulated multicore auxiliary cables, or thermosetting insulated multicore cables can have cores numbered in accordance with British Standards, providing that 1, 2 & 3 are used for phase conductors, and 0 for the neutral conductor. MICC cables shall be identified at their terminations by tapes, sleeves or discs of the appropriate colour.

514-06-01, Table 51A

Red yellow and blue may be used as phase colours for cables up to final circuit distribution boards, but single-phase circuits from a final distribution board shall be coloured red and black.

Table 51A

Bare conductors and switchboard busbars shall, where necessary, be identified by the application of tape, sleeves or discs in accordance with Table 51 A, as specified above.

514-06-03, 514-06-04

Flexible cords

Every core of a flexible cable or cord shall be identified throughout its length, Brown for phase, Blue for neutral, Green/yellow for the protective conductor. (See cable colours.) Coded marker sleeves shall be used where phase rotation needs to be indicated.

514-07-01, 514-07-02

Excessive tensile and torsional stresses to conductors and connections shall be avoided in a flexible wiring system installation.

522-08-06

Flexible cords shall not be used for fixed wiring unless the installation complies with the requirements of the British Standard, such as complying with all the requirements for a fixed wiring installation, as well as with the additional requirements for flexible conductors.

521-01-04

Provision shall be made for portable equipment to be fed from an adjacent and accessible socket-outlet, due allowance being made for the length of the flex fitted to the portable equipment *(usually 1.5m). Cookers rated over 3 kW are not considered portable.*

533-01-07

Where a flexible cord supports, or helps to support a luminaire, the maximum weight that can be placed on the cord is as follows:

Conductor size mm^2	Maximum weight
0.5	2 kg
0.75	3 kg.
1.0	5 kg.
1.5	5 kg.
2.5	5 kg.
4.0	5 kg.

Table 4H3A

Chapter 35

Selection and erection of equipment

General
To prevent danger, proper materials shall be used; in addition, equipment shall be properly constructed, installed and protected, and shall be capable of being maintained, inspected and tested. Equipment must also be suitable for the maximum power it is required to handle.
130-01, 130-02-01, 130-02-02

Standard of equipment
All equipment shall comply with the current edition of the applicable British Standard or Harmonized Standard.
511-01-01

Where a British Standard takes into account a CENELEC Harmonization Document, then equipment manufactured to a foreign standard based on the same Harmonization Document can be used, providing it is no less safe than if manufactured in accordance with the relevant British Standard.
The same comments apply to equipment manufactured to an IEC Standard.
511-01-01, Preface

Location of equipment
Equipment shall be located so that any heat generated by the equipment will be dissipated, and not present a fire hazard to the adjacent building materials, or to materials likely to be placed in the same location.
422-01-01

Where electrical equipment has a surface operating temperature that could cause a risk of fire, could be harmful to adjacent materials, or could produce harmful effects, one of the following methods of installation shall be utilised:

1) the equipment shall be mounted on a support, or placed in an enclosure which is capable of withstanding the temperature generated by the equipment, without being set on fire, or creating a harmful effect. Any support used shall limit the transmission of heat from the equipment by having a low thermal conductance, or
2) the equipment is screened by material which will withstand and not be affected by the heat generated by the equipment, or
3) the equipment is fixed so as to allow the safe dissipation of heat, and at such a distance from other materials as to avoid fire and harmful effects.

422-01-02

Provision shall be made for portable appliances and luminaires to be fed from an accessible and convenient adjacent socket-outlet.
553-01-07

Where equipment in one location contains more than 25 litres of flammable liquid, precautions shall be taken to contain the liquid, and to stop burning liquid or the gases of combustion spreading to other parts of the building.
422-01-05

Enclosures for cables shall be suitable for the extremes of ambient temperature to which they will be exposed. Where a non-metallic or plastics box is used for suspension of a luminaire, it shall be suitable for the suspended weight and the expected temperature when the luminaire is working. **522-01-01, 522-01-02, 422-01-01**

A means of switching off shall be provided for equipment having electrically heated surfaces that can be touched. **421-01-01**

Suitability of supply

All equipment shall be suitable for the nominal voltage of the supply, the design current, and the current likely to flow under short-circuit, overload, or earth fault conditions. **512-01, 512-02**

Equipment designed to work at a given frequency shall match the frequency of the supply, *(e.g., frequency of contactor coils)*, and the power demanded by the circuit, *(e.g., h.p. of motor suitable for the power output required)*. **512-03, 512-04**

Compatibility and external influences

Equipment installed shall not cause harmful effects on other equipment, or on the supply, and shall be suitable for the environmental conditions likely to be encountered.

Where the equipment is not suitable for the environmental conditions it shall be provided during erection with suitable protection , which must not interfere with the equipment's operation.

The degree of protection that is given shall take into account the simultaneous occurrence of different external influences. **512-05, 512-06-01,**

512-06-02, 331-01

Examples of the characteristics likely to have a harmful effect on other electrical equipment or services include:

1) transient overvoltages,

2) rapidly fluctuating loads,

3) starting currents,

4) harmonic currents from fluorescent lighting loads and thyristor drives,

5) surge currents of transformers,

6) mutual inductance,

7) d.c. feedback,

8) high frequency oscillations,

9) earth leakage currents,

10) additional connections to earth for equipment in order to avoid interference with its operation.

INF

Where equipment that is installed is likely to affect the supply, the supplier of electricity shall be consulted. An assessment of any harmful effects equipment is likely to have on any other equipment or services shall be made. **331-01**

Accessibility and maintainability

With the exception of joints that are allowed to be inaccessible, all equipment shall be accessible with sufficient working space to enable safe operation, maintenance, inspection and testing to be carried out.

130-07, 513-01,529-01-02

The quality and frequency of maintenance shall be assessed, so that the reliability of equipment chosen is appropriate to its intended life so that the protective measures for safety remain effective during the intended life of the installation and so that any periodic maintenance or inspection and testing can be carried out safely.

341-01

Voltage drop

The voltage drop in the installation shall be such that the voltage at the terminals of fixed equipment is not less than that value required by the British Standard for such equipment.

Where no British Standard exists for the equipment, then the voltage drop within the installation must not exceed a value appropriate to the safe functioning of the equipment.

Where the supply to the premises is provided in accordance with the Electricity Supply Regulations 1988 as amended, the above requirements are deemed to be complied with if the voltage drops between the origin of the installation and the fixed equipment or socket-outlets does not exceed 4% of the nominal supply voltage. With a 400/230 V supply, this is 9.2 V single-phase, and 16.0 V three-phase: with a 415 V / 240 v supply this is 16.6 v and 9.6 v respectively. When calculating voltage drop, starting currents or inrush currents can be ignored.

525-01-01, 525-01-02

Protection against undervoltage

Devices for protection against undervoltage shall comply with the following requirements.

535-01

Where a reduction in voltage, or loss and subsequent restoration of the supply could cause danger, suitable precautions shall be taken, such as provision of no-volt or undervoltage releases.

451-01-01

Every motor shall be provided with a means of preventing the automatic restarting of the motor if there is a drop in voltage, or if the supply fails, and if the unexpected automatic restarting would cause danger. This does not apply to those instances where there is a brief interruption of the supply, and where failure of the motor to restart would cause a greater danger. In this case other safety precautions would be required.

552-01-03

Where damage to equipment or to the installation may be caused by a drop in voltage but where the drop in voltage will not cause danger, then one of the following arrangements shall be chosen:

a) precautions against any foreseen damage shall be taken, or
b) it shall be confirmed with those responsible for the operation and maintenance of the installation that the foreseen damage is an acceptable risk.

451-01-02

Providing equipment can withstand a brief reduction or loss of voltage without danger, then a suitable time delay may be incorporated in the operation of an undervoltage protective device controlling the equipment.

The instantaneous disconnection by control or protective devices must not be affected by any delay in the opening or closing of contactors, and the characteristics of the undervoltage protective device shall be compatible with the requirements for the starting and use of the associated equipment.
451-01-03, 451-01-04, 451-01-05

The re-closure of a protective device shall not be automatic if it is likely to cause danger.
451-01-06

Moisture and corrosion

Wiring systems shall not be exposed to rain, water, condensation, or steam, unless suitably designed to withstand these. When exposed to the weather, or in damp situations, all materials used shall be corrosion-resistant, enclosures shall be damp-proof, and every cable and joint shall be protected against the effects of moisture.
522-03-01, 522-05-01

Precautions are to be taken to maintain electrical continuity when aluminium conductors are connected to brass or other metals with a high copper content, or to any other dissimilar metals that are in contact with each other.
522-05-03

Where metalwork is installed in a corrosive atmosphere, it shall be protected to withstand corrosion. Non-metallic materials shall not be placed in contact with materials that will cause chemical deterioration of the wiring.
522-05-01

Provision shall be made for the harmless escape of any water which may accumulate in a wiring system due to condensation. Where the equipment is subject to waves, mechanical protection shall be provided.
522-03-01, 522-03-02

Protection shall be given against indirect contact, and against the hazards that are likely to arise when a wiring system is located in close proximity to non-electrical services.
528-02-01

Unless a wiring system is suitably protected, and derated where appropriate, it shall not be installed in the vicinity of a service which produces heat, smoke, fumes, or condensation likely to be detrimental to the wiring.
528-02-02, 528-02-03

Where non-electrical services and electrical services are installed close to each other, it shall be so arranged that work being carried out on either service will not cause damage to the other.
528-02-04

Careful selection is needed to avoid any harmful influence between the electrical installation and other services; additionally, where equipment uses a.c or d.c. current, or a different frequency or voltage and are contained in a common enclosure, the cables and equipment of each type of current, frequency or voltage should be effectively segregated to avoid any detrimental influence between the different circuits or services.
515-01

Electromagnetic compatibility

Bearing in mind the intended level of maintaining continuity of service, equipment selected for installation must be suitable, in terms of its immunity levels, for the electromagnetic influences present. Similarly, the selected equipment must not cause electromagnetic interference; where this is likely, measures shall be taken to minimize the effects of emissions. Reference should be made to the particular product standard or to BS EN 50081: Electromagnetic compatibility. Generic emission standard, and BS 50082: Electromagnetic compatibility. Generic immunity standard.
515-02

Joints and terminations

Where Band I and Band II circuits are terminated at a common box, accessory, switch plate or box, the conductors and terminals of each category shall be effectively segregated by a partition. The partition shall be earthed if it is made from metal.
528-01-07

Joints shall be accessible for inspection unless they are compound filled, encapsulated, or made by welding, brazing or compression lugs, and contained in a suitable enclosure. Apart from the exceptions listed above and the cold tail connection made between a cold tail and a heating element, joints shall be accessible for inspection.
526-04

Every cable termination or joint shall be:
   a) electrically and mechanically sound,
   b) protected against mechanical damage and vibration,
   c) shall not cause harmful mechanical damage to the conductor, or place a strain on the connection.
Mechanical clamps and compression lugs must retain all the wires of the conductor.
526-01

Terminations shall be of the correct size and type for the conductor to be used, and suitably insulated for the nominal voltage present. They must take into consideration the temperature attained by the conductor connections, and its affect on conductor insulation. Where connections or joints are soldered, they must take into consideration mechanical stress, creep, and temperature rise under fault conditions.
526-02-01

Note: Where soldered connections are used, then the 'k' factor used in fault calculations shall be based on the k factor for soldered joints if less than the conductor. (Use k for lead from K1A or K1B.)
INF

All terminations and joints in conductors shall be made in the appropriate enclosure manufactured to British Standards. Any enclosure used shall provide adequate mechanical protection against external influences.
526-03-02, 526-03-01

Agricultural installations

In agricultural buildings and areas accessible to livestock, electrical equipment shall be of Class II construction, or protected by a suitable insulating material, and suitable for all the environmental conditions likely to occur.
605-11

Where cables are liable to attack by vermin, flora or mould growth they shall either be of a suitable type or suitably protected.
522-10

Outside installations

Where cables are installed in positions where they may be exposed to direct sunlight or other ultra-violet radiation, they shall be of a type that is resistant to ultra-violet light. *Note: insulated and sheathed PVC cables are suitable since the inner cores are protected against direct sunlight, as are single core cables installed in conduit.*
522-11

Dusty atmospheres

Enclosures for conductors, joints, and terminations shall be protected against the ingress of foreign bodies, protection being given in accordance with the IP codes.

Where the dust conditions are onerous, additional precautions shall be taken to stop accumulated dust reducing the heat dissipated from the wiring system.
522-04-01, 522-04-02

Motors

Where motors are subject to frequent starting, the effect of the temperature rise on equipment and conductors caused by the starting current shall be taken into account.
552-01-01

Motors exceeding 0.37 kW shall be provided with overload protection.
552-01-02

Every motor shall be provided with a means of preventing the automatic restarting of the motor if there is a drop in voltage, or if the supply fails, and if the unexpected automatic restarting would cause danger. This does not apply to those instances where there is a brief interruption of the supply, and where failure of the motor to restart would cause a greater danger. In this case other safety precautions would be required.
552-01-03

If the reversal of a motor can cause danger precautions shall be taken to stop the motor reversing. Precautions shall therefore be taken when reverse-current braking is used or reversal can occur with the loss of one phase.
552-01-04, 552-01-05

Every motor or electrically powered equipment shall be provided with a means of switching off which shall be accessible, easily operated, and so placed so as to prevent danger.
130-06-02, 476-01-02

Transformers

The common terminal of an auto-transformer connected to a circuit having a neutral conductor shall be connected to the neutral conductor.
555-01-01

A linked switch shall be provided for disconnecting a step-up transformer from all live conductors of the supply. Additionally, a step-up transformer shall not be used in an IT system.
555-01-02, 551-01-03

Switchboards and switchgear

The current-carrying capacity and the temperature limits of busbars, busbar connections, and bare conductors which form part of a switchboard, shall comply with BS 5486 and BS EN 60439-1.
INF

The nominal current rating of low voltage equipment such as switchgear, protective devices, and accessories is normally based on the conductors connected to the equipment not exceeding a working temperature of 70 °C. Where the operating temperature of conductors exceeds 70 °C, the manufacturer should be consulted since the current rating of equipment may need to be reduced by an amount specified by the manufacturer of the equipment.
512-02

Lighting

Every switch used to control discharge lighting shall be designed in accordance with BS 3676 and/ or BS 5518. Alternatively, its nominal rating shall be twice the total steady current of the circuit. *Where used for both discharge lighting and tungsten lighting, its nominal rating shall be twice the total steady current of the discharge lighting plus the total current of the tungsten lighting.*
553-04-01, INF

Final circuits for discharge lighting shall be capable of carrying the total steady current, harmonic currents, and any fault current likely to flow.
512-02

Each lighting point shall terminate in an accessory manufactured to British Standards such as: a ceiling rose, a luminaire supporting coupler, a batten lampholder, a direct connecting luminaire, a suitable socket-outlet or a connection unit.
533-04-01

Lampholders, luminaires and ceiling roses

The maximum voltage allowed for a filament lamp lampholder is 250 V. BC and SBC lampholders shall comply with , and have the temperature rating T2 specified in, BS 5042.

The outer contact of ES lampholders shall be connected to the neutral in TN and TT systems, as will lampholders mounted on a track system.
553-03-02, 553-03-03, 553-03-04

Luminaire supporting couplers shall not be used for the support or connection of other equipment.

The weight of a lighting fitting suspended from a pendant luminaire shall not damage the flexible cable or its connections.

Where an extra-low voltage luminaire has no facilities for connecting a protective conductor, it shall only be used as part of a SELV system
553-04-04, 554-01

Ceiling roses shall not be used in circuits exceeding 250 volts, nor shall they have more than one outgoing flexible cord, unless designed for multiple pendants.
553-04-02,553-04-03

Socket-outlets

With the exception of administration buildings such as site offices and toilets, every socket-outlet used on a construction site shall comply with BS EN 60309-2.

604-12-02

Socket-outlets mounted on a vertical wall shall be at a height sufficient to minimise the risk of mechanical damage to the socket-outlet, plug, or cable, during insertion and withdrawal of the plug.
553-01-06

With the exception of 12 V SELV circuits and shaver sockets complying with BS 3535, there shall be no socket-outlets or provision for connecting portable equipment in a room containing a fixed bath or shower.
601-10-02

Where socket-outlets rated up to 32 A are fitted in an installation, and where it is possible that they may be used to supply portable equipment for use outdoors, they shall be protected by a 30 mA RCD protective device which will disconnect within 40 ms a residual current of 150 mA.
471-16-01

Immersion heaters
Every heater of liquid shall be provided with a thermostat or other means of preventing a dangerous rise in temperature.
554-04

It is recommended that where immersion heaters are fitted to storage vessels in excess of 15 litres capacity, they shall be supplied from their own separate circuit.
INF

Instantaneous water heaters having uninsulated elements
The metal water pipe through which the water supply to the heater is provided shall be solidly and mechanically connected to all metal parts (other than live parts) of the heater or boiler. There must also be an effective electrical connection between the metal water pipe and the main earthing terminal, independent of the cpc.
554-05-02

The supply to the heater shall be through a double pole linked switch. The switch shall be either separate from, and within easy reach of the heater, or incorporated in the heater. The wiring from the heater to the switch shall be direct, without using a plug and socket-outlet.

Where the heater is installed in a room containing a fixed bath or shower, the switch shall be out of reach of a person using the bath or shower: an insulated cord operated pull switch would be acceptable.
554-05-03

The installer shall check or confirm that no single pole switch , fuse, or non-linked circuit-breaker is installed in the heater circuit between the origin of the installation and the heater, before the heater is connected.
554-05-04

Cables
All cables and conductors shall either be suitably constructed, or mechanically protected to withstand any mechanical damage to which they may be subjected in normal conditions of service.
522-06-01

Building structures

The cable support and protection system employed where a building is likely to have structural movement shall allow for the movement of conductors, so that they are not damaged.

Where the building structure is flexible or unstable, a flexible wiring system shall be installed.
522-12

Chapter 36

Socket-outlet circuits

General

Every installation shall be divided into circuits to avoid danger in the event of a fault, and to facilitate safe operation, inspection, testing, and maintenance.
314-01-01

The installation shall be broken down into circuits, such that the number of final circuits installed, and the number of points connected to each final circuit will ensure compliance with the regulations for current-carrying capacity of conductors, overcurrent protection, isolation, and switching.
314-01-03

Where more than one final circuit is installed, each circuit shall be connected to a separate way in the distribution board, and shall be electrically separate from every other final circuit, *e.g., neutrals must not be borrowed from other circuits.* This is to prevent an isolated circuit from being indirectly energised.
314-01-04

BS 1363 13 A socket-outlets

General requirements

A ring or radial final circuit, with or without spurs, can feed an unlimited number of socket-outlets, or feed permanently connected equipment, providing any immersion heater fitted to a storage vessel having a capacity exceeding 15 litres, or permanently connected heating appliances used as part of a space heating installation, are supplied from their own separate circuits.

The area covered by a circuit shall be determined by the known or estimated load, which must not exceed the nominal rating of the circuit protective device, and the floor area covered shall not exceed the area given in Table 1.

The number of socket-outlets shall be such that every portable luminaire and appliance can be fed from an adjacent and conveniently accessible socket-outlet.
INF, 553-01-07

Table 1 already takes into account diversity within the circuit, so no further diversity shall be applied. Where more than one ring final circuit is installed, the load shall be distributed among them. The cable sizes given in Table 1 may need to be increased if the ambient temperature is likely to exceed 30 °C, or if more than two circuits are bunched or grouped together.

Overload protection

A ring final circuit is deemed to comply with the overload requirements if:

a) the circuit is protected by a 30 A or 32 A protective device complying with BS 88, BS 1361, BS 3036, fuses or BS 3871 or BS EN 60898 MCBs, and

b) the accessories of the ring circuit comply with BS 1363, and

c) copper conductors with a minimum cross-sectional area of 2.5 mm^2 or, 1.5mm^2 if mineral-insulated cables are used, and

d) the current-carrying capacity I_z of the cable is not less than 0.67 times the rating of the protective device used.

433-02-04

Determination of cable size

When more than two ring final circuits are grouped together, or when the ambient temperature exceeds 30 °C, or when the conductor is enclosed in thermal insulation, the cable size required has to be calculated. The cable size is determined by calculating the current-carrying capacity required for the conductors from the following formula:

$$I_t = \frac{I_n \times 0.67}{G \times A \times T}$$

where G is the factor for grouping, A the factor for ambient temperature and T the factor for length in thermal insulation. Only the correction factor applicable is placed in the formula. No derating factor is used if the cables are in contact on one side only with thermal insulation, since current-carrying capacity tables are provided for this condition. No further diversity is allowed, since this has already been taken into account in multiplying I_n by 0.67. No derating for rewirable fuses is required.

Spurs off the ring

The total number of fused spurs allowed is unlimited, but the following minimum cable sizes shall be used if a fused spur feeds socket-outlets.

1.5 mm^2 for PVC insulated copper conductor cables.

1 mm^2 for MICC copper conductor cable.

Fused spurs are connected to the ring circuit through a fused connection unit, the fuse of which must not exceed 13 A or the rating of the cable taken from the fused spur.

The number of non-fused spurs must not exceed the total number of socket-outlets and stationary equipment connected directly in the ring circuit.

A non-fused spur can only feed one single or one twin socket-outlet, or one permanently connected item of equipment.

The supply for a non-fused spur can be taken from the origin of the circuit in the distribution board, or from a junction box, or from a socket-outlet connected directly to the ring circuit, the conductors used being the same size as the ring circuit.

Permanently connected equipment shall be controlled by a switch or switches, separate from the equipment, in an accessible position, and protected by a fuse not exceeding 13 A, or a circuit-breaker not exceeding 16 A.

INF

BS 4343 (BS EN 60 309-2) 16 A socket-outlets

The requirements can be summarised in a list:

a) the circuit shall be a radial circuit,

b) the overcurrent protective device for the circuit shall be 20 A, and the maximum demand for the circuit, having allowed for diversity, must not exceed 20 A,

c) the number of socket-outlets is unlimited,

d) the circuit conductor size is determined in the same way as for other circuits, the conductor size being obtained from the tables in Part 3, after derating factors for grouping, ambient temperature, contact with thermal insulation, and the type of protective device used if applicable, have been taken into account,

e) the current-carrying capacity of the cable shall be not less than the nominal rating of the overcurrent protective device protecting the circuit,

f) the current rating of socket-outlets used shall be 16 A, and of a type appropriate to the number of phases, circuit voltage, and earthing arrangement applicable.

INF

BS 196 socket-outlets

The requirements are listed as follows:

a) the circuit can be a radial final circuit or a ring final circuit with spurs,

b) the number of socket-outlets is unlimited,

c) the overcurrent protective device shall have a nominal rating not exceeding 32 A, and the maximum demand for the circuit after diversity has been taken into account must not exceed the nominal rating of the protective device, or 32 A.

d) no diversity is to be allowed for permanently connected equipment. Such equipment shall be assumed to be working continuously,

e) the total current taken by a fused spur must not exceed 16 A,

f) the circuit conductor size is determined in the same way as for other circuits, the conductor size being obtained from the tables in Part 3, after derating factors for grouping, ambient temperature, contact with thermal insulation, and the type of protective device used if applicable, have been taken into account,

h) the current-carrying capacity of the cable shall be such that it is not less than 0.67 times the rating of the overcurrent protective device for ring circuits, or not less than the nominal rating of the overcurrent protective device protecting the circuit for radial circuits,

i) the current-carrying capacity of the conductor for a fused spur is determined by the total load on the spur, which load, in any event, must not exceed 16 A,

j) non-fused spurs are not allowed, and a fused spur is to be connected to a circuit through a fused connection unit,

k) the rating of the fuse in the fused spur unit must not exceed 16 A or the current rating of the cable used for the spur,

l) permanently connected equipment shall be locally protected by fuse or circuit-breaker whose rating does not exceed 16 A, and controlled by a switch or switches, separate from the equipment, in an accessible position,

m) only 2 pole and earth contact plugs, with single pole fusing on the live pole, are to be used in a circuit where one pole of the supply is earthed,

n) only 2 pole and earth contact plugs with double pole fusing are to be used if the circuit has neither pole earthed: *e.g. the supply from a double wound transformer having the mid point of its secondary winding earthed.*

See paragraphs on SELV, FELV, and reduced voltage systems.

Shock protection when U_0 is 230 V

Where the socket-outlet is to be used for hand held equipment indoors, and where the supply voltage to the socket-outlet circuit is 230 V, the regulation for protection against indirect contact can be satisfied if the circuit disconnection by the protective device is achieved in 0.4 seconds when an earth fault occurs. Where the socket-outlet is used for hand held equipment outdoors, the circuit shall be protected by an RCD with r

ated residual operating current, $I_{\Delta n}$, of not more than 30 mA.

413-02-08, 413-02-09

Where a socket-outlet feeds an item of fixed equipment indoors, and where it can be guaranteed that the socket-outlet cannot be used for hand-held equipment, then the disconnection time can be 5 seconds. Where the socket-outlet is installed outdoors, the disconnection time shall be 0.4 seconds.

413-02-09, 413-02-13

When the supply voltage is 110 volts between phases, and 55 volts to the earthed midpoint for single-phase supplies, or 63.5 volts to the earthed neutral point for three-phase supplies, protection against indirect contact is provided if the circuit protective device disconnects the supply in 5 seconds when an earth fault occurs.

471-15-06

Where the nominal voltage to earth is 230 V r.m.s., a.c., and protection is given by an overcurrent protective device, the disconnection times of 0.4 or 5 seconds will be achieved if the phase earth loop impedance at every outlet on the circuit complies with the appropriate Tables ZS1 to ZS 9 in Part 3, or Table 41 B1/B2 and 41D in the regulations.

413-02-10

Except when the voltage is 230 V, where the voltage to earth is in the range 230 V to 277 V, the impedance values of Tables 41B1/B2, and 41D, and ZS1 to ZS 6 shall be adjusted by multiplying the Z_S value from the table by the nominal voltage to earth, and dividing the result by 230 V

INF

The earth fault loop impedance values given in Tables ZS 1 to ZS 7 are applicable for conventionally normal body resistance. Where the body resistance is expected to be reduced, the values of Z_S shall be reduced appropriately.

471-08-01

Where RCDs are used to comply with the disconnection times of 0.4 or 5 seconds, then the residual operating current in amperes multiplied by Z_S in ohms must not exceed 50 V.

413-02-16

Bathrooms and shower rooms

Unless the circuit is a 12 V SELV system or a shaver socket detailed below, socket-outlets shall not be installed in a room containing a fixed bath or shower, and there shall be no provision for connecting portable equipment.

Where a shower is installed in a room that is not a bathroom, socket-outlets shall be at least 2.5 metres from the shower cubicle.

Providing a shaver socket has been manufactured to BS 3535: Part 1, it can be installed in a bathroom for the use of electric shavers. In this case, the protective conductor of the final circuit feeding the shaver socket shall be connected to the earth terminal of the shaver socket.

601-09-01, 601-10-01, 601-10-02, 601-10-03

Providing socket-outlets are part of a SELV system which is derived from a safety source with a nominal voltage not exceeding 12 V, they can be installed in a room containing a fixed bath or shower.

The safety source for the SELV system shall be out of reach of a person using the bath or shower, the socket-outlets must not have any accessible metal parts, and shall be insulated or protected with barriers or enclosures against direct contact.

601-03-01, 601-10-01

Equipment not allowed
No electrical equipment shall be installed in the interior of a bath tub or shower basin.
601-02-01

Equipment to be used outside
Unless the circuit is a SELV system or a reduced low-voltage circuit, where socket-outlets rated up to 32 A are fitted in an installation and where it is possible that they may be used to supply portable equipment or hand-held for use outdoors, they shall be protected by a ($I_{\Delta n} \leq 30$mA) RCD, having an operating time not exceeding 40 ms, when subjected to a residual current of 150 mA.
471-16-01, 471-08-04, 412-06-02

SELV and PELV systems
The plugs used on SELV circuits shall be incapable of insertion into any socket-outlet of another voltage in the same premises. The socket-outlets must also prevent plugs of other voltages from being inserted into them, and shall not have a protective conductor connection.
411-02-10, 471-14-02

FELV systems
The socket-outlets used in the same premises for functional extra-low voltage shall not accept plugs used for other voltage systems.
471-14-06

Reduced low- voltage systems
A protective- conductor contact shall be provided in plugs, socket-outlets, and cable couplers used on a reduced low-voltage system. Additionally, they shall not be interchangeable with plugs, socket-outlets or cable couplers of other voltages used in the same installation.
471-15-07

Protective conductors
Except for socket-outlets connected to a SELV or PELV circuit, a protective conductor shall be provided at every socket-outlet.
471-08-08

The earthing terminal of each socket-outlet shall be connected by a separate protective conductor to an earthing terminal in the associated box or enclosure when the protective conductor is formed by conduit, trunking, ducting, or the metal sheath and/or armour of a cable.
543-12-07

Switching
Plugs and socket-outlets shall not be selected for emergency switching.
537-04-02

Other than for emergency switching, plugs and socket-outlets up to 16 A a.c can be used for switching; if the socket-outlet exceeds 16 A, it can be used to switch a.c., circuits, providing it has the appropriate breaking capacity for the use intended. Socket-outlets must not be used on d.c. for switching.
537-05-04, 537-05-05

Caravan site installations.

The electrical supply point to a caravan pitch shall be located not more than 20 m from any point which it is serving on the pitch .
608-13-01

The caravan supply point shall contain an IPX4 BS EN 60309-2 16A socket-outlet, with its key position at 6h. The socket-outlet shall be placed at a height between 0.8 m and 1.5 m from the ground. There shall be not less than one socket-outlet for each caravan pitch. Socket-outlets with a higher current rating shall be provided where the demand of the caravan pitch exceeds 16 A.
608-13-02, 608-13-03

Each socket-outlet shall be protected individually by an overcurrent protective device, or in groups of not more than three socket-outlets, by a 30 mA RCD complying with BS 4293, BS EN 61008-1 or BS EN 61009-1 which will operate within 40 ms with a 150 mA earth fault. The protective conductor must not be bonded to the PME earth terminal.
Where the supply is PME, each RCD shall be connected to an earth electrode, and the installation shall be considered to be a TT installation; the requirements for TT systems are applicable for protection against indirect contact and the sum of the resistance of the earth electrode and protective conductors R_A multiplied by the residual current of the RCD ($I_{\Delta n}$) shall not exceed 50 V.
608-13-04, 608-13-05

Socket-outlets shall incorporate an earthing contact when they are not supplied by an individual winding on an isolation transformer.
608-08-02

Unless the demand exceeds 16 A single-phase, every plug and socket-outlet used to connect a caravan to a caravan site installation shall be of the two pole and earthing contact type to BS EN 60309-2, with the key position at 6h.
608-07-01

Installations forming part of a TT system

Where automatic disconnection by RCD is used for protection against indirect contact, every socket-outlet circuit shall satisfy the equation $Z_s\, I_{\Delta n} \leq 50V$
471-08-06

Requirements for plugs and socket-outlets

With the exception of SELV socket-outlets, every plug and socket-outlet shall be of the non-reversible type, with provision for connecting a protective conductor. It must not be possible for any pin of a plug to make contact with live parts, whilst any other pin is completely exposed, nor shall it be possible for the pin of a plug to be engaged with a live contact in any socket-outlet which is not of the same type as the plug top. A plug top which contains a fuse shall be non-reversible.
553-01-01, 553-01-02

The position of socket-outlets mounted on a vertical wall or surface shall minimise the risk of damage to the socket-outlet, plug, or the flexible cable, whilst the plug is being inserted or withdrawn.
553-01-06

With the exception of toilets and offices, every socket-outlet and cable coupler used on a construction site shall comply with BS 4343 (BS EN 60 309-2).
604-12-02

Plugs and socket-outlets shall conform to the appropriate British Standard; BS 1363 for 13 A plugs with BS 1362 fuses and socket-outlets shuttered 2 pole and earth, BS 546 for 2, 5, 15 and 30 A fused or non-fused plugs with BS 646 fuses if applicable and 2 pole and earth socket-outlets, BS 196 for 5, 15, 30 A fused or non-fused plugs with 2 pole and earthing contact socket-outlets, BS EN 60309-2 for industrial type plugs and socket-outlets

553-01-03

Plugs and socket-outlets shall not be used for instantaneous water heaters having immersed and uninsulated heating elements.

554-05-03

In domestic installations, socket-outlets shall be shuttered, and should preferably comply with BS 1363 for an a.c. installation.

553-01-04

In single-phase a.c., or two-wire d.c circuits, a plug and socket-outlet may be used, even though it does not comply with BS specifications 1363, 546, 196 or 60309-2, providing the nominal operating voltage does not exceed 250 V, and the socket-outlet is used for the following purposes:

   a)   special circuits used to distinguish their function, or circuits having special characteristics in which the socket-outlet is used to prevent danger, or

   b)   shaver sockets incorporated into shaver supply units or lighting fittings complying with BS 3535 or, of a type complying with BS 4573, when used in a room other than a bath or shower room, or

   c)   electric clock socket-outlets designed specifically for that purpose, and incorporating a cartridge fuse not exceeding 3 A in the plug top, which complies with BS 646 or BS 1362.

553-01-05

A conveniently accessible socket-outlet shall be provided where portable equipment is likely to be used, taking into consideration that the length of flex on portable equipment, *which is usually only 1.5 m long.*

553-01-07

Cable couplers

With the exception of Class II equipment, or SELV equipment, cable couplers shall be non-reversible, and shall be provided with a connection for a protective conductor. The cable coupler shall also be arranged so that the shrouded pins of the coupler are connected to the supply, the exposed pins being connected to the equipment to be used.

553-02

High earth leakage current equipment

Where socket-outlets are used for equipment having a high leakage current see Chapter "Data processing equipment".

TABLE 1
Copper cable sizes and maximum floor area allowed
for BS 1363 socket-outlet circuits to comply with overload regulations

Type of final circuit	Overcurrent device allowed		Minimum conductor size in mm^2		Maximum floor area in
	Rating	Type	PVC	MICC	square metres
Ring	30/32 A	Any	2.5	2.5	100
Radial	30/32 A	Cartridge fuse or circuit breaker	4	2.5	50
Radial	20 A	Any	2.5	1.5	20

Chapter 37

Safety service supplies

Requirements for safety services

Where an electrical supply is provided for safety services, it must be capable of supplying the maximum demand for the duration needed for the safety services and shall have a suitable fire resistance where it is required to operate in fire conditions; the necessary fire protection can be provided by the construction of the equipment or during erection. The source of the safety service can be: a primary cell, battery, an independently operated generator, or an independent supply from a separate source which will not fail at the same time as the normal source of supply.
561-01-01, 561-01-02, 562-01-01

Installation of equipment

The electrical supply for safety services and all switchgear and controlgear must be installed as fixed equipment in a suitable location accessible only to skilled or instructed persons and must not be affected by the failure of the normal service. The location chosen for the electrical safety services supply must be adequately ventilated to prevent the hazardous penetration of exhaust gases, smoke or fumes into areas occupied by people.

Where a single electrical supply is used for the safety service it shall not be used for any other purpose, but where more than one source of supply is available they may be used for standby systems, providing that the capacity of the sources is such that, in the event of the failure of one of the sources, the remaining sources are capable of starting and operating the safety services.

The requirements of installing the equipment in a location accessible only to skilled or instructed persons or for the equipment being used for another purpose shall not apply to equipment supplied individually by a self contained battery *(such as the self contained emergency lighting luminaire)*.
562-01-02, 562-01-03, 562-01-03,

562-01, 04, 562-01-05, 562-01-06

563-01-05

The supplies for electrical services shall be so installed that periodic inspection, testing and maintenance can be carried out.
561-01-04

Protection

It is preferred that the protection provided for protection against indirect contact does not disconnect the circuit by the first phase to earth fault. *(Protection can be provided by ensuring the touch voltage does not exceed a safe value.)*

Where the system is IT an earth monitoring system is required to give visible and audible warning of a first fault.
561-01-03

It is permissible not to install overload protective devices at a point where a reduction occurs in the current-carrying capacity of a conductor due to:
  a) a change in cross-sectional area,
  b) a change in the method of installation

c)   a change in the type of conductor,
d)   a change in the type of cable,
e)   a change in the environmental conditions.

No overload protective device need be provided where the overload device protects the smallest conductor in the circuit or, because of the characteristics of the load, an overload is unlikely or, where the electricity supplier agrees that their protective device provides protection between the origin of the installation and the main switch for the installation.
563-01-03, 473-01-01, 473-01-04

Protective devices giving protection against overload, short-circuit and earth fault current have to be selected so that an overload or fault in one circuit does not interfere with the normal operation of any other safety services circuit. *(Correct selection of devices so that discrimination is assured.)*

Additionally, where equipment is supplied from dual supplies, a fault in one of the circuits shall not affect the correct operation of the other circuit and decrease the protection against electric shock.
563-01-04, 564-01-01

Identification
All alarms, indicators and control devices have to be clearly identified as well as switchgear and controlgear.
563-01-06

Wiring
The wiring cables used for safety services shall have the correct fire resistance for the areas through which the circuit passes. Additionally, each safety services circuit must be independent of all other circuits and not be affected by electrical faults or modifications to other electrical services or systems. Similarly, faults or alterations to the electrical services or its circuits shall not affect other electrical systems.
563-01-01, 563-01-02

Safety services operating in parallel
Each electrical safety services supply must comply with the requirements, for overload, fault protection and protection against electric shock.

Where the sources of electricity are not capable of being operated in parallel, they shall be interlocked both mechanically and electrically so that parallel operation will not occur.

Where the sources of electricity are capable of operating in parallel, protection against fault current and electric shock shall be taken into account for each source operating independently or for all sources in parallel. Additionally, precautions are required to limit any circulating currents in the neutrals of the various sources of supply, particularly from third, sixth and ninth harmonics.
565-01, 566-01

Chapter 38

Swimming pools

General

The International Electrotechnical Commission produced an international standard for swimming pools in 1983. This standard has now been incorporated into the British Standard 7671.

Swimming pools can be large or small: they can be indoors and out: they can be private, or open to the public. Most of the swimming pools installed in the gardens of domestic properties are in the open, and sunk into the ground. There are a few domestic pool installations which are installed in a building, (so as to make the pool usable all the year round) and a few that are in the open, but built above ground level. The requirements are to be applied to all types of swimming pool installations.

A swimming pool looks harmless enough, but in practice, it is more dangerous than a bathroom, as far as electrical services are concerned. In the first place, a bathroom is usually carpeted and the floor itself has a certain amount of insulation; secondly, most of the equipment in the modern bathroom is made up of insulating material, such as plastic baths, wash hand basins etc. Additionally, over the years, stringent rules have been applied to the installation of electrical equipment in the bathroom. The swimming pool area, on the other hand, is in direct contact with the ground: the area round it is usually wet, and people are walking round the pool with bare feet, so more care is needed with the electrical services.

Extent of requirements

Except where modified by the following requirements, the general requirements are applicable and in the absence of any reference to a Regulation, Section or Chapter means that the general wiring regulations should be referred to.

600-01, 600-02

Scope

The requirements for swimming pools apply to swimming pools and paddling pools and the zones surrounding the pools where, as mentioned above, the body is in direct contact with earth and where there is a risk of electric shock.

602-01-01

Classification of Zones

To make it easier to understand the requirements of the standards or regulations, the area round the pool is divided into zones. The IEC classify these areas as Zone 0, Zone 1, and Zone 2. BS 7671 classifies them as Zone A, Zone B and Zone C as shown in Figures 37.1 and 37. 2.

Zone A, covers the internal pool area, including any accessible recesses, and the internal area of a chute or flume.

Zone B, is an area expected to be accessible to persons and extends from the pool edge for a distance of 2 metres in the horizontal plane.

Zone C extends from Zone B for a further 1.5 metres, i.e., from 2 m to 3.5 m from pool edge.

Zone D (3) is the area outside the Zones A, B, and C.

Zones A, B and C extend upwards for 2.5 metres *(this being the limit of arm's reach)*; where diving boards, water chutes or spring boards are installed, the zone is extended by the height of the board above the datum and 1.5 metres either side of the equipment installed, as shown in Figure 37.1.

602-02

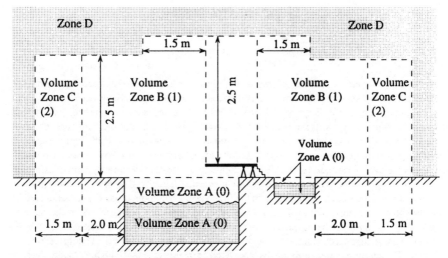

NOTES: The dimensions are measured taking account of walls and fixed partitions.
IEC 364-7-702 Zones are 0, 1 & 2, Wiring Regulations Zones are A, B & C.

Figure 37.1 - Zone dimensions of swimming pools and paddling pools

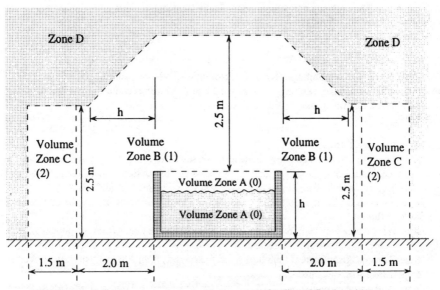

NOTES: The dimensions are measured taking account of walls and fixed partitions.
IEC 364-7-702 Zones are 0, 1 & 2, Wiring Regulations Zones are A, B & C.

Figure 37.2 - Zone dimensions for basins above ground

Electrical equipment installed in Zone A

Zone A (0), is the pool itself. The only electrical services which are allowed to be taken to the pool are for those appliances which are directly associated with the pool: *these are usually pool lights.* The degree of protection for the equipment should be IPX8. This means that they should be suitable for total immersion in water under a specified pressure. They must also be designed to resist the chemicals in the water.

602-05-01, 602-08-02

The electrical supply to each pool light must be from its own SELV transformer, or an individual winding on a multi-secondary winding transformer; in each case, the secondary's open circuit voltage must not exceed 18 V. The electrical supply to any other type of equipment must be from a SELV source situated in Zone D; it must comply with the SELV requirements, and be on its own circuit and must not exceed 12 volts a.c. r.m.s. or 30 V d.c. Any surface wiring should be of Class II construction *(i.e. double insulated)* without any metallic covering or exposed earthing or bonding conductor. No accessible metal junction boxes or other switchgear or accessories are permitted in the Zone.

602-04-01, 602-06-01,
602-06-03, 602-07-01

All SELV circuits and equipment must be protected against direct contact by using enclosures or barriers giving either IP2X or IPXXB protection or by insulation that will withstand a voltage of 500 V d.c for one minute.

602-03-01

Electrical equipment installed in Zone B

No switchgear, controlgear or accessories must be installed in Zone B, but current-using equipment specifically designed for use in swimming pools can be installed. Where it is impossible to install socket-outlets outside the Zone B area they are allowed, but see the requirements under socket-outlets.

602-07-01, 602-08-02

The supply to the current-using equipment should be from a SELV source situated outside Zones A, B and C; the source having a nominal voltage not exceeding 12 V a.c. r.m.s. or 30 V d.c.

Any fixed equipment installed in Zone B must have been manufactured specifically for swimming pools. The degree of protection required for such equipment is IPX4, unless water jets are used for cleaning, when the degree of protection required is IPX5.

Where floodlights are installed, each floodlight shall be supplied from its own transformer, or from an individual winding on a multi-secondary transformer, having an open circuit voltage not exceeding 18 volts.

602-04-01, 602-08-02
602-06-03

Any surface wiring should be of Class II construction *(i.e. double insulated)* without any metallic covering or exposed earthing or bonding conductor *(No steel conduit, steel trunking or bare MICC cable).* No accessible metal junction boxes are permitted in the Zone.

602-06-01

Electrical equipment installed in Zone C

Switchgear, controlgear, accessories and metal junction boxes are allowed in Zone C. However, the degree of protection for enclosures is IP2X for indoor pools and IP4X for outdoor pools unless water jets are likely to be used in which case protection must be to IPX5. Appliances in Zone C may be Class I or Class II construction.

With the exception of instantaneous water heaters manufactured in accordance with British Standards, equipment, accessories, socket-outlets and switches (unless switches are cord operated) have to be protected by one of the following methods:

1. Individually by electrical separation,
2. SELV with a nominal voltage not exceeding 50 V, or
3. 30 mA RCD which will disconnect within 40 ms with a residual current of 150 mA.

A shaver socket-outlet complying with BS 3535 is also allowed in Zone C.

602-05-01, 602-07-02
602-08-03

Protection against electric shock

Protection by placing out of reach, obstacles, non-conducting location or earth-free local equipotential bonding shall not be used in Zones A, B and C.

In Zones A and B protection against electric shock shall only be by SELV, the voltage of which must not exceed 12 V a.c. r.m.s. or 30 V d.c. and protection against direct contact shall be by barriers or enclosures giving protection to IP2X or IPXXB or insulation which will withstand 500 V d.c. for one minute.

602-03-01, 602-04-02

Lighting

In Zones A and B each floodlight shall be supplied from its own transformer or a separate winding on a multi-winding transformer the open circuit voltage of which must not exceed 18 V. Protection being either IPX8 (Zone A)

In Zone C lighting circuits shall be protected by electrical separation, SELV or an RCD which will disconnect within 40 ms with a residual current of 150 mA.

In Zone D the normal requirements of the British Standard apply.

602-04-01, 602-08-03

Socket-outlets

Socket-outlets are prohibited in Zone A. Socket-outlets are only allowed in Zone B if they cannot be installed in either Zone C or Zone D and only then if they comply with the following:

(a) they are installed more than 1.25 m from the edge of Zone A (*3.25 m from pool edge)* and
(b) installed at a height from the floor of 300 mm, and
(c) the socket-outlets comply with BS EN 60309-2, and
(d) the circuit is protected by either an RCD complying with a British Standard which will disconnect the circuit within 40 ms with a residual current of 150 mA, or by electrical separation with the isolating transformer installed in Zone D.

Socket-outlets installed in Zone C must comply with BS EN 60309-2 and the circuit must be protected individually by electrical separation, or SELV, or an RCD which will trip within 40 ms with a fault current of 150 mA, or it is a shaver socket-outlet complying with BS 3535. The earthing terminal of socket-outlets must be included in the equipotential bonding.

Protection shall be:
(a) in Zone B, IPX4 or IPX5 if water jets are used,
(b) in Zone C, IPX2 for indoor pools, IPX4 for outdoor pools or IPX5 if water jets are used.
602-07-01, 602-07-02,
602-08-01, 602-05-01

Electric heating

Electric heating units embedded in the floor are allowed in Zones B and C provided that either the units have an earthed metallic sheath connected to the local supplementary bonding conductors or the units are covered by a metallic grid connected to the same local supplementary bonding conductors.

Non-electrical metallic pipe heating buried in the floor of Zones A and B will require bonding to the local supplementary bonding conductors.
602-08-04

Local supplementary bonding

Local supplementary bonding is required to connect all extraneous-conductive-parts and exposed-conductive-parts together in Zones A, B and C, which includes the earth terminal of socket-outlets. Additionally, if the floor in Zones A, B or C has a metal grid in it *(reinforcement in a concrete floor)* it too shall be included in the supplementary bonding. SELV circuits must not be included in the supplementary bonding.
602-03-02

Size of supplementary bonding

Where there are no exposed-conductive-parts in Zones A, B and C, the minimum size of supplementary bonding conductor is 2.5 mm^2 if it is sheathed or otherwise mechanically protected, or 4 mm^2 if is not mechanically protected.

Where exposed-conductive-parts are in Zones A, B and C, then two sets of rules have to be complied with. One of the rules determines the actual size of the supplementary bonding conductor needed*, and the other specifies the minimum size allowed.

$$\text{Resistance of the bonding conductor} = \frac{12\,V}{I_a}$$

where I_a is the current which will disconnect an overcurrent protective device within 5 s, or is the residual current $I_{\Delta n}$ of an RCD.

The minimum size is determined from 547-03, and is illustrated in Figure 3.
602-03-02, 547-03

Additional requirements

All other requirements of the British Standard should be met, such as the requirements for overloads, fault currents, voltage drop, and installation design.

---

* The British Standard does not give a reduction in voltage for determining the resistance of bonding conductors. Having a voltage appearing between exposed-and extraneous-conductive-parts of 50 V is clearly undesirable. Since SELV circuits in zones A and B are limited to 12 V, this seems a reasonable value of voltage to use in the formula.

Where SELV circuits are used, whatever the nominal voltage, protection against direct contact has to be provided by barriers or enclosures affording at least IP2X or IPXXB protection, or insulation capable of withstanding a test voltage of 500 V d.c. for 1 minute. The measures of protection by non-conducting location, placing out of reach, obstacles and earth-free equipotential bonding, are not allowed.

600-01, 600-02

Exposed-conductive-part to exposed-conductive-part

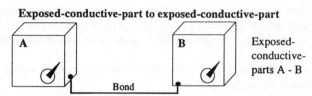

Exposed-
conductive-
parts A - B

Same conductance as the smallest cpc feeding A or B, providing that it is mechanically protected: otherwise minimum size is 4 mm².

Exposed-conductive-part C to extraneous-conductive-part

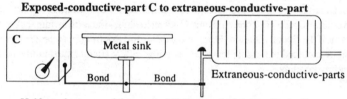

Extraneous-conductive-parts

Half conductance of the cpc in 'C' if sheathed or mechanically protected: otherwise minimum size required is 4 mm².

Extraneous-conductive-part to extraneous-conductive-part

Figure 37.3 - Minimum size of supplementary bonding conductors

Equipment in Zone D (outside Zones A, B & C)

Zone D is outside the Zones A, B and C and the normal requirements apply, *but boilers and pumps for swimming pools are usually placed in their own room, or in the garage for outdoor pools on private property, they are, therefore, in contact with the pool water. The pump should be one manufactured for swimming pools, and made of plastics, so that there is no contact between the water in the pump and the metal parts of the motor; for safety the supply to the pump is taken through a 30 mA RCD. (If the seals fail water can enter the motor without it disconnecting.)*

In domestic installations, the pool boiler generates its own millivolts for the control of the boiler. The only precaution needed is with time switches installed to ensure the boiler switches off before the pump. There must be electrical separation between the supply mains to the timer and the contacts connected to the boiler wiring.; this separation should be the same as that for a safety isolating transformer manufactured to BS 3535.

In many domestic installations the pool water is heated from the domestic central heating boiler (or a separate one) the danger here is the importation of voltages from the domestic installation into the swimming pool area, either from the hot water pipes or the cold water feed to the boiler.

Where reinforcement is installed in the concrete it can be used as the bonding conductor, providing the various sections of the reinforcement are welded or brazed together; it should be noted that mechanical clamps are not allowed for joints when buried. Connection to the reinforcement can be by a copper-clad steel earth rod one end of which has the copper ground off and then welded to the reinforcement. The earth rod is long enough to protrude through the concrete after it has been poured.
INF.

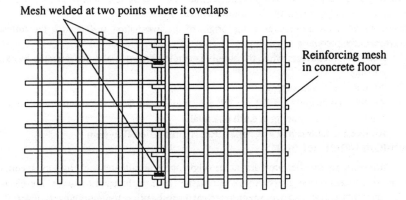

Figure 37.4 - Bonding of reinforcing mesh in swimming pool floor

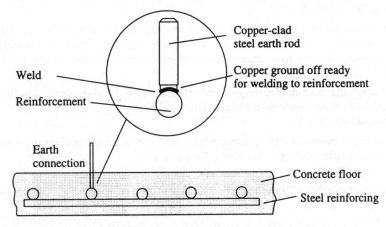

Figure 37.5 - Making electrical connection to steel reinforcing

Chapter 39

Switching

General requirements

A main linked switch or linked circuit-breaker shall be provided for every installation. This switch or circuit-breaker, which shall be installed as close to the origin as possible, shall be capable of interrupting the supply on load. The same device must also be capable of being used for the isolation of the supply, by complying with the requirements for isolation.

A main switch or circuit-breaker at the origin of the installation shall switch the following conductors of the incoming supply:

1. Both live conductors when the supply is single-phase a.c. and not operated by skilled persons.
2. All poles of a d.c., supply,
3. All phase conductors in a T.P., or a T.P. & N., in a TN-S or TN-C-S system where the neutral can be regarded as being at earth potential.
4. All live conductors in a T.P. , or a T.P.& N., in a TT or IT system.

460-01-02, 461-01-01, 461-01-04

Where the electricity supplier provides, at the origin of the installation, switchgear that complies with the requirements and gives express permission for it to be used to isolate the electrical installation from the origin up to the main switch of the installation, the requirements of 460-01-02 above are met.

476-01-01

One switching device is allowed to isolate a group of circuits.

461-01-01, 476-01-02

Where more than one source of supply feeds an installation, each source shall be provided with a main switch, together with a warning notice advising that all switches have to be operated to switch off the installation. Alternatively, an interlock system can be used instead of the warning notice.

460-01-02

A switch shall be provided for every circuit and final circuit, and shall be capable of interrupting the circuit on-load, and under any foreseeable fault conditions.

314-01-02, 476-01-02

Where it is necessary for safety reasons, switches shall be provided to enable circuits, or parts of the installation, to be switched independently from each other.

476-01-02, 314-01-02

A means of switching off the supply under load conditions shall be provided for every distribution circuit and final circuit.

476-01-02

Where the requirements require all the live conductors to be disconnected, it must not be possible to disconnect the neutral before the phase conductors are disconnected. It is essential that the neutral be re-connected at the same time as the phase conductors, or before the phase conductors are connected.

530-01-01

An isolator or switch shall be provided to cut off the supply voltage from every installation, every item of equipment or circuit, as may be necessary to prevent danger. The device provided shall be suitably placed for ready operation.

130-06-01

Single pole switches

Single pole switches shall only be inserted in the phase conductor.

130-05-01, 530-01-02

Appliances

A means of interrupting the supply on load shall be provided for every appliance or luminaire not connected to the supply by a plug and socket-outlet; the switching device must be suitable for the most onerous conditions imposed on it in normal use. The means of interrupting the supply shall be in a readily accessible position, and so placed that the operator is not put in danger.

476-03-04, 537-05-01

The means of interrupting the supply to an appliance or luminaire may be mounted on, or incorporated in, the appliance or luminaire.

476-03-04

One means of interrupting the supply may be used where two or more appliances are installed in the same room.

476-03-04

Heating and cooking appliances

A means of interrupting the supply on load shall be provided for every appliance which may give rise to a hazard in normal use. The device provided shall be so placed in an accessible position that it can be operated without putting the operator in danger.

476-03-04

Bathrooms

Switches and control devices shall be mounted so that they are inaccessible to a person using a fixed bath or shower.

The only switches or controls that are allowed to be accessible in a room with a fixed bath or shower are: shaver socket-outlets manufactured in accordance with BS 3535, the switches or controls of instantaneous water heaters and shower pumps manufactured to British Standards, insulating cords of cord-operated switches which comply with BS 3676, switches which form part of a SELV system with a nominal voltage not exceeding 12 V r.m.s., a.c., or d.c., and switches or controls operated by mechanical actuators with insulating linkages.

601-08

The safety source for the SELV system shall be out of reach of a person using the bath or shower; the switches must not have any accessible metal parts, and shall be insulated or protected with barriers or enclosures against direct contact.

601-03

No electrical equipment shall be installed in the interior of a bath tub or shower basin.

Installations carried out on the surface shall not use metal conduit, metal trunking, a cable with an exposed metallic sheath , or an exposed earthing or bonding conductor.

601-02, 601-07

Motors

Every motor shall be provided with an efficient means of switching off; this shall be readily accessible, easily operated, and so placed as to prevent danger.

130-06-02, 476-02-03

Every motor shall be provided with a means of switching off for mechanical maintenance. The device to be suitably located and clearly identifiable.

462-01-01, 462-01-02

Precautions to be taken to prevent the unintentional or inadvertent energisation of the circuit where the circuit is not under the control of the person working on it.

462-01-03

Transformers

A linked switch that disconnects all the live conductors shall be provided for disconnecting a step-up transformer from the supply.

551-01-03

Discharge lighting (functional switching)

Every switch which is controlling only discharge lighting must have a nominal current rating of at least twice the steady current of the circuit, unless it has been manufactured to the appropriate section of BS 3676, and is marked accordingly.

Where a switch controls a circuit containing filament lamps as well as discharge lighting, it shall have a nominal current rating of the current for the filament lamps, plus twice the steady current of the discharge lighting.

INF

Where the open circuit voltage of discharge lighting exceeds low voltage a switch is required in addition to the control switch for isolating the circuit. Alternatively, a self-contained luminaire containing an interlock which disconnects the supply before access is made to live parts, or a lockable switch or one with a removable handle or lockable distribution board is used to isolate the circuit.

476-02-04

Plugs and socket-outlets (functional switching)

Except for d.c. circuits a plug and socket-outlet may be used as a switching device providing its rating does not exceed 16 A. A socket-outlet having a rating exceeding 16 A may be used for switching a.c. circuits, providing it has the required breaking capacity for the use intended. Plugs and socket-outlets should not be selected for emergency switching.

537-05-05, 537-04-02,
464-01-04, 537-05-04

Fireman's switch

A fireman's emergency switch shall be provided in the low voltage circuit for interior and exterior discharge lighting which exceeds low voltage. (See emergency switching.)

476-03-05

Switches prohibited

A switch must not be installed in a protective conductor or PEN conductor. The only exception to this requirement is where an installation receives its supply from more than one source. *(See Isolation - General requirements.)*
543-03-04, 460-01-03

A switch shall not be installed in the outer conductor of a concentric cable and fuses, links and off-load isolators must not be used for functional switching.
546-02-06, 537-05-03

The neutral conductor shall not be switched on its own, the poles of any switch used being linked.
464-01-02

Semiconductor devices

Semiconductor devices can be used for control and functional switching, providing they are suitable for the nominal voltage of the installation, the design current (full load current), the frequency of the supply and the power requirements of the equipment. The device must not cause harmful effects on other equipment, nor impair the supply; it shall be suitable for any external influences likely to be encountered. They are allowed to control the current without opening the corresponding poles.
512-06, 537-05-02

Accessibility

Every switch which requires operation or attention during normal use shall be so positioned as to give adequate working space and safe access to the equipment for operation, inspection and maintenance.
130-07, 513-01, 529-01-02

Control switching (functional)

476-01-02 requires every circuit and final circuit to be provided with a means of switching, similarly 464-01-03 requires all electrical equipment that requires to be controlled to be controlled by a functional switching device, although one functional switching device may control several items of equipment which operate at the same time.

Where different parts of a circuit need to be independently controlled, a functional switching device shall be provided for each part. A functional switching device does not have to switch all live conductors in a circuit, but must not be placed exclusively in the neutral conductor. The switch should be selected for the most demanding duty it has to perform.
476-01-02, 464-01-01, 464-01-02, 464-01-03, 537-05-01

Change-over switches (functional)

Where change-over switches are used to switch from alternative sources of supply such as from the mains to a generator, all live conductors must be switched. The switch shall be so constructed that it cannot connect both sources of supply in parallel, unless it has been specifically designed for this purpose. Similarly, unless the design requires it, no provision is to be made for the isolation of protective conductors or PEN conductors.
464-01-05

Chapter 40

Trunking

Material

Trunking shall be manufactured to BS 4678. Where this standard does not apply, the material of insulated trunking shall have the ignitability characteristic "P" as specified in BS 476 Part 5.
521-05

Corrosion & exposed to weather

Precautions are to be taken in damp situations against corrosion that can be caused by metal trunking coming into contact with building materials containing chemicals, of which magnesium chloride, lime, and acidic woods are examples.
522-05-01

Metal trunking and associated fixings exposed to the weather, or installed in damp situations, shall be made from corrosion resistant material, and shall not be in contact with metals likely to cause electrolytic action.
522-05-01, 522-05-02

Entries into trunking shall be so placed or protected that the ingress of water is prevented. Trunking exposed to the weather shall be of the weatherproof type; it is important to make certain that entries are not made in the top face of the trunking and that any entry is made waterproof, by, *for example, flanged coupling and lead washer if the entry is for conduit.*
522-03-01

Heat barriers

Vertical trunking shall be equipped with internal barriers to prevent the air at the top of the trunking becoming excessively hot. *As a recommendation, the distance between barriers shall be either the distance between floors, or 5 metres whichever distance is least.*
422-01

Fire barriers

Where trunking passes through walls or floors, the hole round the trunking shall be made good to the same degree of fire resistance as that of the wall or floor. Where the wall or floor has a designated fire resistance, it shall be internally sealed to maintain the fire resistance of the floor or wall. The sealing material shall be compatible with the wiring system, permit thermal movement of the cable without reducing the quality of the seal, be removable without damaging the cables and have the same degree of resistance to environmental conditions as the wiring system. Where the fire resistance required is for not more than one hour, then a sealing system type tested to the method given in BS 476 Part 23 complies with this requirement.
527-02

Trunking supports

Trunking shall be so supported, and of such a type, that it is suitable for the risk of the mechanical damage which it is likely to incur during normal use. Alternatively, it shall be protected against such mechanical damage.
522-08-04, 522-08-05

Spacing of supports for trunking

The following table does not apply to lighting suspension trunking, or trunking using special strengthening couplers.

Trunking size	Maximum distance between supports in metres			
	Metal		Insulating	
mm × mm	Horizontal	Vertical	Horizontal	Vertical
16 × 16	0.75	1.0	0.5	0.5
25 × 16	0.75	1.0	0.5	0.5
38 × 16	0.75	1.0	0.5	0.5
25 × 38	1.25	1.5	0.5	0.5
38 × 38	1.25	1.5	0.5	0.5
50 × 38	1.75	2.0	1.25	1.25
50 × 38	·1.75	2.0	1.25	1.25
50 × 50	1.75	2.0	1.25	1.25
75 × 50	3.0	3.0	1.5	2.0
75 × 75	3.0	3.0	1.75	2.0
Larger sizes	3.0	3.0	1.75	2.0

Trunking as a protective conductor

The cross-sectional area of trunking used as a protective conductor shall be not less than that given in the formula of 543-01-03, or by Table 54G. Sizing the trunking by calculation has to be carried out if the phase conductors have been sized to take account of short-circuit currents and the earth fault current is going to be less than the short-circuit current; in this event the phase conductors should also be checked again to ensure they are protected with the lower fault current.
543-01-01

Where the trunking is common to several circuits, its cross-sectional area shall be determined by either of the following criteria:

1. Using the formula of 543-01-03, using the most unfavourable values of fault current and disconnection time for each of the circuits it contains, or
2. Using the conductor with the largest cross-sectional area with Table 54G.

543-01-02

Protective conductors shall be protected against mechanical and chemical deterioration, and against electrodynamic effects.
543-03-01

When trunking is used as a cpc, the joints do not need to be accessible. (*In such cases bonding across joints to be certain of permanent and reliable continuity is advisable.*)
543-03-03

Busbar trunking

When used as a protective conductor, the electrical continuity of enclosed metallic busbar trunking systems shall be protected against mechanical, chemical, and electrochemical deterioration. They shall have a cross-sectional area not less than that given by the formula in 543-01-03, by Table 54G or it is verified by test in accordance with BS EN 60439-1, and the enclosure shall allow connection of other protective conductors at every tap-off point.
543-02-04

Trunking shall not be used as a PEN conductor.
543-02-10

Circuits for fire alarms and emergency lighting

Circuits for fire alarms and emergency lighting must be segregated from other ciruits and from each other in accordance with BS 5839: Part 1 and BS 5266: Part 1 respectively.

Band I and Band II

A trunking enclosing Band I and/or Band II circuits must not also enclose HV circuits unless every cable is insulated for the highest voltage. Alternatively, HV cables must be separately compartmentalised or, if a multicore cable is used, every conductor must be insulated for the highest voltage present in the cable, or enclosed within an earthed metallic screen of current-carrying capacity of no less than the larger or largest conductor within the screen.

Where trunking encloses Band I and Band II circuits every cable must be insulated for the highest voltage. For multicore cables in trunking, Band I cores must be insulated, either individually or collectively, for the highest voltage present. Alternatively, Band I cores must be separated from cores of Band II circuits by an earthed metal screen of current-carrying capacity of no less than the larger or largest core of the Band II circuit. Circuits of Band I and Band II in separate compartments of a trunking need no further consideration in this respect.

Circuits of Band I and Band II terminating in a common box switchplate or similar must be segregated by an effective partition, which must be earthed if metal.
528-01

Where controls or outlets for Category 1 and Category 2 circuits are mounted in or on common enclosures, the cables and connections of the two Categories shall be partitioned by rigidly fixed screens or barriers.
528-01-07

Termination of trunking.

At the termination of trunking, non-sheathed cables or sheathed cables that have had the outer sheath removed, shall be enclosed in fire resistant material, unless they are terminated in a box, accessory, or luminaire complying with British Standards.

The building structure may form part of the enclosure referred to above.
526-03-02, 526-03-03

Cables in trunking

Any type of cable installed in a vertical trunking shall be supported at regular intervals, *usually at 5 metre intervals.*
INF

CHAPTER 40

The internal radius of every trunking bend shall comply with the minimum bending radius for cables

Conductor material	Conductor insulation	Overall diameter	Internal radius of bend given by diameter of cable × factor below
Copper or	Rubber or	Less than 10 mm	2
aluminium	PVC	Between 11 and 25 mm	3
	circular or	Greater than 25 mm	6
	stranded		

522-08-03

Socket-outlets in trunking
The earthing terminal of each socket-outlet shall be connected by a separate protective conductor to an earthing terminal in the trunking, when the socket-outlet is mounted in the trunking. Socket-outlets mounted on trunking should be mounted at a height to reduce the risk of damage to the socket-outlet or plug top.
543-02-07, 533-01-06

Accessibility
Adequate access has to be assured to allow the cables to be placed in and withdrawable from the trunking.

The trunking system must be completely erected before cable are drawn into the trunking.
522-08-02

The radius of every bend or junction formed in trunking must be such that the cables are not damaged.
522-08-03

Calculations
When calculating the thermal capacity of trunking or calculating the earth fault loop impedance the impedance of the trunking should be taken not its resistance, since the reactance of trunking is high compared to the resistance.

The impedance of trunking is given in Table ZCT 1 and the cross-area of trunking in PCA 1, in Part 4.

The k factor for steel trunking is 47 for 70 °C PVC cables, 44 for 90 °C PVC cables, 54 for 85 °C cables and 58 for 90 °C thermosetting cables (XLPE). The assumed initial temperatures are respectively: 50 °C, 60 °C, 58 °C and 60 °C(See table K1A and K1B in Part 4.
INF, Table 54E

Wiring capacity of trunking
See Table CCT 1 in Part 3.

Chapter 41

Voltage

Nominal voltage

The declared nominal voltage for electricity supplies to customers in Great Britain is: 230 V a.c. single-phase and 400 V a.c three-phase.

These voltages can vary between the limits of +10 % - 6 %, which means that the voltage of the supply is allowed to fluctuate between:

a) 216 V and 253 V single-phase, and

b) 376 V and 440 v three-phase.

Electricity Supply Regulations 1988 amended

Open circuit voltage u_{oc}

The open circuit voltage, as far as Uoc is concerned for calculating Zs, is the voltage between phase and the earthed neutral point of the electrical source transformer and no load.

Voltage limits of the Wiring Regulations BS 7671

Extra-low voltage

A voltage not exceeding 50 V a.c or 120 V ripple free d.c between conductors or between conductors and earth.

Low voltage

A voltage exceeding:

a) extra-low voltage but not exceed 1000 V a.c or 1500 V d.c between conductors, or

b) 600 V a.c or 900 V d.c between phase conductors and earth.

High voltage

A voltage exceeding 1000 V a.c. or 1500 V d.c between conductors, or 600 V a.c. or 900 V d.c between phase conductors and earth.

Definitions

Reduced low voltage

A system in which the voltage between phase and neutral and phase and earth is limited to a maximum of 110 V r.m.s. for single phase supplies (55 V from phase to an earthed mid point) and to 110 V r.m.s. between phases and 63.5 V from phase to earthed neutral.

Definitions, 471-15-02

Voltage drop

The voltage at the terminals of fixed electrical equipment has to be not less than the minimum specified in the appropriate British Standard for the fixed equipment.

Where no minimum operating voltage for the equipment is specified in the British Standard the voltage at the equipment has to be such as not to impair the safe functioning of the equipment.

The requirements specified above are deemed to be satisfied if the electrical supply is provided in accordance with The Electricity Supply Regulations 1988 amended *(i.e. the voltage is supplied within the tolerances specified in those regulations as mentioned above)* and if the voltage drop between the origin of the installation up to the fixed equipment or the terminals of a socket-outlet

in a final circuit does not exceed 4 % of the nominal voltage of the supply.
525-01-01, 525-01-02

The voltage drop specified above is applicable under normal load conditions; starting currents and inrush currents to equipment can be ignored providing the voltage variations are within the limits specified in British Standards, or the manufacturers' recommendations.

Diversity can be taken into account when calculating voltage drop.
525-01-02, INF

Warning notices

Where it would not normally be expected that equipment or an enclosure contains a voltage exceeding 250 V the equipment shall be so arranged that a notice, which is clearly visible, warning of the maximum voltage present, is seen before access can be made to live parts.

Where separate items of equipment or enclosures have a voltage in excess of 250 V between them, and are within arm's reach of each other, a notice warning of the voltages present must be securely fixed in such a position, that it warns anyone of the voltages present, before they gain access to live parts.

Access to live parts in switchgear, control panels or equipment in which different nominal voltages exist should be clearly marked indicating the different voltages present.
514-10-01

Lampholders and ceiling roses

Lampholders for filament lamps and ceiling roses are not to be used in circuits which normally operate at a voltage exceeding 250 V.
553-03-02, 553-04-02

Lighting

Where an extra-low voltage luminaire does not contain the provision for the connection of a protective conductor it must be connected only to a SELV circuit.
554-01-01

The requirements of BS 559 must be complied with for the construction and the selection and erection of high voltage (see definitions) discharge lighting and high voltage electric signs.
55-02-03

Water heating

Where high voltage is connected to an electrode water heater or an electrode boiler protection shall be by means of an RCD which will disconnect the circuit if the earth leakage current exceeds 10 % of the full load current of the heater or boiler, except that where at any instance a higher earth leakage current is essential to ensure operational stability of the heater or boiler the earth leakage current can be increased to 15 %. Alternatively, a time delay can be incorporated in the RCD so that it does not trip unnecessarily if an imbalance of short duration occurs.
554-03-04

Protection by electrical separation

The voltage of an electrically separated circuit shall not exceed 500 V.
413-06-02(iv)

Assessment of general characteristics

When assessing the general characteristics of an installation the nominal voltage of the installation

must be ascertained.
313-01-01(i)

Operational conditions

All items of equipment installed in an installation shall be suitable for the nominal voltage of the installation taking into account the minimum and maximum voltages likely to occur in normal service. Equipment must be insulated for the nominal voltages between phases in an IT system.
512-01-01

Undervoltage protection

The following requirements shall be complied with when selecting devices for the protection against undervoltage.
535-01-01

Where there could be a loss of the electricity supply and subsequent restoration of the supply, or a reduction in voltage which could cause danger, the control of every motor shall be so arranged to stop the automatic restarting of the motor after it has stopped due to a failure of the supply or a reduction in voltage, unless the failure of the motor to restart could cause a greater danger.

This requirement does not preclude the motor being automatically stopped and started at intervals when it is part of an overall control scheme where safety precautions take into account the stopping and automatic restarting of motors.
451-01-01, 552-01-03

Where damage to equipment or to the installation may be caused by a drop in voltage, and where the drop in voltage will not cause danger, then one of the following arrangements shall be chosen:

a) precautions against any foreseen damage shall be taken, or
b) it shall be confirmed with those responsible for the operation and maintenance of the installation that the foreseen damage is an acceptable risk.

451-01-02

Providing equipment can withstand a brief reduction or loss of voltage without danger, then a suitable time delay may be incorporated in the operation of an undervoltage protective device controlling the equipment.

The instantaneous disconnection by control or protective devices must not be affected by any delay in the opening or closing of contactors, and the characteristics of the undervoltage protective device shall be compatible with the requirements for the starting and use of the associated equipment.
451-01-03, 451-01-04, 451-01-05

The re-closure of a protective device shall not be automatic if it is likely to cause danger.
451-01-06

PART TWO

ELECTRICAL INSTALLATION DESIGN

Explanations and worked examples

of the requirements

Contents - index

Chapter 1

Introduction

Scope of the requirements
BS 7671 covers all general commercial, industrial, agricultural and horticultural installations operating at voltages up to 1000 volts; it also covers construction sites, street lighting, caravans and their sites.

The general rules apply to hazardous areas, but do not include the extra precautions that have to be observed with such installations as outlined in BS 5345.

The British Standard does not cover installations for electric railways, motor cars, ships, mines and quarries, aircraft, off-shore installations, or the generation, transmission and distribution of energy to the public.

Object of the requirements
The main object is to provide protection of persons, property and livestock from hazards associated with the electrical installation. This is achieved by giving the requirements for protection against fire, shock, burns and for injury from mechanical movement of electrically actuated equipment, where such equipment is controlled by electrical devices intended to prevent such accidents.

It is not intended to instruct untrained persons, or to provide for every conceivable circumstance that may arise. The regulations are contained in the Fundamental Requirements for Safety, the remaining chapters giving the methods and practices that will in general satisfy these Fundamental Requirements for Safety.

Compliance with the Fundamental Requirements for Safety will generally satisfy the statutory requirements applicable to electrical installations so far as the fixed installation is concerned.

Legality of the requirements
BS 7671 is not mandatory but is accepted as satisfying the requirements of the Electricity Supply Regulations 1988, as amended, and the Electricity at Work Regulations 1989 for general installations up to 1000 V for the fixed installation . If a contract specifies that the work will be carried out in accordance with BS 7671, then this would be legally binding, and the British Standard then becomes a legal requirement of the contract.

Health and Safety at Work Act.
Electrical installations must be carried out to a suitable standard to satisfy the Health and Safety at Work Act. Your attention is drawn to Section 6.(3) of that Act 'It shall be the duty of any person who erects or installs any article for use at work in any premises where that article is to be used by persons at work, to ensure, so far as is reasonably practicable, that nothing about the way in which it is erected or installed makes it unsafe, or a risk to health, when it is being used, cleaned or manufactured by a person at work.'

Installations carried out to BS 7671, should satisfy the Health and Safety at Work etc. Act as to being of a suitable standard for the fixed installation. However, other management procedures will be required to provide compliance with the Health and Safety at Work Act and other legislation.

CHAPTER 1

Electricity at Work Regulations 1989

There is insufficient space available to give a full explanation of these regulations in this book. The following outline gives an indication of their requirements. A fuller explanation of these regulations and how to comply with them is given in the handbook 'Electricity at Work and Related Regulations' by the author.

The Electricity at Work Regulations 1989 (EAW) were made under the Health and Safety at Work Act 1974 and reinforce the requirements of the Health and Safety at Work Act with respect to the use of electricity at work. They are also applicable to mines and quarries.

The 1989 regulations apply to employers, the self employed and employees, all of whom will be responsible where matters are within their control, these duties being in addition to those imposed by the Health and Safety at Work Act.

A summary of the responsibilities called for under the 1989 regulations are given against the relevant regulation number as follows:

4(1)   to ensure at all times that the electrical system is of such construction as to make it safe,

4(2)   the electrical system is maintained so that it remains safe,

4(3)   that no person is in danger whilst using, operating, maintaining, or working near such a system,

4(4)   that any equipment provided for the purpose of protecting persons at work, on or near electrical equipment, shall be suitable for that purpose, shall be maintained and shall be properly used,

5)   that the system has sufficient strength and capability for any short-circuit currents, overload currents and voltage surges etc., that can foreseeably occur,

6)   the equipment is suitable for any environmental conditions to which it may reasonably and foreseeably be exposed,

7)   all conductors that can give rise to danger to be insulated, protected or suitably placed so as not to cause danger,

8)   precautions to be taken either by earthing or other suitable means to prevent danger when any conductor (other than a circuit conductor) becomes charged *(has a voltage on it)*; the earthing conductor being capable of carrying any fault current without danger,

10)   that every joint and connection in a system is mechanically and electrically suitable so as to prevent danger,

11)   that where necessary, protection is provided to disconnect any excess current before danger can occur,

12)   that provision is made for cutting off the supply or isolating equipment as may be necessary to prevent danger,

13)   when equipment and conductors are made dead, suitable precautions must be taken to ensure that they do not become electrically charged,

14)   that no person shall be allowed to work on or near any live conductor where a danger could arise, unless:

a) it is unreasonable in all the circumstances for the equipment or conductor to be dead, and

b) it is reasonable in all the circumstances for that person to be at work on or near it while it is live, and

c) suitable precautions are taken to prevent injury,

15)   that there is adequate working space, access and lighting for working on or near an electrical system,

16) that no person is engaged in any work activity unless they are competent for the duties they have to perform.

The regulations apply to all voltages from the lowest to the highest available. They apply to all electrical systems whenever manufactured, purchased, or taken into use, even if the manufacture pre-dates the regulations. Existing equipment can still be used, even if it has been made to a standard that has since been modified; its replacement will only become necessary when it becomes unsafe, or falls due for replacement.

The EAW Regulations are applicable to the design of an installation as well as to the operation and maintenance of it. They also apply where any person is working near an electrical system, and as such , they are applicable to managers, mechanical and civil engineers and to any other person who has personnel working with or near electricity, who are under his control.

By definition, in the EAW Regulations, portable electrical tools, portable electrical equipment and test instruments become part of the system. So that any person using such equipment is covered by the regulations, as is any person in charge of other personnel using such equipment.

In practice almost everyone is working near or using electricity at work, the regulations therefore cover such persons as typists, computer operators, photocopier operators and labourers digging trenches on a building site, to name but a few.

General

BS 7671 only takeS into account established equipment, materials and methods of installation, but does not preclude the use of new materials, inventions or designs, providing the degree of safety is not less than that required by that British Standard.

Any departures from BS 7671 should be noted on the Completion Certificate, when the contract is finished.

Licensing or other authorities exercise control over certain premises, and whereas BS 7671 is still applicable to those premises, the additional requirements of the appropriate authority must be determined.

Voltage Bands

There are now two defined bands of voltage, namely Band I and Band II. Band I embraces installations which, for operational considerations, the voltage is limited, such as in telecommunication and bell systems, control and signalling circuits. Also covered under this Band are installations that, by virtue of applying certain conditions, protection against electric shock is achieved. Installations operating ELV can normally be defined as Band I, though circuits which occassionally exhibit voltages higher than ELV, such telecommunication circuits, can also be so defined. Band II covers all other voltages above Band I and below high voltage and would be used to describe voltages to domestic, commercial and industrial premises (excluding HV).

Voltage ranges

The voltage ranges covered by BS 7671, are:

Extra-low voltage	Not exceeding 50 V a.c., or 120 V ripple free d.c., between conductors or between any conductor and Earth.
Low voltage	From 50 V a.c., to 1000 V a.c., between conductors or from 50 V a.c., to 600 V a.c., between any conductor and Earth. From 120 V d.c., to 1500 V d.c., between conductors or from 120 V d.c., to 900 V d.c., between any conductor and Earth.

CHAPTER 1

It follows from the above that:

High voltage   Exceeds 1000 V a.c. between conductors and 600 V a.c between any conductor and Earth.
Exceeds 1500 V d.c between conductors and 900 V d.c. between any conductor and Earth.

Chapter 2

Assessment of the installation

Before a start can be made on the design, alteration or carrying out of an installation , an assessment has to be made of the installation, which involves the characteristics which are applicable to that installation.

Maximum demand
The first requirement is to work out the total electrical load that will be connected to the installation, and then to decide what diversity may be applicable . This is best worked out from the knowledge one has of the installation, or from experience. For instance: there may be two groups of motors that cannot possibly run at the same time, such as the re-circulation pumps and storm water pumps in a sewage works. In this instance, only the set of motors that has the largest demand needs to be included in the maximum demand for the installation.

Where no information is available, tables can be used as a guide to make an assessment of the diversity that will be applicable to that type of installation.

When assessing the maximum demand, starting currents or in-rush currents of equipment can be ignored.

Supply characteristics
The next step is to determine how the maximum demand is going to be supplied. Most installations will obtain their electrical supply from the local supplier. The first approach is to check with the supply company that they can give a supply and the type of supply they will provide. This will include determining the number and types of live conductors, i.e. single-phase or three-phase four wire, and the type of earthing arrangement, i.e., the type of system of which the installation will be part. This is important since the type of protection required is determined by the type of system. For instance, if the system is TT then the installation will require its own earth electrode, and to satisfy the requirements for protection against indirect contact an RCD will probably be required at the mains position, also all socket-outlets in the installation must be supplied through an RCD, such that, with the exception of special locations, $Z_S I_{\Delta n} \leq 50$.

The supply company must also be asked for the supply characteristics to enable the protection within the installation to be designed. These characteristics are still required even if the supply is from a private source such as a generator. The characteristics required can be listed as follows:

a) nominal voltage,
b) nature of current and frequency - (a.c., or d.c., 50 Hz or 60 Hz etc.),
c) the single-phase and three-phase prospective short circuit current at the origin of the installation,
d) type and rating of the overcurrent protective device at the origin of the installation,
e) the earth fault loop impedance external to the origin of the installation.

The single-phase and, where appropriate, three-phase values of prospective short-circuit required by item (c) are required to enable the correct type of protective devices to be installed within the installation.

Purpose and structure of the installation
Having determined the maximum demand and the characteristics of the supply, there are two items that have still to be considered.

The first is the purpose for which the installation is intended. This will include such items as: the type of building structure: whether the building will be used for handicapped people: are there any hazardous areas: are corrosive substances being used? The purpose of the installation is required to be ascertained to enable the correct selection of equipment and wiring materials to be made.

The next stage is to determine the distribution arrangement: in other words, the structure of the installation. This includes the earthing arrangement, i.e. the type of system. It also involves dividing the installation into circuits, so as to minimize inconvenience and danger in the event of a fault, and to enable the operation, maintenance, inspection and testing of the installation to be carried out without danger.

Separate circuits have to be provided for those parts of the installation that need to be separately controlled to avoid danger. This is to make certain that other circuits remain energised when there is a fault on one circuit. Consideration has to be given as to what would happen when a protective device operates. For instance, what would happen if all the lights went out due to a fault in a power circuit. Could there be an accident due to the lights going out? This means checking that there is discrimination between protective devices.

Each final circuit has to be connected to a separate way in a distribution board, so that it is electrically separate from every other final circuit. Consequently where the final circuits are single-phase and are taken from a three-phase and neutral distribution board, each circuit must have its own neutral conductor.

Environmental conditions

The environmental conditions applicable to the installation have to be taken into account when selecting equipment and wiring materials. Consideration must be given to the ambient temperature, mechanical stresses, corrosive substances and any other conditions which can foreseeably be seen to affect the installation when new and in the future.

Compatibility

Consideration has to be given to what effects the installation is likely to have on other equipment and services, this being important to comply with the Electromagnetic Compatibility Regulations. For instance, installing mains cables in close proximity to telemetry cables can affect the signals carried by those cables, or creating a loop with the cables for two-way switching, which sets up a magnetic field, affecting computer equipment. Starting currents of large motors can cause a dip in voltage which will affect electronic control circuits. The electricity supplier will also need to be informed of any equipment which is likely to affect their supplies, such as large motors or welding sets thus affecting other customers connected to the same service.

In assessing compatibility with other electrical equipment and other services, including non-electrical services, account must be taken of starting currents, harmonic currents, rapidly fluctuating loads, high frequency oscillations, transient over-voltages, earth leakage currents and the need for additional connections with earth, where appropriate.

Maintainability

When selecting the protective devices for the installation, consideration has to be given to the frequency and quality of maintenance that the installation will receive throughout its intended life.

This is required so that the protective measures taken to provide safety and the reliability of the equipment selected remain effective throughout the intended life of the installation. In addition, consideration has to be given to any periodic inspection, testing, maintenance of, and repairs to, the installation that may be necessary during its intended life. Information on the frequency that the installation should be tested should also be determined at this stage.

Chapter 3

Selecting protective devices

The next stage in the installation design is to select the type and rating of protective devices for each final circuit and distribution circuit. There are several types of protective device available, each type having its own particular characteristics. The various types available are listed as follows:
  i) cartridge fuses manufactured to BS EN 60269-1 and BS 88; known as HRC or HBC fuses,
  ii) cartridge fuses manufactured to BS 1361; generally known as domestic or service fuses,
  iii) semi-enclosed fuses manufactured to PS 3036; also known as rewirable fuses,
  iv) circuit-breakers manufactured to BS EN 60898 and BS 3871 ; known as MCBs,
  v) moulded case circuit-breakers manufactured to BS EN 60947-2; known as MCCBs.
The BS 3871 MCB standard was withdrawn in July 1994, but the MCBs to that British Standard will still be available for about five years, at which point they will be superseded by BS EN 60898 MCBs.

Each type of device has advantages and disadvantages over the other types, and a brief outline of these is given later. Whichever type of device is selected, it is better to use that type of device throughout the installation, since discrimination has to be achieved between successive devices that are in series. For instance, the operation of a protective device in a final circuit from a distribution board, should not cause operation of the protective device in the circuit feeding that distribution board. If it did, all the other circuits in the distribution board would become de-energised, causing loss of those services.

When selecting the current rating of the protective device for a fixed load circuit,such as a lighting circuit or a heating circuit, the current rating of the protective device must not be less than the total load connected to that circuit.

When selecting a protective device for a circuit that has an initial in-rush or starting current, such as a motor or transformer, the protective device must take into account the starting current or in-rush current, otherwise it will operate unwantedly. In many cases the protective device is not providing overload protection, but fault current protection. A table for selecting HRC fuses for motors is given in Part 4. The selection of protective devices has also to take into account fault currents, which is covered in more detail in Chapter 13.

Advantages and disadvantages of protective devices

BS 88 industrial fuses
BS 88 fuses which are generally called HRC (High Rupturing Capacity) or HBC (High Breaking Capacity) fuses, are made with two different utilisation categories in the U.K., which are distinguished by a two letter code introduced by IEC 269. The category known as 'aM' is also available in other countries, but not normally used in the U.K. The aM category fuse will interrupt a fault current but not an overload current safely.
The first letter indicates the breaking range of the fuse link:
  g   indicates full range breaking capacity,
  a   indicates partial range breaking capacity,
The second letter indicates the utilisation category:
  G   denotes general application including the protection of motor circuits,
  M   indicates protection of motor circuits or circuits with an inrush current.

The two utilisation classes associated with British fuse links are gG and gM, both of which have a full range breaking capacity. The European style aM motor fuse referred to above is also seen in the British market associated generally with imported equipment. Care is, therefore, needed when designing an installation in which imported equipment is involved.

Confusion often exists as to the rating of a British Standard motor fuse due to the label having two numbers as shown in Figure 3.1. If the Type T characteristics in Part 4 are looked at, it will be seen that the 32M50 fuse has the same characteristic as the 50 A fuse. The 32M50 is in practice a 50 A fuse element in a 32 A fuse size body. The lower number therefore indicates the maximum continuous current rating of the fuse, whilst the larger number indicates the fuse element size for short duration overload. The larger number should also be used for determining the energy let-through of the fuse under fault conditions.

The normal gG fuses are also suitable for motor protection so gM motor fuses are only required where the distribution board would need to have larger rating fuse carriers to cater for the starting current of the motor. In Figure 3.1 this would mean using a 63 A distribution board if a 50 A gG fuse was used, whereas by using a 32M50 fuse it will fit in a 32 A fuse carrier.

Fuses are available at a lower breaking capacity and voltage rating for domestic installations.

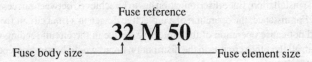

Fuse reference

32 M 50

Fuse body size ⸻⸻⸻⸻⸻⸻⸻⸜ ⸝⸻⸻⸻⸻⸻⸻Fuse element size

Figure: 3.1 - Motor fuse label

The gG and gM fuses used for industrial installations will interrupt all fault currents up to 80,000 A with all fuse ratings from 2 A to 1250 A, making them more than adequate for the normal industrial installation, since it is only on extremely large installations that the fault current will exceed 80,000 A.

The advantage the BS 88 industrial fuse has over other devices comprises: its breaking capacity, so that prospective short-circuit currents are restricted to levels well within the overload capability of cables; that when it has operated and is replaced, the circuit is protected by a brand new device; proven non-deterioration over many years; it is often used by motor starter manufacturers to provide ASTA certified type 2 coordination to IEC 947-4 .

The disadvantage is that it cannot be controlled: the fuse element is either whole or broken, and once operated it cannot be reset, it must be replaced manually by an authorised person. Additionally, it cannot be operated by remote control and cannot indicate that it has operated. (Some fuses can be fitted with an indicating device.) One further disadvantage, of the equipment rather than the fuse, is that fuse wire can so easily be used in the fuse carriers.

BS 1361 fuses (to be revised to BS 88 Section 3-1)

Fuses to BS 1361 are going to be superseded by BS 88 Section 3-1 , generally referred to as a cartridge fuse for domestic and similar premises. They are made in two types: Type 1 rated at 240 V for domestic installations, and having a breaking capacity of 16,500 A; Type II rated at 415 V for industrial installations and service cut-outs, having a breaking capacity of 33,000 A. The advantages and disadvantages are the same as for the BS 88 fuse. However, these types have a lower certified breaking capacity (33,000 A) and a limited range of ratings, the maximum fuse rating available being only 100 A.

BS 3036 fuse

The BS 3036 fuse which is usually called the rewirable fuse, or semi-enclosed fuse, is only rated at 240 V and should only be used in domestic installations. Its breaking capacity is limited, in the case of Type S1 fuses, to 1,000 A, Type S2 to 2,000 A and Type S3 to 4,000 A. It suffers the same disadvantages as the previous fuses, with the additional disadvantages of a significantly lower breaking capacity, and of only interrupting the fault current when it passes through zero. Additionally, when used for overload protection the current-carrying capacity of the circuit conductors has to be adjusted by a factor 0.725. The maximum rating of fuse available is 100 A.

BS 3871 Type 1 MCBs

These circuit-breakers (whose standard is now withdrawn and replaced by BS EN 60898 -see later) have a range of breaking capacities, from M1 - 1000 A to M9 - 9000 A, but devices with a breaking capacity up to 16 kA can be obtained. These are used in domestic and industrial installations. They have a low breaking capacity compared to the BS 88 fuse. They have the advantage that the circuit can be re-energised quickly by pressing the switch on the circuit-breaker. They do have a further disadvantage, in that, they are only usually designed to break a fault current at their rated breaking capacity twice, so when they have operated twice due to a fault at their rated breaking capacity they should be changed. They will, of course, break a lower fault current more often, but unfortunately there is no indication on the breaker rating plate of the value of fault current interrupted.

BS EN 60898 MCBs

BS EN 60898 Circuit Breakers for Household and Similar Installations, still referred to as MCBs, are manufactured in a range of breaking capacities of 1.5 kA, 3 kA, 4.5 kA, 6 kA and 10 kA; they can go up to 15 kA, 20 kA and 25 kA. The 1.5 kA MCB is only for incorporation in equipment associated with socket-outlets or switches for household and similar applications.

The values stated above are the ultimate breaking capacity I_{cn} of the MCB. The test carried out on the MCB for I_{cn} is O-t-CO (open, time interval, close/open) at various power factors, this means that the MCB will disconnect the circuit twice at its rated breaking capacity. Having disconnected the circuit twice at its rated breaking capacity the MCB should be changed.

The service breaking capacity I_{cs} must be the same as I_{cn} for MCBs up to 6 kA, but, subject to a minimum breaking capacity, can be less for higher rated MCBs. I_{cs} is 0.75 Icn for MCBs rated between 6 kA and 10 kA subject to a minimum of 6 kA. I_{cs} is 0.5 Icn for MCBs rated over 10 kA subject to a minimum of 7.5 kA. The test for I_{cs} is O-t-CO-t-CO (open, time interval, close/open, time interval, close/open). Thus if the fault current does not exceed I_{cs} the MCB can be put back into service. This will be discussed further after considering moulded case circuit-breakers.

The advantages and disadvantages of the BS EN 60898 MCB are the same as the BS 3871 MCB.

BS EN 60947-2 MCCBs

The BS EN 60947-2 moulded case circuit-breaker, referred to as an MCCB, is manufactured with a range of breaking capacities and can be equipped with different ratings of overloads. MCCBs are often used in conjunction with the BS EN 60898 MCBs to give back-up protection for the lower rated MCBs in final circuits.

They have the same disadvantage as MCBs, that is, they are designed only to break their ultimate rated breaking capacity I_{cu} twice. They also have a service breaking capacity I_{cs} which is a lower breaking capacity than I_{cu}, but the test O - t - CO -t - CO means that the circuit breaker can be put back into service without loss of performance providing the fault current does not exceed I_{cs}. Some manufacturers make I_{cs} equal to I_{cu} and in this case the MCCB must be changed if I_{cu} is exceeded twice.

They have the advantage that they are easily switched back on again once the breaker has cooled down after a fault, and can be operated by remote control and can be equipped with interlocks. They can also be made to give remote indication showing whether the breaker is closed or open. Time delays can be fitted, so that discrimination with lower rated devices is achieved. Additionally, the rating of overload protection in the breaker can be changed.

These different breaking capacities are confusing, but are used in the selection of the MCCB or MCB. The fault current at the point the MCCB or MCB is installed must not exceed the rated breaking capacity I_{cu} (I_{cn}) of the device (there are exceptions to this rule).

The fault level at the end of the cable protected by the MCCB or MCB will be less due to the resistance of the cable, so a check should be made that the service breaking capacity I_{cs} is not exceeded at the end of the cable.

If the breaking capacity I_{cs} is likely to be exceeded then a circuit breaker with a larger ultimate breaking capacity I_{cu} (MCCB) and I_{cn} (MCB) can be installed so that the service breaking capacity I_{cs} is higher; this allowing more interruptions before the breaker has to be changed.

Overload discrimination

Having selected the protective devices for the size and type of load, the protective device rating in the section board has to be determined. The current rating of the protective device in section board 'A' in Figure 3.2 will be determined by the total connected load on the final distribution board 'B',

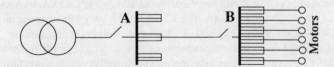

Figure 3.2 - Distribution diagram

reduced by any diversity that is allowed. Where the final distribution board contains equipment that has a peak current at start-up such as the starting current of motors, the number of items that may start at the same time has to be determined, together with the length of time the starting currents will persist. These starting currents are then added to the total current load on the final distribution board, to give the total current demand at start-up. It is now necessary to check the protective device characteristic for the total current demand and the time the starting currents will persist. This is to ensure that the protective device will not operate on start-up, as shown in Figure 3.3.

As shown in Figure 3.3, where the total current intersects the characteristic, read off the time on the time axis. This must be greater than the time taken for the motors to start, otherwise the protective device will operate.

Overload devices should be selected so that the current causing disconnection does not exceed 1.45 times the nominal rating of the protective device. It therefore follows (except for the above illustration) that $2A \times 1.45$ will always be less than a $4A \times 1.45$, and no problem with discrimination will occur between devices of the same type used solely for overload protection.

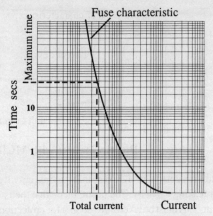

Figure 3.3 - Selecting protective device

Fault current discrimination

Although protective devices may discriminate with each other under overload conditions, they may not do so under fault current conditions, and so a check has to be made to determine that they discriminate under the latter conditions.

As far as fuses are concerned, this means checking that the total energy (I^2t) let-through of the smaller device does not exceed the pre-arcing energy (I^2t) of the larger device, for discrimination to be effective. See Figure 3. 4. As a rough guide, with BS 88 industrial fuses the rating of the major fuse should be 1.6 times the rating of the minor fuse.

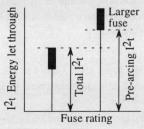

Figure 3.4 - HRC fuse discrimination

The total I^2t let-through of the lower rated device should be less than the minimum pre-arcing value of the higher rated device. Since the British Standard allows a certain amount of tolerance within which the design of the fuse characteristic can be made by the fuse manufacturer, it may be prudent to allow a margin of safety between the two values to avoid any nuisance blowing which may occur between fuses from different manufacturers. The characteristics given in Figure 3.4 are no longer used, instead the manufacturers now give the actual pre-arcing I^2t and total I^2t let-through for each device. Where circuit-breakers are concerned, it is a case of considering their characteristics, to ensure that the fault current flowing through the smaller device is less than the current needed for the instantaneous trip of the higher rated device as shown in Figure 3.5.

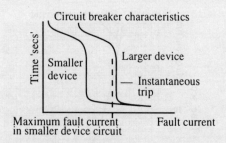

Figure 3.5 - Discrimination with MCBs

Delays can be incorporated in an upstream MCCB so that the final circuit protective device protects the circuit up to its rated breaking capacity. The upstream device then provides protection for all higher levels of fault current. This must be taken into account for protection of cables and equipment.

Some MCCBs used for back-up protection work on the principle that they open the circuit and then re-close, this causes a reduction in the fault current giving the final circuit device time to operate. If the fault current persists the back-up MCCB then operates again. This form of discrimination can switch off computers with subsequent loss of data.

Discrimination between RCDs

Discrimination between RCDs is achieved by using a time delay in the upstream RCD, this allows sufficient time for the RCD protecting the final circuit to disconnect the circuit before the main RCD operates.

The speed of operation has to be 0.04 s with five times the rated tripping current or where a time delay is used the speed of operation has to be the time delay plus 0.04 s.

With the rated tripping current applied to each pole in turn the RCD should trip within 0.2 s or plus any time delay incorporated in the device.

Type of MCB to be used

MCBs are supplied in different types. Types 1, 2 , 3 and 4 to BS 3871 and types B, C and D to BS EN 60898. Each type of MCB has a different current band for instantaneous operation and are used for different types of load detailed as follows:

Type 1 - $2.7I_n$ to $4\,I_n$, general circuits where the load does not give a high inrush current,

Type 2 - $4\,I_n$ to $7\,I_n$, general circuits with a moderate inrush current from the load,

Type 3 - $7\,I_n$ to $10\,I_n$, for motor or similar circuits,

Type 4 - $10\,I_n$ to $50\,I_n$, circuits with very high inrush current, such as X-ray equipment,

Type B - $3\,I_n$ to $5\,I_n$, general circuits with a moderate inrush current,

Type C - $5\,I_n$ to $10\,I_n$, for motor or similar circuits

Type D - $10\,I_n$ to $20\,I_n$, for high inrush currents, such as X-ray equipment.

Instantaneous tripping will occur within 0.1s within the range of currents which are multiples of the nominal rating of the device I_n. For example, a 16 A Type B MCB will disconnect a fault in 0.1s when the fault current is in the range of 3×16 and 5×16, i.e., between 48 A and 80 A. These multiples can also be used to check discrimination between devices in series.

Temperature effect on protective device ratings

When designing an installation for an area with a high ambient temperature, such as installations carried out in the Middle or Far East, the protective device rating has to be reduced. The amount of reduction is dependent upon the rating of the protective device and the ambient temperature, details of which should be available from the manufacturer.

Effect of altitude

Altitude also has an effect on the ratings of devices and manufacturers should be consulted when designing such installations.

Overload and fault current protection

Protective devices have to be selected to give overload and fault current protection, details of which are given in the following chapters.

Chapter 4

Overload selection

Overload selection

All circuits in the installation will require protection against fault currents, but only those circuits where an overload can occur will require overload protection.

An overload is caused by a circuit carrying more current than that for which it was designed, as in the case of a 3 kW load being connected to a circuit designed for 1 kW, or a motor being overloaded.

An overload occurs in a circuit which is a healthy circuit, whereas a short-circuit is a fault condition, which occurs between live conductors, (which includes the neutral) in conductor connections, or in the equipment. In general, live conductors must be protected by one or more devices for the automatic interruption of the supply, in the event of an overload, or fault current.

The devices installed must break any overload current flowing in the circuit, before such a current causes a temperature rise detrimental to the conductors, their insulation, joints, terminations, or any material surrounding or in contact with the conductors.

Relationship between I_n, I_b, I_z, I_{tab} and I_2

The nominal current rating of the protective device is designated by I_n, the design current (full-load current of the circuit) by I_b and the current-carrying capacity of the conductors by I_z. The calculated value of current- carrying capacity required for the conductor is given by I_t and the current-carrying capacity of the conductor given in the tables by I_{tab}.

The nominal current rating of the protective device, or the setting of the overload device, must be not less than the design (full-load) current of the circuit, i.e.

$$I_n \text{ must be greater than or equal to } I_b$$

The nominal current rating or setting of the overload device must not be greater than the current-carrying capacity of the smallest conductor in the circuit, i.e.

$$I_z \text{ must be greater than or equal to } I_n$$

There is a further requirement, that the current causing the effective operation of the protective device, referred to as I_2, must not exceed 1.45 times the current-carrying capacity of the smallest conductor in the circuit, i.e.

$$I_2 \text{ must not exceed } 1.45\, I_z$$

Fusing factor

The factor 1.45 referred to in BS 7671 is the fusing factor of the device obtained from the equation:

$$\text{Fusing factor} = \frac{I_2}{I_n}$$

which is:

$$\text{Fusing factor} = \frac{\text{Current causing operation of the protective device}}{\text{Nominal rating of protective device}}$$

Devices that have a fusing or tripping factor in excess of 1.45 do not comply with the above requirements, and compensation has to be made for the increase in temperature rise of the conductors, caused by the higher overload current which will flow through the circuit, before it is interrupted by the protective device.

Semi-enclosed fuse factor

One device which does not comply with the above requirement is the BS 3036 fuse, more commonly known as the 'rewirable fuse', or more correctly, 'semi-enclosed fuse', because it has a fusing factor of 2.

The requirements can be used to determine the degree of compensation that must be applied to the cables. This compensation allows for the higher overload current that will flow before the device interrupts the circuit, i.e.

$$I_2 = 1.45\, I_z \tag{1}$$

From the previous equations: fusing factor for BS 3036 fuse = 2

$$\text{Therefore } 2 = \frac{I_2}{I_n}$$

Substituting for I_2 in equation (1) as follows:

$$2\, I_n = 1.45\, I_z \qquad \text{i.e. } I_n = 0.725\, I_z$$

This is more conveniently expressed in the form:

$$I_z \text{ minimum} = \frac{I_n}{0.725}$$

This shows that the cable size has to be increased when using semi-enclosed fuses as overload devices.

HRC fuses referred to as BS 88 fuses, cartridge fuses referred to as BS 1361 fuses, and MCBs, all comply with the requirement that the fusing current does not exceed $1.45\, I_z$, providing the current-carrying capacity of any conductor in the circuit is not less than the nominal rating of the protective device. The MCCB is not included in the list because its overloads can be changed.

The fusing factor for the HRC fuse is often referred to as having a fusing factor of 1.5, but this is in an open test rig, and when installed in an enclosure, its fusing factor can be considered to be 1.45.

Confusion between factors

Confusion often arises between the two factors of 1.45 and 0.725. The factor 1.45, is the amount by which the nominal rating of the protective device has to be multiplied to give the overload current that must flow, in order to disconnect the circuit.

Technically speaking, the 0.725 factor is the amount by which current-carrying capacity of the smallest conductor in the circuit has to be multiplied, to give the nominal rating of the protective device, when its fusing factor is 2.

Since the application of the 0.725 factor in this way is impractical, the nominal rating of the protective device is divided by 0.725 to give I_t. It is commonly referred to as the derating factor for semi-enclosed fuses (S). (S is a symbol used in a formula given later.)

Motor overloads

Fuses and circuit-breakers are not the only means of protecting a circuit against overload: the most common alternative is the overload device in a motor starter. The rules are still the same, i.e.

$$I_n \text{ greater than or equal to } I_b$$
$$I_z \text{ greater than or equal to } I_n$$
$$I_2 \text{ less than or equal to } 1.45\, I_z$$

The last item does not usually pose a problem, but since most starter overload devices have a variable setting, the maximum setting must be used in the calculations, since the device can always be turned up to the maximum setting; particularly at a later date.

CHAPTER 4

Cables in parallel

Conductors are allowed to be installed in parallel from the same protective device, providing the conductors are of the same type, have the same cross-sectional area, are the same length, follow the same route, and have the same disposition, and there are no branch circuits throughout their length. The sum of the current-carrying capacities of all of the cables in parallel is then equal to the nominal rating of the protective device. This can be expressed as follows: (See Figure 4.1.)

$$I_t = \frac{I_n}{\text{Number of cables in parallel}}$$

The above cannot be applied to ring circuits since they are not parallel circuits.

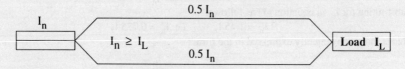

Figure 4.1 - Protection of conductors in parallel

Positioning overload devices

Overload protection devices should be placed at the point where a reduction occurs in the current-carrying capacity of the conductors. When considering where to install overload protective devices, the circuit must be looked at so as to determine the factors that will cause a reduction in current-carrying capacity. For instance, if there is a change in the cross-sectional area of the conductor, or a change in the method of installation, overload devices may be required at the point of change. Similarly, if there is a change in the type of cable or conductor, overload protection may be required.

The environmental conditions can affect the current-carrying capacity of a conductor: for example, a change in the ambient temperature along the route the conductor takes, or contact with thermal insulation. Grouping with other cables will also affect the current-carrying capacity of the conductor.

Overloads along the run of a conductor

Under certain circumstances overload devices can be installed along the run of a conductor, as shown in Figure 4.2.

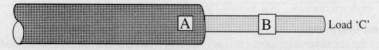

Figure 4.2 - Overload device installed along the run of a conductor

The overload device can be installed at point 'B', providing there are no branch circuits or outlets for the connection of current-using equipment between the points 'A' and 'B'. A typical example of this arrangement is a motor circuit where the starter with its overload device is installed adjacent to the motor. However, this arrangement is not allowed in locations in an installation where there is an abnormal fire risk or risk of explosion and special requirements apply, such as a hazardous area. Neither is this arrangement allowed with an IT system unless the conductors concerned are protected by an RCD or all the equipment and conductors form part of a Class II installation.

Omission of overload devices

Under certain circumstances overload protection devices may be omitted as shown in Figure 4.3.

Overload devices need not be provided at point 'B', if the protective device at point 'A', protects the conductors between 'B' and the load 'C'.

Figure 4.3 - Overload devices can be omitted in certain circumstances

No overload protective device is required where the unexpected opening of the circuit would give rise to a greater danger than the overload condition. In these situations audible alarms warning that the overload has occurred should be installed. Overload devices are not therefore required in the secondary circuit of current transformers, exciter circuits of rotating machines, lifting magnets, and fire extinguishing circuits. With current transformers and exciters the open-circuiting or the sudden opening of the circuit, can cause a dangerously peaking high voltage, which could also damage the equipment.

An overload protective device is not required if the characteristics of the load are not likely to cause an overload in the conductors. For example, where the load is constant, as in the case of a heating load, an overload is unlikely to occur. The protective device is only providing short-circuit protection and not overload protection; in this case, the semi-enclosed fuse factor 0.725 is not applicable.

An overload protective device is not required at the origin of an installation where the supplier of electricity provides overload protection between the origin of the installation and the main switch of the installation, for example the tails to a consumer unit.

The above is only applicable to an IT system if the conductors are protected by an RCD or the equipment and conductors form part of a Class II installation.

Protection of neutral conductors (3½ core cables)

Where the cross-sectional area of the neutral conductor is normally at least equal to that of the phase conductors in a TN or TT system, overcurrent detection is not required in the neutral conductor.

However, if the neutral cross-sectional area is less than that of the phase conductors, overcurrent detection shall be provided for the neutral conductor, unless the reduced neutral is protected against fault current by the phase conductor protective devices and the normal current carried by the neutral conductor is considerably less than its current-carrying capacity.

Where overcurrent detection is provided in the neutral it must disconnect the phase conductors, but not necessarily the neutral conductor itself.

Sizing conductors

Having determined the types and ratings of the protective devices the following chapters show how the designer should determine the distribution and circuit conductors cross-sectional areas.

Chapter 5

Voltage drop

Where the overcurrent protective overcurrent device is providing overload protection, the conductors of the circuit should be sized so that they have a current-carrying capacity not less than the rating of the protective device I_n. Where the circuit has an inrush current, such as a circuit supplying a motor, the protective device is only providing fault current protection and the cables will have been sized to the normal load current I_b.

The next logical step is to check that the voltage drop in the circuit complies with the requirements, because the voltage drop may require larger conductors to be installed, which may make the application of other correction factors inapplicable.

An explanation of the requirements

BS 7671 requires that the voltage drop up to the terminals of fixed equipment, shall ensure that the voltage at the equipment is greater than the lowest operating voltage specified for the equipment in the relevant British Standard. Where no such standard exists, then the voltage drop within the installation should not exceed the value appropriate to the safe functioning of the equipment.

Where the electrical supply complies with the Electricity Supply Regulations 1988 amended, the above requirement can be satisfied by ensuring that the voltage drop from the origin of the installation to the fixed equipment does not exceed 4% of the nominal supply voltage. The nominal voltage in the UK is now 230 V single-phase and 400 V three-phase.

When calculating the voltage drop in an installation the starting conditions, or inrush currents can be ignored. For instance, it is the full load current taken by a motor and not the starting current that is used in the voltage drop calculation.

If diversity has been applied, then the current load after diversity has been taken into account is used and not the total full load current. For example, the total connected load on a distribution board may be 100 A, but after diversity is taken into account, the load could be 90 A, 90 A would therefore be used in the voltage drop calculation.

Extra-low voltage

The same rule applies to extra low voltage circuits. Care is needed however, when applying the 4% voltage drop, to make certain that the voltage drop in the installation does not exceed a value appropriate to the safe functioning of the associated equipment in normal service.

Additional precautions have to be taken where extra-low circuits are supplied from a transformer and it is important to know the regulation of the transformer.

$$\text{Formula 1 - Regulation} = \frac{U_{oc} - U_{fl}}{U_{oc}}$$

Where: U_{oc} is the open circuit secondary voltage and U_{fl} is secondary voltage at full load. If the regulation was 5% then the maximum voltage drop allowed would be 0.6 V.

This is important when sizing conductors for ELV lighting because consideration has to be given as to what would happen if lamps in the circuit failed.

Motor circuits

Although the full load current of the motor is used to calculate voltage drop, consideration has to be given to the voltage drop due to the starting current for motors or similar circuits, otherwise trouble may be encountered.

For example: the supplier of electricity is allowed a tolerance on the supply voltage of + 10% - 6%. If the nominal supply voltage is 400 volts, and is supplied at -6% and a further 4% voltage drop, allowance is made within the installation based on a supply voltage of 400 V; then the voltage at the motor terminals when the motor is started will be 400V - 6% - 4%, giving a final voltage of 400V - 24 V - 16 V = 360 volts. The voltage available for the motor is only 360 V, before any voltage is dropped due to the starting current. Further reductions in voltage could seriously affect the motor's ability to reach its final speed, or to deliver the necessary driving torque. This could cause a breach of the requirements. Particular care is needed where electronic circuits are involved particularly programmable logic controllers (PLCs).

Basis of voltage drop tables

The voltage drop in mV/A/m in the tables in BS 7671 gives the single-phase voltage drop for single-phase circuits, i.e., the value given allows for the voltage drop in the phase and neutral conductors. The three-phase voltage drop for three-phase three wire or four wire circuits, gives the line voltage drop between phases.

The voltage drops given in the tables are based on the circuit conductors working at the maximum permitted operating temperature of the conductor, and at a load power factor the same as that for the cable. This leads to a larger voltage drop for the circuit, resulting in shorter lengths of circuit cables. Therefore, the voltage drop tables give the resistance, reactance and impedance of the cables for cable sizes larger than 16 mm². The tables in BS 7671 give the voltage drop per amp per metre, for each size of cable, whether used on single-phase or three-phase. The tables in Part 3 use a factor based on the length of the circuit multiplied by the tabulated current-carrying capacity of the conductor I_{tab}.

Correcting for load power factor

Where the load has a power factor less than unity a correction can be made to the mV/A/m value given in the tables to give a more accurate value of voltage drop.

For cables up to 16 mm² the figure in the volt drop tables is multiplied by cos Ø, i.e. the power factor. For cables larger than 16 mm² the voltage drop is given by the following formula:

Formula 2   Voltage drop $= \dfrac{(\cos Ør \ + \ \sin Øx) \ mV/A/m \times L \times I_b}{1000}$

where 'r' is the resistance and 'x' the reactance figure taken from the volt drop table for the cable size concerned, ' Ø' is the power factor angle of the load, L is the cable length and I_b is the design or full load current (f.l.c.). This can be rearranged to give the length allowed.

Formula 3   Length allowed $= \dfrac{\text{Voltage drop available} \times 1000}{(\text{Cos } Ør \ + \ \text{Sin } Øx) \, I_b}$

In general, the voltage drop calculation should first be made by using the mV/A/m impedance value given in the tables, the above formula being used only if the calculation gives a voltage drop slightly higher than the desired value, or a conductor length less than that required.

Table SPF 1 'Sine of Power Factor' in Part 4 gives the sine of the power factor angle from

CHAPTER 5

0.1 to 0.99, to enable the sine of the power factor to be easily obtained; it can of course, be calculated from the formula:

Formula 4 $\text{Sine } \emptyset = \sqrt{1 - \cos^2 \emptyset}$

where: $\cos \emptyset$ is the power factor

Correction for conductor operating temperature

Where a conductor is not carrying its tabulated current rating I_{tab}, its temperature will not reach the operating temperature allowed for the conductor. This means that a correction for the lower temperature can be made, as the voltage drop tables are based on the allowed operating temperature for the conductor.

Since reactance is not affected by temperature, it is only the resistive element that needs correction. This can be done by using the following formula to calculate the actual temperature of the conductor and then to calculate its resistance;

Formula 5 $t_f = t_p - \left(G^2 A^2 - \dfrac{I_b^2}{I_{tab}^2} \right) \left(t_p - t_a \right)$

Where: t_p is the maximum permitted operating temperature given in the tables.

t_a is the ambient temperature,

A is the ambient temperature correction factor,

G is the correction factor for grouping,

I_b is the design or full load current of the circuit and,

I_{tab} is the tabulated current rating of the cable from the tables.

The correction factor is then given by:

Formula 6 $C_t = \dfrac{230 + t_f}{230 + t_p}$

Alternatively, the correction factor for operating temperature is obtained by the following formula:

Formula 7 $C_t = \dfrac{230 + t_p - \left(G^2 A^2 - \dfrac{I_b^2}{I_{tab}^2} \right) \left(t_p - 30 \right)}{230 + t_p}$

This formula cannot be used where the protective device is a BS 3036 fuse, or where the actual ambient temperature is less than 30 °C.

Where the conductors are 16 mm² or less the resistance given in the voltage drop tables is multiplied by C_t to give the revised mV/A/m volt drop; for larger cables, the impedance given in the voltage drop tables is multiplied by C_t.

Combining the temperature and power factor factors

Where the load for the circuit has a power factor, then correction for the conductor operating temperature can be combined with the formula for correcting for the load's power factor. Thus for cables up to 16 mm² the revised voltage drop is obtained from the following formula:

Formula 8   Voltage drop = $C_t \cos \emptyset \, (\text{mV/A/m}) \times L \times I_b \times 10^{-3}$

Where the cables are larger than 16 mm² the following formula is used:

Formula 9   Voltage drop = $(C_t \cos \emptyset r + \sin \emptyset x) \times L \times I_b \times 10^{-3}$

The correction factor is not applied to the reactive component, since temperature has no effect on reactance. Thus for very large cables where the reactance is greater than the resistance such that: $\frac{x}{r} \geq 3$, no correction need be made for conductor temperature.

Calculating voltage drop

When corrections for operating temperature or load power factor are not important, there are two ways that the correct size of cable can be checked as being suitable for voltage drop. The first method is to obtain the mV/A/m for the size of cable for the correct current-carrying capacity, and then calculate the voltage drop as follows:

Formula 10 Voltage drop = $\dfrac{I_b \times L \times mV/A/m}{1000}$

The above formula can be rearranged to give the length allowed for a particular cable size.

Formula 11 Length allowed = $\dfrac{\text{Voltage drop allowed} \times 1000}{I_b \times mV/A/m}$

The problem with this method is that if the voltage drop is larger than that allowed for the circuit, the calculation has to be repeated with different cable sizes until the voltage drop is within the amount allowed.

The second method is to re-arrange the voltage drop formula, so that the mV/A/m is the unknown quantity as follows:

Formula 12 mV / A/ m = $\dfrac{\text{Voltage drop allowed} \times 1000}{I_b \times L}$

The cable size is then found by looking down the impedance column in the volt drop table until the value is either the same as, or less than, the figure calculated. This method, therefore, only requires one calculation.

There is another method, which is to use the Factor from the CR tables in Part 3.

Formula 13 Factor = $L \times I_b$

This is fully explained in 'How to use the Tables' in Part 3. A brief explanation of how it works is as follows: The load is multiplied by the length of circuit, and a factor that is equal to or larger than the figure calculated for load multiplied by current, is chosen from the appropriate current rating table. This then gives the size of cable which is suitable for 4% voltage drop. It makes calculations quick and easy, and does not involve using 10^{-3}.

Again the formula can be adjusted to give the actual cable length allowed.

Formula 14 $\dfrac{\text{Factor}}{I_b} \times \dfrac{\text{Available volt drop}}{\text{Allowed volt drop}} = L$ (Actual length allowed)

Voltage drop in a ring final circuit

Strictly speaking, voltage drop cannot be worked out for a ring final circuit at the design stage, since the designer has no control over what the user may plug into the circuit in the future. The length of the circuit is determined by the maximum floor area allowed for the circuit.

As an approximation, the average length of a ring circuit for voltage drop can be worked out by taking the worst and best current distribution for the circuit. This is how the total length is determined for this example. Since the electricity suppliers are not changing the voltage supplied to the consumer, the example is worked out on the basis of the voltage drop allowed at 240 V using a 30 A protective device.

First, calculate the pessimistic length. This will occur if all the load were at the mid point of the ring.

A current equal to half the rating of the protective device will flow in each half of the ring circuit; therefore, using table CR7, mV/A/m =18.

$$\text{\textbf{Formula 11}} \quad \text{Length of half ring circuit} = \frac{9.6 \text{ V} \times 1000}{15 \text{ A} \times 18 \text{ }\Omega} = 35.56 \text{ m}$$

The total length of ring circuit will therefore be equal to $35.56 \times 2 = 71.12$ m

Secondly, the optimistic length will be when each socket outlet carries an equal amount of current, and there are equal lengths of cable between each item. Assume 30 socket outlets. There will be 15 sockets in each leg of the ring circuit, and each socket will have a 1 A load; therefore, 9,600 mV = 18 mV $\times$ L$_S$(15A+ 14A + 13A + 12A + 11A + 10A + 9 A + 8 A + 7 A + 6 A + 5 A + 4 A + 3 A + 2 A + 1A), where L$_S$ is the length between socket-outlets and to distribution board.

$$9,600 \text{ mV} = 18 \text{ mV} \times L_S \times 120: \therefore L_S = \frac{9600}{18 \times 120} = 4.44 \text{ m / cable section}$$

There are 31 equal sections of cable, so the total circuit length = $4.44 \times 31 = 137.64$ m.

$$\text{Average ring circuit length} = \frac{71.12 + 137.64}{2} = 104.38 \text{ m}$$

This indicates a maximum ring circuit length of 100 metres, providing the floor area is not exceeded, or the actual loading on the ring is not known. A 32 A protective device would give an average of 98 m. See summary concerning other checks that will be required.

Voltage drop in a radial socket-outlet circuit

This is quite easy to determine. The rating of the protective device is used as the full load current I$_b$ of the circuit and the voltage drop calculation is then carried out by using Formula 10 or 14.

$$\text{\textbf{Formula 10}} \quad \text{Voltage drop} = \frac{I_b \times L \times \text{mV/A/m}}{1000}$$

Worked examples

Example 1 - Single-phase voltage drop.

A 3 kW fan heater is to be wired in twin and CPC cable clipped direct to a non-metallic surface. If the length of cable run is 25 metres, what size of cable will be required for 9.6 V voltage drop, if the nominal supply voltage is 240 V ?

Working (first method)

$$\text{Current taken by fan heater} = \frac{3000 \text{ W}}{240 \text{ V}} = 12.5 \text{ A}$$

Voltage drop allowed = 240 V $\times$ 4% = 9.6 V. Table CR7 column 10, 1.0 mm^2 cable rated at 15 A appears large enough. Volt drop/A/m = 44 mV from column 11 of same table.

$$\text{\textbf{Formula 10}} \quad \text{Voltage drop} = \frac{12.5 \text{ A} \times 25 \text{ m} \times 44\text{mV/A/m}}{1000} = 13.75 \text{ V}$$

13.75 volts drop is too much. Try 1.5 mm^2. rated at 19.5 A, mV/A/m = 29

$$\text{Voltage drop} = \frac{12.5 \times 25 \times 29}{1000} = 9.06 \text{ V}$$

which is satisfactory. Cable size achieved with a minimum of two calculations.

Working (second method)

Current taken by fan heater as before = 12.5 A.

Formula 12 $\quad mV/A/m \quad = \quad \dfrac{9.6V \times 1000}{12.5 \times 25} \quad = \quad 30.72 \, mV / A / m$

Now look in column 11 of table CR 7 for a mV drop equal to, or less than 30.72. The nearest lower value is 29 mV/A/m for 1.5 mm^2 cable, so this is the size to choose for voltage drop purposes. A check must now be made that this cable size has the correct current-carrying capacity, and whether any derating factors that could affect the cable size are applicable.

Working (third method using factor from CR7)
Formula 13 - length × load = 25 × 12.5 = 312.5.
Now select a factor from column 13 equal to, or larger than 312.5. This gives a cable size of 1.5 mm^2 as before, but is easier to work out.

Working (fourth method)
An even quicker method is to look at Table INST 2 in Part 3, which gives the cable size for a 3 kW load as 1.5 mm^2, irrespective of protective device type. This table also shows that the cable size is also suitable for protection against indirect contact, short-circuit protection, and thermal protection of the CPC, in addition to voltage drop.

Example 2 - Three-phase voltage drop.
A motor with a full load current of 15 A, is to be wired in PVC single core cable in conduit installed on the surface. If the supply voltage is 415 V, and the length to the motor is 30 metres, what size cable will be required for voltage drop purposes?

Working
Voltage drop allowed 4% of 415V = 16.6 V

Formula 12 $\quad mV/A/m \quad = \quad \dfrac{16.6 \times 1000}{15A \times 30m} \quad = \quad 36.89 \, mV$

From Table CR 1 column 7, 1.5 mm^2 is required, with a mV/A/m of 25.
Proof: Voltage drop = 15 A x 30 m x 25 ÷ 1000 = 11.25 V.

Working (using the factor method)
Formula 13 - Length 30 m × load 15 A = 450. Select a factor from table larger than 450. Column 9 next larger factor = 664 for 1.5 mm^2 cable. This is the cable size required for voltage drop.

Example 3 - Motor circuit(finding length allowed).
Suppose a motor with a full load current of 20 A, wired in 2.5 mm^2 PVC cable enclosed in conduit, is to be moved to a new location, and suppose the voltage drop up to the motor distribution board is 3 volts. How far can we move the motor without having to change the existing 2.5 mm^2 cables, if the supply voltage is 415 V, the existing length of run is 24 metres, and the total voltage drop allowed is 16.6 volts?

Working

Formula 11 $\quad$ Length allowed $\quad = \quad \dfrac{\text{Voltage drop allowed} \times 1000}{I_b \times mV/A/m}$

From CR1 column 7 mV/A/m for 2.5 mm^2 cable is 15 mV

CHAPTER 5

$$\text{Length allowed} \ = \ \frac{(16.6 - 3) \times 1000}{20 \times 15} \ = \ 45.33 \text{ m}$$

Maximum length to which cables can be extended is 45.33 - 24 = 21.33 metres.

Working (alternative method)
Using the factor method explained in Part 3.

$$\textbf{Formula 14} \quad \frac{\text{Factor}}{I_b} \ \times \ \frac{\text{Available volt drop}}{\text{Allowed volt drop}} \ = \ \text{Actual length allowed}$$

From CR1 for 2.5 mm^2 cable has a cable factor = 1107 (Column 9)

$$\frac{1107}{20} \ \times \ \frac{(16.6 - 3)}{16.6} \ = \ 45.34 \text{ m.}$$ As before the circuit can be extended by 21 m

Although extending the circuit 21 metres would satisfy the voltage drop requirements, a check would have to be made that the following were still satisfactory, with the added impedance in the circuit:

1. The phase earth loop impedance of the circuit for protection against indirect contact,
2. The protective conductor was still satisfactory,
3. Short circuit protection was still given by the protective device,
4. The voltage drop due to the starting current, when the motor was started, would allow the motor to develop its correct starting torque.

Correction factor examples
So far, only simple voltage drop calculations have been considered, but the occasion will arise when the voltage drop is limiting the correct length of cable being installed by just a few metres; under these circumstances it can be economical to carry out the additional calculations.

Example 4 - Power factor in motor circuit.
A three-phase motor with a full load current of 150 A, and a power factor of 0.8, is to be fed by a 150 metre PVC SWA & PVC armoured cable clipped direct to a non-metallic surface. What size cable would be required if the maximum voltage drop allowed in the cable is 16.6 volts?

Working
Trying the simple method first:
From table CR9 Column 6, a 50 mm^2 cable will carry 151 A, and from column 9 the factor is 20494

$$\textbf{Formula 13} \quad \text{Actual length allowed} \ = \ \frac{\text{Factor}}{I_b} \ = \ \frac{20494}{150} \ = \ 136.6 \text{ m}$$

This is less than the 150 metres required, so try compensating for the power factor.

Power factor = 0.8. From Table 'Sine of Power Factor' Part 4, when power factor is 0.8, sin $\emptyset$ = 0.6.
From Table CR9 column 7 'r' = 0.8 and from column 8 'x' = 0.14 .
Substituting these in the formula from Part 3:

$$\textbf{Formula 3} \quad \text{Actual length allowed} \ = \ \frac{\text{Voltage drop available} \times 1000}{(\text{Cos } \emptyset r + \text{Sin } \emptyset x)\, I_b} \ = $$

$$\frac{16.6 \text{ V} \times 1000}{(0.8 \times 0.8 + 0.6 \times 0.14) \times 150} \ = \ 152.85 \text{ m}$$

By taking the power factor into account, the cable is suitable as far as voltage drop is concerned.

Example 5 - Correcting for both power factor and conductor temperature.

A PVCSWA&PVC armoured cable is to be installed from an HRC fuse in a distribution board to a 415 V 55 kW motor, along with 5 other cables fixed to a perforated metal cable tray, where the cable sheaths will be touching. If the cable length is 100 metres, and the setting of the overloads in the starter are: minimum 90, mid point 100 A, maximum 110 A, and the power factor of the load is 0.75, what size cable would be required to satisfy voltage drop, if the ambient temperature is 30 °C, the voltage drop in the 3 phase feeder cable up to the distribution board is 9 V, and the total voltage drop allowed is 4%?

Working

In this case the cable is grouped, so this will determine the size of cable. From table GF1 Part 4, the derating factor for 6 cables is 0.74. The voltage drop allowed is 415 V $\times$ 4% = 16.6 V From Table MC1 Part 4 , 55 kW at 415 V gives motor current of 98 A; but since the overloads can be turned up to the maximum value, this should be used for the voltage drop calculation. Therefore :-

$$I_t = \frac{110}{0.74} = 148.65 \text{ A}$$

From table CR9 column 14, 50 mm^2 cable is required. I_{tab} = 163 A, and the factor is 20494, but is this cable size satisfactory? Try the simple calculation first, using Formula 14.

$$\textbf{Formula 14} \quad \text{Length allowed} = \frac{20494}{110} \times \frac{(16.6 - 9.0)}{16.6} = 85.3 \text{ m}$$

This is too short, so try compensating for the load's power factor.
From CR9, for 50 mm^2 cable 'r' = 0.8 and 'x' = 0.14
From SPF 1 'Sine of Power Factor' table Part 4 power factor 0.75 = sin Ø 0.6614

$$\textbf{Formula 3} \quad \text{Actual length allowed} = \frac{\text{Voltage drop available} \times 1000}{(\cos Ø r + \sin Ø x) I_b} =$$

$$\text{Actual length allowed} = \frac{7.6 \times 1000}{\left(0.75 \times 0.8 + 0.6614 \times 0.14\right) 110} = 99.76 \text{ m}$$

This is still too short, so investigate whether using the actual operating temperature of the conductors will allow a longer cable. (In practice this figure would be acceptable.)
Use alternative formula for C$_t$.

$$\textbf{Formula 7} \quad C_t = \frac{230 + t_p - \left(A^2 G^2 - \dfrac{I_b^2}{I_{tab}^2}\right)(t_p - 30)}{230 + t_p}$$

$$C_t = \frac{230 + 70 - \left(1^2 \times 0.74^2 - \dfrac{110^2}{163^2}\right)(70 - 30)}{230 + 70} = 0.988$$

Use this value in Formula 9 but rearranged similar to formula 3:

CHAPTER 5

$$\text{Length allowed} = \frac{7.6 \times 1000}{\left(0.988 \times 0.75 \times 0.8 + 0.6614 \times 0.14\right) 110} = 100.8 \text{ m}$$

This is now greater than the length required for the cable.

Summary

This chapter has shown various ways of calculating the voltage drop in a circuit. Whichever method is used, a check must always be made that the cable size chosen will have the correct current-carrying capacity for the load and the conditions under which the cable is installed. This is important because with the 4% voltage drop allowance quite often the cable size is suitable for voltage drop but its current-carrying capacity is not suitable for the load.

It is also necessary to check that the circuit complies with the other requirements of BS 7671, such as; shock protection, earthing, short circuit current overloads and any derating factors that may be applicable.

Chapter 6

Sizing conductors

Having sized the conductors for the load, the overload requirements and voltage drop, the method of installation and environmental conditions must now be considered to check that the current-carrying capacity of the conductors is still satisfactory.

Correction factors

The object of correction factors is to ensure that the temperature of the conductor does not rise to a temperature which is greater than the temperature the conductor insulation can withstand, when carrying its full rated current, and takes into consideration the temperature rise of the conductor during an overload.

The current-carrying capacity of conductors given in the current rating tables is based on the heat generated by the current flowing through the conductor being dissipated from the conductor, so that the conductor operating temperature is as stated in the current rating tables.

When conductors are grouped with other conductors carrying current, or are installed in an ambient temperature higher than 30 °C, or are in contact with thermal insulation, the rate of flow of heat from the conductor is reduced.

The diagram on the left in Figure 6.1, shows that when the conductor is carrying its full rated current, the rate of flow of heat out of the conductor leaves the conductor temperature at 70 °C. The diagram on the right shows that where the conductor is surrounded by insulation, the rate of flow of heat from the conductor is reduced; this reduction will raise the conductor temperature above 70 °C. The unit of heat is the Joule, which is Watt seconds, but Watt seconds are equal to I^2Rt. Since time is constant and the current I, is the current needed for the load, it can be seen that the only item that can be changed is the resistance of the conductor. The resistance of the conductor is reduced the larger the conductor is made. Correction factors should not be ignored, because there will be cases where their application indicates the need to install a larger conductor.

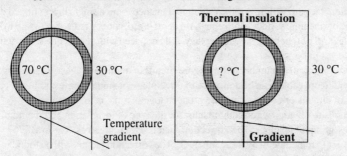

Figure 6.1 - Rate of flow of heat from a conductor

Circuits to equipment may be grouped together, be subjected to an ambient temperature higher than 30 °C, or be in contact with thermal insulation. Any of these conditions will necessitate applying the appropriate correction factor, so that the conductor's operating temperature does not exceed the value given in the current rating tables.

There are four correction factors to be taken into account: grouping, ambient temperature, contact with thermal insulation, and type of protective device used. These correction factors are often referred to as C_g, C_a, C_i and C_4 and can be easily remembered by the word G.A.T.S., detailed as follows:

G – Grouping Factor = C_g.

A = Ambient temperature = C_a.

T = Thermal insulation = C_i.

S = Semi-enclosed fuse = C_4, F_4, C_d or C_f.

The first three conditions listed above, can interfere with the natural flow of heat from the conductor, thus raising the conductor's temperature above its designed working temperature. The last factor is concerned with the type of protective device used, and the current which causes the effective disconnection of the circuit.

The first three correction factors are designed to increase the conductor size so that with the restricted flow of heat the conductor temperature does not exceed the designed working temperature considering each correction factor in turn.

Grouping factor (G)

Not all cable types need correction for grouping, the exception being non-sheathed mineral insulated cables which are not exposed to touch or in contact with combustible material. With the exception of non-sheathed MICC cables in Table CR 16 Part 3, correction factors for grouping should be applied to all the CR tables in Part 3 of this Handbook.

The first step is to determine the method of installation of the cables:

a) enclosed in conduit, trunking, or bunched and clipped direct,

b) in a single layer clipped direct to, or lying on, a non-metallic surface,

c) multicore cables in a single layer on a perforated metal cable tray,

d) single core cables in a single layer on a perforated metal cable tray,

e) multicore cables in a single layer on a ladder support system,

f) MICC cables installed on cable tray.

Care has to be exercised when choosing the correction factor for multicore cables clipped direct, since the correction factor is different if the cables are bunched, or in a single layer clipped direct, or just lying on a non-metallic surface. With regards to items b, c, d, and e, it is also necessary to know whether the cables will be touching, or whether they will be spaced with at least one cable diameter between them.

The next step is to count the number of single core circuits or multicore cables grouped together. The circuits or multicore cables can be a mixture of three-phase or single-phase circuits or cables; it is the number of circuits or cables which are grouped together that is important.

The third step is to obtain the correction factor for the type of installation and the number of circuits or cables grouped together. The correction factor is now given as 'Cg in BS7671; it is however, still easier to remember it by the symbol 'G'. The correction factors will be found in table GF 1 in Part 4 of the Handbook.

The fourth step is to look at the type of protective device for the circuit, since the formula used to determine the calculated current-carrying capacity required for the conductor I_t, is dependent upon the type of protective device and whether the device is giving overload protection.

Ambient temperature (A)

Ambient temperature is the temperature of the immediate surroundings of the equipment and cables, before such equipment or cables contribute to the temperature rise, in that location.

Two tables are provided in BS 7671: one table (4C2) is for BS 3036 fuses i.e. rewirable fuses, and the other table (4C1) is for all other types of protective device. The correction factors for rewirable fuses are the least onerous, because the rewirable fuse element is not enclosed, enabling it to dissipate heat more easily than the other types of device. Tables CR 1 to 17 in Part 3 contain the ambient temperature correction factors at the bottom of the table. When the protective device is being used for short-circuit protection, including BS 3036 fuses, the correction factor is taken from the table headed 'For other protective devices' in the CR tables in Part 3, or Table 4C1 in BS 7671.

Thermal insulation (T)

Where cables are likely to be in contact with thermal insulation, the current-carrying capacity of the cable has to be increased to allow for the reduction in heat dissipated from the cable.

Tables are provided giving the current-carrying capacities of cables installed in a thermally insulated wall or ceiling, where one side of the cable is in contact with a thermally conductive surface. Where cables are totally enclosed in thermal insulation, the current-carrying capacity of the cables has to be increased by a factor dependent upon the length the conductor is enclosed in the thermal insulation.

Length in mm conductors totally enclosed in thermal insulation	Derating factor
50	0.89
100	0.81
200	0.68
400	0.55
500 or over	0.5

The correction factor is applied to the current-carrying capacity taken from the 'open and clipped direct column' in the table for the type of cable being used.

Semi-enclosed fuse (S)

The tables in BS 7671 give the full thermal ratings of the conductors. They are based on the current needed to operate the device, not exceeding 1.45 times the nominal rating of the device.

Where the current required to operate a protective device exceeds 1.45, then a calculation is needed to determine the correction factor to be applied to the cable rating.

A typical example of the calculation has been given in Chapter 4. When a rewirable fuse is used for overload protection, the figure of 0.725 is used to derate the current-carrying capacity of the conductors. It is easier to apply the correction factor to the device's nominal rating as follows.

$$I_t = \frac{I_n}{S}$$

Where I_n is the nominal rating of the protective device, and S is the calculated correction factor. The correction factor for semi-enclosed (rewirable fuse) is only applied when the fuse is providing overload protection, i.e., it is not applied when the fuse is providing only short-circuit protection. The instances where it is not applied is where the fuse is used for a motor circuit and the overload protective devices are in the motor starter, or the circuit is supplying a load which is unlikely to cause an overload: such as a resistive load.

Application of correction factors

The formula to be used depends upon whether the protection device is providing overload and fault current protection or just fault current protection. Additionally, there are several alternative ways of determining the cable size for grouped cables, depending upon how they are installed, the type of circuit and the load current. All of which are explained as follows.

Circuits subject to overload

In this instance, the protective device is providing overload protection and more than one circuit can be overloaded at the same time. The minimum tabulated current-carrying capacity required. I_t, is determined by dividing the nominal rating of the fuse I_n by the correction factors.

$$\text{Formula 15} \quad I_t \text{ to be not less than } \frac{I_n}{G \times A \times T \times S}$$

where: G is the correction factor for grouping, A is the correction factor for ambient temperature, T the correction factor for thermal insulation and S the correction factor for a semi-enclosed fuse and I_n is the overcurrent protective device rating.

Circuits not subject to overload

When the circuit conductors are either not required to carry overload current, or because of the characteristics of the load, they are not likely to carry overload current, then I_n is replaced by I_b in formula 15; where I_b is the full load current (design current) of the circuit.

$$\text{Formula 16} \quad I_t \text{ to be not less than } \frac{I_b}{G \times A \times T}$$

Note the omission of the correction factor S from formula 16. An example of the application of this formula is the heating circuit, such as, an immersion heater, where its load is constant; it is therefore unlikely to carry overload current. Thus the 15 A rating of the semi-enclosed fuse does not have to be divided by 0.725 to determine the I_t for the cable when feeding a 3 kW immersion heater.

Selecting cable size

Having calculated I_t, the cable size is chosen from the single circuit column of the appropriate current rating table; the cable chosen must have an actual tabulated (I_{tab}) current-carrying capacity equal to, or greater than, the I_t calculated from the equation concerned. Single-phase circuits are chosen from the single-phase column and three-phase circuits from the three-phase column.

Grouped circuits not subject to simultaneous overload

Protective device not a semi-enclosed fuse

Although any circuit in the group can carry overload current, providing it can be guaranteed that only one of the circuits will be carrying overload current at any one time, the following calculation can be used to obtain the calculated current-carrying capacity I_t required for the conductors. This formula must not be used for socket-outlet circuits, since overloads cannot be guaranteed to be confined to one circuit.

Two calculations have to be made, and the one giving the highest calculated current-carrying capacity I_t required for the conductors, is the one that is used to size the cables.

First calculation

Formula 16 I_t to be not less than $\dfrac{I_b}{G \times A \times T}$

Second calculation:

Formula 17 I_t to be not less than $= \dfrac{1}{A \times T} \sqrt{I_n^2 + 0.48I_b^2 \left(\dfrac{1 - G^2}{G^2} \right)}$

Protective device is a semi-enclosed fuse (BS 3036)
Sizing the cables is carried out in exactly in the same way as for the other types of protective device and again two methods are available.

Method 1 - Circuits subject to simultaneous overload
Formula 1 is used and includes the derating factor 0.725 for S, as follows.

Formula 15 I_t to be not less than $\dfrac{I_n}{G \times A \times T \times 0.725}$

Again having worked out I_t the cable size is chosen from the single circuit column for either single-phase or three-phase in the appropriate table.

Circuits not subject to simultaneous overload
Again the higher value of I_t, obtained from either formula 16 or 18, is used to size the cables.

Formula 16 I_t to be not less than $\dfrac{I_b}{G \times A \times T}$

Formula 18 I_t to be not less than $\dfrac{1}{A \times T} \sqrt{1.9I_n^2 + 0.48I_b^2 \left(\dfrac{1 - G^2}{G^2} \right)}$

where: G is the correction factor for grouping, I_b is the design or full load current for the circuit, A is the ambient temperature correction factor and T the thermal insulation correction factor.

As before, formula 15 is used when more than one circuit can be overloaded at the same time; formulae 16 and 18 are used when it can be guaranteed that not more than one circuit will be carrying overload current at any one time; multiple socket-outlet circuits and motor circuits are not covered by this arrangement.

Socket-outlet circuits
The problem with socket-outlet-circuits, is that the designer of the installation has no knowledge of what the customer may plug into the socket-outlets in the future. The designer cannot therefore assume that the socket-outlet circuits will not be overloaded. Neither can he assume that simultaneous overloads will not occur.

Ring socket-outlet final circuits
For ring final circuit socket-outlets to be deemed to comply with the overload requirements they must:
a) be protected by a 30 A or 32 A device complying with BS 88, BS 1361, BS 3036, BS 3871 or BS EN 60898 (note a lower rating protective device can be used), and
b) the accessories used must comply with BS 1363, and
c) copper conductors with a minimum cross-sectional area of 2.5 mm^2, or 1.5mm^2 if mineral

insulated must be used, and

d) the current-carrying capacity I_z of the cable must not be less than 0.67 times the rating of the protective device used.

The correct way to derate for socket-outlet ring circuits when more than two circuits are grouped together is to use the following formula.

Formula 19 $I_t = \dfrac{\text{Rating of overcurrent protective device} \times 0.67}{G \times A \times T} = \dfrac{I_n \times 0.67}{G \times A \times T}$

No further diversity can be allowed, since this has already been taken into account by reducing the protective device size by 0.67. No correction for semi-enclosed fuse is required.

Radial socket-outlet final circuits

As far as radial socket-outlet final circuits are concerned, the overcurrent protective device nominal current rating is divided by the correction factor to obtain the current-carrying capacity, I_t, required for the cable, i.e., use formula 15.

Formula 15 I_t to be not less than $\dfrac{I_n}{G \times A \times T \times S}$

Lightly loaded cables

One of the most useful relaxations to the corrections for grouping is when circuits are lightly loaded. The relaxation works on the principle that the restriction of heat loss from the fully loaded cable by the lightly loaded cable is counteracted by the conductor in the lightly loaded cable acting as a heat sink, since its temperature is only just above ambient temperature. This relaxation will be found most useful in trunking runs which carry large numbers of control circuits, each of which carry only a few amperes.

If it is known that a cable will not carry more than 30% of its grouped rating, that cable can be excluded from the number of circuits or cables grouped together when counting the number of circuits or cables that are grouped together for derating factor purposes. This means they are not included, when counting the remaining cables or circuits that are grouped together.

In these circumstances it is necessary to count all the circuits or cables grouped together and work out the current-carrying capacity required, I_t, for the lightly loaded circuit or cable. This is done by using Formula 15 and then selecting the cable size required from the tables.

Having worked out the cable size for the lightly loaded circuit, the following calculation can be made to see whether it can be excluded from the number of circuits grouped together, when determining the correction factor for the remaining circuits.

Formula 20 Grouped current-carrying capacity % $= \dfrac{100 \times I_b}{I_{tab} \times G}$

If the full load current (design current I_b) of the lightly loaded conductor does not exceed 30% of the cables grouped current-carrying capacity, it need not be counted when working out the number of cables grouped together for the rest of the group, for determining the 'new' grouping factor for the rest of the cables in the group.

Each circuit can be considered in turn so that the correction factor gradually becomes less onerous as each circuit is excluded from the group.

The cable size worked out using the grouping factor for the lightly loaded cable is the size of cable that has to be used for the lightly loaded circuit. An example of this calculation is given in the next chapter.

Circuits and cables for which correction for grouping is not required

Where the horizontal distance between adjacent cables exceeds twice their overall diameter, no correction factor for grouping need be applied. When considering this exemption use the larger diameter of two adjacent but dissimilar sized cables, as shown in Figure 6.2.

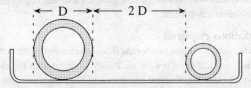

Figure 6.2 - Obviating the need to derate for grouping

Mixing cables with different maximum permitted operating temperatures

Where cables having different maximum permitted operating temperatures are grouped together, the current rating has to be based upon the lowest operating temperature of any cable in the group. This requires a correction factor for the higher operating temperature cables C_{mc} (correction for mixed cables).

Formula 21 $C_{mc} = \sqrt{1 - \dfrac{t_p - t_f}{t_p - t_a}}$ where $C_{mc} = \dfrac{I_b}{I_{tab}}$

Using this formula the following correction factors for C_{mc} are obtained.

Maximum Permitted operating temperature of conductor in group	Correction factor C_{mc} to be applied to the I_{tab} of the higher operating temperature conductor in the group				
	60 °C	70 °C	85 °C	90 °C	150 °C
60 °C	1	-	-	-	-
70 °C	0.866	1	-	-	-
85 °C	0.739	0.852	1	-	-
90 °C	0.707	0.816	0.957	1	-
150 °C	0.5	0.577	0.736	0.707	1

This correction factor is in addition to the correction factor for grouping. The easiest way to use it is to combine it with the correction factor for grouping when sizing the conductor with the higher maximum permitted operating temperature. For example, if a 90 °C cable is grouped with 70 °C cables, the I_t required becomes:

Formula 22 $I_t = \dfrac{I_n}{G \times C_{mc}} = I_t = \dfrac{I_n}{G \times 0.816}$

General rules

The correction factors are only applied to the section of the cable run where they are applicable, because the transmission of heat from the cables will only be affected by Grouping, Ambient and Thermal insulation, over the length of cable where the particular factor is applicable. The factor for semi-enclosed fuses affects the whole length of the cable run since it is installed at the source of the circuit.

The notes below the tables in BS 7671 state that the factors are applicable to groups of one size of cable, the reason being that smaller cables in the group could become overheated. Since no other guidance for the correction of grouped circuits or cables is available, use the derating factors given in BS 7671 as a guide when small conductors are grouped with larger conductors, but err on the safe side when choosing the conductor size.

Additional calculations required

Having determined the size of conductors for the load, protective device nominal current rating and any correction factors applicable, it is now necessary to check that the circuits comply with the requirements for indirect contact, thermal protection of the protective conductor and short-circuit protection. These subjects are covered in detail in the following chapters.

Where a more accurate or economical design is needed, adjustments can be made to take into account the actual operating temperature of the conductor and, in consequence, a revised k factor for the revised operating temperature. The formulae are given here since they are applicable to each subject.

Determination of final conductor temperature

The actual operating temperature of a conductor can be obtained by using part of the formula already given in the regulations. However, this formula is not applicable when t_a is less than 30 °C, or when BS 3036 fuses are used as the protective device.

Formula 5 $\quad t_f = t_p - \left(G^2 A^2 - \dfrac{I_b^2}{I_{tab}^2} \right) \left(t_p - t_a \right)$

where: t_f is the final conductor operating temperature, t_p is the maximum conductor temperature allowed and t_a is the ambient temperature; which in BS7671 is taken as 30 °C for the U.K.

This formula can be used to determine the initial temperature of a conductor for determining a revised 'k' factor, or the resistance of the conductor at its actual operating temperature.

Resistance of conductor at another temperature

Having calculated the conductor's actual operating temperature; the resistance of the conductor can be worked out using the resistance at a base temperature of 20 °C.

Formula 23 $\quad R_f = R_{20} \left(1 + \alpha_{20} \left(t_f - 20 \right) \right)$

where: R_f is the conductor resistance for t_f,

$\quad\quad R_{20}$ is the resistance of the conductor material at 20 °C,

$\quad\quad \alpha_{20}$ is the approximate resistance temperature coefficient of 0.004 per °C at 20 °C for copper and aluminium,

$\quad\quad t_f$ is the actual conductor operating temperature from the previous formula.

An extension of the above formula when dealing with two resistances and two temperatures is:

Formula 24 $\quad \dfrac{R_1}{R_2} = \dfrac{1 + \alpha t_a}{1 + \alpha t_b}$

Alternatively, the following formula, which may be more convenient, can be used:

Formula 25 $R_f = \dfrac{230 + t_f}{230 + t_0} \times R_0$

where: t_0 is the temperature of the resistance R_0,
     R_0 is the resistance of the material at temperature t_0.

However, the following formula can be used to determine the resistance of the conductor at the average of operating temperature and insulation limit temperature:

Formula 26 $t_1 = \dfrac{t_f + t_L}{2}$ **Formula 27** $R_{av} = \dfrac{(230 + t_1)\, R_{20}}{250}$

where: t_L is the limit temperature of the conductor's insulation; R_{av} is the resistance of the conductor at the average temperature; R_{20} is the resistance of the conductor material at 20 °C; t_f is the conductor operating temperature; t_1 is the average of conductor operating temperature and conductor's insulation limit temperature.

Part 4 contains tables giving conductor resistance at 20 °C, for copper and aluminium.

Determination of k for conductor operating temperature

When the actual operating temperature of a conductor is less than its maximum permitted operating temperature, the value of k used in calculations for the thermal capacity of protective conductors and live conductors can be revised.

IEC standard 724 gives a generalised form of the adiabatic temperature-rise formula which can be used to develop a formula to give the value of k for the actual operating temperature of a conductor:

Formula 28 $k = \sqrt{\dfrac{Qc\,(\beta + 20)}{\sigma_{20}}\, \log_e \left(\dfrac{t_L + \beta}{t_f + \beta} \right)}$

This can be simplified to:

Formula 29 $k = K \sqrt{\log_e \left(\dfrac{t_L + \beta}{t_f + \beta} \right)}$

where: t_L is the limit temperature for the conductor's insulation, and
     t_f is the actual operating temperature of the conductor.

The other constants in the equation, including K, being obtained from the following table:

Material	K $As^{1/2}/mm^2$	β (°C)	Qc $J/°C\,mm^3$	σ_{20} $\Omega\,mm$
Copper	226	234.5	3.45×10^{-3}	17.241×10^{-6}
Aluminium	148	228	2.5×10^{-3}	28.264×10^{-6}
Lead	41	230	1.45×10^{-3}	214×10^{-6}
Steel	78	202	3.8×10^{-3}	138×10^{-6}

CHAPTER 6

Fault current

Separate chapters are devoted to fault current protection and sizing protective conductors; but since the final circuit temperature has been covered in this chapter, it seems the appropriate place to discuss final conductor temperature in relation to fault currents.

Calculations to determine fault current should use the resistance at the average temperature of the initial conductor temperature and the limit temperature for the conductor's insulation because you cannot be sure otherwise that the conductors are protected with all the different types of protective devices available. In general, the permitted operating temperature of the conductor is assumed to be the figure given in the current rating tables.

When a conductor is not carrying its rated current-carrying capacity, its operating temperature is less than that given in the tables; this means that the average temperature will be lower, so the conductor resistance will be lower; this improves protection against indirect contact, voltage drop, but increases the fault current.

In large installations, the supply to the final circuit will be through several distribution cables and items of switchgear; each distribution cable getting progressively smaller as it approaches the final circuit. The temperature rise on conductors, caused by a short-circuit or earth fault in the final circuit, will progressively become less, as the distribution cables become larger, as they approach the source of electrical power; even to the point of having negligible effect on conductor temperature.

For calculation purposes, the formula for final conductor temperature, could be used to determine the temperature a distribution cable conductor would reach, when carrying full load current and the additional current caused by a fault in the final circuit.

The same is true when a fault occurs in a distribution cable; but the temperature rise effect on the other distribution cables, with which it is in series, back to the source of power, is dependent upon its position relative to the source of energy.

In practice, and to reduce the number of calculations necessary, assume that all the conductors are raised in temperature by the fault current. It is then only necessary to carry out the additional calculations of adjusting for temperature and k factor if it is a borderline case that the conductor may not be protected. In such circumstances, conductors which are more than one section board from a final distribution board can be considered not to be affected by an increase in temperature due to a fault in a final circuit. This is illustrated in Figure 6.3.

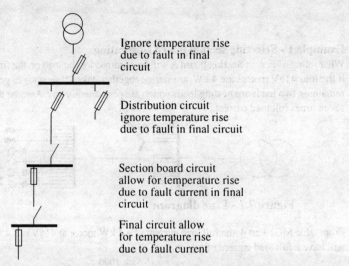

Ignore temperature rise
due to fault in final
circuit

Distribution circuit
ignore temperature rise
due to fault in final circuit

Section board circuit
allow for temperature rise
due to fault current in final
circuit

Final circuit allow
for temperature rise
due to fault current

Figure 6.3 - Cables affected by temperature rise under fault conditions

Chapter 7

Examples of sizing conductors

Example 1 - Selecting section board fuse rating.

What rating of fuse in Section board A will be required for the load on the final distribution board if the four 415V motors are 4 kW, are started together, take 10 seconds to get up to speed and the remaining two loads are heating loads which are permanently on? Assume that starting current is seven times full load current.

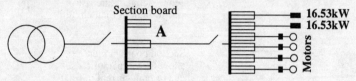

Figure 7.1 - Line diagram

From table MC1 Part 4 the full load current of a 4 kW motor at 415V is 8.4 A. Each heating load will have a full load current of:

$$I_{HL} = \frac{16.53 \times 1000}{415 \times 1.732} = 23 \text{ A}$$

The total connected load will be ($23 \times 2 + 8.4 \times 4$) = 79.6 A, which would require an 80 A fuse in section board 'A'. The motors all start together, so the current for which the fuse in the section board 'A' has to cater for upon start-up will be ($23 \times 2 + 8.4 \times 4 \times 7$) = 281.2 A.

It is now necessary to consult the fuse characteristics for the 80 A HRC TIA fuse.

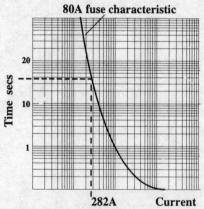

Figure 7.2 - Checking fuse will not blow

The total start-up load of 282 A is checked against the fuse characteristic to ensure that the point at which it touches the fuse characteristic is greater than 10 seconds; if it is, then the fuse rating is satisfactory: if it isn't, then a larger fuse would have to be selected.

Example 2 - Grouping of cables in conduit, with HRC fuses.

Three circuits, each with a full load current of 32 A, are to be wired in PVC enclosed in conduit from HRC BS 88 40 A fuses. If the ambient temperature is 30 °C and the circuits are not in contact with thermal insulation, what size cables will be required,

    a)   if the circuits are feeding 3-phase 3-wire BS EN 60309-2 socket-outlets, and

    b)   if the circuits are feeding single-phase heating loads ?

Working problem (a)

The correction factors for ambient temperature and thermal insulation will both be 1.

From table GF 1 Part 4, for 3 circuits enclosed in conduit the correction factor is 0.7

The circuits are feeding socket-outlets, so formula 15 has to be used.

Formula 15 $I_t = \dfrac{I_n}{G \times A \times T \times S} = \dfrac{40\,A}{0.7 \times 1 \times 1 \times 1} = 57.14\,A$

From column 6 of table CR1 Part 3, 16 mm^2 cable is required with an I_{tab} of 68 A

Working problem (b)

Since the loads are heating and simultaneous overloading of the circuits will not occur, formulae 2 and 3 can be used.

Formula 16 $I_t = \dfrac{I_b}{G \times A \times T \times S} = \dfrac{32\,A}{0.7 \times 1 \times 1} = 45.71\,A$

Formula 17 $I_t = \dfrac{1}{A \times T} \sqrt{I_n^2 + 0.48 I_b^2 \times \left(\dfrac{1 - G^2}{G^2}\right)}$

therefore $I_t = \sqrt{40^2 + 0.48 \times 32^2 \times \left(\dfrac{1 - 0.7^2}{0.7^2}\right)} = 45.95\,A$

The worst condition for I_t is formula 17; I_{tab} from CR1 Part 3 must not be less than this value. From column 2 table CR 1 Part 3, a 10 mm^2 cable is required with an I_{tab} of 57 A

Example 3 - Grouping with BS 3036 fuse.

Use the same example as number 2, but circuits are now protected by a semi-enclosed fuse and enclosed in conduit.

Working problem (a)

As before, correction factor is 0.7; but since a rewirable fuse is being used, formula 15 has to be used. **Formula 15** $I_t = \dfrac{I_n}{G \times A \times T \times S} = \dfrac{40\,A}{0.7 \times 1 \times 1 \times 0.725} = 78.818\,A$

From Table CR1 column 6, a 25 mm^2 cable is required with an I_{tab} of 89 A

Working problem (b)

This time formulae 16 and 18 have to be used.

Formula 16 $I_t = \dfrac{I_b}{G \times A \times T} = \dfrac{32\,A}{0.7 \times 1 \times 1} = 45.71\,A$

Formula 18 $\quad I_t = \sqrt{1.9 \times 40^2 + 0.48 \times 32^2 \times \left(\dfrac{1 - 0.7^2}{0.7^2}\right)} = 59.595\ A$

Formula 18 gives the larger value for I_{tab}, and from column 2 table CR 1 Part 3, 16 mm^2 cable is required.

Example 4 - Overload protection not required (single-phase).

A 3 kW immersion heater is supplied from a 15 A rewirable fuse in twin and CPC cable, the voltage of the supply being 240 V; it is grouped with one other cable on the surface, the ambient temperature is 30 °C and thermal insulation is not involved. What cable size will be required?

Working

From GF1 the derating factor is 0.8. 3 kW load is 12.5 A at 240 V.

The circuit is not likely to carry overload current (Regulation 473-01-04(ii)); use formula 16.

Formula 16 $\quad I_t = \dfrac{I_b}{G \times A \times T} = \dfrac{12.5\ A}{0.8 \times 1 \times 1} = 15.625\ A$

From CR 7 Part 3, a 1.5 mm^2 cable is required with an I_{tab} of 19.5 A. Note if the cable had not been grouped then a 1.0 mm^2 cable with an I_{tab} of 15 A would have been satisfactory.

Example 5 - Overload protection not required (three-phase).

A three-phase heater with a full load current of 21 A is protected by a 30 A HRC BS 88 Part 2 fuse. The circuit is to be wired in PVC single core cable (6491X) which will be installed along with four other motor circuits in conduit. If the cables are not in contact with thermal insulation and the ambient temperature is 50 °C, what size cable will be required ?

Working

In this example the cable will not be required to carry overload current, since the load is fixed. It cannot be guaranteed that simultaneous overload will not occur, since the heating circuit is mixed with motor circuits, but advantage can be taken of Regulation 473-01-04; so use formula 16.

From table CR1, the ambient temperature factor is 0.71, and from table GF1 the grouping factor for five circuits is 0.6. Therefore:

Formula 16 $\quad I_t = \dfrac{I_b}{G \times A \times T} = \dfrac{21\ A}{0.6 \times 0.71 \times 1} = 49.296\ A$

From column 6 table CR1, cable size required is 10 mm^2, with an I_{tab} of 50 A

Example 6 - Socket-outlet ring final circuit

Four socket-outlet ring final circuits are to be installed in conduit using PVC cable from a 30 A MCB. What minimum size of cable must be used to comply with the overload requirements?

Working

From GF 1; Part 4, grouping factor is 0.65. Use formula 19.

Formula 19 $\quad I_t = \dfrac{I_n \times 0.67}{G \times A \times T} = \dfrac{30\ A \times 0.67}{0.65 \times 1 \times 1} = 30.92\ A$

From CR1 column 2 Part 3; 4 mm^2 cable is required.

Example 7 - Radial socket-outlet final circuits in conduit

Two three-phase socket-outlet final circuits are to be installed from a 32 A BS 88 fuse in PVC single core 6491X cable installed in the same conduit. If the ambient temperature is 50 °C and the conduit length between draw-in boxes is 4 m with one bend between draw-in boxes what cable and conduit size must be used?

Working
From GF1 Part 4; grouping factor is 0.8 and from CR1 Part 3; ambient temperature factor is 0.71.

Formula 16 $\quad I_t = \dfrac{I_n}{G \times A \times T} = \dfrac{32 \text{ A}}{0.8 \times 0.71 \times 1} = 56.338 \text{ A}$

From CR1 column 6 Part 3, cable size required is 16 mm^2.
Number of 16 mm^2 cables enclosed in conduit is $2 \times 3 = 6$
From Table 1 of Table CCC1 Part 3; factor for 16 mm^2 cable is 145. Total factor $=145 \times 6 = 870$
From Table 2 of CCC1 Part 3; conduit size for 4 m between draw-in boxes with one bend between draw-in boxes is 38 mm (Factor 973), i.e., 1.5 mm conduit required.

Example 8 - Ambient temperature correction with BS 3036 fuse.

A three-phase circuit to be wired in PVC single core 6491X cable in its own conduit in a boiler house where the ambient temperature is normally 45 °C. It will be protected by a 20 A rewirable fuse. If it is not grouped or in contact with thermal insulation what cable size is required?

Working
In this example, both G and T will equal 1. The value of S is 0.725 from MISC 1 Part 4, but the value of A is found from table CR1 for rewirable fuses, and is 0.91, since 6491X cable is a general purpose PVC cable; use formula 15.

Formula 15 $\quad I_t = \dfrac{I_n}{G \times A \times T \times S} = \dfrac{20A}{1 \times 0.91 \times 1 \times 0.725} = 30.315 \text{ A}$

From column 6 of table CR1, cable size required is 6 mm^2 with an I_{tab} of 36 A

Example 9 - Thermal insulation; cable in contact on one side only.

A house is to be fitted with a 7 kW shower heater, using twin and cpc cable for the supply, which will not be grouped with other cables, but will be installed along the side of a joist with thermal insulation installed between the joists. If the ambient temperature is 30 °C and the voltage is 240 V, what current rating will be required for a Type B MCB and what size of cable will be required?

Working
First step: Method of installation: contact on one side with thermal insulation.
Second step: Full-load current of the heater: $= 7000 \text{ W} \div 240 \text{ V} = 29.17 \text{ A}$; therefore a 30 A MCB can be used.
Third step: No calculation is required, since tables are now given for cables in contact with thermal insulation on one side only.
Fourth step: Table CR8 Part 3 column 2; 6 mm^2 cable required with an I_{tab} of 32 A
Fifth step: Factor from column 5, CR8 is 1315. The maximum length the circuit can be is:

Formula 13 $\quad$ Length allowed $= \dfrac{\text{Factor (from table)}}{I_b} = \dfrac{1315}{29.17 \text{ A}} = 45.08 \text{ m}$

CHAPTER 7

Alternative 1, using volt drop/A/m from Table CR8. Vd/A/m = 7.3 mV.

Formula 11 Length allowed $= \dfrac{9.6 \text{ V} \times 1000}{29.17 \text{ A} \times 7.3 \text{ mV}} = 45.08 \text{ m}$

Alternative 2, Part 3 Table INST 1: 30 A MCB maximum length to comply with regulations is 45 m.

Example 10 - Thermal insulation; cable totally enclosed.

The same problem as example 9, but cable is installed on the surface and then passes through 200 mm of thermal insulation where it passes through the ceiling ; work out the cable size required.

Working

The full load current of the heater will be 29.17 A; the MCB size 30 A.
First step: derating factors applicable; G = 1, A = 1, T = ?, S = 1.
In this case the thermal insulation factor is 0.68 from MISC 1 Part 4.

Formula 16 $I_t = \dfrac{I_b}{G \times A \times T} = \dfrac{29.17 \text{A}}{1 \times 1 \times 0.68} = 42.89 \text{ A}$

Note: Since the load is resistive the MCB is not providing overload protection, so I_b is used not I_n
   The size of cable is now selected from CR7 column 10 for cables 'clipped direct', since they have already been derated by the above calculation. In this case, the cable size required is 6 mm^2, with an I_{tab} of 46 A
From INST 1 Part 3. Cable length allowed 45 m for a 7 kW resistive load for TN-C-S system.

Example 11 - Lightly loaded conductors.

This example illustrates the application of the note 2 of Table 4B1/2.
Two three-phase three-wire motor circuits are each protected by a 50 A HRC fuse, and are to be installed in the same trunking as 16 single-phase control circuits, each of which has a design current I_b of 1.5 A , and each of which is protected by a 2 A HRC fuse. What size cables will be required if all other derating factors = 1 ?

Working

Total number of circuits grouped together = 2 + 16 = 18.
From table GF1 Part 4, Grouping factor G = 0.39.
Consider the lightly loaded conductors first: fuse type HRC so use formula 16.

Formula 16 $I_t = \dfrac{I_n}{G \times A \times T} = \dfrac{2 \text{ A}}{0.39 \times 1 \times 1} = 5.128 \text{ A}$

From table CR1 Part 3, 1.0 mm^2 cable could be used having an I_{tab} of 13.5 A.
If the control cables will not carry more than 30% of their grouped current rating, they can be ignored when considering the three-phase motor circuits.

Formula 20 Percentage of grouped rating $= \dfrac{100 \, I_b}{I_{tab} \times G} = \dfrac{100 \times 1.5}{13.5 \times 0.39} = 28.49\%$

Control cables can be ignored so grouping factor for two three-phase circuits = 0.8.

Formula 16 $I_t = \dfrac{I_n}{G \times A \times T} = \dfrac{50 \text{ A}}{0.8 \times 1 \times 1} = 62.5 \text{ A}$

From Table CR1, column 6, 16 mm^2 cable required. Without this relaxation 50 mm^2 cable required.

Example 12 - cables with different maximum permitted operating temperatures

A 90 °C four core XLPESWA&PVC armoured cable is to be installed on a cable tray alongside three 70 °C PVCSWA&PVC armoured cables. The cables are in a single layer with sheaths touching. If the protective device is a 63 A fuse, what size of cable is required and what action should be taken concerning the existing PVC cables ?

Working

Four cables grouped together; from GF1 Part 4 correction factor is 0.77.

From table below Formula 21 in previous chapter C_{mc} factor for XLPE cable is 0.816.

Formula 22 $\quad I_t = \dfrac{I_n}{G \times C_{mc}} = I_t = \dfrac{63\ A}{0.77 \times 0.816} = 100.267\ A$

From CR10 column 4 Part 3; 25 mm^2 cable required with I_{tab} of 131 A.

Action concerning the existing PVC armoured cables would be to recalculate their current-carrying capacity to determine they are still large enough now that there are four cables in the group.

Example 13 - Actual operating temperature

Due to voltage drop a four core 50 mm^2 70 °C PVCSWA&PVC armoured cable with an I_{tab} of 163 A is installed for a load of 63 A; no correction factors apply. Calculate: a) the actual operating temperature of the cable and b) the revised mV/A/m voltage drop impedance if the resistive mV/A/m is 0.8 and the reactive mV/A/m is 0.14. (Note: figures obtained from Table CR9 Part 3.)

Working (a)

Since there are no correction factors for grouping and ambient temperature both G and A equal 1.

Formula 5 $\quad t_f = t_p - \left(G^2 A^2 - \dfrac{I_b^2}{I_{tab}^2} \right) \left(t_p - t_a \right)$

Formula 5 $\quad t_f = 70 - \left(1 - \dfrac{63^2}{163^2} \right) (70 - 30) = 36\ °C$

Working (b) Use Formula 25.

Formula 25 $\quad \dfrac{\text{Revised resistive mV/A/m}}{\text{Tabulated resistive mV/A/m}} = \dfrac{230 + t_f}{230 + t_p}$

Revised resistive mV/A/m $= 0.8\ \text{mV/A/m} \times \dfrac{230 + 36}{230 + 70} = 0.709\ \text{mV/A/m}$

Revised mV/A/m impedance $= \sqrt{0.709^2 + 0.14^2} = 0.7226\ \text{mV/A/m}$

Example 14 - revised k factor

The actual operating temperature of a 70 °C PVCSWA&PVC armoured cable is 36 °C, calculate the revised k factor for this cable.

Working

Use formula 29 from previous chapter.

Formula 29 $\quad k = K \sqrt{\log_e \left(\dfrac{t_L + \beta}{t_f + \beta} \right)}$

CHAPTER 7

From the Table in Chapter 6, K equals 226 and β is 234.5
From Table K1A Part 4, Limit temperature for a PVC insulated copper conductor is 160 °C

$$k = 226 \times \sqrt{\log_e \left(\frac{160 + 234.5}{36 + 234.5} \right)} = 138.828$$

The k factor is increased from 115 to 138.8

Example 15 - Combined factors.

So far, circuits have been considered where the correction factors affected the whole circuit, but in practice the correction factors may only be applicable to a portion of the circuit they affect. This is illustrated in this example.

Two three-phase loads 'A' and 'B' in Figure 7.3, are to be supplied by PVCSWA&PVC cable, the circuits being fed from a distribution board with BS 3036 fuses; the normal ambient temperature is 30 °C. The cables start off clipped to a perforated cable tray with their sheaths touching. The cable for load 'A' passes through a cavity wall where it is totally enclosed in thermal insulation for 50 mm. The cable for load 'B' is installed into a boiler house where the ambient temperature is 60 °C Ignoring voltage drop, indirect contact and short circuit current, what size cables are required ?

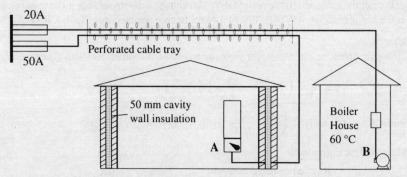

Figure 7.3 - Combining the derating factors

Working

General Formula :   **Formula 15** I_t to be not less than $\dfrac{I_n}{G \times A \times T \times S}$

Load 'A' cable

G = grouping = 0.86 from table GF1.
A = ambient = 1 for 30 °C
T = thermal insulation = 0.89 from MISC 1 Part 4.
S = type of protective device = 0.725, from MISC 1 Part 4.

Condition 1 - Length on cable tray

$$I_t = \frac{50}{0.86 \times 1 \times 1 \times 0.725} = 80.19 \text{ A}$$

Condition 2 - Length in contact with thermal insulation

$$I_t = \frac{50}{1 \times 1 \times 0.89 \times 0.725} = 77.48 \text{ A}$$

The cable size is selected for the most onerous case, i.e., condition 1 - 80.19 A.
From Table CR 9 column 14, 16 mm^2 cable required with an I_{tab} of 83 A

Load 'B' cable
G = grouping = 0.86 (as before).
A = ambient = 0.69 from table CR9.
T = thermal insulation = 1.

Condition 1 - Length on cable tray

$$I_t = \frac{20}{0.86 \times 1 \times 1 \times 0.725} = 32.07 \text{ A}$$

Condition 2 - Length in boiler house

$$I_t = \frac{20}{1 \times 0.69 \times 1 \times 0.725} = 39.98 \text{ A}$$

Condition 2 is the worst case so this is used. From Table CR 9 column 14 the cable size required is 6 mm^2 with an I_{tab} of 45 A.

Comments

It should be noted that the cable is only sized to the worst condition and the correction factors are only applied to that section of the conductor they affect.

Where long runs of large cables are involved, it may be economic to change the size of cable in each section according to how it is affected by the particular correction factor; the saving in conductor material and labour costs would have to be weighed against the cost of jointing material and the cost of installing the joints. This arrangement would satisfy the overload requirements, providing the conductor with the lowest current-carrying capacity was protected by the overcurrent device; care would be needed with fault current protection.

It should also be noted that in practice (except where conductors are in contact with thermal insulation on one side only) it is not necessary to carry out separate calculations for each section of cable; the cable can be sized by taking the correction factors that will most increase the cable size.

A further point to note is that the ambient temperature correction factor for BS 3036 rewirable fuses is only used when the fuse is giving overload protection. When the fuse is giving only short-circuit protection, the ambient temperature correction factor is obtained from the list of correction factors 'for other protective devices' in the CR Tables, or Table 4C1 in BS7671.

Having sized the circuit conductors, it will now be necessary to check that the circuit provides protection against indirect contact, that the thermal capacity of the protective conductor is satisfactory, that the overcurrent protective device protects the circuit conductors against short-circuit current and that the live conductors are still protected when an earth fault occurs.

Discrimination with RCDs

Discrimination between an incoming supply RCD and circuit RCDs can be achieved by having a time delay on the incoming supply RCD. In general RCD protection should be placed in the circuit where the protection is required. For TT installations, install a time delayed RCD on the mains.

Chapter 8

Protection against electric shock

A person can receive an electric shock in two ways; firstly by coming into contact with live parts (direct contact) and secondly by touching metallic parts that have become live due to a fault (indirect contact). Protection against electric shock is covered by three Sections in Chapter 41 of BS 7671.

Section 411 - Protection against both direct and indirect contact.

Section 412 - Protection against direct contact.

Section 413 - Protection against indirect contact.

Protection against electric shock can be achieved by application of the requirements in Section 411, or by the combined application of the requirements in Sections 412 and 413. Looking at each of these in turn;

Section 411 concerns the following methods of protection:

1. Protection by SELV,
2. Protection by limitation of discharge energy.

The rules for item 1 are covered in Part 1 of the handbook, so there is little point in repeating them again in this chapter.

Protection by limitation of discharge energy means that the equipment incorporates a means of limiting the current which can pass through the body of a person or animal to a value lower than the shock current. Any circuits using this method of protection must be separated from other circuits; the separation being equivalent to that specified in BS 7671, for SELV systems.

Protective low voltage (PELV) is the term used when a circuit complying with the SELV requirements is connected to earth at one point only and is not suitable for protection against direct contact.

Functional extra low voltage is the term used, where extra low voltage is used, but where compliance with all the SELV requirements is not possible. In these circumstances measures have to be taken to provide protection against electric shock.

Protection against direct contact

Four basic protective measures for protection against direct contact are given in BS 7671, and are illustrated by the following diagram:

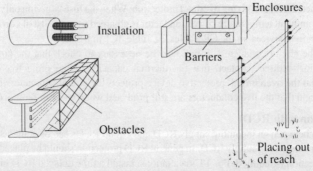

Insulation

Enclosures

Barriers

Obstacles

Placing out of reach

Figure 8.1 - Protection against direct contact

Insulation

Insulation is the basic insulation of cables and parts required in every installation. BS 7671 requires that the insulation can only be removed by destruction, and that it will withstand any electrical, mechanical, thermal and chemical stresses to which it may be subjected whilst it is in service.

Barriers or enclosures

Where protection by barriers or enclosures is used to protect against direct contact, then the degree of protection must be at least IP2X or IPXXB, which is the Index of Protection for the standard finger 80 mm long and 12 mm in diameter. Where the opening in equipment has to be larger than IP2X or IPXXB to enable maintenance to be carried out, precautions must be taken so that there can be no unintentional touching of live parts, and that persons are warned of the proximity of live parts within the enclosure. Where a top surface of equipment is readily accessible, the degree of protection is more stringent, and must be at least IP4X.

During the maintenance of an installation, or the carrying out of new works, enclosures may have to be opened and barriers removed. BS 7671 gives four alternative methods of safety against direct contact; these are detailed as follows:

1. The opening of the enclosure, or the removal of a barrier, must only be possible by using a key or tool. This means that access is limited to skilled persons who should know the dangers and take the necessary precautions,
2. Opening an enclosure, or removing a barrier can only be carried out after the supply to live parts has been disconnected. The supply may only be restored after the barriers have been replaced or the equipment reclosed. Isolators interlocked with doors are examples of this type of protection,
3. An intermediate barrier having a degree of protection of IP2X or IPXXB is provided to prevent contact with live parts, the barriers being removable only by using a tool.

Care has to be taken when selecting distribution boards to comply with the requirements for protection against direct contact. Quite often, access to the interior of distribution boards is not limited to skilled or instructed persons; quite frequently, barriers are not replaced after work has been completed, and many boards have knurled screws for fixing the door, allowing anyone access to the interior.

Obstacles

Obstacles need little explanation. They are intended for use only where access is limited to skilled and instructed persons. They should be securely fixed but can be removed without using a key or tool; as such, they do not prevent intentional contact with live parts.

Arm's reach

Similarly, placing out of reach does not stop intentional contact with live parts, but BS 7671 does define the limit of arm's reach. (See definitions in Part 1.) The limit must, however, be increased in areas where long or bulky metallic objects are handled.

Where a barrier or obstacle limits a person's movement, such as a handrail, the limit of arm's reach in the horizontal plane starts at the obstacle unless the degree of protection is greater than IP2X or IPXXB (i.e., IP3X or larger number)

Residual current devices (RCD)

Although a residual current device can reduce the risk of injury from electric shock, a residual current device cannot be used as the sole means of protection against direct contact. For the risk of injury to be reduced, the live parts must either be insulated, protected by barriers, enclosures, or obstacles, or by placing the conductors out of reach. Additionally, the tripping current of the RCD must not exceed 30 mA, and must operate within 40 ms., with a test current of 150 mA, when used as supplementary protection against direct contact.

Protection against indirect contact

When the primary insulation fails, a fault which can give rise to danger occurs, even though the live electrical parts and conductors cannot be touched. BS 7671 therefore gives requirements that try to limit the degree of danger.

Five methods of protection against indirect contact are given:
1. Earthed equipotential bonding and automatic disconnection of the supply,
2. Using Class II equipment or equivalent insulation,
3. Non-conducting location,
4. Earth-free local equipotential bonding,
5. Electrical separation.

The last three items in the list would normally be covered by a detailed specification; they do not arise very often, and are therefore beyond the scope of this handbook.

As far as the second item is concerned, a protective conductor has to be installed to all Class II items used in an installation such as switches and ceiling roses. Other areas where Class II is used exclusively for protection against indirect contact would be covered by a detailed specification and will not arise very often, so they will not therefore be dealt with here.

One area where double insulation will be encountered is with portable tools and equipment; in this case a protective conductor is not installed to the tool or the equipment.

The most common form of protection against indirect contact used in this country is item one in the list, so this will be looked at in detail.

Earthed equipotential bonding and automatic disconnection

The basic rule requires the characteristics of the protective devices, the earthing arrangements for the installation, and the impedances of the circuits concerned to be co-ordinated, so that during an earth fault, the voltage appearing between simultaneously exposed-and extraneous-conductive-parts occurring anywhere in the installation shall be of such magnitude and duration as not to cause danger.

This basic rule co-ordinates two subjects: equipotential bonding and automatic disconnection. They are more easily understood if considered separately.

Equipotential bonding

To minimise the magnitude of the voltage that can appear between exposed-and extraneous-conductive-parts, BS 7671 requires that main equipotential bonding conductors are installed between the extraneous-conductive-parts and exposed-conductive-parts and the main earthing terminal of the installation.

The items to be bonded include, water pipes, central heating pipes, gas installation pipes, other service pipes, metallic ducting, exposed metallic parts of the building structure and the lightning protective system. These are illustrated by 1 to 7 in Figure 8.2.

The bond to the gas installation pipe, water pipe, or any other metallic service should be made as near as possible to the position the service enters the building. Where a meter is installed, or there is an insulating insert in the service pipe, the connection has to be made on the consumer's side of the insert or meter, before there are any branches off the pipe. It is recommended that the bond should be within 600 mm of the meter outlet, or the entry point of the service into the building.

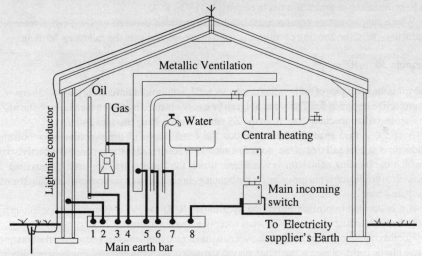

Figure 8.2 - Main equipotential bonding

Where the supply is not PME, the cross-sectional area of the bonding conductors should be half that of the main earthing conductor, the minimum size allowed being 6 mm². The maximum size need not exceed 25 mm² if the bonding conductor is a copper conductor, or a conductor of a different material having an equivalent conductance to the copper conductor.

If the supply is PME, then the PME regulations apply, and the minimum size will be as given by the local electricity supplier, as shown in the following table.

Equivalent copper cross-sectional area of supply neutral conductor	Minimum size of copper main equipotential bonding conductor
Up to 35 mm²	10 mm²
36 mm² to 50 mm²	16 mm²
51 mm² to 95 mm²	25 mm²
96 mm² to 150 mm²	35 mm²
Over 150 mm²	50 mm²

Supplementary bonding
If there are any metallic parts going out of the building into the ground, they should have been included in the main equipotential bonding, isolated parts being bonded to extraneous-conductive-parts connected to the main bonding conductors with the same size of bonding conductor as the main equipotential bonding conductor.

The only area left for supplementary bonding (except for special situations) is where a voltage can appear between an exposed-conductive-part and an extraneous-conductive-part.

The principle is that if the current can be disconnected within the times specified in BS 7671, then there is no need for a supplementary bond.

CHAPTER 8

Where the disconnection time cannot be met, then either local supplementary equipotential bonding has to be installed, or protection has to be provided by an RCD.

Where supplementary equipotential bonding is provided between exposed-and extraneous-conductive-parts, the bonding conductor is sized in accordance with the following formula:

Formula 30 $R = \dfrac{50}{I_a}$

where R is the resistance of the bonding conductor and I_a is the minimum current that will disconnect the protective device in 5 s. The value of I_a can be easily obtained by taking the design value of Z_S for 5 second disconnection time from the ZS tables, and dividing this into 240 V.

A check is then made to determine that the conductance of the supplementary bonding conductor is at least half that of the protective conductor connected to the exposed-conductive-part to which the bonding conductor is connected. If not, the larger conductor must be installed. If mechanical protection is not provided for the bonding conductor, its cross-sectional area shall be not less than 4 mm^2.

If it is decided to provide protection by using an RCD instead of providing supplementary equipotential bonding, then the RCD must comply with the condition: $Z_s \times I_{\Delta n} \leq 50$ V

Rooms containing a fixed bath or shower must have local supplementary equipotential bonding. Where plastic baths, plastic waste pipes, plastic shower trays and no exposed-conductive-parts are present, it will only be necessary to bond all the extraneous-conductive-parts at one point within the bathroom; this point will however have to be accessible for inspection, especially in domestic premises, where the electricity supplier may inspect the installation before connecting the supply.

Where extraneous-conductive-parts are bonded together, the minimum size of conductor, if it is mechanically protected, is 2.5 mm^2; if not, the minimum size is 4 mm^2. If one of the extraneous-conductive-parts is bonded to an exposed-conductive-part, then the conductor must be sized in accordance with the earlier paragraph on supplementary bonding.

Metal framed windows

Metal windows do not need supplementary bonding if the circuits in the vicinity of the windows can be disconnected within the disconnection time specified in BS 7671 for the type of circuit installed. Bonding always needs careful consideration, since it is very easy to introduce a danger where none existed previously.

Installing a supplementary bond onto a metal window frame may create a safe situation inside the equipotential zone, but a possible dangerous one outside the zone. Protection using an RCD would be preferable to installing supplementary bonds onto metal parts that are exposed to the outside of the building, such as the metal framed window .

It must be remembered that the object of bonding is to create an equipotential zone, so that a voltage appearing on the main earth bar will also appear on all exposed- and extraneous- conductive-parts within the zone. This does not mean that a person will not receive an electric shock when main equipotential bonding is carried out.

In Figure 8.3, the voltage $I_f(R_2 + R_3)$ is the voltage appearing on the exposed- conductive-part with respect to the earthed point at the source of supply when a fault occurs. I_f is the fault current, R_2 the resistance of the cpc to the main earth bar, and R_3 is the resistance of the cpc from the earth bar back to the source of supply. It can also be seen from Figure 8.3 that even when the extraneous-conductive-parts are bonded to the main earth bar, a voltage will appear between an exposed-conductive -part and an extraneous-conductive-part when a fault occurs. The bond has only reduced the magnitude of the voltage.

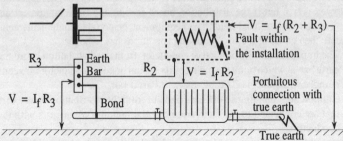

Figure 8.3 - Shock voltage with bonding

If a metal framed patio door is electrically bonded to the main earth bar, a voltage on the main earth bar will also appear on the frame of the patio door. A person touching the frame of the door, would then be touching that voltage. The magnitude of the current flowing through the body would depend upon the type of floor, i.e., its insulation value, and the type of footwear worn by the individual. The resistance of the ground outside the window would be zero; a window cleaner, for instance, would therefore be subjected to a higher touch voltage.

Automatic disconnection

Since a voltage will exist between exposed-and extraneous-conductive-parts when a fault occurs, it is essential to remove this voltage as quickly as possible; this is the object of automatic disconnection.

The actual disconnection time is dependent upon the environmental conditions, and whether a person or livestock is likely to be in contact with exposed- conductive -parts at the instant of the fault. The basic formula used to ensure that a protective device disconnects in a specified time is:

Formula 31 $\quad Z_s = \dfrac{U_{oc}}{I_a}$

where U_{oc} is the open circuit voltage and I_a is the current for the specified disconnection time.

Where a circuit supplies a socket-outlet, hand-held Class I equipment, or portable equipment intended for manual movement, then the disconnection times allowed are:

Voltage to earth U_o	Disconnection time 't' seconds
120	0.8
230	0.4
277	0.4
400	0.2
Over 400	0.1

The disconnection time of 0.4 s is also applicable for voltages within the range of 207 V to 253 V. For any other intermediate voltage the next higher voltage in the table is used.

These disconnection times are only suitable for normal environmental conditions. The disconnection times allowed are therefore modified for areas having poor environmental conditions, such as bathrooms and construction sites.

Where a circuit feeds fixed equipment, or when reduced low voltage is used, the disconnection time allowed is 5 s, but if the distribution board feeding the fixed equipment also has circuits supplying socket-outlets, hand-held Class I equipment, or portable equipment intended for manual movement, then either of the two following conditions has to be met.

1. The impedance of the protective conductor from the distribution board back to the main earth bar, at which point the main equipotential bonding conductors are connected, should comply with table 41C (Table PCZ 10 Part 3) or,

2. equipotential bonding shall be carried out from the distribution board to all the services that are connected to the main equipotential bonding conductors with the same size of conductors.

As far as condition '1' is concerned, the impedance of the cpc can be found for any size and type of protective device by using the following formula:

Formula 32 $\quad Z_{cpc} = \dfrac{50 \times Z_s}{U_{oc}} \ \Omega$

where Z_s is the earth loop impedance for 5 s disconnection and U_{oc} is the open circuit voltage to earth. The value of Z_S can be obtained from the ZS tables in Part 3, when U_{oc} is 240 V.

BS 7671 states that the figures given in the tables are for the 'gG' type fuses; for the type 'gM' or other types of fuses, the manufacturer should be consulted.

Calculations

It should be noted that although the declared nominal voltage is now 230 V, the electricity suppliers are not changing the voltage of their supplies. Therefore, when carrying out calculations the actual voltage present should be used and not the theoretical voltage of 230 V.

There is a close relationship between automatic disconnection and the sizing of protective conductors, since the same conductors are involved, i.e. the phase earth loop impedance Z_s. This chapter should therefore be read in conjunction with the chapter on circuit protective conductors.

The calculation of phase earth loop impedance is carried out in the same way as that for sizing the circuit protective conductors.

Formula 33 $\quad\quad Z_s = Z_e + Z_{inst}$

where Z_e is considered to be the earth loop impedance external to the circuit under consideration, and Z_{inst} is the impedance of the phase and cpc conductors of the circuit.

Since reactance only needs to be taken into account when the conductor size exceeds 35 mm^2, impedance in the above formula can be replaced by resistance for small cables.

BS 7671 now state that the operating temperature of the conductors can be used in calculations, but this is only applicable when the characteristics in Appendix 3 and the Z_s values given in BS 7671 are used. Where there are no values of Z_s in BS 7671 for the protective device used or manufacturer's information is used, calculations have to be carried out at the design temperature. This could result in a messy design, therefore, the values of resistance or impedance used should be based on the design temperature of the circuit conductors. The value of Z_s will also have to be worked out for the temperature at which the installation will be tested.

In practice it is better to work out the size of the protective conductor and Z_s. Then check with the tables that Z_s complies with the value required for the protective device being used, and the disconnection time required for the circuit. In certain cases, the size of the cpc is determined by the thermal constraints on the cpc, and not by the limits laid down in the tables for shock protection.

This accounts for the difference between the PCZ tables and the values for 5 s in the ZS tables.

Adjusting the value of Z_s

The values of Z_s given in BS 7671 are for an open circuit voltage U_{oc} of 240 V, i.e., a nominal voltage U_0 of 230 V. Where the installation is supplied at 240 V, i.e., its nominal voltage is 240 V, the open circuit voltage U_{oc} will be 250 V. An adjustment can be made to the Z_s values given in the tables for the different open circuit voltage U_0. The Z_s values can also be adjusted for any other open circuit voltage U_0 as follows:

Formula 34   Revised Z_s = Z_s from tables $\times \dfrac{\text{Actual } U_{oc}}{240 \text{ V}}$

Temperature effect on Z_s

It must be remembered that automatic disconnection is concerned with phase earth faults. The fault current will therefore raise the temperature of the phase and protective conductors. The calculated value of Z_s must, therefore, take into consideration the resistance of the conductors at the temperature created by the fault current in those conductors . Otherwise, the increase in conductor resistance due to the fault current will reduce the fault current flowing, thus increasing the disconnection time for the circuit. This could be more than that allowed by BS 7671 as shown in Figure 8.4.

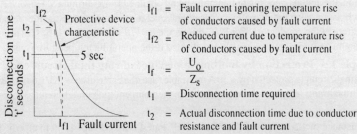

I_{f1} = Fault current ignoring temperature rise of conductors caused by fault current

I_{f2} = Reduced current due to temperature rise of conductors caused by fault current

I_f = $\dfrac{U_0}{Z_s}$

t_1 = Disconnection time required

t_2 = Actual disconnection time due to conductor resistance and fault current

Figure 8.4 - Effect of conductor temperature rise on fault current

BS 7671 does not contain the characteristics or the Z_s values for all types of protective devices. For these instances and where manufacturer's characteristics are used, the impedance of the conductors should be calculated on the average temperature attained by the conductor. The average temperature is obtained by taking the average of the assumed initial temperature (or actual operating temperature from Formula 5) and the limit temperature for the conductor insulation. For example: operating temperature of the conductor 60 °C: limit temperature of PVC insulation 160 °C: average temperature is 60 °C + 160 °C divided by 2, giving 110 °C. The impedance of the conductors at 110 °C is then used in calculations. Since BS 7671 does not contain all the information required, it seems logical to use the above method in calculations throughout an installation. See comment under worked examples on the next page.

When conduit, trunking, or steel wired armoured cables are used as the cpc, it is important that the impedance value of the cpc is used, since the reactance has a significant influence on the overall impedance, although the reactance is not affected by temperature.

CHAPTER 8

Socket-outlet final circuits

There is a relationship between the effects of an electric shock on a person, the magnitude of the shock voltage, and the speed at which the current is disconnected. Thus a person can tolerate a relatively high shock current through the body, providing that the circuit is disconnected fast enough; this is the basis of Regulation 413-02-12.

The principle of Regulation 413-02-12 (Table 41C) is to limit the resistance (impedance) of the protective conductor so that the voltage appearing on exposed-conductive-parts under fault conditions is limited to a safe value, or if the voltage is a higher value, the circuit is disconnected faster.

The application of Regulation 413-02-12 is achieved by determining that the resistance (impedance) of the protective conductor, back to the point at which equipotential bonding is carried out, does not exceed the value given in table PCZ 10 for the appropriate protective device, or the value given in Table 41C in BS 7671.

In practice, where the circuit is single-phase and the voltage U_0 is 240 volts, it may be found that voltage drop constraints limit the circuit length more than the impedance of the protective conductor.

Worked examples

There is a general consensus of opinion that it is wrong to just take the resistance of conductors at their maximum permitted operating temperature and that the resistance should always be taken at the average of the actual operating temperature and the limit temperature of the conductor's insulation, the actual operating temperature being obtained from Formula 5. However, a simplistic approach is to take the maximum permitted operating temperature as the actual operating temperature of the conductors. For PVC this would be $(70 + 160) \div 2 = 115\,°C$. The examples are based on this premise, but can easily be adjusted if the operating temperature is used.

Example 1 - Protection by limiting impedance of ring final circuit cpc.

A ring final circuit is protected by a 30 A rewirable fuse, the cable used is 2.5/1.5 twin & CPC, what can the maximum length of the circuit be from the main equipotential bonding?

Working

The resistance of 1.5 mm^2 CPC from table RC1 Part 4 is 16.698 Ω / 1000 m at 115 °C.
From Table PCZ 10 in Part 3, maximum value CPC impedance for 30 A rewirable fuse is 0.43 Ω

$$\text{Maximum circuit length} = \frac{0.43 \times 4}{0.016698} = 103 \text{ m}$$

Example 2 - Protection by limiting impedance of radial final socket-outlet circuit cpc.

A three-phase 415 V radial final socket-outlet circuit is to be protected by a 32 A Type B MCB, and wired in 4 mm^2 PVCSWA&PVC copper cable installed open and clipped direct. What can the maximum length of the circuit be?

Working

From Table PCZ 10, maximum allowed value of impedance for CPC is 0.31 Ω. From RC1 Part 4, the impedance of the phase conductor is 6.362 Ω per 1000 metres at 115 °C. From ELI 1 Part 4, the phase earth loop impedance of the cable is 15.22 Ω per 1000 metres. The impedance of the cable armour will be $15.22 - 6.362 = 8.858\ \Omega$ / 1000 m. = 0.008858 Ω / m. This result is the same as given in ELI 1 for the armour; the calculation was, therefore, unnecessary.

$$\text{Maximum circuit length} = \frac{0.31}{0.008858} = 34.99 \text{ m}$$

From CR9, Factor = 1747. I_n 32 A. The maximum length for voltage drop will therefore be:

Fomula 13   Maximum length for voltage drop = $\dfrac{1747}{32\,A}$ = 54.59 m

Shock protection constraint limits circuit length not the voltage drop.

Example 3 - Protection by limiting phase earth loop impedance.

A 6 mm^2 twin & cpc cable is to be installed to a cooker unit containing a socket-outlet; the protective device is a 30 A rewirable fuse, the circuit length is 28 m, and Z_e up to the distribution board is 0.35 Ω. Will it satisfy the requirements for indirect contact ?

Working
A 6 mm^2 cable has a cpc of 2.5 mm^2 (Table PCA 1 Part 4).
The resistance per metre of 2.5/1.5 twin and cpc = 0.01448 Ω at 115 °C (from ELI 6, Part 4).
Formula 33 $Z_S = Z_e + Z_{inst}$.
$Z_S = 0.35 + 28 \times 0.01448 = 0.755\,\Omega$.
From Table ZS 3 Part 3, for 30 A fuse 0.4 second disconnection Z_S allowed = 1.14 Ω.
Therefore circuit is suitable for shock protection.

Example 4 - Protection by limiting impedance of cpc.
Same problem as example 3.

Working
From PCZ 10, maximum impedance for cpc and 30 A rewirable fuse is 0.43 Ω. From PCA 1, size of cpc is 2.5 mm^2. From RC1 Part 4, resistance of 2.5 mm^2 cable is 0.010226 Ω/m at 115 °C.

Maximum length from equipotential bonding = $\dfrac{0.43}{0.010226}$ = 42.04 m

Maximum length from INST 1 Part 3 for 7 kW load is 45 metres. 28 m therefore satisfactory.

Example 5 - Protection by limiting phase earth loop impedance.
A 2.5 mm^2 circuit is to be installed in 20 mm heavy gauge conduit to fixed equipment from a 20 A, HRC fuse. If it is 30 m long, will it comply with the shock protection regulations if Z_e up to the distribution board is 0.8 Ω ?

Working
The resistance per metre of 2.5 mm^2 cable at 115 °C is 0.010226 Ω from Table RC1 Part 4. The impedance per metre of 20 mm conduit is 0.0047 Ω (Table ZCT 1 Part 4). Now use formula 33.
$Z_s = Z_e + Z_{inst}$, therefore, $Z_S = 0.8 + 30\,(0.010226 + 0.0047) = 1.247\,\Omega$.
From Table ZS1 A, Z_S allowed for 20 A HRC fuse disconnection time 5 s is 3.04 Ω; therefore circuit complies. (Note: resistance of joints in conduit can be ignored.)

Example 6 - Determination of protective device for three-phase circuit
A four core 50 mm^2 PVCSWA&PVC armoured cable is to be installed to a remote building. If the protective device is to be 60 or 63 A, the cable length is 100 m and Z_e at the mains is 0.35 Ω, what protective devices could be used to give protection against indirect contact?

Working
From Table ELI 2 Part 4 the design value of earth loop impedance for the cable is 2.17 Ω / 1000 m.

therefore, $Z_{inst} = 100 \times \dfrac{2.17}{1000} = 0.217\,\Omega$

$Z_s = 0.35 + 0.217 = 0.567\,\Omega$. Z_s for protective device must not exceed this value.

Disconnection time allowed is 5 s.

HRC fuse: from Table ZS 1A Part 3, Z_s allowed for 63 A fuse is 0.86 Ω. Therefore satisfactory.
BS 1361 fuse: from ZS 2, Z_s allowed for 60 A fuse is 0.73 Ω. Therefore satisfactory.
BS 3036 fuse: from ZS 3, Z_s allowed for 60 A fuse is 1.17 Ω. Therefore satisfactory.
Type 1 MCB: from ZS 4, Z_s allowed for 63 A MCB is 0.95 Ω. Therefore satisfactory.
Type B MCB: from Table ZS7, Z_s allowed for 63 A is 0.76 Ω. Therefore satisfactory.

Note: since I_b is less than I_{tab} the temperature at which the resistance of the conductors is taken can be adjusted by using formula 5 given previously and calculating a revised average temperature. It may then be found that other MCBs may be satisfactory.

Example 7- Circuit using MICC cable

A 12 metre length of light duty1.0 mm^2 PVC MICC cable is to supply a 1000 W tungsten halogen light fitting installed outside from a 6 A Type 2 MCB. If Z_e is 0.8 ohm will the circuit be suitable for indirect contact ?

Working

From ELI 5 Part 4, earth loop impedance is 30.7 Ω / 1000 m which is 0.0307 Ω / m
$Z_{inst} = 12\ m \times 0.0307 = 0.3684\,\Omega$.

Formula 33 $Zs = Z_e + Z_{inst} = 0.8 + 0.3684 = 1.168\,\Omega$

From ZS 5 Part 3, Z_s allowed = 5.71
Indirect contact protection is provided.

General

It does not matter whether the cable is MICC, PVCSWA&PVC, or whether trunking and conduit are used, the calculation is always worked out the same way. It saves time if the cpc is sized first, then the value of Z_S obtained for sizing the protective conductor is checked against the tables for protection against indirect contact.

Using the operating temperature given in the current-carrying capacity tables to obtain the average of operating temperature and the conductor's insulation limit temperature will give a lower fault current than will occur if the conductor is not carrying its rated current.

Calculations based on the operating temperature given in the current rating tables, will err on the safe side. The formulae given earlier in Chapter 6 Sizing Conductors can be used to determine the resistance of a conductor, when the operating temperature is less than that given in the current rating tables.

As far as the distribution cables are concerned, rather than trying to work out the actual temperature of the distribution cables when a fault occurs in a final circuit, it is quicker and simpler to assume that all cables rise in temperature to their maximum value. It is then only necessary to consider the lower temperatures of the distribution cables if the circuit does not comply with the values of Z_s allowed by the tables.

The temperature of the final distribution cable will depend upon the load it is carrying and the magnitude of the fault current. The fault current can be added to the load on the distribution cable to determine its temperature.

This is only applicable to the distribution cable up to the final distribution board, see Figure 6.3 on page 241, all others being at their actual operating temperature.

Using the Z_s values from the tables for U_0 240 V (U_0 230 V) will err on the side of safety when the nominal voltage U_0 and the open circuit voltage U_0 is higher than that given in the ZS tables.

Chapter 9

Circuit protective conductors

Whereas the requirements for equipotential bonding and automatic disconnection of the supply are concerned with the phase conductor and circuit protective conductor impedances being low enough to provide protection against indirect contact, Chapter 54 in BS 7671 is concerned with the protective conductor having sufficient thermal capacity to withstand any fault current flowing through it. This is also a requirement of Regulation 8 in the Electricity at Work Regulations 1989.

To provide compliance with the Electricity at Work Regulations it is felt that the resistance of conductors should be taken at the average of their actual operating temperature and the limit temperature of the conductor's insulation. For convenience the maximum permitted operating temperature of the cable, as given in the CR table in Part 3, can be used instead of the actual operating temperature when working out the average temperature.

Options available
BS 7671 gives two options for the determination of the size of the cpc; it can either be calculated or determined from Table 54G.

Phase earth loop
It is important to understand what an earth fault is, and the path the current takes when such a fault occurs. Figure 9.1 shows a TN-S system, starting at the source of energy, with the feeder cables up to the origin, and from the origin to the end of a final circuit. Meters and fuses etc., have been omitted for clarity.

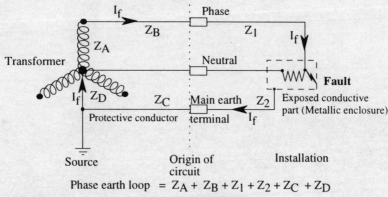

Phase earth loop $= Z_A + Z_B + Z_1 + Z_2 + Z_C + Z_D$

Figure 9.1 - Phase earth loop impedance

The equipment is shown with a phase to earth fault, which for the purpose of calculation is always assumed to be of negligible impedance, i.e., there is no resistance between the phase conductor and the exposed-conductive-part.

At the instant of the fault, current will flow through the phase winding of the transformer, along the phase conductor up to the fault, and then along the protective conductor back to the transformer. This is called the phase earth loop, and the magnitude of the current is only limited by the impedance of the transformer phase winding, the phase conductor, and the impedance of the protective conductor.

These are represented in the diagram by Z_A, Z_B, Z_1, Z_2, Z_C, and Z_D. The phase earth loop impedance Z_s (sometimes just referred to as the earth loop impedance) is obtained from the addition of all the individual impedances in the loop; i.e., the system impedance:

$$Z_S = Z_A + Z_B + Z_1 + Z_2 + Z_C + Z_D$$

In the diagram, impedances Z_A, Z_B, Z_C, and Z_D are all external to the installation and are referred to as Z_e.

Small conductors

If the conductors in the installation do not exceed 35 mm^2 the reactance of the conductors can be ignored, so only the resistance of the conductors is used. For conductors that do not exceed 35 mm^2, the system impedance Z_S is therefore equal to $Z_e + R_1 + R_2$, where R_1 is the resistance of the installation's phase conductor, and R_2 is the resistance of the installation's circuit protective conductor. Where the conductors exceed 35 mm^2, then $Z_S = Z_1 + Z_2$.

The installation impedances Z_1 and Z_2, or the resistance R_1 and R_2, can be referred to as Z_{inst}, representing the phase earth loop within the circuit or installation. The system impedance is then: **Formula 33:** $Z_S = Z_e + Z_{inst}$.

BS 7671 talks of protective conductors, but where the conductor is used for a particular circuit, it is then known as the circuit protective conductor, which is abbreviated to cpc.

Relationship between earth fault and short-circuit

There may appear to be no connection between an earth fault and a short-circuit fault, but a relationship does exist between them. When a fault to earth occurs, the fault current not only flows down the protective conductor, it also flows in the phase conductor. When the phase conductor has been sized to allow for the prospective short-circuit current, it is possible for the earth fault current to be less than the prospective short-circuit current. In these circumstances, the protective conductor has to be sized by using the formula. Additionally, since the protection of the phase conductor is based on the fault current being not less than a certain minimum value, as explained in the Chapter on Fault Currents, a check must therefore be made to determine that the phase conductor is still protected with the smaller earth fault current.

The formula used for calculating that the thermal capacity of the protective conductor is satisfactory is the same as the formula used for checking the thermal capacity of the live conductors. Although at first sight the formula for sizing the protective conductor looks different, it is in fact the same as that used for checking that the live conductors are protected against short-circuit current. The formula has just been re-arranged so that the area of the conductor is the unknown quantity. Consequently, where the disconnection time obtained from the protective device characteristic for a given value of I_f, is 0.1 seconds or less, then the I^2t (not I_f x disconnection time) from the manufacturer's I^2t characteristics for the type of protective device being used must not be less than $k^2 S^2$. In this case the protective conductor can be sized from $I^2 t \le k^2 S^2$

Calculating the size of circuit protective conductor

To use the formula for determining the size of the circuit protective conductor (abbreviated to cpc), the values of I_f, t and k have to be known.

Formula 35 $\quad S \ge \dfrac{\sqrt{I^2 t}}{k}$

Looking at each one of these values in turn, starting with k. The value of k is determined from the initial temperature of the conductor at the start of the fault and the limit temperature of the insulation with which the conductor is in contact. It is important to check that the assumed initial temperature is correct for the protective conductor that is to be used. For instance, if the protective conductor is to be a PVC insulated copper conductor, installed in plastics trunking along with other circuits, then its initial temperature will rise to the working temperature of the other conductors in the trunking. In the absence of more detailed information, it has to be assumed that the working temperature of the other conductors will be the maximum permitted operating temperature of those conductors. The assumed initial temperature of the cpc could be 70 °C, even though it is not carrying current.

In the case of armoured cable, the armour is in the temperature gradient from the core of the cable running at 70 °C, to the ambient temperature at 30 °C; thus its initial temperature is assumed to be 60 °C. The final temperature of the conductor is taken as the limit temperature of the conductor's insulation, which for PVC is 160 °C. Tables are provided in BS 7671 giving the value of k in different situations: they are also given in Table K1A and K1B in Part 4.

The fault current 'I' in phase to earth faults is called I_f, and to determine I_f the phase to earth voltage is divided by Z_S.

Formula 36 $I_f = \dfrac{U_{oc}}{Z_s}$

The difficulty here is that Z_S includes the impedance of the cpc which, of course, in the majority of cases has not yet been sized. This problem does not occur on the very small installations, such as house wiring, since twin and cpc cables are used and the electricity supplier will declare Z_e at the origin of the installation. Using twin and cpc cables means that the size of the cpc is known and since the conductors will not exceed 35 mm^2, all that is required is the resistance of the phase and circuit protective conductors. This resistance should be at the average of the actual operating temperature of the conductor and the limit temperature for the conductor's insulation.

Determination of 't'

The value of 't' in the formula is the time it takes the protective device to disconnect a circuit with a given earth fault current flowing, as illustrated in Figure 9.2. A vertical line is drawn upwards from the current axis on the protective device characteristic for the value of I_f, and where it touches the characteristic for the protective device being used, a horizontal line is then drawn from this point to the time axis. The disconnection time is read off the time axis at the point where the horizontal line crosses it. This is the value used in the formula. Of course, if the time is 0.1 seconds or less, then the I^2t in the formula becomes the value from the manufacturer's I^2t characteristics.

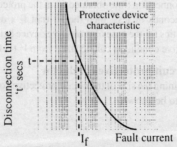

Figure 9.2 - Finding disconnection time 't'

Determination of 'k'

The value of 'k' is obtained either from table K1A or K1B in Part 4, or from the tables in Chapter 54 in BS 7671. Select the initial temperature of the conductor at the start of the fault for the type of conductor insulation. Then select the value of ' k' from the appropriate column, depending upon whether the protective conductor is bunched with live conductors, installed separately, is a sheath or armour of a cable, or is conduit or trunking.

Resultant cable size 'S'

The value of 'S' obtained will rarely be a standard cable size, so the next larger cable has to be used.

Where one cpc is used for several circuits, the values of I_f and t have to be worked out for each circuit, and the values that will give the largest size S are then used.

Impedance tables

Part 4 of the handbook contains tables giving the resistance, reactance, and impedance of cables up to 1000 mm^2. The tables give this information at the design temperature for calculations.

The design temperature is based on the average of the permitted maximum operating temperature of the conductor and the limit temperature for the conductor's insulation. Information is also given for an ambient temperature at 20 °C and for the permitted maximum operating temperature of the conductor.

The tables cover aluminium conductors with either PVC or XLPE insulation (RA1 and 2), copper conductors with PVC insulation (RC1 and 2), and copper conductors with XLPE insulation (XLPE 1). It is therefore unnecessary to have to calculate the impedance of the conductors at the design temperature, since this has already been done in the tables. Figures are also given at 20 °C to enable the maximum fault current to be determined, when sizing switchgear etc.

Calculations for larger installations

For the larger installations, the exact size of the protective conductor required will not initially be known. The distribution cables should be sized for the load, protective device rating, de-rating factors and voltage drop as explained in previous chapters. This will determine the cross-sectional area of the protective conductor. The suitability of the protective conductor can then be checked by selecting the appropriate information from Part 4.

The earth loop impedance for armoured and MICC cables can then be obtained from the earth loop impedance tables ELI 1 to 5 in Part 4 . For twin and cpc cables the resistance for each conductor can be obtained from ELI 6. For cables enclosed in conduit or trunking the resistance / impedance of the phase conductor can be obtained from RC 1 and the impedance of conduit or trunking from ZCT 1.

Where the circuit conductor exceeds 35 mm^2 and is close to the substation in a factory , or other type of premises, the impedance of the supply and of the conductors has to be used. For example, using the characters in figure 9.1.

Formula 37 $\quad Z_s = \sqrt{(R_A + R_B + R_C + R_D)^2 + (X_A + X_B + X_C + X_D)^2}$

where R is the resistance of each conductor in the phase earth loop, and X is the reactance.

Table 54G

Calculating the size of cpc can be avoided by using table 54G. It cannot, however, be used for twin and cpc cables larger than 1 mm^2, since the cpcs in the larger cables are smaller than the phase conductor. In any event the earth loop impedance is needed for protection against indirect contact.

Where the phase conductor exceeds 35 mm², the table will invariably give a non standard size for the cpc, in which case the next largest standard size conductor must be used.

The size of the cpc given in the second column of the table is based on that conductor being made from the same material as the phase conductor. Where the cpc is made from a different material from that of the phase conductor, its size is determined by the ratio of the k factor for the live conductor divided by the k factor for the protective conductor material, as detailed in the third column.

MINIMUM SIZE OF PROTECTIVE CONDUCTORS		
Area of phase conductor S in mm² S	Area of protective conductor in mm² being the same material as the phase conductor	Area of protective conductor in mm² for material different to phase conductor
S not exceeding 16	S	$S \times \dfrac{k_1}{k_2}$
S from 16 to 35	16	$16 \times \dfrac{k_1}{k_2}$
S exceeding 35	$\dfrac{S}{2}$	$\dfrac{S}{2} \times \dfrac{k_1}{k_2}$

The factor k_1 is the factor for live conductors, applicable to the type of conductor and insulation being used. The factor k_2 is the factor for the protective conductor based on the conductor material, conductor insulation, and type of conductor: i.e., whether it is enclosed with live conductors, installed separately, is the armour or sheath of a cable, or conduit and trunking.

Values of k1 / k2 are given in Table M54G in Part 4 for aluminium, steel and lead, based on the worst values for k_2. Table TCCA 1 shows which armoured cables comply with Table 54G.

Using table 54G will in most cases be uneconomic, but it can be used to advantage in certain cases, such as:

1. Aluminium cables with an aluminium strip armour, since the armour has an area, in most cases, at least equal to the area of the phase conductors.
2. MICC multicore cable, because the sheath has an area larger than the phase conductor.
3. Steel conduit, since the cross-sectional area of the conduit is so large relative to the size of conductor that can be installed through it.

One protective conductor can be used for several circuits; its size is then determined by the largest conductor of the circuits to be protected. Thus a number of circuits with conductors of 2.5 mm², 4 mm², and 10 mm² can be protected by one cpc; in this case the cpc size is based on the largest phase conductor, which is 10 mm².

What can be used as a protective conductor

The most common items that are used as protective conductors are: the bare or insulated conductors contained in a cable, the metal sheath, screen or armour of a cable, and steel enclosures such as conduit or trunking.

The protective conductor can be a single core cable installed on its own, or installed along with live conductors, either in plastics or steel conduit or trunking. It can be the metal enclosure of distribution equipment, or busbar trunking, providing it complies with the following paragraph.

Protection of conductors

Where the metal enclosure of switchgear, distribution boards, controlgear, or busbar trunking is used as a protective conductor, it is essential that its electrical continuity is assured, and that its cross-sectional area is suitable for use as a protective conductor.

Protective conductors must be protected against mechanical damage, and against chemical and electrodynamic deterioration. With the exception of ducting, conduit, trunking, or underground cables, all joints must be accessible.

Sizing CPCs by using the PCZ tables from Part 3

Where the cpc size is unknown, it can be found by using the following formula and the PCZ tables:

Formula 38 $R_2 = Z_s - (Z_e + R_1)$ where Z_s is from the appropriate PCZ table

Impedance values Z_1 and Z_2 can be used in place of R_1 and R_2.

Determination of Z_S in a large installation

The value Z_e is referred to as the external phase earth loop impedance. This is satisfactory for a very small installation, such as a house, but is inconvenient when considering large installations since there can be many distribution boards between the origin of the installation and the circuit for which calculations are required.

It is therefore more convenient to consider Z_e as being the earth loop impedance external to the circuit under consideration. This makes calculations easier, since only three components are involved in the calculation, i.e., Z_e, Z_s and Z_{inst}. Knowing any two of these items will enable the third to be calculated. A convenient way of remembering the above is given in Figure 9.3

The three calculations resulting from the diagram are:

Formula 33	Z_s	$=$	Z_e	$+$ Z_{inst}
Formula 39	Z_e	$=$	Z_s	$-$ Z_{inst}
Formula 40	Z_{inst}	$=$	Z_s	$-$ Z_e

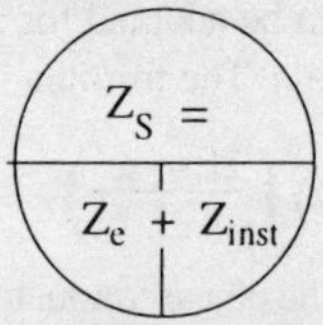

$$Z_S =$$
$$Z_e + Z_{inst}$$

Figure 9.3 - Calculation diagram

Two values of Z_S

When carrying out calculations, it must be remembered that there are two values of Z_s, one that is the maximum allowed for the circuit, (as given in the ZS tables Part 3) whilst the other value is the actual calculated Z_s of the circuit.

When using the notation from Figure 9.3 above to carry out a calculation for a circuit from a sub-distribution board, the Z_e for the circuit will be the Z_S at the busbars of the distribution board, as illustrated in Figure 9.4.

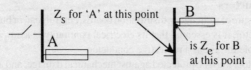

Figure 9.4 - Illustrating how Z_s becomes Z_e

When using the PCZ tables from Part 3 for a small installation, the maximum length of the circuit for the cpc can be found by using the above formula. As Z_S is obtained from the tables for the type and size of protective device and size of protective conductor, the value of Z_e will be known, so Z_{inst} can be found. The maximum length of the circuit as far as the size of protective conductor is concerned, will be;

Formula 41   Maximum length of circuit $= \dfrac{Z_{inst}}{R_1 + R_2 \text{ per metre at } t_1}$

where the temperature 't_1' is the average of the actual conductor operating temperature and the limit temperature for the conductor's insulation. If the maximum permitted operating temperature of the conductor is used for the actual operating temperature, which for 70 °C PVC is 115 °C, R_1 is the resistance of the phase conductor per metre, and R_2 is the resistance of the cpc per metre.

The maximum length obtained meets only the requirement that the thermal capacity of the protective conductor is satisfactory. Protection against indirect contact, overload and short-circuit will have to be complied with, and calculations will still be needed for voltage drop.

Ring socket-outlet final circuits

If a phase to cpc fault occurs part way along a ring circuit the fault current will be higher in the shortest leg of the ring. In the chapters on short-circuit currents and examples, details are given of two methods of determining the fault current for part way along two conductors in parallel. One of the methods uses the total length of the ring and works out the resistance in each leg. The other method uses a formula, in which the symbols can be adapted for a phase to cpc fault in a ring circuit by modifying the symbols used in the formula. The formula would then become:

Fornula 42 $Z_s = Z_e + (R_{ph} + R_{cpc}) \left(\dfrac{2Lx - x^2}{2L} \right)$

where: R_{ph} is the resistance per metre of the phase conductor, R_{cpc} is the resistance per metre of the cpc, L is the length from the fuseboard to the mid point of the ring and x is distance from the fuseboard to the fault.

The total fault current is then determined by:

Formula 36 $I_f = \dfrac{U_{oc}}{Z_s}$

The fault current will divide as the ratio of the conductor lengths since each leg is in parallel, therefore:

Formula 43   The fault current in the longer leg $= I_f \left(\dfrac{x}{2L} \right)$

Formula 44   The fault current in the shorter leg $= I_f \left(\dfrac{2L - x}{2L} \right)$

Having determined the fault current in the shorter leg, the cpc can now be checked to ensure it has sufficient thermal capability, i.e., withstand capacity by using the formula:

Formula 35 $\quad S \geq \dfrac{\sqrt{I^2 t}}{k}$

Residual current devices

BS 4293 clause 8.7 requires the RCD to trip within 0.04 s at five times the rated tripping current of the RCD, i.e., $5I_{\Delta n}$. The minimum fault current must therefore be $5I_{\Delta n}$ to comply with BS 4293.

First check that Regulation 413-02-16 is complied with by **Formula 45:** $\quad Z_s \times I_{\Delta n} \leq 50$ V. Now check the thermal capacity of the protective conductor using:

Formula 47 $\quad I^2 t \leq k^2 S^2$. This formula can also be used to find the maximum tripping time

Formula 46. $\quad t = \dfrac{k^2 S^2}{I^2}$

Conduit and trunking

The problem all engineers have when designing an installation using conduit or trunking as a protective conductor, is what values should they use in the calculations.

Table ZCT 1 in Part 4 has been provided to assist with this problem. Similarly tables ELI 1, 2, 3, 4 and 5 have been provided giving the phase earth loop impedance of armoured cables and MICC cables. Both light duty and heavy duty MICC cables are included in ELI 5. Twin and cpc cables and single core PVC cables enclosed in conduit are given in ELI 6.

As far as conduit is concerned its reactance is quite high relative to its resistance, so that during the short time the fault current is flowing, there will be little, if no increase, in the temperature of the conduit. By calculation it can be shown that if a 20 mm heavy gauge conduit has its temperature increased from 20 °C to 115 °C by a fault current the increase in impedance is only 0.00036 Ω/m. The increase is only some 0.4 Ω per 1000 metres. Conduit when installed forms a network, so that the fault current can divide and take several routes back to the origin of the circuit. The only conduit that will take the full fault current is the length that feeds the actual piece of equipment in which the fault develops, or where a single run of conduit is installed from a distribution board to a piece of equipment.

What designers are concerned with is the thermal capacity of the protective conductor. As far as conduit is concerned, its thermal capacity is greater than the PVC insulated copper cables that can be installed through it in accordance with BS7671. Bearing this in mind, the values given in ZCT 1 can be applied without alteration due to temperature rise by a fault current.

Table ZCT 1 gives the impedance of conduit for fault currents up to and over 100 A, the impedances being higher for fault currents up to 100 A. It is suggested that in the initial calculation the values up to 100 A be used, it is only then necessary to use the values for fault currents over 100 A if this will give a sufficient reduction in earth loop impedance that will enable the circuit to comply with the indirect contact requirements for shock protection, when the fault current exceeds 100 A.

Armoured cables

As far as the armour of cables is concerned, the same comments as those made for conduit apply. In this case, the armour does not have the same relative cross-sectional area, so the effects of reactance are not as dramatic as with conduit. However, it should still be taken into account when carrying out calculations. Table TCCA 1 in Part 4 can be used for checking whether the armour complies with Table 54G, but this will be more restrictive than can be obtained by calculation.

Chapter 10

Circuit protective conductors
examples of calculations

In the following examples, the values of resistance at the design temperature have been taken from the tables in Part 4. The method of calculation will be the same even if the resistance of the conductors is taken at a different temperature to that used in the examples. Additionally, the declared nominal voltage is now 230 V, however, the installation should be designed to the voltage at which it is supplied. Using the values of Z_s given in the tables will err on the side of safety for $U_o = 250$ V.

Example 1 - Conduit as cpc by calculation.

A three-phase circuit feeds a 4 HP 400 V motor and is wired in 1 mm² PVC single core cable enclosed in 20 mm light gauge conduit; the length of the circuit is 35.84 metres, and the protective device in the distribution board is a 20 A BS 88 HRC fuse. The value of Z_s up to the distribution board is 0.86 Ω. Is the conduit satisfactory as a cpc? Supply voltage 230 / 400 V.

Working.

From Table RC 1 Part 4: resistance of one 1 mm² cable at 115 °C = 0.024978 Ω / m.
From Table ZCT 1 Part 4: impedance of light gauge conduit = 0.0054 Ω / m (without joints).
The first task is to determine Z_s for the circuit.
Z_{inst} = 35.84 (0.024978+ 0.0054) = 1.0887 Ω.

Formula 33 $Z_s = Z_e + Z_{inst} = 0.86 + 1.0887 = 1.9487$ Ω.

Formula 36 $I_f = \dfrac{U_{oc}}{Z_s} = \dfrac{240 \text{ V}}{1.9487} = 123.159$ A

The disconnection time for the fault current I_f of (say) 125 A is now obtained from the 20 A HRC fuse characteristic as shown in Figure 10.1.

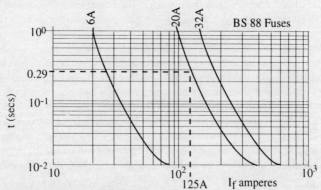

Figure 10.1 - Example 1

From the characteristic, when $I_f = 125$ A, t = 0.29 s. It is now necessary to determine the value of k from Table K1A (Part 4). The conduit is in the temperature gradient between the conductors at 70 °C, and the ambient temperature of 30 °C. This means that its initial temperature will be

50 °C. It is however in contact with the PVC insulation of the cables, whose temperature must not exceed 160 °C i.e. the limit temperature is 160 °C. The value of k therefore is 47.

All the necessary information is now available for the calculation:

Formula 35 $\qquad S \geq \dfrac{\sqrt{I^2 t}}{k} = \dfrac{\sqrt{125^2 \times 0.29}}{47} = 1.432$ mm^2

From PCA 1 Part 4: The cross-sectional area of the conduit is 59 mm^2: therefore it is satisfactory as a circuit protective conductor being some 40 times larger than required.

Example 2 - Conduit as a cpc using Table 54G.

The same problem as example 1, but conduit sized using Table 54G.

Working

From Table K1A Part 4: k for copper for PVC live conductor is 115.

k for conduit for same type of cable is 47.

Size of phase conductor is 1 mm^2. From 54G: cross-sectional area of cpc must be:

$$S \times \frac{k_1}{k_2} = 1 \text{ mm}^2 \times \frac{115}{47} = 2.447 \text{ mm}^2. \quad \text{Conduit therefore suitable.}$$

Looking up the different k values can be avoided by using Table M54G from PCA 1 in Part 4.

Example 3 - Armoured cable using Table M54G, Part 4.

Same problem as example 1, but supply to motor is now a 1.5 mm^2 PVC SWA&PVC steel wired armoured cable.

From Table M54G in PCA 1 Part 4, cross-sectional area required for steel armour is 2.25 times area of phase conductor; therefore area of armour = 1.5 mm$^2 \times 2.25$ = 3.375 mm^2.

From Table ELI 1 Part 4, area of 3 x 1.5 mm^2 PVCSWA&PVC cable armour = 16 mm^2, so it is satisfactory. Note : Smallest armoured cable available is 1.5 mm^2.

Example 4 - Twin and cpc cable by calculation.

A cooker circuit is to be installed from a 30 A rewirable fuse in a consumer unit using 6 mm^2 twin and cpc cable to a cooker control containing a socket-outlet, the length of run being 28 metres, if Z_e to the consumer unit is 0.35 Ω and U_o is 240 V, check that the cpc will be suitable as a circuit protective conductor. (Since U_o = 240 V, U_{oc} = 250 V.)

Working (using formula)

From Table PCA 1 Part 4: size of cpc for 6 mm^2 twin and cpc cable is 2.5 mm^2.

From table ELI 6 Part 4: resistance at 115 °C for 6 mm^2 twin and cpc = 14.48 Ω per 1000 metre.

Z_{inst} = 28 × 0.01448 = 0.405 Ω.

$Z_S = Z_e + Z_{inst}$ = 0.35 + 0.405 = 0.755 Ω.

Formula 36 $\qquad I_f = \dfrac{U_o}{Z_s} = \dfrac{250 \text{ V}}{0.755} = 331.13$ A

The disconnection time for a fault current of (say) 331 A has to be found from the 30 A rewirable fuse characteristic, as shown in Figure 10.2.

From the characteristic, when I_f is 331 A, t = 0.175 s.

The cpc is enclosed in the cable: therefore its initial temperature will be the same as the live conductors i.e. 70 °C, and because the cable is PVC, the final temperature must not exceed 160 °C.

CHAPTER 10

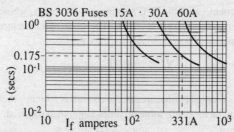

Figure 10.2 - Example 4

The value of k from table K1A Part 4 is 115. Substituting this in the formula:

Formula 35 $\quad S \geq \dfrac{\sqrt{I^2 t}}{k} \quad$ therefore, $\quad S = \dfrac{\sqrt{331^2 \times 0.175}}{115} = 1.2 \text{ mm}^2$

The area of the cpc is therefore satisfactory.
Note: Table 54G cannot be used since the cpc is smaller than the phase conductor.

Example 5 - Using PCZ tables from Part 3.

Same problem as example 4, but using PCZ tables from Handbook for checking cpc.

Working

Z_S for the circuit would be calculated as in the previous working, giving a total Z_S of 0.755 Ω.
The maximum Z_S allowed for the cpc is obtained from the PCZ table for rewirable fuses, by selecting the value for the fuse size and cpc size.
From Table PCZ 3, the value of Z_S allowed for 2.5 mm^2 cpc and 30 A fuse is 2.76 Ω, so the CPC is satisfactory. From Table ZS 3 Part 3 the Z_S allowed is 1.14 Ω for a disconnection time of 0.4 s, so the circuit also complies with shock protection requirements even though the figures used from the tables are for a nominal voltage of 230 V. Correcting for U_{OC} being 250 V will increase the disconnection time allowed, so using the tables for a U_O of 240 V errs on the side of safety.

Formula 34 $\quad$ Revised $Z_s = 1.14 \times \dfrac{250}{240} = 1.187 \, \Omega$

Example 6 - Earth fault in a ring circuit

A ring circuit is wired in 2.5/1.5 twin and cpc PVC cable and is 100 m long. A fault develops 10 m from the distribution board between the phase conductor and the cpc. If Z_e is 0.8 ohms, is the thermal capacity of the cpc satisfactory? The nominal supply voltage is 240 V and the fuse is a 30 A rewirable fuse. (U_o is 240 V therefore U_{oc} is 250 V.)

Working

Resistance per metre of 2.5 / 1.5 cable from ELI 6 is 0.026924 Ω / m.
The length to the mid point will be half the ring length which is 50 m.

Formula 42 $\quad Z_s = Z_e + (R_{ph} + R_{cpc}) \left(\dfrac{2Lx - x^2}{2L} \right)$

therefore, $\quad Z_s = 0.8 + (0.026924) \left(\dfrac{2 \times 50 \times 10 - 10^2}{2 \times 50} \right) = 1.042 \, \Omega$

The total fault current will be:

Formula 36 $I_f = \dfrac{U_{oc}}{Z_s} = \dfrac{250 \text{ V}}{1.042 \ \Omega} = 239.92 \text{ A say } 240 \text{ A}$

This fault current will divide as the ratio of the conductor's length.

Formula 43 I_f in the longer leg $= I_f \left(\dfrac{x}{2L}\right) = 240 \text{ A} \times \dfrac{10}{100} = 24 \text{ A}$

Formula 44 I_f in the shorter leg $= I_f \left(\dfrac{2L-x}{2L}\right) = 240 \text{ A} \times \dfrac{100 - 10}{100} = 216 \text{ A}$

The fuse will see the full fault current of 240 A. The disconnection time is 0.35 s from the rewirable fuse characteristic in Part 4. The k factor from K1 A is 115. The minimum cross-sectional area for the cpc in the short leg can now be found.

Formula 35 $S \geq \dfrac{\sqrt{I^2 t}}{k} = \dfrac{\sqrt{216^2 \times 0.35}}{115} = 1.1 \text{ mm}^2$

The area of the cpc is 1.5 mm^2 so it will withstand the fault current.

Example 7 - RCD protected circuit

A 100 mA RCD with a 2 s time delay is to be installed at the mains position to ensure discrimination with RCDs installed in final circuits. The value of Z_{inst} for the installation is 1 Ω, Z_e is 0.8 Ω, Will the RCD be satisfactory if the earthing conductor is 2.5 mm^2 with 70 °C PVC insulation? $U_o = 230$ V, therefore $U_{oc} = 240$ V.

Working

Formula 33 $Z_s = Z_e + Z_{inst} = 0.8 + 1.0 \ \Omega = 1.8 \ \Omega$.

Now check compliance with Regulation 413-02-16 by using Formula 45: $Z_s \times I_{\Delta n} \leq 50$ V.
$Z_s I_{\Delta n} = 1.8 \ \Omega \times 0.1 \text{ A} = 0.18$ V. This is less than the 50 V requirement.
Now check suitability of protective conductor by using Formula 47: $I^2 t \leq k^2 S^2$.
Value of k for earthing conductor not installed with other conductors is 143.

Formula 36 $I_f = \dfrac{U_{oc}}{Z_s} = \dfrac{240 \text{ V}}{1.8 \Omega} = 133.33 \text{ A say } 134 \text{ A}$
$k^2 S^2 = 143^2 \times 2.5^2 = 127,806 \text{ A}^2\text{s}.$
$I^2 t = 134^2 \times 2 \text{ s} = 35,912 \text{ A}^2\text{s}.$

$I^2 t$ is less than $k^2 S^2$ so earthing conductor thermally protected and RCD satisfactory.

General

Where the sheath of multicore MICC cables and the strip aluminium armour of aluminium cables is used as the protective conductor, no calculations will be required to determine that the sheath or armour has the correct thermal capacity. The sheath and armour quite easily comply with table 54G. The earth loop impedance for such cables will, however, still be required, to enable protection against indirect contact to be worked out.

The thermal capacity of the protective conductor and protection against indirect contact must always be checked, since there are instances where protection against indirect contact is provided but the thermal capacity of the protective conductor is insufficient and vice versa.

Chapter 11

Fault current regulations

Section 434 of BS 7671 is only concerned with fault currents flowing in the live conductors belonging to the same circuit. This would comprise a short between two phases, between all three phases on a three-phase circuit, and between all three phases and neutral, or a short between phase and neutral on a single-phase circuit. Where there is a phase to earth fault, Section 434 is concerned with the protection of the phase conductor, and Section 543 with the protection of the protective conductor. For the purposes of calculation, the fault is considered to be a bolted short of negligible impedance on the load side of the device installed.

The fundamental requirements for safety in BS 7671 call for every installation and circuit to be protected against overcurrent. The device used must operate automatically at a safe current, related to the current rating of the circuit, and must have an adequate breaking capacity. Where circuit-breakers are used for protection, they must also have adequate making capacity. The protective device must be located so that there is no danger from overheating, arcing, or scattering of hot particles.

BS 7671 stipulates that a protective device must be provided to break a fault current occurring in the conductors of a circuit before the fault current can cause danger from thermal and mechanical effects. The thermal effect (I^2t) referred to is the heating effect on the conductors created by the fault current I^2. If the fault current is not interrupted, the heating effect can be such as to melt the conductors through which it is flowing. The mechanical effect is the magnetic field set up round the conductor due to the fault current I^2, which exerts a mechanical stress on the conductors. BS 7671 also states that the nominal current rating of the protective device 'may be greater than' the current-carrying capacity of the conductor being protected.

There is another stipulation, that where a protective device is used for protection against fault currents, the time taken by the protective device to clear the fault must not allow the limit temperature of the conductor's insulation to be exceeded.

Cable size independent of protective device rating
Protecting against fault current, is only protecting against the thermal and mechanical effects that can occur with a fault. Consequently, the cable conductors do not need to be sized to the nominal rating of the protective device. The relationship that I_b must not be greater than I_n, which in turn must not be greater than I_z, is related only to overloads, and has nothing to do with short-circuit protection.

This allows the use of cables with a lower current-carrying capacity than the nominal rating of the protective device when feeding a motor circuit. But the circuit still has to be protected against overload in accordance with Sections 433 and 473 by another device, i.e. an overload incorporated in a motor starter. Similarly, circuits unlikely to carry overload current do not need to be sized to the protective device rating, but to the load, the protection provided being against fault current.

Determination of prospective short-circuit current at relevant point
Section 434 calls for the prospective short-circuit current and earth fault current at every relevant point of the installation to be determined. Additionally, (with a few exceptions covered later) Section 473 calls for the installation of a short-circuit protective device where there is a reduction in the current-carrying capacity of the conductors of the installation.

Such reductions are caused by changes in cross-sectional area, method of installation, type of cable, or environmental conditions.

Breaking capacity lower than fault current

Section 434 (434-03-01) calls for the breaking capacity of the protective device to be not less than the prospective fault current at the point at which it is installed. A protective device that has a breaking capacity lower than the prospective fault current available at the point it is installed, can be used, provided that:

a) there is a protective device on the supply side of the lower rated protective device which has the necessary breaking capacity; and

b) the characteristics of both protective devices are co-ordinated, so that the energy let-through of the supply side device will not damage the lower rated device, or the cable conductors protected by both devices, (see examples for details of how co-ordination is achieved).

Protective device allowed at different position than the relevant point

A protective device can be installed at a point other than at the point of change in current-carrying capacity, (473-02-02) provided that:

a) the distance from the point of change in current-carrying capacity does not exceed 3 m, and

b) the conductors are so erected that the risk of short-circuit, earth fault, and the risk of fire or danger to persons is reduced to a minimum.

In general, the conductors will, however, have to comply with the overload requirements for the circuit concerned.

This relaxation allows the conductors which tap off a busbar chamber, feeding a switch fuse mounted on the busbar chamber, to be smaller than the rating of the busbars. However, the conductors between the busbars and the switch fuse must comply with the overload requirements for the circuit under consideration.

The overload requirements allow the device protecting a conductor against overload to be placed along the run of that conductor, provided that the part of the run between the point where the value of the current-carrying capacity is reduced and the position of the protective device has no branch circuits or outlets for the connection of current using equipment; providing there are no special requirements or recommendations or an abnormal risk of fire or explosion.

The conductors in the busbar chamber should be installed carefully so that a short-circuit will not occur between them, or between the conductors and any other conductor or earth. In practice it is better to connect the tails from the switch further along the busbar chamber, rather than directly below the switch, and preferably at the rear of the busbars so that the cables can be set to the back of the chamber away from the other busbars, eliminating the risk of a short-circuit. The actual connection is then in line with the busbar.

The insulation or sheath on the conductors must be suitable for the maximum operating temperature of the busbars. When determining the size of the conductor to install from the busbar to the switch fuse for overload protection purposes, the correction factors should be taken into account.

It must be remembered that the maximum working temperature of the busbar enclosure becomes the ambient temperature for the conductors to the switch fuse. The nominal rating of the fuse in the switch fuse is taken as I_n in the calculation:

Formula 15 I_t not to be less than $\dfrac{I_n}{G \times A \times T \times S}$

There is a further requirement, (473-02-03) which allows the short-circuit protective device to be installed at a point other than at the point where the current-carrying capacity of the conductors change, provided that there is a protective device on the supply side of the change in current-carrying capacity which will protect the conductors against short-circuit or an earth fault.

In this case, a calculation is required to prove that the conductors are indeed protected, but there is no restriction on the length of the conductor installed. However, protection of the conductors has to be confirmed by the Formula 47: $I^2t \leq k^2S^2$.

This arrangement cannot be used in that part of an installation where special requirements or recommendations apply or where there is an abnormal risk of fire or explosion.

A typical example of the use of this requirement is in a main switchroom, where there is spare capacity in the main switchgear, but the busbar chamber is physically full of switch fuses. In such a situation a cable can be tapped off the busbars and installed to a switch fuse mounted on the switchroom wall. A calculation must then be made to ensure that the main incoming protective device to the switchgear will protect the cable against short-circuit current.

Omission of protective devices

The omission of a device for protecting against a fault current is permitted where conductors connect a generator, transformer, rectifier or battery to a control panel, when a protective device is installed in the panel (473-02-04). One such example of this relaxation, is when the conductors from a transformer are installed to a main l.v. board which incorporates protective devices in the l.v.board.

In this case it is important to remember that the cables are only protected by the protective device on the H.V. side of the transformer; it is for this reason that BS 7671 specifies that the risk of fault current, fire, and danger to persons must be a minimum. This is achieved by installing the transformer tails in smooth ducts between the transformer and the main l.v. switchboard; open type cable ducts are usually provided with covers.

The other occasions where a protective device can be omitted, are at the origin of an installation where the electricity supplier agrees that their protective device protects the tails to the installation's main switch, or where the circuit is a measuring circuit or where the sudden opening of the circuit could cause a greater danger than the fault current.

Calculations redundant

One of the most important requirements in Section 434 is 434-03-02, which states that where a protective device is used for both overload and fault-current protection (as in the case, say of a 5 A fuse protecting a lighting circuit), the conductors on the load side of the protective device will have been sized to the nominal current rating of the protective device in accordance with Section 433. If the protective device has a breaking capacity not less than the prospective fault current available at the point at which it is installed it can be assumed that the conductors on the load side of the device are protected against fault current. However, where conductors are installed in parallel, or the protection is by a non-current limiting type of circuit-breaker, the circuit has to be checked by calculation.

The non-current limiting type of circuit-breaker to which BS 7671 is referring is the zero point circuit-breaker. These circuit-breakers, depending upon the circuit conditions, may not protect the cables. This is because they let through all the energy for at least the first half cycle, and can let it through for several cycles as illustrated in Figure 11.1.

If a circuit-breaker does not have a characteristic similar to those shown in Part 4, it must be assumed to be a non-current limiting type. This means that a calculation is required to ensure the conductors are protected against fault current.

In the majority of cases Regulation 434-03-02 will apply, and it will generally be necessary only to carry out calculations by using the formula of Regulation 434-03-03, where the cable conductors are not sized to the nominal rating of the protective device, as in the case of cables feeding a motor, or where Regulation 473-02-03 is used.

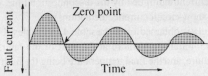

Shaded area is the energy let through protective device

Zero point

Fault current

Time ——→

Figure 11. 1 - Energy let through a non-current limiting circuit breaker

Calculating disconnection time

Where Regulation 434-03-02 is not applicable then Regulation 434-03-03 has to be used to ensure that the live conductors are protected against short-circuit current. This latter regulation gives a formula for determining whether a fault current occurring at any point of a circuit will be interrupted in a time which will not allow the cable conductors to exceed their limiting temperature. The formula given in 434-03-03 is:

Formula 46 $t = \dfrac{k^2 S^2}{I^2}$

What the formula means

In this form, the formula does not convey any particular meaning, but if this formula is re-arranged it becomes:

Formula 47 $I^2 t = k^2 S^2$

Now energy is equal to watts × time = $I^2 Rt$. If R remains constant, or is minute compared with I, then energy becomes proportional to $I^2 t$, and this information can be inserted into the above formula,

$I^2 t$	=	$k^2 S^2$
Thermal energy		Thermal capacity
let-through the	=	of conductor .
protective device.		

It follows that if the conductor is not going to be damaged, then $I^2 t$ must never exceed $k^2 S^2$. This is important because the heat produced by a fault current is considered to be contained within the conductor core of the cable for fault durations up to five seconds.

The equation is commonly called adiabatic (without loss or gain of heat). It gets this name because the equation is considered to remain adiabatic for fault current durations up to five seconds, i.e., it ignores heat loss. This is not strictly correct, since the conductor will be losing some heat, and will be gaining a considerable amount of heat due to the fault current.

The time 't' given by the formula is the time taken for the fault current to raise the conductor's temperature from the maximum operating temperature to the limit temperature for the conductor's insulation. The time 't' is the maximum time the fault current can be allowed to flow, and is compared with the actual disconnection time taken by the circuit's protective device for the fault current I.

Disconnection time 0.1 second or less

Where the disconnection time of the protective device is equal to or less than 0.1 seconds, then $I^2 t$ should be less than $k^2 S^2$ for the conductor.

The I^2t referred to is the energy let-through of the protective device and is obtained from the manufacturer's characteristics for the protective device protecting the circuit.

It is not the square of the fault current multiplied by the disconnection time. (An example is given in the chapter on examples of short-circuit current calculations).

The value of k used in the formula is obtained from the table 43A (Tables K1A and K1B in Part 4). The values are based on the initial temperature at the start of the fault, this being the conductor's actual operating temperature, and the final temperature being the limit temperature of the conductor's insulation.

The value of 'S' in the formula is the conductor's area in square millimetres.

Application of the fault current requirements

The application of the fault current requirements are illustrated in Figure 11.2. The diagram shows a transformer feeding a main switchboard, which in turn feeds two distribution boards.

Regulation 473-02-04 allows the cables to be installed from the transformer to the switchgear without protective devices being installed at the transformer.

Regulation 473-02-02 allows the tails to switch I_{n2} to have a current-carrying capacity less than the busbars, providing they comply with Section 433 for overloads, and do not exceed 3 m in length. No calculations are required for the cable from I_{n2} to DB1, since it complies with 434-03-02: i.e. its current-carrying capacity is not less than the nominal rating of the fuse in I_{n2}, and the fuse has a breaking capacity greater than the 14 kA at the busbars. (Note BS 88 fuses have a breaking capacity of 80 kA.)

The characteristics of the MCBs I_{n6} to I_{n9} must be co-ordinated with I_{n2}, so that the energy let-through I_{n2} is within the withstand capacity of the MCBs, and that both devices protect the circuit cables L1 to L4, in compliance with 434-03-01.

The other distribution board DB2 is fed through an isolating switch and the cable exceeds 3 m in length, so regulation 473-02-03, along with 434-03-03, has to be used to make certain that the cable feeding the distribution board is protected against short-circuit current by I_{n1}.

Regulation 434-03-02 can again be used for the lighting cable fed from I_{n4}, but a calculation will be needed for the cable to the motor, since its current-carrying capacity is less than the nominal rating of the fuse I_{n5}; so 434-03-03 has to be used.

Regulation 435-01-01 requires that the characteristics of the starter are co-ordinated with its protective device I_{n5}; this means that the withstand capacity of the starter must be suitable for the energy let-through I_{n5}.

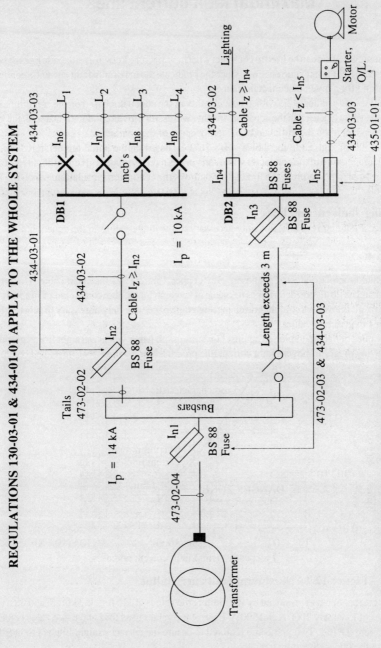

Figure 11.2.- Illustration of where fault current regulations apply

Chapter 12

Maximum fault current lines

The adiabatic equation can be used to construct what have become known as adiabatic lines. A better name for these lines is fault current lines, since they indicate the maximum fault current a conductor can withstand for a given disconnection time.

Such lines only indicate the withstand capability of the conductor based on the conditions used to obtain the value of k used in the equation. In other words, for a given conductor material, the value of k is dependant upon the initial and final temperature of the conductor.

The values of k given in the tables in BS 7671 is based on the initial temperature being the operating temperature of the conductor when carrying its rated current, as given in the current rating tables. The final temperature is the limit temperature of the conductor's insulation, which for 70°C PVC is 160 °C.

Producing fault current lines

To produce a fault current line, it is only necessary to substitute values in the equation:

Formula 46 $\qquad t = \dfrac{k^2 \, S^2}{I^2}$

The size and type of conductor is chosen - say copper 4 mm^2 PVC insulated cable. The type of conductor and its insulation determines the value of k, which for the above cable is 115 (Table K1A). Since the result is always a straight line on log-log graph paper it is only necessary to substitute two values for I to give two values of t.

The values of t are then plotted against I on such graph paper when a straight line is produced. The following diagram shows a fault current line produced for a 4 mm^2 PVC insulated conductor.

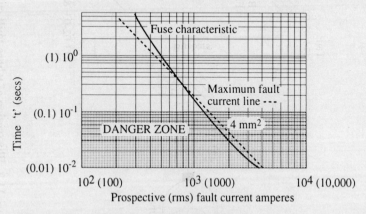

Figure 12.1 - Maximum fault current line

The fault current line was produced by substituting the values of 115 for 'k', 4 for S, and then using two values of I, namely 2000 A and 500 A. Each of the values used for I gave two values of t, viz: 0.0529 s, and 0.846 s. Two points are required to ensure the correct straight line is produced and to confirm the line's direction.

To determine the maximum disconnection time for a fault current, draw a line vertically from the fault current axis for the value of fault current, up to the fault current line; where it touches the fault current line, draw a horizontal line to cut the axis t; this is the maximum time allowed for the circuit to be disconnected.

A family of such fault current lines can be produced for a particular type of cable. In the case above, a group of lines could be produced for 1 mm^2 cable up to the maximum size made for that type of conductor.

The fault current lines are more useful if they are superimposed onto the characteristics of protective devices, since this shows at a glance whether the protective device will effect disconnection (of the faulty circuit) before the fault current reaches the fault current line.

Using clear acetate sheet for a set of lines will enable them to be used with various protective device characteristic, providing the characteristics have been produced on the same size of log-log graph paper. Marking the corners of the log-log paper on the acetate will ensure that the fault current lines are correctly positioned.

Dangers in using fault current lines

There are two hidden dangers when using fault current lines for determining whether a circuit is protected. The first is due to the scale of the characteristics; this makes it impossible to determine the exact position of an odd numbered fault current. The danger occurs when the fault current is very close to the intersection of the fault current line with the protective device characteristic.

The second and most important danger (shown shaded on the previous diagram) is where the disconnection time is equal to, or less than 0.1 s. It is possible for the fault current line to be above the protective device characteristic in the time range of 0 to 0.1 s, visually indicating that the conductor is protected, whereas by checking the I^2t from the manufacturer's characteristics and comparing it with k^2S^2 for the conductor, it can be shown not to be protected. It is for this reason when using a fault current line to check that a conductor is protected against short-circuit current, or an earth fault current, that care has to be taken to check that the disconnection time is not less than 0.1 s; if it is, then k^2S^2 must be compared with the I^2t let- through of the protective device, as given in the manufacturer's literature. (See last page in Part 4.)

General practice

In practice it is better to ignore using fault current lines and use the adiabatic equation to determine the maximum time allowed for a protective device to disconnect the circuit for the calculated fault current. It is then only necessary to read from the protective device characteristic the actual disconnection time for the calculated fault current and compare this value of t with the value of t obtained from the adiabatic equation. If the value of t from the characteristic is less than the value of t calculated using the adiabatic equation the conductors are protected against the calculated fault current.

Where the disconnection time from the characteristic is 0.1 s or less then the I^2t energy let through of the protective device given by the manufacturer is compared with $k^2 S^2$ for the conductor, the former must be less than the latter for the conductors to be protected.

Fault current lines are useful in demonstrating that a minimum fault current is required for the conductors to be protected as explained in the next chapter.

Chapter 13

Short-circuit current calculations

Information required before starting

Before commencing any calculations, there are four fundamental questions that have to be answered, as called for in the assessment of general characteristics:

a) the single-phase and three-phase prospective short-circuit current at the origin of the installation (I_{pn} and I_p) which will usually be at the supply company's fuse;

b) the external earth loop impedance (Z_e) ;

c) the type of protective device installed at the origin (usually the supply company's fuse); and

d) the type of system to be provided.

The type of supply required, and the system of which the installation will form part, will affect the approach to the calculations. If a three-phase four-wire supply is required, then the I_p given by the supply company will be the three-phase prospective short-circuit current.

The Z_e quoted will only give the impedance of the phase and neutral conductors in the TN-C type system, because the protective conductor is combined with the neutral up to the origin of the installation. If the system is TN-S, then the supply company will have to be asked for the phase-neutral impedance external to the installation, since the neutral conductor is a separate conductor from the protective conductor.

Two calculations required

Two calculations will be needed, one to determine whether the equipment (such as switchgear, starters, control panels and distribution boards) have sufficient withstand capacity. The second calculation is to determine whether the circuit conductors are protected.

Since the open circuit voltage is used for phase to earth faults, it is logical that the open circuit voltage should also be used with short-circuit current calculations.

When calculating symmetrical three-phase short circuit currents, the impedances of the neutral and earth conductors are ignored.

The three-phase symmetrical short-circuit current I_p can be determined by the formula:

Formula 48 $\qquad I_p = \dfrac{U_{loc}}{\sqrt{3} \times Z_p}$

where U_{loc} is the open circuit line voltage, Z_p is the impedance of one phase conductor, and $\sqrt{3}$ is 1.732.

To determine the single-phase short-circuit current on a TN-C type system, the voltage U_o is divided by Z_e , but if the system is TN-S, then the following formula should be used:

Formula 49 $\qquad I_{pn} = \dfrac{U_{oc}}{Z_{pn}}$

where Z_{pn} is the impedance of the phase and neutral conductor, and I_{pn} is the prospective single-phase short-circuit current.

To determine the impedance to the short-circuit current:

The following formula for single-phase will give the impedance of the phase and neutral conductor.

The three-phase formula will give the impedance of one phase conductor only, where I_p is the three-phase sym

Single-phase

Formula 50 $\quad Z_{pn} = \dfrac{U_{oc}}{I_{pn}}$

Three-phase

Formula 50b $\quad Z_p = \dfrac{U_{oc}}{I_p}$

Where the fault is only between two phase conductors then I_{pp} will give the impedance of both phase conductors as indicated in the following formula:

Formula 51 $\quad Z_{pp} = \dfrac{U_{loc}}{I_{pp}}$

The main difference between this formula and the previous formula is that the open circuit line voltage is divided by the fault current instead of the open circuit phase voltage.

In all the above cases the impedance must be the total for the circuit from the source of electrical energy (i.e. the generator) to the end of the final circuit, or to any intermediate point at which the prospective short-circuit current is required to be calculated.

For the single-phase short-circuit this means the impedance of the transformer phase winding, the phase conductor up to the fault and the neutral conductor back to the transformer. For the three-phase short-circuit current only the impedance of the transformer phase winding and the phase conductor are used. For a fault between phases the impedance of both phase windings and both phase conductors are used.

Although impedance is always referred to when discussing short-circuit currents, only the resistance needs to be taken into account if the conductors are 35 mm^2 or less, the effect of reactance on these smaller conductors being considered negligible. In fact, the reactance has a decreasing effect on the prospective short-circuit current as the distance from the origin of the installation increases.

Where short runs of very large conductors are involved, the reactance of the conductor can exceed the resistance by a considerable amount; in such circumstances, the resistance of the conductor can be ignored, only the reactance being used in the calculation.

When calculations are made where the reactance is given separately, or is known, the impedance is obtained from the equation:

Formula 37 $\quad Z = \sqrt{R^2 + X^2}$

When considering whether the equipment will be suitable for the prospective short-circuit current that might flow, there are two considerations that have to be made. The first concerns the actual switchgear or equipment, and the second concerns the conductors of the installation.

As far as the equipment is concerned, the maximum fault current that may flow is required, so that switchgear or equipment with the correct withstand capacity can be selected.

On the other hand, to protect the conductors of the installation, the minimum fault current that will flow is required.

Determination of maximum short-circuit current

To determine the maximum prospective short-circuit current that will flow in a three-phase installation, the three-phase symmetrical short-circuit current should be calculated, using the resistance of the conductors at the temperature at which they could possibly be when a fault occurs. For instance, if a cable was installed between buildings, and it was the depth of winter, then the conductor temperature could be the same as the outside temperature. The actual conductor temperature would depend upon whether the conductor was just being energised, or whether it had been loaded for some time. When in doubt as to the temperature, take the lowest value that could occur. The resistance obtained at the selected temperature is then used in the impedance formula given above.

Determination of minimum short-circuit current

To determine the minimum fault current in a three-phase and neutral installation, the phase to neutral short-circuit current should be calculated. The phase to neutral fault current is calculated since this value will be less than the three-phase symmetrical short circuit current.

Where the supply is three-phase three-wire, the minimum fault current will be with a fault between two phases. In this case the open circuit line voltage is divided by the impedance of two phases: from the generator up to the end of the circuit, or to any other point at which the calculation is being carried out.

Formula 51 $\qquad I_{pp} = \dfrac{U_{loc}}{Z_{pp}} = \qquad$ **Formula 52** $\qquad \dfrac{U_{loc}}{2\,Z_p}$

Where the fault is close to the transformer, it could be found that a fault between phases gives a lower fault current than a single phase fault. Additionally, the further the fault is from the transformer it could be found that a phase to earth fault will be less than a phase to neutral fault.

Reason for minimum short-circuit current being used for conductors

The reason for the minimum prospective short-circuit current being taken for the conductors is more clearly seen if the maximum fault current line is superimposed onto protective device characteristics. Figure 13.1 shows the fault current line superimposed onto a fuse characteristic.

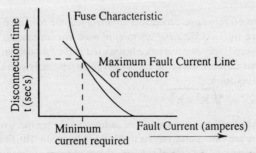

Figure 13.1 - Fault current line and fuse characteristic

Figure 13.1 shows the fault current line crossing the fuse characteristic. The minimum fault current and maximum disconnection time for the conductor is determined by the point at which the fault current line crosses the fuse characteristic.

The point of intersection, therefore, gives the minimum fault current that must flow in the circuit for the conductor to be protected. The fact that a smaller fault current will damage the conductor can be seen by drawing a vertical line to the left of the crossover point. As the line crosses the fault current line before it reaches the fuse characteristic, the current would therefore be allowed to flow too long, and the conductor would be damaged.

If the fault current line is superimposed onto a circuit-breaker characteristic, as in Figure 13. 2, it can be seen that the fault line crosses the characteristic at two places, corresponding to a minimum and a maximum fault current. The fault current in the circuit must therefore lie between these two points if the conductor is to be protected.

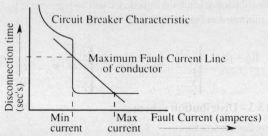

Figure 13.2 - Fault current line and a circuit-breaker characteristic

The fault current line will not cross the protective device characteristic in every case, and therefore as a general rule, if the fault current line is to the right of the characteristic, the conductor is protected; if it is to the left of the characteristic, it is not protected. As previously explained, care has to be taken that the fault current does not exceed the maximum value allowed by the fault current line for circuit breakers. Care has also to be taken that the disconnection time is not 0.1 s or less for the fault current-being considered, because if it is, then the energy let-through characteristics have to be used in the calculations.

Conductor temperature to be used in calculations

To determine the minimum prospective short circuit current is not easy, since the final temperature of the conductor is required. This will depend to some extent on the temperature of the conductor at the start of the fault, and the speed at which the fault is disconnected. Unless other information on what the temperature of the conductor will be at the end of a fault is available, it can only be assumed that the final temperature will be the average between the operating temperature of the conductor and the limit temperature for the conductor insulation.

This is a compromise. The temperature of the conductor at the start of the fault could be its maximum permitted operating temperature; the fault current will raise the conductor temperature, but in all probability, the circuit will be disconnected before the limit temperature of the insulation is reached, due to the time taken by the fault current to raise the conductor's temperature.

The temperature of the conductor will rarely be running at the maximum permitted operating temperature given in the tables, and therefore if a more precise determination is required, a new value of k will need to be worked out, based on the conductor's actual operating temperature; at the same time the conductor's resistance will be required at its actual operating temperature. The average temperature used for the conductor's resistance will therefore need recalculating. It should be mentioned at this point that temperature has no effect upon reactance:

Tables are given in Part 4 for the resistance, reactance and impedance of cables from 1 mm^2 up to 1000 mm^2. These tables give the impedance at various average temperatures as well as at 20 °C.

By using the formulae from Chapter 6, the resistance of a conductor at its actual operating temperature can be calculated by the use of formula 5, and by using formulas 26 and 27 the average temperature can be calculated. Table TCF 1 in Part 4 automatically converts the actual conductor temperature to the average temperature, instead of using formulas 26 and 27.

CHAPTER 13

Checking withstand capacity of switchgear etc.

If the supply is three-phase, the maximum prospective short-circuit current will occur if the fault is across all three phases, since the fault current is only limited by the impedance of the source and one phase conductor. In a distribution scheme the impedance will be represented by $(R_1 + jX_1) + (R_2 + jX_2) + (R_n + jX_n)$ as illustrated in Figure 13.3

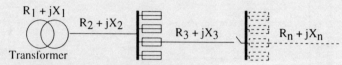

Figure 13.3 - Distribution scheme

The prospective short-circuit current will be:

Formula 53
$$I_p = \frac{U_o}{\sqrt{(R_1 + R_2 + R_3 + R_n)^2 + (X_1 + X_2 + X_3 + X_n)^2}}$$

The damage to the cable or switchgear could occur when the circuit was switched out of service, or when the circuit is lightly loaded; in either event, the conductors could be at ambient temperature, so the resistance values in the calculation ought to be taken at the lowest temperature conditions that can occur, to give the maximum symmetrical short-circuit current. This value is then used to check whether the equipment will withstand the prospective short circuit current.

Checking conductor withstand capacity

Checking whether the conductors are protected is done differently. As already explained, it is the minimum fault current that is required, which means taking the resistance of the conductors at the average of the working temperature and the limit temperature for the insulation. This is, however, not the only criteria; the minimum short circuit current is also dependant upon the type of circuit and type of fault.

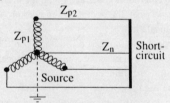

Figure 13.4 - Three-phase and neutral short-circuit

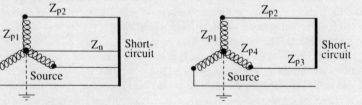

Figure 13.5 - Two-phase short-circuit

In Figure 13.4, a three-phase four wire supply has a short-circuit across all the conductors, so the maximum short-circuit current will be:

Formula 54
$$I_p = \frac{U_o}{Z_{p1} + Z_{p2}}$$

but the minimum short-circuit current will be:

Formula 55
$$I_{pn} = \frac{U_o}{Z_{p1} + Z_{p2} + Z_n}$$

In diagram 13.5, the fault is shown across two phases, but this time it is a three-wire supply: i.e. there is no neutral. In this case the maximum short-circuit current will be the same as in Figure 13.4 above, but the minimum short-circuit current will occur when there is a fault between two phases; the open circuit line voltage is therefore divided by the impedance of all the conductors involved, as given in the following formula:

Formula 56 $I_{pp} = \dfrac{U_{1oc}}{Z_{p1} + Z_{p2} + Z_{p3} + Z_{p4}}$

This formula can be simplified if Z_p is taken as the impedance of one phase up to the point of fault. The formula now becomes:

Formula 52 $I_{pp} = \dfrac{U_{loc}}{2\,Z_p}$

In each formula, the impedances have been shown added together for simplicity of understanding the conductors involved in the circuit. In practice, the resistance of each item and the reactance of each item would be added together and put in the impedance formula similar to that shown in Formula 53 on the previous page.

Conductors in parallel

Where conductors are installed in parallel from one protective device BS 7671 calls for it to be verified by calculation that the conductors are protected against short-circuit current account being taken of what would happen if a fault did not affect all the conductors. Figure 13.6 shows two conductors in parallel with a fault at distance x from the origin of the circuit.

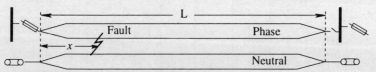

Figure 13.6 - Conductors in parallel

The length to the fault in one leg of the parallel conductors is $2L - x$.
The length to the fault in the other leg of the parallel conductors is x.

Therefore equivalent length $= \dfrac{(2L - x)\,x}{(2L - x) + x} = $ **Formula 57 :** $\dfrac{2Lx - x^2}{2L}$

The phase neutral impedance up to the distribution board will be Z_{pn}, the resistance of the phase conductor R_{ph} and the neutral conductor R_n.
    The total impedance will therefore be:

Formula 58 $Z_{Tpn} = \text{External } Z_{pn} + (R_{ph} + R_n)\left(\dfrac{2Lx - x^2}{2L}\right)$

where: R_{ph} and R_n are in ohms per metre, and L and x are in metres.
    As before the fault current I_{pn} will be U_o divided by Z_{Tpn}.
The current in each leg will, however, divide as the ratio of the conductor resistances.

The fault current in the longer leg will be: **Formula 59** $I_{pn}\left(\dfrac{x}{2L}\right)$

The fault current in the shorter leg will be: **Formula 60** $I_{pn}\left(\dfrac{2L-x}{2L}\right)$

If the cables are larger than 35 mm^2 then reactance has to be taken into account.

Formula 61 Z_{Tpn} = External Z_{pn} + $\sqrt{(R_{ph}+R_n)^2+(X_{ph}+X_n)^2}$ × $\left(\dfrac{2Lx-x^2}{2L}\right)$

If the fault is between phases then the resistance and reactance per metre in Formula 61 are for each phase and the fault current I_{pp} would be U_{loc} divided by Z_{Tpp}.

Calculation sheet

When calculations have to be carried out using R and X, it is advisable to use a table for listing the values of R and X. In this way neutrals and cpcs do not get omitted. The following table is given for guidance. Z_p only has the values of one phase conductor placed in it, the total can always be multiplied by two for a phase to phase fault. The Z_{pn} column has the values of the phase conductor and the neutral conductor placed in it, and the Z_S column has the phase conductor and the cpc conductor values placed in it. The impedance Z does not have to be worked out on each line; it is only required at the point the impedance is required for determining fault current. This will usually be at switchgear and distribution boards, or at the ends of final circuits.

Equipment or location	Z_p			Z_{pn}			Z_S		
	R	X	Z	R	X	Z	R	X	Z

Figure 13.7 - Calculation sheet

Co-ordination between protective devices

A protective device can have a breaking capacity less than the prospective short-circuit current at the point it is installed, provided that on the supply side of that device there is a second further protective device which has the necessary breaking capacity and the characteristics of the two devices are co-ordinated. This is best explained with an illustration.

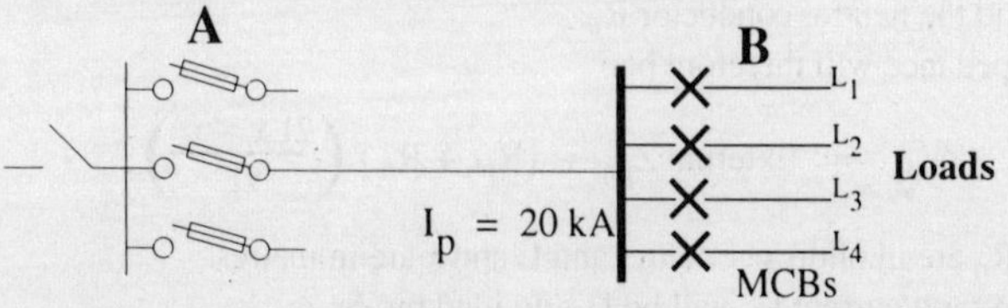

Figure 13.8 - Co-ordination of protective devices

Where it is required to install lower breaking capacity devices at 'B', with a protecting fuse at 'A', the characteristics of the protective devices have to be co-ordinated. The principle is to select a fuse whose pre-arcing time/current characteristic crosses the circuit-breaker time/current characteristic at a point which allows the circuit breaker to operate under all conditions of overcurrent, up to its short-circuit capacity. This leaves the fuse to take over operation at the short-circuit current values in excess of the circuit-breaker's breaking capacity. This involves superimposing the fuse characteristic onto the circuit breaker characteristic to determine the change-over point as shown in the Figure 13.9.

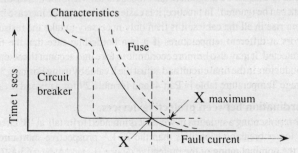

Figure 13.9 - Fuse and circuit-breaker characteristics combined

It must be remembered that all protective devices have a tolerance. The dashed line in Figure 13.9 shows the envelope for both the circuit-breaker and the fuse; it is important that the two characteristics do not overlap.

As can be seen from figure 13.9, the circuit-breaker operates for all fault current values up to the point 'X' on the fuse characteristic. For fault currents in excess of the point 'X' the fuse will operate first, but both devices will recognise the fault, and both devices will start to operate; but because the fuse characteristic goes down to the current axis (i.e. its operating time is less than the circuit-breaker) the fuse will operate first.

This means that the circuit-breaker will not be called upon to break the short-circuit current, and is therefore only required to withstand the peak fault current flowing through it.

The making capacity of a circuit-breaker has to be greater than its breaking capacity, and this fact can be used to determine the value of peak current that the circuit-breaker can withstand flowing through it.

For circuit-breakers, Table 2 of BS EN 60947-2 details how much the making capacity must exceed the breaking capacity for various power factors. For MCBs, the value would have to be obtained from the manufacturers. A typical example for a rated breaking capacity of I_{cn} 4.5 kA to 6 kA at 0.7 power factor from BS EN 60947-2 the minimum making capacity would be 1.5 times I_{cn}.

The value of making capacity for the circuit-breaker can be used in conjunction with the fuse cut-off characteristic which gives the peak current let-through the fuse for various r.m.s. values of short-circuit current.

The method is to find the peak current let-through the fuse for a given fault current from the fuse cut-off characteristic. The peak current can then be divided by the factor by which the making capacity of the breaker must exceed its breaking capacity. The result must not be more than the breaking capacity of the breaker.

Chapter 14

Short-circuit current calculations - examples

The examples given have been kept simple so as to illustrate the method of carrying out the calculations. At the end of Chapter 6 a diagram details those cables where the increase in resistance due to the fault current can be ignored. In practice, it is easier to allow for the increase in resistance due to the temperature rise in all the cables, it is then only necessary to take the resistance of the distribution conductors at different temperatures if it is a borderline case that the final circuit conductors are not protected. It may also be more economic to take into account the actual operating temperature of the conductors in the final circuit and adjust the k value by using Formula 5; the Actual Temperature to Average Temperature table in Part 4; and formula 29 for a revised k value.

Example 1 - Co-ordination between protective devices.

Consider Figure 14.1 representing a situation in which it is intended to install at 'B' circuit-breaker devices which have lower breaking capacities than the 20 kA prospective short-circuit current available at 'B'. Let the nominal rating of the protective device (I_n) at 'A' be a 63 A BS 88 Part 2 fuse. Ignoring any reduction in I_p due to the cable between 'A' and 'B', will the 63 A fuse protect the circuit-breakers and what will be the required breaking capacity of the circuit-breakers?

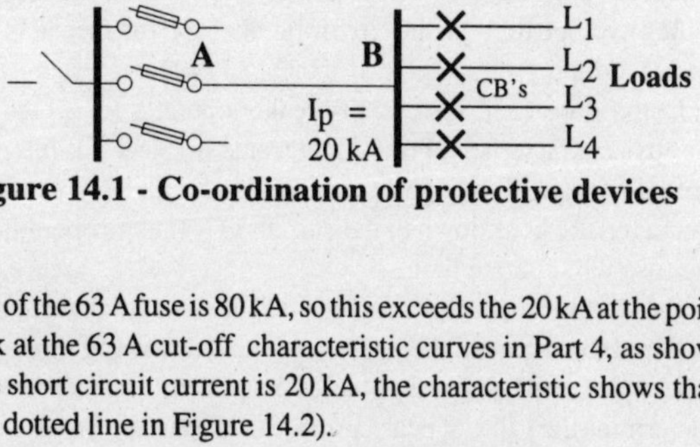

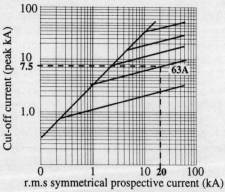

Figure 14.1 - Co-ordination of protective devices

Working

The breaking capacity of the 63 A fuse is 80 kA, so this exceeds the 20 kA at the point at which device is installed. Now look at the 63 A cut-off characteristic curves in Part 4, as shown in Figure 14.2. When the prospective short circuit current is 20 kA, the characteristic shows that the peak cut-off current is 7.5 kA (see dotted line in Figure 14.2).

Figure 14.2 - Fuse cut-off characteristic

The next step is to refer to Table 2 in BS EN 60947-2. This specifies that the making capacity of the circuit-breaker should be greater than its breaking capacity, by a factor given in Table 2. For a breaking capacity of 6 kA, the minimum making capacity is 1.5 times the breaking capacity. Since a large fault current will cause the fuse to operate before the circuit-breaker, the circuit-breaker is not having to break the fault current, but has only to withstand the peak value of fault current flowing through it. As the making capacity of the breaker is greater than its breaking capacity, this can then be used to determine the withstand capacity of the breaker, i.e: 6 kA × 1.5 = 9 kA.

This exceeds the 7.5 kA peak current let through by the fuse, so a 6 kA breaker would be suitable.

A calculation would also be required to make certain, by using 434-03-03, that the conductors feeding L_1 to L_4 are also protected by either the fuse or the circuit-breakers.

Example 2 - Three- phase distribution.

Consider the distribution system shown in Figure 14.3. The supply to the L.V board (A) is from a transformer; the voltage drop up to the board is 0.6 V. A 4 core × 300 mm^2 PVCAWA&PVC strip aluminium armoured feeder cable is then installed to a section board (D); the voltage drop in the cable is 10.5 V. A final circuit to a 5.5 kW three-phase motor (H) is carried out, using 3 core 4 mm^2 PVCSWA&PVC cable. The following information is available:

a)   at 115 °C at 'A' Z_p = 0.0055219 + j0.01786 and Z_{pn} = 0.0059538 + j0.01862, and

b)   at 20 °C at 'A' Z_p = 0.0054 + j0.01786.

The cable from A to D is 161 metres long and is clipped direct. The length of the circuit from board D to motor H is 50 metres, the cable being clipped direct. The distribution boards are suitable for GEC Type T Gg fuses. The object is to check whether the fuses protect the cables against short-circuit current. (Note: for 240 / 415 V supply, U_{loc} = 433 V and U_{oc} = 250 V.)

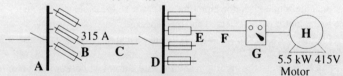

Figure 14.3 - Line diagram for example 2

Working

In this type of problem it is convenient to log the cable parameters as they are worked out on a calculation sheet, or on the drawing; this enables quick reference to their values to be made.

The first thing to remember is that it is the minimum short-circuit current that is required to check that the cables are protected and the maximum to check the withstand capacity of equipment. All the cables are PVC insulated, so their impedances will be taken at 115 °C.

To check the feeder cable 'C', the total impedance up to board 'D' is required. From table RA1 in Part 4, the values per 1000 m for 300 mm^2 cable are 0.138 + j 0.076.

Value for cable 'C' phase = 0.161 (0.138 + j0.076)	0.022218 + j0.012236
Value for cable 'C' neutral =	0.022218 + j0.012236
To this must be added the impedance at the source	
Value of Z_{pn} at Main L.V. board =	<u>0.0059538 + j0.018620</u>
Total	0.0503898 + j0.043092

Calculate Z_{pn} at D

Formula 37 Z_{pn} at D = $\sqrt{0.0503898^2 + 0.043092^2}$ = 0.0663 Ω

Calculate the minimum fault current.

Formula 49 $\quad I_{pn} = \dfrac{U_o}{Z_{pn}} = \dfrac{250 \text{ V}}{0.0663} = 3770 \text{ A}$

From the fuse characteristic in Part 4, look up the disconnection time for 3,770 A for the 315 A fuse, as shown in Figure 14.4.

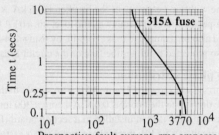

Figure 14.4 - Part 315 A fuse characteristic

For a short circuit current of 3770 A the disconnection time is 0.25 s. This has now to be checked by the maximum disconnection time allowed by the formula of 434-03-03. From K1A Part 4, the value of k for aluminium with a working temperature of 70 °C is 76.

Formula 46 $\quad t = \dfrac{k^2 S^2}{I^2} = \dfrac{76^2 \times 300^2}{3770^2} = 36.57 \text{ s}$

This is greater than the actual disconnection time of 0.25 s, so the cable is protected.

Next the cable to the motor has to be sized and then checked.

From MC1 Part 4, 5.5 kW 415 V motor has a full load current of 11 A, and requires a 25 A GEC Type T 'Gg' fuse, or a GEC 20M25 type NIT fuse.

This is a three-phase circuit, so the minimum short-circuit current will occur when there is a short between two phases; this means that the Z_p is required from the source.

The resistance of one phase of the 4mm^2 cable from table RC1 Part 4 is 6.362 Ω / 1000 m at 115 °C, so the resistance of 50 m = $0.05 \times 6.362 = 0.3181 \Omega$.

Working out the total impedance to the motor:

Value of Z_p at the source	0.0055219 +	j0.01786
Value of Z_p for cable 'C'	0.022218 +	j0.012236
Value of Z_p for cable 'F'	0.3181 +	j0.0000
Total	0.34584 +	j0.030096

Formula 37 $\quad$ Total Z_p of one phase $= \sqrt{0.34584^2 + 0.0301^2} = 0.3471 \Omega$

Formula 52 $\quad I_{pp} = \dfrac{U_{loc}}{2 Z_p} = \dfrac{433}{2 \times 0.3471} = 623.74 \text{ A}$

Now, from the fuse characteristic in Part 4 obtain the disconnection time for the 25 A fuse, in a similar manner to that shown in Figure 14.4. The fault current is past the point at which the characteristic touches the current axis. The disconnection time is less than 0.1 s, so the manufacturer's I^2t let-through has to be compared with k^2S^2, instead of using the formula.

From the I^2t chart on the last page in Part 4, the total I^2t for a type T 25 A fuse at 415 V is $1.85 \times 1000 = 1,850\,A^2$ sec. This must now be compared with k^2S^2 for the cable. From K1A, k is 115, S is 4 mm^2, therefore $k^2S^2 = 115^2 \times 4^2 = 211,600$. This is larger than the energy let-through the fuse, so the fuse protects the cable.

Comment on example two

In the above example the cable sizes were chosen to comply with a total voltage drop of 4%: i.e., 16.6 V three-phase for 415 V supply. Another point worth noting is that each cable has a current-carrying capacity in excess of the size of the fuse. Additionally, the fuses have a breaking capacity greater than the prospective short-circuit current at the point at which they are installed. In these circumstances, Regulation 434-03-02 is applicable, even though the final circuit is to a motor; the calculations were therefore unnecessary.

Example 3 - Maximum short-circuit current.

Calculate what short-circuit current the starter for the motor in example 2 must withstand, if the Z_p at board A at 20 °C is $0.0054 + j0.01786$, and the cable length to the starter is 50 m.

Working

The calculated short-circuit currents in example 2 are only suitable for checking that the cables are protected. For checking that equipment, such as the starter, is protected the maximum prospective short-circuit current is required.

In a three-phase system, the maximum fault current is with a fault across all three-phases. This means that the impedance of one phase conductor only is required, but this must be taken at the appropriate temperature. Since the maximum current is needed, the impedance of the conductors should be taken at the lowest temperature at which the conductors could be when a fault occurs, as the damage could occur when the conductors were not carrying any load current. In the absence of any definite information, take the resistance of the conductors at 20 °C, and those of any transformer involved when the windings are cold.

Z_p at the L.V. board @ 20 °C	$0.0054 + j0.01786$
Z_p for the cable 'C' @ 20 °C = $0.161(0.1 + j\,0.076)$	$0.0161 + j0.012236$
Z_p for the cable 'F' @ 20 °C = 0.05×4.61	$\underline{0.2305 + j0.000000}$
Total Z_p for the cable 'C' @ 20 °C	$0.252 + j0.030096$

Formula 37 $\quad$ Total $Z_p = \sqrt{0.252^2 + 0.030096^2} = 0.2538\ \Omega$

Formula 50 $\quad I_p = \dfrac{U_o}{Z_p} = \dfrac{250}{0.2538} = 985\ A$

It would now require checking that the starter would withstand a fault current of some 1000 A. Note that the fault current is some 360 A larger than that calculated for the protection of the cables.

Example 4 - Socket-outlet ring circuit.

A ring circuit is to be installed using 2.5/1.5 twin and CPC cable. The electricity supplier advises that Z_e will be 0.35 ohms, and the system is TN-C-S. The cable is to be installed clipped direct, and will not be grouped with other cables, or installed in contact with thermal insulation. The ring circuit is 50 m to the mid point and is protected by a 30 A rewirable fuse. Determine whether the circuit

CHAPTER 14

is protected if a phase neutral fault develops 10 m from the fuse in one leg of the circuit.

Working

There are two methods available for working out this problem. One is to use Formula 58 for conductors in parallel. The second way is to use a practical method.

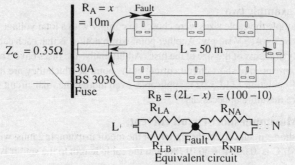

Figure 14.5 - Ring circuit with fault.

First, using **Formula 58** $\quad Z_{Tpn} = \text{External } Z_{pn} + (R_{ph} + R_n)\left(\dfrac{2Lx - x^2}{2L}\right)$

Resistance per metre of R_{ph} and $R_n = 0.010226\ \Omega$ at 115 °C from Table RC1 Part 4.
As the system is TN-C-S, Z_e will give the phase/neutral impedance external to the installation.

Formula 61 $\quad Z_{Tpn} = 0.35 + (0.010226 \times 2)\left(\dfrac{2 \times 50 \times 10 - 10^2}{2 \times 50}\right) = 0.534\ \Omega$

Secondly; using the practical way of determining Z_{Tpn}.

The resistance R_{LA} $\quad = \quad 10 \times 0.010226\ \Omega \quad = \quad 0.10226\ \Omega.$
The resistance R_{LB} $\quad = \quad 90 \times 0.010226\ \Omega \quad = \quad 0.92034\ \Omega.$

These resistance are in parallel therefore :

$$\text{Phase resistance} \quad = \quad \frac{R_{LA} \times R_{LB}}{R_{LA} + R_{LB}} \quad = \quad \frac{0.10226 \times 0.92034}{0.10226 + 0.92034} \quad = \quad 0.92\ \Omega$$

A similar calculation can be carried out for the neutral, but since this is the same size as the phase conductor, it will have the same resistance; i.e. 0.92 Ω.

These two resistances are in series: therefore $Z_{pn} = 0.92 \times 2 = 0.184\ \Omega.$

Total $Z_{Tpn} = Z_e + Z_{pn} = 0.35 + 0.184 = 0.534\ \Omega$, as before.

The fault current I_{pn} can now be determined, assuming U_o is 240 V and U_{oc} is 250 V.

Formula 49 $\quad I_{pn} = \dfrac{250\ V}{0.534} = 468.16\ A$

As far as the ring circuit conductors are concerned, the fault current of 468.16 A will divide by the ratio of the conductor lengths of each leg of the ring circuit, since each leg is in parallel, i.e.,
The fault current in the longer leg will be:

Formula 59 $\quad I_{pn} \left(\dfrac{x}{2L} \right) \; = \; 468 \times \dfrac{10}{2 \times 50} \; = \; 46.8\ A$

Similarly, the short-circuit current in the shorter leg of the ring circuit will be:

Formula 60 $\quad I_{pn} \left(\dfrac{2L - x}{2L} \right) \; = \; 468 \times \dfrac{(2 \times 50) - 10}{2 \times 50} \; = \; 421.2\ A$

It is now important to establish that the energy let-through of the fuse will not damage the conductors.

The fuse will see the total fault current of 468 A the disconnection time t being obtained from the protective device characteristic. For the purposes of this example it will be assumed that the disconnection time is 0.2 s.

The maximum disconnection time allowed for the shorter leg i.e., the one carrying 421.2 A, can be found by using the formula $I^2 t = k^2 S^2$, to check that it is within the disconnection time of the protective device.

Formula 46 $\quad t \; = \; \dfrac{k^2\, S^2}{I^2} \; = \; \dfrac{115^2 \times 2.5^2}{421.2^2} \; = \; 0.466\ s$

This time is greater than the time taken for the protective device to disconnect the circuit so the conductors are protected.

The actual disconnection time for the example was less than 0.1 s, so the energy let-through of the fuse would be used, which from the BS 3036 fuse characteristic in Part 4 is 36, 300 A^2 sec.

Therefore, using **Formula 47** $\quad I^2 t \leq k^2 S^2$:

$I^2 t = 36, 300$, and

$k^2 S^2 = 115^2 \times 2.5^2 = 82,656.25\ A^2 s$.

The value of $k^2 S^2$ is greater than $I^2 t$ which indicates that the conductors will withstand the energy let-through of the protective device.

Energy let-through of a rewirable fuse

When the disconnection time is 0.1 s or less, the manufacturer's $I^2 t$ energy let-through of the protective device has to be compared with $k^2 S^2$ for the conductor. Rewirable fuses are not energy limiting devices, since they only interrupt the fault current as it passes through zero. This means that the arcing time and energy are, in general, insignificant in relation to pre-arcing time and energy. For a rewirable fuse, a sufficiently accurate value of $I^2 t$ let-through energy can be obtained by squaring the current at 0.1 s, and multiplying this by 0.1. This would then be the value to compare with $k^2 S^2$. However, the rewirable fuse characteristics (BS 3036) in Part 4 give the total $I^2 t$ energy let-through of the different ratings of fuses at 2 k up to 30 A..

Chapter 15

Isolation and switching

Types of switching

As a result of international agreements, four types of switching have been introduced, namely:

- Isolation
- Switching off for mechanical maintenance
- Emergency switching
- Functional switching

Each of these will be looked at in turn in the following chapters.

Isolation

Isolation is used to prevent danger. Such a danger exists when work has to be carried out on parts that are live in normal use, or with mechanical equipment driven by electrical means.

When work has to be carried out on such equipment, it is essential to make certain that isolation from every source of electrical energy is complete. For example, in damp situations, motors are quite often provided with internal heaters to limit condensation within the motor. The circuits are designed so that the heater is switched on when the motor is switched off. In these circumstances the isolator at the motor must isolate both the power supply to the motor and the supply to the heating unit in the winding of the motor, if danger is to be avoided.

Another example of dual supplies going into equipment is with control circuits which derive their supply from a source of power different from that of the equipment. In this instance, isolation is not complete until the control circuits have also been isolated.

The requirements of BS 7671 for isolation and switching, are given in Chapter 46, Section 476 and 537. Compliance with these requirements is also likely to comply with the Electricity at Work Regulations. BS 7671 does not, however, lay down any rules for the management of safe systems of working, being only concerned with the hardware of the installation.

Object and purpose

The object of isolation is to make certain that normally live parts are made dead before work commences. The purpose of isolation is to enable work to be carried out on parts that are normally live, in complete safety. One of the precautions to be taken is to make certain the supply cannot be restored by a remote switch or control circuit. The electrical equipment to be worked on should not be accepted as safe until it has been proved dead by testing. Any test instrument used must have fusible leads. Testing to prove that a circuit or piece of equipment is dead is considered to be live working, so training is required before personnel are allowed to test.

Isolation by non-electrical personnel

Isolation can also include the isolation of equipment that is permanently being taken out of service, and can be used to enable non-electrical work to be carried out. However, where non-electrical personnel use isolators to enable mechanical work to be carried out, they should be taught that the load must be switched off first, since isolators are not designed to switch load on or off. It is better to have the isolator interlocked with a load breaking device, so that the load is switched off before the isolator's contacts open. This is best achieved by using isolators with additional contacts which break first and make last; these are then wired into the stop circuit of the controlling contactor, which disconnects the load before the isolator contacts start to open. This arrangement also has the advantage that the equipment can only be restarted at the starter in the proper manner. Alternatively, use an isolating switch.

Where required

Isolation, in the form of a linked switch or linked circuit-breaker, capable of switching the full load current of the installation, has to be provided as near as possible to the origin of every installation. With the exception of three-phase TN-S and TN-C-S systems where a solid neutral link is allowed, all live conductors must be disconnected.

An isolator or switch near the origin of the installation, can be used to switch a group of circuits, and can be the only isolator installed to comply with the BS 7671. An example is the consumer unit installed in a house.

Where there are more than one source of supply to an installation, a switch is required for each source, together with a notice warning that all switches must be operated to isolate the installation. Where one of the sources of electricity requires its earth to be independent, a switch may be inserted between the neutral point and earth, providing that the switch is linked so that it connects and disconnects the live conductors at the same time as the earth connection.

To comply with BS 7671, isolators are required at switchgear to enable maintenance to be carried out. Where the system is TT or IT then all the phase and neutral conductors must be switched. The isolator must isolate all phase conductors in a TN-S or TN-C-S system, the neutral not being required to be switched where it can be reliably regarded as being at earth potential.

The instances where the neutral can be reliably regarded as being at earth potential are:

a) a balanced three-phase load with very little current in the neutral,

b) the incoming supply from a generator, the distribution complying with (a),

c) with a dedicated supply.

Where the supply is single-phase the main switch must switch the phase and neutral if it is going to be operated by unskilled persons, for example in a domestic installation.

BS 7671 also requires every circuit and final circuit to have a means of switching the supply on load, although one switch may switch a group of circuits. They will also be required at every motor, and for high voltage discharge lighting.

Isolators can be placed remotely from the equipment or circuits they control, providing that the means of isolation can be secured in the open position. This means preventing inadvertent or accidental re-closure whilst being used to isolate the equipment or circuits they control. This requires their being secured against inadvertently or unintentionally being returned to the 'on' position by other personnel, or through mechanical shock or vibration.

Any locks or handles within the same installation used to make certain that the supply cannot be inadvertently switched on, must be unique to that installation. This is best achieved by each lock having a different type of key, one of which is kept by the manager, and the other by the skilled operative. An alternative arrangement, when the primary isolator is remote, is to have another isolator adjacent to the equipment.

Warning notices required

Where an isolator is not capable of cutting off all the supplies to a piece of equipment, a notice must be displayed, warning that the remaining live parts must be isolated elsewhere, and it is more meaningful if the label specifies where. Alternatively, an interlocking arrangement may be used so that all circuits are isolated before access can be gained to the equipment. Except where previously mentioned, isolators must not be placed in protective conductors, nor in protective earthed neutrals.

CHAPTER 15

Identification

The isolator which controls each circuit or installation has to be identified so that persons know what the isolator controls. Identification can be by the position of the isolator so that it is absolutely clear what the isolator controls, or the isolator can have a permanent durable label.

Requirements for isolators

Since isolators are intended to stop an electrical supply getting to parts that are normally live whilst work is being carried out on those parts, they have to comply with certain requirements. They must have sufficient poles to disconnect all live conductors including neutrals, unless the system is TN-S or TN-C-S and the neutral is at earth potential, in which case only the phase conductors need to be disconnected.

They must also have a specified isolating distance between poles when open, and the position of the contacts shall be either externally visible, or clearly and reliably indicated only when the specified isolating distance has been achieved on each pole. This precludes the use of semi-conductor devices or micro-switches as isolators.

Isolators must be selected and installed so as to prevent unintentional re-closure by mechanical shock or vibration, or by inadvertent or unauthorised operation by personnel.

Where the neutral of an isolator or switch in a three-phase four-wire TN-S and TN-C-S system is a solid link, it must either be only removed by using a tool, or accessible to skilled persons only. Where it is a joint, it must be in an accessible position and disconnection must be by means of a tool.

What can be used as an isolator

Switches and circuit-breakers can be used as isolators, providing they comply with the rules for isolators. Plugs and sockets can also be used providing they are suitable for breaking current whilst being disengaged.

Using lock off stop buttons that are mounted next to the equipment is not a safe means of isolation, since it would be possible for someone to tamper with the remote starter (pushing the contactor in would energise the cables), or a fault could energise the main circuit.

Solid links or fuse links can be used for isolation, but will require a management procedure to make certain that the load is switched off first to ensure that no danger will arise.

The difference between an isolator and a switch, or an isolating switch, is often misunderstood. Isolators are not designed to make or break load or fault currents; the load should therefore be switched off before the isolator is operated. On the other hand, a switch or an isolating switch is designed to make and break load current. Some high voltage isolators are designed to break a fault current for the period of time it takes the feeder circuit-breaker to disconnect the fault.

Where an isolator is used in conjunction with a circuit-breaker for maintenance, it must either be interlocked with it, or accessible only to skilled persons.

Fundamental requirements for safety

There are three fundamental requirements in Chapter 13 of BS 7671 that have to be taken into account. Regulations 130-06-01 and 130-06-02 are generally taken into account in Chapter 46, Sections 476 and 537, but account should also be taken of 130-07-01; this calls for every piece of equipment which requires operation or attention by a person to be installed so that adequate and safe means of access and working space are afforded for such operation or attention. The Electricity at Work Regulations also have a similar requirement in Regulation 15. Additionally, Regulations 12 and 13 of these Regulations, should also be taken into account.

Chapter 16

Mechanical maintenance

What is meant by mechanical maintenance

Mechanical maintenance includes the replacement, refurbishment or cleaning of lamps and non-electrical parts of equipment, plant and machinery. Confusion arises between 'isolation' and 'switching off for mechanical maintenance', because 'isolation' is a general term which has normally been used when maintenance has to be carried out on equipment. As a result of the international discussions, however, a distinction has been drawn between 'isolation' and 'switching off for mechanical maintenance'.

The difference is, isolation enables work to be carried out on normally live parts in safety, whereas switching off for mechanical maintenance allows work to be carried out on non-electrical parts without the risk of those parts being moved or energised by electrical means. Thus isolation is for the use of electrically skilled or instructed persons, to enable them to be able to work in safety.

General requirements

Where there is a risk of physical injury, some means of switching off for mechanical maintenance has to be provided. The devices used for switching off for mechanical maintenance must be suitably placed, readily identifiable, and convenient for their intended use.

Suitably placed means that the switch should be operable safely without personnel having to stretch over moving equipment to reach the switch.

Labelling

In process plants, it is quite common for the switches to be grouped adjacent to each other. In this case, each switch should be labelled so that no confusion arises as to which switch to use. 'Convenient for their intended use,' implies having the switch local to the equipment it controls, otherwise personnel will not be bothered to walk the distance to switch off whilst they make some small adjustment to the machinery. Any labels should be of a permanent and durable type.

Where required

A device for switching off for mechanical maintenance has to be provided for every motor, electrically heated surfaces which can be touched, and electromagnetic equipment, the operation of which could cause a mechanical accident. The device should be inserted in the main supply circuit.

BS 7671 allows a device to be inserted in the control circuit, providing precautions are taken to make sure the same degree of safety as that if the device were inserted in the main supply circuit. This, however, can be dangerous, and it can be very difficult to ensure such safety. This method should only be used where specified in a British Standard, and only then as a last resort.

A fundamental requirement is that the switch must be under the control of the person carrying out the mechanical maintenance. This means that precautions have to be taken to prevent the equipment from being unintentionally or inadvertently reactivated. Padlocking the switch in the OFF or OPEN position is one method of achieving this requirement.

Devices such as lock off type stop buttons inserted in the control circuit should not be used as this practice can be dangerous, since a fault could start up the equipment; therefore this mode of switching off is not recommended.

The replacing and cleaning of lamps is considered to be mechanical maintenance, and the local lighting switches can be used for switching off; this naturally precludes two way switches, since one of the switches can never be under the control of the operator, however, usually standard domestic lighting switches do not comply with the device requirements for switching off for mechanical maintenance. It is also going to be very difficult to comply with this requirement in industrial and commercial premises when all the lighting switches are grouped in one position, since BS 7671 call for precautions to be taken against the unintentional or inadvertent re-activation of the switch. Another factor which has to be taken into account is whether the lighting switches comply with the requirements laid down for isolating devices given in the paragraph of type of device below.

Type of device

The type of device used for switching off for mechanical maintenance can be:
- a switch,
- plug and socket-outlet not exceeding 16 A,
- isolator, providing there is a means of switching off the load before the isolator is operated,
- circuit-breakers, providing the manufacturer confirms that the circuit-breaker is suitable for this duty,
- control switches operating contactors (not recommended).

Each of the devices listed above must comply with certain requirements:
- they should be inserted in the main supply circuit,
- should be manually operated,
- have an externally visible contact gap, or
- clearly and reliably indicate the OFF or OPEN position only when each pole is open, and
- be capable of cutting off the full load current of the circuit they control.

Switches for mechanical maintenance are provided for the use of non-electrical personnel. In general, it is far safer to provide isolating switches (instead of isolators) for mechanical equipment, to enable maintenance to be carried out. Additionally, non-electrical personnel should be trained in the procedures of switching off for mechanical maintenance and testing to make certain the circuit really is dead.

BS 7671 allows one type of device to be used for different functions, providing it complies with the requirements for each of the functions for which it is used. This means that a switch can be used to perform the functions of: isolation, switching off for mechanical maintenance, or emergency switching; the switch must, however, comply with the requirements for each function.

Chapter 17

Emergency switching

Definition

Emergency switching is defined as an operation intended to remove, as quickly as possible, danger, which may have occurred unexpectedly, and is provided for use by any person. People should therefore be advised of the location of emergency switching devices and the devices should be sited so that they will always be available for use.

Where required

Every fixed or stationary appliance that can give rise to danger and is not connected to the supply by a plug and socket which is designed for the most onerous use intended, shall be provided with a means of interrupting the supply on load. The means of interrupting the supply can be incorporated in the appliance, but if it separate from the appliance it shall be so placed that no danger to the operator of the equipment shall occur.

Emergency switching devices are operated when danger or a hazard is occurring, and as such they can be operated by anyone; they should therefore be installed at every point at which it may be necessary to disconnect the supply rapidly to prevent or remove the hazard that is occurring.

Where a machine driven by electrical means may give rise to danger, an emergency switch shall be provided; it shall be readily accessible and easily operated by the person in charge of the machine. Where more than one means of manually stopping the machine is provided, and a danger could arise by the unexpected restarting of the machine, arrangements have to be made to prevent the restarting of the machine.

In process plants, or where conveyors are used, additional emergency switching will be required at other hazardous points. In the case of open conveyors, these will be required throughout the conveyor's length. This is best achieved by the use of a trip wire running the full length of the conveyor, and in the case of long conveyors, with intermediate emergency switching points along the length of the conveyor. When installing trip wires, it is important to ensure that the switch is operated by a pull on the wire in either direction. A trip wire with a single switch at one end would not be acceptable.

Operation of emergency switches

Emergency switches should preferably be manually operated, and should be of the latching type of device, or be capable of being restrained in the off position; the handles or buttons should be coloured red.

The means of interrupting the supply must be capable of cutting off the full load current of the installation, acting directly on the appropriate supply conductors, so that a single action will cut off the supply. Where there is a risk of electric shock, means must be provided to interrupt all live conductors, except that in TN-S and TN-C-S system the neutral need not be interrupted.

Where devices such as circuit-breakers and contactors are operated by an emergency switch or stop button which is remote, they shall be arranged to open on the de-energisation of the coils.

Where emergency switching could result in a greater danger arising, for instance, from the disconnection of safety services, then it is permissible for the emergency switching to be carried out only by skilled or instructed persons.

As stated earlier, emergency switches can be operated by anyone. Although not a requirement of

BS 7671, staff should be instructed as to the location and purpose of emergency switching devices.

Emergency stopping

Where mechanical movements caused by electrical means can give rise to danger, a means of emergency stopping has to be provided. Emergency stopping is only concerned with stopping mechanical movement, and can be achieved by energising a circuit to stop the mechanical movement, such as energising a circuit to feed d.c., into the motor stator at the same time as the a.c., mains supply to the motor is cut off. This is known as 'd.c., injection breaking'. It is important to note that the de-energising of the main circuit and the energising of the braking circuit is achieved by one initiation of the emergency stopping device.

Colour of emergency switches

Emergency switches should preferably be coloured red.

Fireman's switch

A fireman's emergency switch has to be provided in the low voltage circuit for exterior discharge lighting and for interior discharge lighting operating at a voltage above 1000 volts. A fireman's switch is not required for a portable discharge lighting unit or a sign which does not exceed a rating of 100 Watts, if it is fed from a readily accessible socket-outlet.

Exterior and interior lighting

Covered markets, arcades and shopping malls are all considered to be exterior installations, but temporary installations in permanent exhibition halls are considered to be interior installations.

For exterior installations the switch should be mounted outside the building, whilst for interior installations the switch should be mounted in the main entrance to the building. In both cases the switch position can be agreed with the local Fire Authority. Switches should be installed in a conspicuous position, and where more than one switch is installed, each switch must be marked to indicate the installation it controls.

In general, switches for discharge lighting should be mounted adjacent to the discharge lighting fitting, at a height not exceeding 2.75 metres from the ground, unless the local Fire Authority agree the switch position. The switch can be placed in another position, but a nameplate, which is clearly readable from the ground, has to be placed next to the discharge lighting, advising where the switch is located, and the switch must be clearly labelled to indicate which lighting it controls. (See Part 1 for details of label.)

Colour of fireman's switch

The switches should be coloured red and be labelled 'FIREMAN'S SWITCH'. They should have the ON and OFF positions clearly marked, so that they can easily be read by a person standing on the ground. The OFF position of the switch must be at the top.

What cannot be used

The installation design should not select a plug and socket-outlet as an emergency switching device. It must be recognised, however, that the first reaction of a person, if the TV catches fire, or a person is receiving an electric shock from a portable tool, will be to pull the plug out of the socket-outlet.

Chapter 18

Other means of switching

Difference between switching and interrupting the supply

A main switch or circuit-breaker has to be provided for every installation, and every circuit and final circuit has to be provided with a means of *interrupting* the supply on load, or in any foreseen fault conditions. There is a difference between switching and *interrupting* the supply; *interrupting* does not necessarily mean using a mechanical switching device; the *interruption* could be brought about electronically.

Functional switching

Functional switching is the device used to control a circuit or group of circuits in normal use. The device used may not comply with the requirements given previously for switches or isolators. Thus micro-gap switches and semi-conductor devices are functional switches.

Where a circuit is required to be controlled independently from other parts of the installation, such as the lighting circuits from a distribution board, it has to be provided with a functional switching device. The switching device does not have to switch all the live conductors, but it must never be installed solely in the neutral conductor.

A plug and socket-outlet up to a rating of 16 A can be used on an a.c. supply (but never on d.c.) as a controlling switch.

Where functional switching is used to change-over the supply from different sources the switches must switch the phase and neutral conductors and not allow the sources to be switched in parallel, unless the installation has been designed for the sources to be operated in parallel. Additionally, PEN or protective conductors must not be isolated unless the installation is designed for such isolation.

Circuits and appliances

A group of circuits may be switched by a common device, such as a switch controlling a multiple lighting circuit.

Every circuit and final circuit shall be provided with a means of switching for *interrupting* the supply on load. A plug and socket-outlet with a rating not exceeding 16 A may be used as a switching device. Only with an a.c., supply may a socket exceeding 16 A be used as a switching device, and only then, if it has a breaking capacity appropriate to its intended use.

Cooking appliances

Every fixed or stationary cooking appliance has to be controlled by a switch which is separate from the appliance. The switch must be in a position where it can be operated without putting the operator of the switch in danger. A danger can occur should a fire start due to cooking oil or fat overflowing on the hot plates of the cooker. The switch should, therefore, be in a position where it will be clear of any flames emanating from the cooker and where it can be operated by persons without them being placed in danger from the fire.

One switch can be used to control more than one appliance in the same room, but consideration has to be given to the total load on the switch, its terminal capacity, and, if danger is to be avoided, the location of the switch relative to the equipment.

This arrangement is usually adopted where split level cookers are installed in domestic kitchens.

CHAPTER 18

Control circuits

Care is needed when designing control circuits to ensure that a fault between a control circuit conductor and any other conductive part does not short out the control or safety devices for the circuit as illustrated in Figure 18.1.

The correct connections are illustrated in Figure 18.2, which illustrates that a fault on the neutral side of the coil does not render the stop button or limit switch inoperative. A fault on the fuse side of the coil would, of course, blow the fuse.

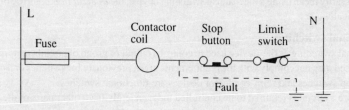

Figure 18.1 - Incorrect connections for contactor

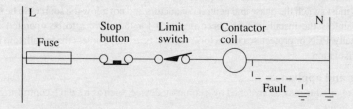

Figure 18.2 - Correct connections for contactor

Chapter 19

Selection and erection of equipment

This chapter should possibly be considered at the same time as the assessment of the installation's characteristics, but having designed the installation, the type of equipment, cables accessories etc., have to be chosen. Chapters 51 and 52 in BS 7671 are to assist in the correct choice of equipment. The British Standard just refers to equipment, but this term is an abbreviation of electrical equipment and includes all the materials used.

Compliance with standards

BS 7671 calls for all equipment to comply with the appropriate British or Harmonised Standard. Equipment in this context includes all materials used such as: conduit, cable, switchgear, distribution boards, trunking and wiring accessories. A harmonised standard is one with an EN in the number, such as BS EN 60898.

This does not preclude the use of equipment manufactured to a foreign standard or an IEC standard, but the designer or specifier then has to verify that any differences between these standards and the British or Harmonised Standards does not reduce the safety of the installation.

Operational conditions

The operational conditions cover: voltage, current, frequency, power, compatibility, external influences and accessibility and should have already been determined under the assessment of general characteristics.

Voltage

Make sure all the equipment to be used is suitable for the nominal voltage present. This includes the voltage between phases where three-phase supplies and equipment are concerned.

Current

All equipment has to be suitable for the design current and any fault currents that are likely to occur.

Where cables are used that operate at a temperature in excess of 70 °C, ensure that the switchgear, distribution boards and any other utilisation equipment is suitable for the higher cable temperature.

Frequency

The frequency of equipment has to be appropriate for the frequency of the supply. This is particularly important when designing installations in other countries.

Power

Where equipment is selected for its power characteristics then it should be checked that it is suitable for the power demanded by the equipment, for example, selecting a motor to drive a particular machine.

Compatibility

Compatibility means making sure that the characteristics of the equipment installed does not interfere with the correct operation of other equipment. It also includes taking into consideration the Electromagnetic Compatibility regulations, where the radiation from cables can interfere with the correct operation of computer systems.

CHAPTER 19

External influences
Includes selecting equipment that is suitable for the environmental conditions that will be present in the area where the equipment is installed and takes into consideration: ambient temperature, humidity, condensation, presence of water, the likelihood of thermal insulation being installed at some future date, vibration, corrosive or polluting substances, dust, ignitable gases, solar radiation, vermin or mechanical damage.

Cables
In addition to external influences, cable selection should take into consideration the maximum permitted operating temperature of the cable and what effect heat generated by the equipment will have on the cables entering the equipment. Where the temperature in the equipment exceeds that of the cable, then either cables of the correct temperature rating must be installed or a sleeve giving protection at that temperature must be placed over each individual conductor entering the equipment.

Where single core cables are installed they must not be completely surrounded by ferrous material and where single core cables with a non-magnetic armour are installed the armour should be bonded at least at one end of the run and preferably at both ends to limit the touch voltage developed on the armour under fault conditions.

Consideration has to be given to the mechanical protection of cables and any limit on the bending radius of the cable. Where they are buried underground they must have mechanical protection and be buried at a depth where any further normal work carried out in that area will not damage the cables; they should then be covered by marking tape or cable tiles to give warning of the presence of the cable. MICC cables when buried underground should have a PVC sheath to protect the cable from corrosion.

Cables installed through timber joists should be installed at least 50 mm from the top and bottom of the joist, or have a metal plate installed over them that will stop the penetration of a nail.

Conduit and trunking
In addition to external influences, where conduit is totally concealed it should be installed so that the only draw-in boxes are those exposed at the points of utilisation. Screwed conduit should have a full thread so that it fully enters the conduit box and so that the conduits butt-together at couplings; the outside ends of the joints should be given a coat of paint afterwards to stop dampness corroding the screw threads.

Where galvanised trunking is used there is no need for earth tags across the joint, providing the sections and accessories are bolted together properly. Where painted trunking is used a copper link should be installed across all joints.

Consideration should be given to expansion and contraction on long runs of conduit and trunking, particularly if they are of plastics material.

Accessibility
Access to switchgear and distribution boards must be provided. Additionally, every joint and connection has to be accessible unless it is made by welding, brazing or a compression tool, or the joint is one that is filled with compound, as in the case of armoured cables underground.

Fire risk
Consideration must be given to stopping the spread of fire as detailed in Part 1 'Heat and fire' and for enclosing cables, not tested for the resistance to the propagation of flame, in non-combustible material complying with the ignitability characteristic P in BS 476 Part 5

Chapter 20

Inspection, testing and certification

The fundamental requirements for safety in the regulations call for inspection and testing to be carried out. The inspection and testing has to be carried out on completion of an installation, on completion of an alteration, or on completion of an addition to an installation. Moreover, inspection and testing has to be carried out before the installation is put into use. BS 7671 does not specify any size of alteration or addition, so no matter how small an alteration or addition is, inspection and testing is required.

BS 7671 is concerned with providing safety, especially from fire, shock, burns and injury from mechanical movement actuated by electrical means. The object is to check that all the fundamental requirements for safety are complied with. To assist in compliance with these requirements for safety, Parts 3 to 6 in BS 7671 give methods and practices which will satisfy these safety requirements.

BS 7671 specifies that the tests carried out shall present no danger to persons, property, or equipment. This means that the installation has to be proved before the application of any mains voltage necessary to complete the tests, and that the necessary steps have been taken to ensure that any equipment that can be damaged by insulation testers is disconnected before the tests are carried out.

Obviously the only person or persons who should carry out an inspection and test are those who are competent to do so. This also means that they must have a thorough understanding of BS 7671 requirements and their application, as well as the instruments they have to use.

When to start inspection
On the larger contract, the inspection and testing should start at the same time as the contract, and continue as the contract proceeds. Quite often the contract can be broken down into areas. As each area is completed, the inspection and testing can be completed; for example, each floor in a multistorey building. This may be impractical for the small contract, so the inspection and test will have to be carried out on completion of the work.

Care is needed when inspecting and approving areas of a contract, that the installation which joins all the areas together, or branch circuits feeding separate areas, do not get omitted from the inspection and testing schedule.

Information to be provided
The person carrying out the inspection and test has to be provided with the details of the assessment of general characteristics. These details will comprise:
  a) the maximum demand of the installation,
  b) the number and type of live conductors, i.e., whether the supply is single-phase, three-phase three-wire, or three-phase four-wire etc., or d.c. two wire or three wire,
  c) the type of earthing arrangement: i.e. of which system is the installation part, is it, TN-S, TN-C-S, or TT? The system is unlikely to be TN-C, but where the supply is obtained from the customer's own transformer, it is possible to find an IT system,
  d) the nominal voltage of the supply and its frequency,
  e) the prospective short-circuit current at the origin of the installation. In a three-phase four wire supply, both the single-phase and three-phase value will be required,

f) the phase earth loop impedance external to the installation,

g) the type and rating of the overcurrent protective device installed at the origin of the installation by the supplier,

h) whether any supplies for safety services are provided, and whether they are automatic or non-automatic. The characteristics of the source of the safety service will also be required in the same way that the characteristics of the main supply is provided,

i) a legible diagram, chart or table showing in particular: the type and composition of each circuit. This means an indication of the outlet points or equipment served by the circuit, and the number and size of conductors, and the type of wiring used (PVC single core cable in conduit, for instance),

j) the information to enable the identification of the devices that perform the functions of protection, isolation, and switching, together with details of where they are located,

k) a description of how the installation's exposed and extraneous conductive parts are protected against the persistence and the magnitude of earth fault voltages appearing on the metalwork.

Checking design

The information provided either by the person who designed the installation, or who carried out the installation has to be checked as listed above. In checking the above, it is important to check the following items, since these are the foundation stones on which the installation was built.

- Prospective short circuit current at the origin.
- Size and type of protective device at the origin.
- Phase earth loop impedance external to the installation.
- Type of earthing arrangement (i.e. system type).
- Is the supply suitable for the installation's maximum demand?

As far as the prospective short-circuit current is concerned, where the supply is three-phase, it is important to obtain the single-phase as well as the three-phase short-circuit current. Both values are required, since the single-phase value is needed to check that conductors are protected, and the three-phase value is required to check that the switchgear, starters etc., have the correct withstand capacity.

Inspection: visual and physical

Visual inspection has to precede testing in order that danger cannot arise when tests are carried out. The visual inspection is made with the supply disconnected. This is a good safety measure, since equipment will be opened and live conductors touched during the visual inspection.

A visual inspection of the installation has to be made to establish that the equipment complies with British Standards. On large contracts this is best done by inspecting equipment, cables, conduit etc., for British Standard labels before the installation is carried out (see Figure 20.1).

Figure 20.1 - British Standard Kite and Safety Mark

If the designer (installer or specifier) has stated the equipment complies with a foreign standard or IEC code, it should be verified that the designer (installer or specifier) certifies that it complies with the safety requirements of BS 7671.

Visual inspection also means physically checking the installation to make certain that the electrical equipment (which includes cables etc.) has been correctly selected and erected, and not damaged during erection.

Items that should be checked whilst carrying out a visual inspection include:

a) the connection of conductors: are they properly made and tight, and include all the strands of the conductor in the lug or termination, i.e., have strands been removed to enable the connection to be made?

b) that conductors are properly identified, and are properly protected against mechanical damage, and that single pole switches have been installed only in the phase conductors,

c) that the number of cables enclosed in conduit or trunking specified in the design has not been exceeded, and that burrs have been removed from conduit etc.,

d) that conductors have the correct current-carrying capacity, and circuits comply with the voltage drop allowed for the supply nominal voltage,

e) where conductors have been de-rated for grouping that the correct formula from Appendix 4 has been used. For instance, the formula used when simultaneous overload cannot occur has not been used for socket-outlet circuits,

f) that the correct connections have been made at accessories and equipment and lampholders, such as the phase conductor being connected to the centre contact of an E.S. lampholder,

g) that where equipment passes through a fire barrier, the holes round the equipment have been filled to the same degree of fire resistance as the wall, floor or ceiling through which the equipment passes, and that internal fire barriers have been installed in the equipment,

h) that fixed equipment that causes a focusing or concentration of heat has been installed a sufficient distance from any other object or building material so that it will not cause a heat build-up in the material that is likely to cause a fire,

i) where forced air heating systems or appliances producing hot water or steam are installed, that the safety devices called for in Section 424 are installed and functioning properly,

j) that protection against direct contact has been made, and where RCDs are used for supplementary protection, they will disconnect within 40 ms with an earth fault of 150 mA,

k) a check should be made to ensure that all barriers have been installed, and that enclosures comply with BS 7671. Holes in equipment should not exceed IP2X generally, or IP4X in top surfaces,

l) that the methods of protection against indirect contact comply with BS 7671. In particular, that main equipotential bonding conductors have been installed, and are the correct size. That where disconnection cannot be made within the times specified in the regulations either supplementary bonding conductors have been installed or RCDs, and that the resistance of the supplementary bonding conductor is 50 / Ia and the minimum sizes given in Section 547-03 are complied with. Where the protective conductor impedance has been limited to a value given in Table 41C, Table PCZ 10 Part 3, that the value is correct for the circuit,

m) where distribution boards contain circuits that have to be disconnected within 5˙s and 0.4 s, that either the protective conductor impedance back to the main earth bar (to which the equipotential bonding conductors are connected) is limited in accordance with Table 41C (PCZ 10), or equipotential bonding is carried out at the distribution board to the same extraneous conductive parts as the main equipotential bonding and with the same size of conductor,

n) that Category I, II and III circuits have been installed in accordance with BS 7671,

o) there will be no problems concerning compatibility and no mutual detrimental influence between the materials and equipment installed,

p) all protective devices and overloads and undervoltage devices are of the correct rating and setting for the duty they have to perform,

q) that isolators, switches and where required emergency stop buttons, have been installed in the correct locations,

r) equipment selection and protective measures taken is in accordance with the external influences applicable to the area where the equipment is installed.

s) a check has to be made that equipment which requires labelling has been properly labelled: for example: labels attached to the earth clamps of bonding conductors,

t) that distribution boards, circuits, switches and terminations are correctly identified and that the necessary information in the form of diagrams, charts and instructions is provided for the installation,

u) that there is sufficient access and working space at equipment, and that danger or warning notices are installed at the appropriate places. Posting of notices also includes those required for the periodic testing of RCDs, and notice of the frequency of a periodic inspection and test of the installation.

Testing

Before any tests are carried out it is essential that the operator is fully familiar with and competent to use the instruments required. It is also essential that the instruments used comply with British Standards, and have their calibration checked regularly. Since some of the tests are to be made with the mains voltage on, it is important to make certain that those particular instruments are equipped with fusible leads.

Sequence of tests

Once the visual inspection is complete, tests have to be carried out. The initial tests being carried out with the electrical supply disconnected; the final tests with the installation energised. These can be summarised in a list.

The following tests have to be carried out.

1. Continuity of protective conductors. This involves every protective conductor including main and supplementary bonding conductors.
2. Continuity of ring final circuit conductors.
3. Insulation resistance.
4. Insulation applied on site.
5. Protection by separation of circuits.
6. Protection against direct contact by barriers or enclosures.
7. Insulation of non-conducting floors and walls.
8. Polarity.
9. Earth electrode resistance.
10. Earth fault loop impedance.
11. Operation of residual current operated devices.

12. Prospective short-circuit current test.

13. Functional testing of the equipment, safety devices, limit switches etc.

Tests 1 to 8 do not need the supply connected. Test 9 can be carried out before the installation is energised , but it is usually done at the same time as the earth loop impedance or the RCD tests are made.

Protective conductors

BS 7671 calls for every protective conductor to be separately tested and checked that it is electrically sound and correctly connected. This requirement means that every circuit protective conductor, main equipotential bonding conductor, and supplementary bonding conductor are tested. It is recommended that the test instrument has an open circuit voltage between 4 V and 24 V, d.c or a.c., and a short-circuit current of at least 200 mA.

Where the alternative form of protection against indirect contact has been adopted, i.e., limiting the resistance of the protective conductor in line with Table 41C (PCZ 10) then the resistance of such protective conductors will also have to be measured and checked for compliance with BS 7671.

Additionally, where supplementary bonding has been installed the resistance of the bond will be required to ensure that the resistance complies with 50 / Ia and the minimum sizes specified in Section 547.

Non-ferrous protective conductors

The test can be made with an ohmmeter as described above. The resistance reading obtained can be checked against the installation design to make certain that there are no high resistance joints.

Providing the phase conductor is the same length and follows the same route as the protective conductor, the ratio of the area of the phase conductor to the protective conductor can be used to determine the resistance of the protective conductor; the phase conductor can be used as the return conductor to the ohmmeter, and by temporarily connecting the phase conductor to the earth terminal at the distribution board, readings can then be taken at the end of the circuit.

Where a radial circuit has several outlet points, this test will combine some of the polarity tests which are required later in the test schedule.

The resistance of the protective conductor obtained by this method is given by the formula:

Formula 62 $$R_{pc} = R_0 \times \frac{A_p}{A_p + A_{pc}}$$

Where: A_{pc} = Area of protective conductor.
   A_p = Area of the phase conductor.
   R_0 = The ohmmeter reading.
   R_{pc} = Resistance of protective conductor.

To obtain the phase earth loop impedance by measuring the circuit's resistance with an ohmmeter, connect the phase conductor to earth at the consumer unit, then measure the resistance at each accessory or outlet point between phase and earth.

There is an alternative to the above method. This has to be used when checking the continuity of bonding conductors. In this case it involves using long test leads. One lead of the continuity tester is connected to the main earthing terminal, whilst the other test lead is connected to various exposed conductive parts of the circuit, or accessory earth terminals. This resistance obtained includes the resistance of the test leads, so this must be measured and deducted from the readings obtained.

When measuring the resistance of bonding conductors, the bonding conductors must be disconnected whilst the test is carried out, otherwise parallel paths can be included in the test giving an incorrect result. This test will also be checking in part the polarity of the conductors and connections.

Ferrous protective conductors

Where the protective conductor is a steel enclosure, conduit, or the steel wire armour of cables then tests can be made with a low resistance ohmmeter in a similar manner to checking the main equipotential bonding conductors.

If it is felt that the result from the ohmmeter test is not satisfactory then a test using a phase earth loop impedance tester can be used. In practice it is easier and quicker to use the phase earth loop impedance tester rather than the previous method described. If this test is not satisfactory then installation should then be rigorously subjected to a physical inspection to determine the problem and then re-tested after rectification work carried out.

There is a further test recommended if both the above tests are not satisfactory. The continuity test is made by passing a current 1.5 times the design (full load) current of the circuit through the phase conductor and steel CPC. The test current need not exceed 25 A, and the test voltage must not exceed 50 volts a.c., but must be at the frequency of the supply.

The test is shown in Figure 20.2, the resistance of the test leads T_A and T_B are also required, the resistance of the protective conductor being:

Formula 63 $\qquad R_{pc} = \dfrac{U_o}{I} - (R_{TA} + R_{TB})$

Test instruments have been developed by the instrument manufacturers to carry out this test, but if the procedure given under paragraph two above is used this test should be unnecessary.

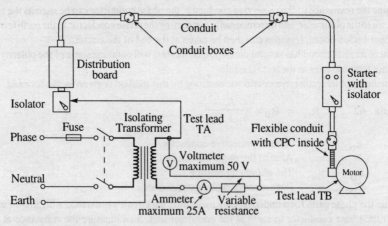

Figure 20.2 - Test 3 - Continuity of ferrous protective conductor

Ring circuit continuity

The continuity of ring circuit conductors has to be proved. This will include the protective conductor where it is installed in the form of a ring, for instance: when twin and CPC cable is used.

Test Method 1 (Figure 20.3)

It is first important to identify the outward leg conductors and the inward leg conductors of the ring. Connect the phase conductor of the outward leg to the neutral of the inward leg at the distribution board. Now take the resistance reading between the inward phase conductor and the outward neutral conductor (i.e. the phase and neutral conductors that have not been connected together).

The next stage is to connect these remaining phase and neutral conductors together. Now measure the resistance at each socket-outlet between the phase and neutral conductor; this should be approximately the same at each socket-outlet. If the connections at the distribution board are wrong, the resistance reading at each socket will get smaller the further you move away from the mid point of the ring.

Now repeat the above, but this time use the circuit protective conductor instead of the neutral conductor. Repeat the tests at each socket-outlet, but this time between phase and CPC. Keep a record of these test results, since they give the total resistance of the phase and CPC round the ring. The phase earth loop impedance will be a quarter of the value at the mid point socket-outlet.

Test method 2 (Figure 20.4)

There is another method of proving a ring circuit. The first test is made before the ends of each conductor are joined together, and the resistance of the loop is obtained for each conductor R_p, R_n and R_E The next test is made with the conductors of each loop joined together, and a shorting-out plug put in the socket nearest the mid point of the ring circuit.

The resistance is obtained between the phase conductor and the neutral, and the phase conductor and the CPC. The second set of resistance should then be as follows:

For the phase and neutral loop:

Formula 64 $$R = \frac{R_p}{4} + \frac{R_n}{4} = \frac{R_p}{2} \text{ or } \frac{R_n}{2}$$

For the phase to CPC loop:

Formula 65 $$R = \frac{R_p}{4} + \frac{R_E}{4}$$

The main difference between this test and the previous one is that the testing is all done at the distribution board, but it does require knowledge of which socket-outlet is near the mid point.

Insulation resistance

The soundness of the conductor's insulation has to be checked, since leakage currents will cause a deterioration and ultimate breakdown of the insulation.

The test has to be made before the installation is permanently connected to the supply, with a d.c., test voltage not less than that given below.

> **Extra-low voltage circuits (SELV and PELV)**
>
> Supplied from an isolating transformer to BS 3535. Test voltage 250 V d.c., minimum insulation resistance 0.25 megohm.
>
> **Low-voltage 50 V to 500 V**
>
> Test voltage 500 V d.c., minimum insulation resistance 0.5 megohm
>
> **Low-voltage above 500 V**
>
> Test voltage 1000 V d.c., minimum insulation resistance 1 megohm.

The test equipment used should be capable of maintaining the test voltage indicated above when a load of 1 mA is placed on the instrument.

It should be remembered that these test voltages can destroy electronic equipment, such as electronic starter switches in fluorescent fittings and electronic central heating controllers, electronic amplifiers in RCDs, and temperature sensing equipment. All these items as well as other electronic equipment, neon indicators and capacitors (such as those installed for power factor correction) should be temporarily disconnected.

Continuity test on socket-outlet ring circuit Method 1

OPEN CIRCUIT TEST

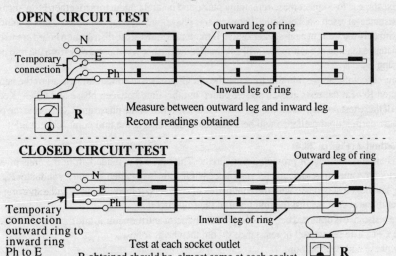

Measure between outward leg and inward leg
Record readings obtained

CLOSED CIRCUIT TEST

Temporary
connection
outward ring to
inward ring
Ph to E

Test at each socket outlet
R obtained should be almost same at each socket

Figure 20.3 - Testing ring circuit at each socket outlet

Continuity test on socket-outlet ring circuit Method 2

OPEN CIRCUIT TEST

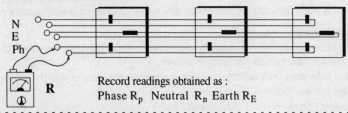

Record readings obtained as :
Phase R_p Neutral R_n Earth R_E

CLOSED CIRCUIT TEST

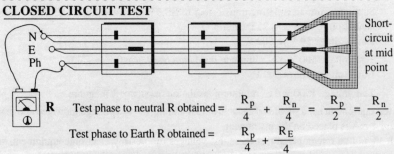

Short-circuit at mid point

Test phase to neutral R obtained = $\dfrac{R_p}{4} + \dfrac{R_n}{4} = \dfrac{R_p}{2} = \dfrac{R_n}{2}$

Test phase to Earth R obtained = $\dfrac{R_p}{4} + \dfrac{R_E}{4}$

Figure 20.4 - Testing ring circuit from distribution board

The first test is made between the earthing conductor and all live conductors, with all switches closed, all protective devices in place (or circuit breakers switched on), and all circuits connected. The test can be carried at the mains position with the phase and neutral conductors connected together, the test being made between these conductors and earth. It is only necessary to carry out the test at other distribution boards in large installations where, due to the long length of the cables involved, the capacitance of the conductors gives a poor reading. The only exception to this test is the TN-C system where the neutral is also the protective conductor and is therefore treated as an earth conductor.

The minimum insulation resistance allowed is given in the above table, but the installation is considered satisfactory if the above values or better are obtained. It is not necessary to remove lamps for this test, since both phase and neutral are being tested to earth. The TN-C system being the exception.

The next test is between the live conductors of the installation, this is best carried out to a pattern as illustrated in the following chart.

TEST BETWEEN	
Conductor	Conductors connected together
Red phase	Yellow & Blue phase and Neutral
Yellow phase	Red & Blue phase & Neutral
Blue phase	Red & Yellow phase & Neutral
Neutral	Red, Yellow & Blue phases

This test is not carried out on circuits containing electronic devices, reliance being placed on the test between the phase and neutral conductors connected together and the protective conductor. It would be wise to check with the manufacturer that this test will not damage the electronic components.

As this test is carried out between live conductors, it is necessary to disconnect equipment and remove lamps connected between the conductors; again, the minimum insulation resistance required is as given in the table above.

Earth electrode test
There are several ways in which the resistance of an earth electrode can be measured; some are easier than others, but more dangerous.

The easy way, but perhaps the most dangerous way, is to use a phase earth loop impedance tester. All other protective conductors are disconnected from the earth rod. The phase earth loop impedance tester is then connected between the phase conductor (through an HRC fuse) at the origin of the installation and the earth rod. The reading obtained is taken as the resistance of the earth rod. Extreme care is required when using this method of testing, since the test is being carried out at mains voltage. The connection from the tester should be made to the earth rod first, and everyone kept away from it. Connection is made to the mains last, and the instrument operated immediately, then disconnected from the mains.

A much safer way of checking the earth rod resistance is by using a Null Balance Tester (see Figure 20.5). The operating principal of the Null balance tester is a hand driven generator which passes an a.c., current through the earth electrode to a test probe T_1; the voltage between E and T_2 is measured and deflects a galvanometer.

This voltage is balanced by an equal but opposite voltage produced across an adjustable resistance within the instrument. The resistance in the instrument is adjusted in steps, until a balance

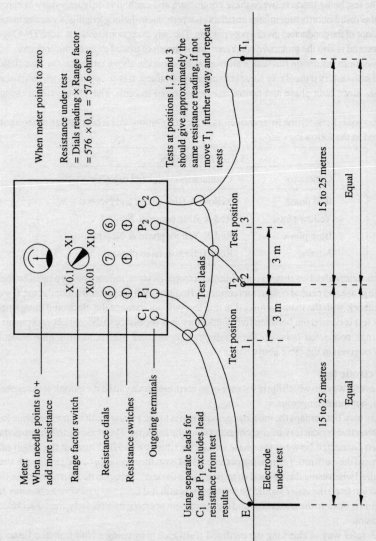

When meter points to zero

Resistance under test
= Dials reading × Range factor
= 576 × 0.1 = 57.6 ohms

Tests at positions 1, 2 and 3
should give approximately the
same resistance reading, if not
move T₁ further away and repeat
tests

T_1

Test position
3

15 to 25 metres

Equal

C_2

P_2

Test position
2

T_2

3 m

3 m

Test leads

C_1 P_1

Meter
When needle points to +
add more resistance

X 0.1   X1
X0.01 X10

⑤ ⑦ ⑥
⊕ ⊕ ⊕

Range factor switch

Resistance dials

Resistance switches

Outgoing terminals

Test position
1

Using separate leads for
C₁ and P₁ excludes lead
resistance from test
results

Electrode
under test

E

15 to 25 metres

Equal

Figure 20.5 - Null Balance Tester - used for measuring the resistance of an earth rod.

is reached between the internal voltage and the voltage between E and T_2, at which point no current flows in the potential circuit, the galvanometer reads zero, and the resistance on the instruments resistance dials gives the electrode resistance.

Adjustment of the resistance is made by looking at the meter, if it deflects to the plus sign then more resistance is added and if it deflects to the minus sign the resistance is reduced.

Three readings are obtained with the test probe T_2 in three positions, the first position being midway between the earth under test and probe T_1

The second reading is made with the test probe T_2 three metres nearer the earth under test, and the third reading is obtained with T_2 three metres further away from the earth under test. The arrangement and connections are shown in Figure 20.5.

The resistance should be practically the same in all three tests; the earth electrode resistance is then the average of the three readings obtained. If a large difference between each of the test readings is obtained, the tests should be repeated with probe T_1 moved further away from the earth under test.

Polarity

Polarity cannot be determined by inspection, and a test is therefore essential to check that the phase conductors are connected to the correct terminals, and that switches are only in phase conductors. The tests made àre carried out in the same way as those for checking protective conductors.

Earth fault loop impedance

Two tests have to be carried out: the first at the origin of the installation to determine the value of Z_e, and the second at the end of each final circuit to determine Z_s. The first test is made as near as possible to the origin of the installation: the main earthing terminal is a convenient point.

The requirement is for the main equipotential bonding conductors to be disconnected from the main earthing terminal for the first test. As a precaution the main earthing conductor should also be disconnected from the main earthing terminal at the same time otherwise, when the test is made, all of the installation's exposed conductive parts will rise to mains voltage. However, the extraneous conductive parts could be at earth potential since the equipotential bonding conductors have been disconnected. A person who is touching an exposed and extraneous conductive part at the same time the test is made is likely to receive an electric shock. See Figure 20.6.

If the supply for the test is taken from a spare fuseway in the main distribution board, the resistance of the tails can be subtracted from the reading obtained. Taking the supply through a fuse in the main distribution board is a safety precaution, in case the instrument, which will be hand held, becomes faulty.

For the second test, the main equipotential bonding conductors are re-connected to the main earth terminal, since they will contribute to the value of Z_s. The values of Z_s are then checked against the tables for shock protection, to verify that they comply with the limits laid down for shock protection.

It must be remembered that the design calculations for shock protection will have been based on a conductor temperature which allows for the temperature rise caused by the phase earth fault current I_f. This temperature will be higher than those prevailing at the time of test. The test results should therefore be reduced by an amount determined by the temperature at the time of test, and the temperature allowed in the design calculations.

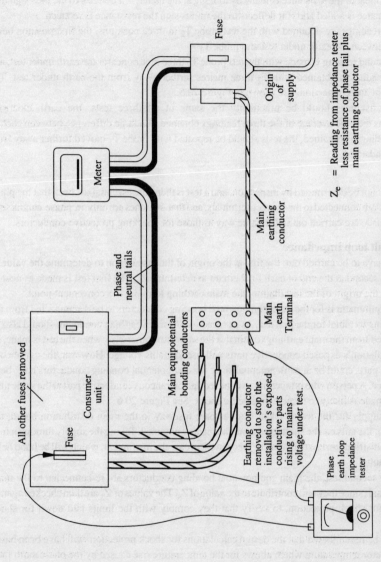

Z_e = Reading from impedance tester less resistance of phase tail plus main earthing conductor

Figure 20.6 - Determination of Z_e at the origin of the installation

Tables ZS 1 to 11 in Part 3 of the handbook give the Z_s values at various temperatures when copper conductors are used as CPCs, such as twin and CPC cable. On TT installations, the resistance of the earth electrode has to be included in the values of Z_s.

Where the alternative form of protection against indirect contact has been adopted, i.e., limiting the resistance of the protective conductor in line with Table 41C (PCZ 10) then the resistance of such protective conductors will also have to be measured and checked for compliance with the regulations.

Residual current devices

Where protection against indirect contact is provided by a residual current device a test of its effectiveness has to be made. Residual current devices are equipped with a test button which tests the device is operating at the correct sensitivity, and that the electrical and mechanical elements of the device are functioning. The test button does not test the circuit, protective conductors, or any earthing conductors or earth .

The device should be operated first by the test button to check that the device is not faulty before tests on the installation are made.

Principle of operation

For simplicity, consider the operation of a single-phase unit (see Figure 20.7). The live conductors of the circuit or installation are passed through a current transformer in the RCD (A). The current flowing down the phase conductor induces a current in the transformer in one direction, the current flowing back through the neutral conductor induces a current in the transformer in the opposite direction to that produced by the phase conductor. In a healthy circuit, the currents induced into the transformer by the phase and neutral currents are equal but opposite, and cancel out.

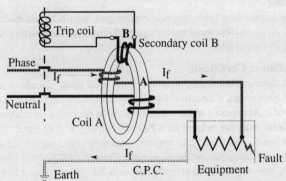

Figure 20.7 - RCD operation

When an earth leakage occurs, some of the current flowing through the phase conductor flows back to the source of energy through the protective conductor. The current flowing back through the neutral is therefore less than that flowing through the phase conductor, causing an imbalance in the current transformer (B), which generates a voltage across the trip coil of the RCD. When the earth leakage current is between 50% and 100% of the rated residual operating current of the RCD it will trip, disconnecting the circuit or the installation.

The RCD does not provide either overload or short-circuit current protection, and does not limit the magnitude of the fault current.

CHAPTER 20

RCD testing

The loads normally supplied by the RCD are disconnected during the test. The test is then made on the load side of the RCD, between the phase conductor and the circuit protective conductor, so that a suitable test current flows.

The first test is made with tripping current set at 50% of the rated tripping current of the RCD for a period of two seconds; during this period the RCD should not trip.

With a test current equal to the residual operating current of the RCD, an RCD manufactured to BS 4293 (BS EN 61008), the RCD should open in less than 0.2 s.

If an RCD has been installed to reduce the risk of shock from direct contact, a residual test current of 150 mA should cause the RCDs breaker to open in less than 40 ms. The test current must not be applied for a period exceeding 50 ms.

Having completed the test, the effectiveness of the test button on the RCD should again be checked to ensure that everything is satisfactory.

Where an S type RCD with a time delay is used, it should not open in less than 0.13 s, but should open within 0.5 s. The maximum time permitted for the test current is 0.5 s.

Where an RCD manufactured to BS 4293 (BS EN 61008) incorporating a time delay is used, it should trip within 50% and 100% of the rated time delay plus 0.2 s. The period of test should not exceed 2 s and care should be taken to ensure the protective conductor does not rise to a voltage exceeding 50 V.

Verification should also be made that the RCD has the correct withstand capacity for possible short-circuit or earth fault currents, and that discrimination between RCDs or between RCDs and other devices will be achieved where this is required to prevent danger.

Functional testing

A functional test is required for limit switches, control gear, interlocks and motors to ensure they are functioning correctly and that they have been properly mounted and that no control switches are mounted in the neutral conductor.

Electrical Installation Certificate

When a new installation is complete, or an alteration or addition is made to an installation is complete, and has been tested, a Electrical Installation Certificate has to be filled in; examples of which will be found in Appendix 6 of the regulations and at the end of this chapter. The Electrical Installation Certificate must not be used for a Periodic Inspection of an installation.

The Electrical Installation Certificate requires signing by the designer of the installation, the installer and the person who carries out the inspection and testing, all of whom should be competent persons, to certify that the installation complies with BS 7671. This Electrical Installation Certificate is required even for alterations or additions. The Electrical Installation Certificate must also be accompanied with a complete set of test results.

The certificate and test results should be at least in duplicate, the original copy being given to the person ordering the work and the copy being retained by the contractor. Where the inspection and test is carried out by an independent organisation, or several organisations are involved, such as, consulting engineers, area health authorities and local hospital engineers, then multiple copies will be required.

Any defects found in the installation must be rectified and the installation re-tested before an Electrical Installation Certificate is issued.

The results of tests should be carefully kept, so that if a problem occurs in the future, a properly completed Electrical Installation Certificate will help to prove that the installation was in good condition when handed over to the customer.

Where tests are carried out by maintenance staff in a factory or large commercial premises, recording the results enables a comparison with previous tests to be made; this can be helpful in preventing a fault developing, which can avoid a possible factory or production line shut-down.

The larger installation requires test schedules for each type of test; this is best achieved by producing a test schedule which also details the circuits installed from a distribution board. If these test schedules are numbered, it is only necessary to record the number on the Schedule of Test Results, and attach the test schedules to it.

Since each distribution board has to be labelled with details of the circuits it supplies with ratings and type of protective device, it is a good idea to have a distribution board chart in the same format as the Schedule of Test Results.

Samples of a Electrical Installation Certificate and distribution board test schedule are illustrated on the following pages.

Additionally an inspection notice of the form outlined in Regulation 514-12-01 has also to be fixed at the origin of the installation.

IMPORTANT

This installation should periodically be inspected and tested, and a report on its condition obtained, as prescribed in BS 7671, (formally the IEE Wiring Regulations for Electrical Installations) published by the Institution of Electrical Engineers.

Date of last inspection ....

Recommended date of next inspection ....

Minor Electrical Installation Works Certificate

Amendment No 2 to BS 7671 introduced for the first time a Minor Electrical Installation Works Certificate which is basically a 'short form' version of the Electrical Installation Certificate. The new form is intended only for minor electrical work, such as an addition or an alteration to an existing installation, which does not include the provision of a new circuit. In completing the certificate, the signatory is confirming that the subject work complies with BS 7671: 1992, except for any designer-sanctioned departures. A separate certificate is required for each modified circuit.

Periodic inspection and testing

When a periodic inspection is carried out, only those parts requiring more careful examination need to be partially dismantled. The first requirement is to check the installation to see whether any alterations or additions have been carried out. If they have, then the next step is to ask to see the Electrical Installation Certificates and test results for the alterations or additions. If these are not available, then the inspection of that part of the installation should be thorough and as though the installation had only just been completed.

As far as the rest of the installation is concerned, careful inspection and testing is required to see if any deterioration due to environmental conditions, or to mechanical damage has occurred. The installation should then be checked to ensure connections are still tight, barriers are still in place, and protective devices are of the correct size etc. The inspection and testing must be carried out so that no danger to persons, livestock or property will occur, particular attention being given to circuits containing electronic equipment.

The results of the inspection and test have to be recorded on a Periodic Inspection Report signed by the person carrying out the inspection (or a person authorised to act on their behalf) and given to the person ordering the inspection and test to be carried out.

The report will detail any dangerous conditions found and will contain all the test results obtained. The extent and limitations of the inspection and test should be agreed with the client and recorded in the "Extent and limitations" box on the Periodic Inspection report form. Needless to say, the inspection and testing must be carried out by a competent person.

The frequency that a periodic inspection and test has to be carried out is dependent upon the type of installation, how it is used and maintained; the list in the following section being used as a guide for the frequency of testing.

Periods of inspection

The frequency of the periodic inspection and testing has to be determined by the type of installation, the environmental conditions and the quality and frequency of maintenance that the installation is going to receive.

For installations having a normal environment, where maintenance is regularly carried out, the following periods between inspections are offered for guidance.

Periods between inspection	
Type of premises or installation	Maximum time between inspections - in years
Domestic installations	10
Commercial, educational and hospitals	5
Industrial	3
Cinemas, leisure complexes, restaurants, hotels, places of public entertainment, theatres, churches over 5 years old	1
Agricultural, horticultural, caravans and motor caravans	3
Caravan and motor caravan sites	1
Emergency lighting	3
Fire alarms, launderettes, petrol filling stations	1
Temporary electrical installations	3 months
Construction sites	3 months

Alterations and additions

Electrical Installation Certificate has to be provided along with test and inspection results when either an alteration or addition to an installation has been made. Any defects found in the alteration or addition must be rectified before a certificate is issued. Any defects found in the electrical installation related to the alteration or addition, have to be notified in writing to the person ordering the work. It must also be verified that the alteration or addition will not impair the safety of the existing installation.

COMPLETION CERTIFICATE   (ELECTRICAL)   Number:_____

Client's name / title _____   New installation / Alteration / Addition*

Address of installation tested _____   (*Delete what is not appropriate)

System of which installation is part (Type of Earthing)	TN-C	TN-S	TN-C-S	TT		IT
Characteristics of the supply	Nominal voltage		Frequency Hz		N° of phases	
Type of protective devices used in the installation (Tick box)	BS 3036	BS 1361	BS 88	MCBs	MCCBs	RCDs
Origin of the installation (Tick box)	Trans. L.V. Box		Electricity supplier's fuse			
Prospective short-circuit current at origin (Give value kA)	Three-phase	Single-phase		Three-phase	Single-phase	
State whether: measured, calculated or other means	Calculated	Measured	Other	Calculated	Measured	Other
Phase earth loop impedance at the origin (Give value Ze Ω)						
Protective device at origin - Number of poles:	BS type	Rating		BS type		Rating
Residual operating current of RCD at origin in mA (If fitted)						
Main linked switch or linked circuit breaker - N° of poles:	BS type	Rating		BS type		Rating
Main equipotential bonding conductors (Give material and CSA)	Material	CSA mm²		Material		CSA mm²
Maximum demand per phase (Give value in amperes)	Red phase		Yellow phase		Blue phase	
Earth Electrode - Type	Location	Method of measurement		resistance		Ω
Method of protection against indirect contact	Earthed equipotential bonding and automatic disconnection of supply					
	Other (specify)					

Tests carried out Date:			Tests recorded on test sheet numbers			
Continuity of main equipotential bonding conductors and size						
Continuity, resistance and size of supplementary bonding conductors						
Continuity of protective conductors						
Continuity of socket-outlet ring circuit conductors						
Insulation resistance live conductors to earth						
Insulation resistance between live conductors						
Polarity checked						
Earth fault loop impedance of each circuit						
Operation of residual current devices and operating time and current						
Checks of I_b, I_n, I_z with respect to G.A.T.S (C_g, C_a, C_i, C_4)						
Condition of switches, plugs, socket-outlets, flexible cable and cords						
Segregation of Category 1, 2 and 3 circuits checked						
Equipment checked for being correct for environmental conditions						
Special measures taken for protection against direct and indirect contact						
Protection against direct contact by site applied insulation (R 713-05)						
Protection by separation of circuits (R 713-06)						
Protection against direct contact by barrier or enclosure (R 713-07)						
Protection against indirect contact by non-conducting location (R 713-08)						

Extent of installation covered by this certificate

Comments on existing installation where an alteration or addition is carried out (R 743-01-01)

Deviations from BS 7671 The IEE Wiring Regulations (R120-02, 120-05)

The following portable equipment has / has not been tested

I /We being the person(s) responsible (as indicated by my / our signatures below) for either the design, construction or inspecting and testing the above installation, CERTIFY that the work for which I/we have been responsible is to the best of my / our knowledge and belief in accordance with BS 7671 (The IEE Wiring Regulations 16th Edition) amended to.............. except for any departures, if any stated in this certificate. The extent of liability of the signatory is limited to the work described above.

For the DESIGN of the installation.   Name (Block letters)_____ Position _____ Signature_____

For and on behalf of _____ Address _____ Date: _____

For the CONSTRUCTION of the installation. Name (Block letters)_____ Position _____ Signature_____

For and on behalf of_____ Address _____ Date: _____

For the INSPECTION and TEST of the installation. Name (Block letters)_____ Position _____ Signature_____

For and on behalf of _____ Address _____ Date: _____

This installation should be further inspected and tested on or before :-

The tests outlined on this sheet have been extracted from BS 7671 IEE Wiring Regulations 16th Edition   © TEM 1995

Figure 20.8 - Electrical Installation Certificate

PERIODIC INSPECTION REPORT (ELECTRICAL)   Number: _____

Client's name	Address

Purpose for which the report is required:

DETAILS OF THE INSTALLATION

Occupier		Address			

Description of premises:	Domestic	Commercial	Industrial	Other

Estimated age of the Electrical Installation:	years.

Any alterations or additions	Yes	Age	No		Not apparent

Date of last inspection		Records available	Yes ☐	No ☐	Records held by:

System of which installation is part (Type of Earthing)	TN-C	TN-S	TN-C-S	TT	IT

Characteristics of the supply	Nominal voltage	Frequency Hz	N° of phases

Type of protective devices used in the installation (Tick box)	BS 3036	BS 1361	BS 88	MCBs	MCCBs	RCDs

Origin of the installation (Tick box)	Transformer L.V. Box	Electricity supplier's fuse

Prospective short-circuit current at origin (Give value kA)	Three-phase Single-phase	Three-phase Single-phase

State whether: measured, calculated or other means	Calculated Measured Other	Calculated Measured Other

Phase earth loop impedance at the origin (Give value Ze Ω)		

Protective device at origin - Number of poles:	BS type Rating	BS type Rating

Residual operating current of RCD at origin in mA (If fitted)		

Main linked switch or linked circuit breaker - N° of poles:	BS type Rating	BS type Rating

Main equipotential bonding conductors (Give material and CSA)	Material CSA mm²	Material CSA mm²

Earth Electrode - Type	Location Method of measurement resistance Ω

Method of protection against indirect contact	Earthed equipotential bonding and automatic disconnection of supply ☐
	Other (specify)

Extent of electrical installation covered by this report:

Limitations:

RECOMMENDATIONS

The following were found unsafe and require immediate attention:

SUMMARY OF THE INSPECTION	Date of inspection:	Results recorded on test sheet numbers		
Continuity of main equipotential bonding conductors and size				
Continuity of supplementary bonding conductors				
Continuity of protective conductors				
Insulation resistance live conductors to earth				
Insulation resistance between live conductors				
Earth fault loop impedance of each circuit				
Operation of residual current devices and operating time and current				
Polarity checked (At same time as earth loop impedance or RCD test)				
Checks for correct protective devices installed on each circuit				
Checks that all barriers are installed in switchgear, distribution boards etc.				
Condition of switches, plugs, socket-outlets, flexible cable and cords				
Checks that Segregation of Category 1, 2 and 3 circuits still correct				
Equipment checked for being correct for environmental conditions				
General condition of the installation:				

Test sheet results are marked as follows: ☐1☐ Requires urgent attention ☐2☐ Requires improvements ☐3☐ Requires further investigation
☐4☐ Does not comply with the latest BS 7671 (the IEE Wiring Regulations) but not unsafe.

The following portable equipment has / has not been tested

To the best of our knowledge and belief we confirm that the details recorded above and on the attached test schedules and the recommendations given are an accurate assessment of the condition of the installation within the limits specified above.

INSPECTED BY: Name (Block letters) _____ Position _____ Signature _____

For and on behalf of _____ Address _____ Date: _____

REVIEWED BY: Name (Block letters) _____ Position _____ Signature _____

For and on behalf of _____ Address _____ Date: _____

This installation should be further inspected and tested on or before :-

The tests outlined on this sheet have been extracted from BS 7671 IEE Wiring Regulations 16th Edition   © TEM 1995

Figure 20.9 - Periodic inspection report

DISTRIBUTION BOARD OR SWITCHGEAR REF:				TEST CERTIFICATE NUMBER :	
Insulation resistance at busbars					
R-Y-B & N to Earth	R to Y-B-N	Y to R-B-N	B to R-Y-N		

Maximum demand	R	Y	B
Design value			
Actual value			

D.B.chart fitted	
Z_S at busbars	
I_P at busbars	

		Signatures	Weather conditions
Date of test			
Person making test			
Witness			

Instrument serial numbers

Circuit				Protection			Cable						Equipment			Continuity of ring circuit					Continuity	Phase earth loop impedance Z_s					Voltage
											Insulation resistance					Open circuit test			Closed circuit test				Test				
											Phase to																
Nº	Phase	I_b	Checked	I_n	Type	Checked	Method of installation	Applicable derating factors	I_{tab}	Checked	Phase	Earth	Polarity checked	Single pole switches in phase conductors	RCD's	P Ω	N Ω	CPC Ω	P - N Ω	P - CPC Ω	Protective conductors	Design Z_s	Z_s	@ °C	Z_s allowed @ C	Position of test	Calculated Volt drop
	R																										
1	Y																										
	B																										
	R																										
2	Y																										
	B																										
	R																										
3	Y																										
	B																										
	R																										
4	Y																										
	B																										
	R																										
5	Y																										
	B																										
	R																										
8	Y																										
	B																										
	R																										
9	Y																										
	B																										
	R																										
10	Y																										
	B																										
	R																										
11	Y																										
	B																										
	R																										
12	Y																										
	B																										

Figure 20.10 - Distribution board test sheet

DISTRIBUTION BOARD REF:

D.B. Size

Feeder cable size

Feeder cable type

Maximum demand allowed

Final circuit Volt drop allowed

Maximum I_p

Minimum I_p

Parameters at Bus-Bars

Z_p / Z_{pp}

Z_{pn} / Z_s

Circuit			Description	Protection			Cable		Length	Voltage drop	Installation method	Protective Conductor		Impedances		Total Load		
N°	Phase	I_b		I_n	Type	I_z	Size	Type				Type	Z_s	Z_p or Z_{pp} or Z_{pn}	Z_s	Red	Yellow	Blue
1	R1																	
	Y1																	
	B1																	
2	R2																	
	Y2																	
	B2																	
3	R3																	
	Y3																	
	B3																	
4	R4																	
	Y4																	
	B4																	
5	R5																	
	Y5																	
	B5																	
6	R6																	
	Y6																	
	B6																	
7	R7																	
	Y7																	
	B7																	
8	R8																	
	Y8																	
	B8																	
9	R9																	
	Y9																	
	B9																	
10	R10																	
	Y10																	
	B10																	
11	R11																	
	Y11																	
	B11																	
12	R12																	
	Y12																	
	B12																	

Figure 20.11 - Distribution board chart

MINOR ELECTRICAL INSTALLATION WORKS CERTIFICATE

(REQUIREMENTS FOR ELECTRICAL INSTALLATIONS - BS 7671 (IEE WIRING REGULATIONS)
To be used only for minor electrical work which does not include the provision of a new circuit

---

PART 1: Description of minor works

1. Description of the minor works

2. Location/address

3. Date minor works completed

4. Details of departures, if any, from BS 7671 : 1992 (as amended)

PART 2: Installation details

1. System earthing arrangement (where known)   TN-C-S ☐   TN-S ☐   TT ☐

2. Method of protection against indirect contact

3. Protective device for the modified circuit   Type............................   Rating............A

Comments on existing installation, including adequacy of earthing & bonding arrangements: (see Regulation 130-09)

PART 3: Essential Tests
Earth continuity satisfactory ☐

Insulation resistance:
  Phase/neutral............................ M Ω

  Phase/earth............................ M Ω

  Neutral/earth............................ M Ω

Earth fault loop impedance...................................... Ω

Polarity satisfactory ☐

RCD operation (if applicable). Rated residual operating current ....mA and operating time of ....ms

---

PART 4: Declaration

I/We CERTIFY that the said works do not impair the safety of the existing installation, that the said works have been designed, constructed, inspected and tested in accordance with BS 77671 : 1992 (IEE Wiring Regulations), amended to .... and that the said works, to the best of my/our knowledge and belief, at the time of my/our inspection, complied with BS 7671 : 1992 except as detailed in Part 1.

Name: ....   Signature: ....

For and behalf of: ....   Position: ....

Address: ....   Date: ....

....

PART THREE

ELECTRICAL INSTALLATION TABLES

Contents Part 3

Note: When a BS 3036 (semi-enclosed/rewirable) fuse is used only for fault current protection, i.e., when the load is not likely to cause an overload, as is the case with immersion heaters, electric radiators etc., tables CSG 1 to CSG 20 can be used for determining the size of conductor. When the BS 3036 fuse is used for overload protection, then tables CSG 21 to 24 should be used.

How to use the tables

Introduction

Part 3 of the handbook has been produced to assist in the sizing of cables without having to carry out complicated calculations. They can also be used as a quick method of checking an installation during an inspection of an installation.

Basis of tables

Since the Regional Electricity Companies are not changing their nominal voltage U_0 to 230 V, most installations will, as before, be operating at a U_0 of 240 V i.e., U_{oc} of 250 V. The voltage drop in the CR tables is therefore based on 240 V. The values of Z_s in the ZS tables are based on U_0 of 230 V, i.e., U_{oc} of 240 V, to keep them in line with the regulations, this means that for U_0 of 240 V they err on the side of safety.

Sizing conductors

This subject has been covered in detail in Part 2 of the handbook. As an outline to using the tables in Part 3, the order of working out the size of conductor are listed as follows:

1. Work out the cable size required from the CR tables to satisfy voltage drop.
2. Determine whether:
   a) the circuit is being protected against overload and fault current, or
   b) only against fault current.
3. Check that the current-carrying capacity of the cable size chosen is suitable for either:
   (a) of item 2, i.e., I_n, or (b) of item 2, i.e., I_b.
4. Check that the current-carrying capacity of the cable size chosen is still suitable after taking into consideration the derating factors for: grouping, ambient temperature, thermal insulation, and a BS 3036 (rewirable) fuse if this type of fuse is used for item 2a.
5. Check that the circuit has protection against indirect contact.
6. Check that the thermal protection of the protective conductor is achieved.
7. Check that the live conductors are thermally protected against fault current.

Installation tables

The INST 1 and 2 tables give the maximum lengths of cable that can be installed for resistive loads to comply with BS 7671. These tables can be used for conduit and MICC cable, providing the live conductors are of the same size as those given in the table.

No allowance has been made in the tables for derating factors. Where circuits are grouped the appropriate cable size should be selected from the CSG tables, the cable length given in the INST table will still be applicable for the cable size chosen from the CSG table.

The INST 3 table enables the voltage drop to be worked out at any lighting point on a circuit wired using the loop in method of wiring. It is based on ten lighting fittings each loaded at 100 watts and can therefore be easily adjusted for other loads.

The INST 4 table covers the maximum length of radial and ring circuits. Since voltage drop cannot be determined by the designer, as he has no control over what the client may plug into the socket-outlets, the ring circuit voltage drop has been taken as the average of the minimum and maximum voltage drop, the maximum length being where the socket-outlets are equally distributed round the ring, and each socket is equally loaded. The minimum length of the ring circuit is based on all the load being placed at the mid point. For both ring and radial circuits the load has been taken to be equal to the size of the protective device rating. The maximum floor area allowed for each ring circuit is 100 square metres.

CR current rating tables

The tables give the maximum current-carrying capacity for single circuits based on an ambient temperature of 30 °C, when carrying the tabulated current (referred to as I_{tab}) given in the table. The maximum permitted operating temperature of the conductors will be that shown at the top of each table, and the ambient temperature correction factors for the most common form of conductor insulation are given at the bottom of each table, both for BS 3036 fuses, and for other types of protective device.

Voltage drop calculations using the CR tables

A factor comprising length × amps is given to enable a quick assessment of the cable size required for a voltage drop of 4% with U_0 240 V and U_{0l} of 415 V these being standard electricity supplies.

Using the factor (length × amps) based on U_0 240 V.
1. I_b is the design, or full load current (f.l.c.) taken by the circuit in amperes.
2. L is the length of the circuit in metres.
3. 'Allowed Vd' is 4% voltage drop, i.e. 9.6 V for single-phase, or 16.6 V for three-phase.
4. 'Available Vd' is the amount of voltage drop available after deducting voltage drop in feeder cable to distribution board.

This means that only two numbers have to be remembered: 9.6 V for single-phase and 16.6 V for three-phase.

Determination of cable size for voltage drop when 9.6 V or 16.6 V available for circuit
Multiply length of circuit L by full load current of equipment I_b to give factor required from the CR tables.
Formula 66: Factor (from CR table) $\geq L \times I_b$.
Select a factor equal to, or larger than, the value calculated. This gives the cable size required. The current-carrying capacity for the cable size chosen should now be checked to ensure that the cable is capable of carrying the circuit's full load current I_b.

Factor for cable size when 'Available Vd' is less than ' Allowed Vd'.
Allowed volt drop (9.6 V or 16.6 V) minus volt drop in feeder cable = Available volt drop.

Formula 67: Factor (for CR table) $\geq \dfrac{\text{Length} \times I_b \times \text{Allowed Vd}}{\text{Available Vd}}$

Now look in a suitable table, and choose a Factor equal to, or larger than the Factor calculated.

Determination of maximum circuit length when conductors are not carrying I_{tab}.
Where the full voltage drop allowance either for single-phase or three-phase is available,

Formula 68: L (length allowed for circuit) $= \dfrac{\text{Factor}}{I_b}$

Circuit length when f.l.c. is less than I_{tab} and Available Vd. is less than the Allowed Vd.
Where both conditions (2) and (3) are applicable, the formula can be combined.

Formula 69: L (cable length allowed) $= \dfrac{\text{Factor}}{I_b} \times \dfrac{\text{Available Vd}}{\text{Allowed Vd}}$

Finding the actual volt drop in a circuit for a given size and length of cable.

Formula 70   Vd (voltage drop in circuit) $= \dfrac{L \times I_b \times \text{Allowed Vd}}{\text{Factor}}$

Factor for any other percentage voltage drop

Where the percentage voltage drop allowed is different from 4%, divide the percentage voltage drop required by four. This ratio can be applied to any factor in the CR tables.

Formula 71   Table Factor $\times \dfrac{\% \text{ Vd required}}{4}$ e.g., for 2.5 % Vd $= \dfrac{2.5}{4} = 0.625 \times$ Factor

Determination of impedance mV/A/m

There may be occasions when an engineer requires to know the mV/A/m for a given cable; this can be obtained as follows:

Formula 72   Single-phase circuits :   mV/A/m $= \dfrac{9.6 \text{ V} \times 1000}{\text{Factor}}$

Formula 73   Three-phase circuits :   mV/A/m $= \dfrac{16.6 \text{ V} \times 1000}{\text{Factor}}$

Power factor

Adjustments can be made for those circuits affected by power factor by using the 'r' and 'x' values from the CR tables. The sine of power factor can be obtained from SPF 1 'The Sine of Power Factor' table in Part 4.

Formula 74   Vd (voltage drop) $= \dfrac{(\text{Cos } \varnothing r + \text{ Sin } \varnothing x) \times L \times I_b}{1000}$

Formula 75   L (circuit length allowed) $= \dfrac{\text{Available Vd } \times 1000}{(\text{Cos } \varnothing r + \text{ Sin } \varnothing x) I_b}$

Allowance for conductor temperature

Where the ambient temperature is not less than 30 °C; where the protective device is not a BS 3036 fuse; where the conductor is not carrying its full rated current; and where the conductors are larger than 16 mm^2, allowance can be made for the reduction in voltage drop due to the resistance of the conductor being lower than its stated maximum permitted operating temperature.

Formula 76:   L (circuit length allowed) $= \dfrac{\text{Available Vd } \times 1000}{(C_t \text{ Cos } \varnothing r + \text{ Sin } \varnothing x) I_b}$

For cable up to 16 mm^2, the increase in length can be determined by increasing the Factor in the CR tables, as given in the following formula.

Formula 77:   New Factor to use in formula $= \dfrac{\text{Factor}}{C_t}$

The calculation of C_t is given in Part 2, in the Chapter on Voltage drop.

ZS tables: earth loop impedance tables

These tables are provided for checking that a circuit will comply with the requirements for protection against indirect contact.

When designing a circuit, the Z_S values in the row headed 'Design' should be used. To be absolutely safe, the average of the conductor operating temperature plus the limit temperature of the conductor's insulation should be used when selecting the resistance (or impedance) of the phase and protective conductors. Tables are provided in Part 4 giving the earth loop impedance of various types of cable. The value Z_S is obtained from:

Formula 33 $Z_S = Z_e + Z_{inst}$

where Z_e is the earth loop impedance up to the origin of the circuit, and Z_{inst} is the phase earth loop impedance of the circuit up to the point being checked.

When testing an installation, the conductors will not be operating at the temperature used when designing the circuit: i.e., 115 °C in the case of PVC, so the values used from the tables are those listed under testing for the ambient temperature ruling at the time of the test.

The values in the tables are based on $U_{oc} = 240$ V, for any other open circuit voltage divide the figure in the table by 240 and multiply by the actual open circuit voltage. U_{oc} of 240 V equals U_0 of 230 V as given in the regulations.

PCZ tables: protective conductor tables

Having worked out the Z_S for the protection against indirect contact, that value can now be used to check whether the circuit protective conductor is suitable for the thermal stresses caused by the fault current, by using the PCZ tables.

Look in the column for the size of protective device protecting the circuit, and then in the row for the protective conductor size. If the Z_S is less than the figure in the table, the protective conductor is satisfactory. If the Z_S is larger than the figure in the table, re-calculate Z_S using a larger protective conductor until the circuit Z_S is less than that given in the table.

When MCBs are used as the protective device, the circuit Z_S must not be less than the minimum value given in the last column.

Although the PCZ tables are based on the thermal capacity of copper conductors, they can be used when conduit or trunking is used as a protective conductor, since the thermal capacity of the conduit is larger than that of the copper conductor which can be installed through it, in accordance with the regulations. The suitability of conduit can easily be checked by using Table M54G in Part 4. Where MICC is used, the area of the sheath is given in Table ELI 5.

SCZ tables: short-circuit protection

The SCZ tables are used to check that the live conductors are suitable for the thermal stresses caused by a short-circuit.

The impedance of the phase and neutral conductor Z_{pn} is worked out at the design temperature, and then checked with the value given in the table under the protective device rating, and opposite the conductor size. If the calculated value of Z_{pn} is less than the table value the conductor is protected

Where three-phase three-wire circuits are concerned, an adjustment has to be made to the calculated impedance. The impedance of two phase conductors should be divided by 1.732 before comparing it with the value in the table. The reason for this adjustment is that the table is based on the single-phase voltage of 240 volts.

In TN-C-S systems, the declared Z_e at the origin of the installation can be used to give the external phase/neutral impedance. This value can then be used for calculating the minimum short-circuit current. In order to check the thermal capability of the conductors, the Z_e is added to the phase and neutral impedance up to the point of fault, to give the total Z_{pn} .

When checking breaking capacity (maximum fault current) in a system, the external Z_{pn} will be equal to 240 V divided by the prospective short-circuit current at the origin of the installation. Thus, for domestic premises, the external Z_{pn} $\Omega = 240$ V $\div$ 16,000 A $= 0.015$ Ω This figure is then used in the calculations to give the total phase/neutral impedance, for determining the maximum short-circuit current, to check whether equipment has the correct breaking capacity.

CSG tables: cable sizes required when circuits grouped together

The CSG tables are provided to give the cable size required for different types of grouping, and different methods of installation. All that is required, is to look under the column for either I_n or I_b (either single-phase or three-phase) for the number of circuits or cables grouped together in the left hand column, to obtain the cable size.

Where the load is not likely to cause overload current (resistive load), the design current I_b is used instead of the rating of the protective device I_n, when derating for grouping.

Derating for BS 3036 fuses is only required if the load is not resistive and they are used for overload protection. Where they are used for overload protection, Tables 21 to 24 have been provided and take into account the derating factor 0.725.

CCC 1 and CCT 1 tables: cable capacity of conduit and trunking

The details of how to use the tables CCC 1 and CCT 1 are given with the tables.

Worked examples

To illustrate the use of the tables, the following worked examples are given by using the tables from Parts 3 and 4.

Example 1 (Heating load)

A single-phase circuit is to supply a fixed heating load of 6 kW; the cable is to be twin and CPC cable clipped direct to a surface along with three other cables, with their sheaths touching. If the circuit length is 25 metres and is to be supplied from a BS 1361 fuse, what size of fuse and cable is required, if the system is TN-C-S, and Z_e is 0.35? The ambient temperature is 30 °C, and the cables will not be in contact with thermal insulation.

Working

Fixed load: so use INST 2, with a disconnection time of 5 s.

From INST 2, the BS 1361 fuse size required is 30 A, full load current is 25 A, but the load is a heating load, so I_b can be used for derating. From CSG 5, the cable size required for 4 circuits grouped, and an I_b of 25 A is 6 mm^2.

From INST 2, the maximum length for 6 mm^2 cable for a TN-C-S system is 52 metres; this is twice the length required for the circuit, so the cable is suitable.

Example 2 (Motor circuit)

A 75 kW DOL 415 V motor with a power factor of 0.7 is to be supplied from an HRC fuse board 100 metres from the motor. The installation is to be carried out with an XLPE armoured cable installed on a cable tray, with one other cable. The starter will be adjacent to the motor; the overload settings will be 130 A minimum 160 A maximum. If the length of the cable is 100 metres, and the three-phase voltage drop up to the distribution board is 4 volts, determine that the circuit complies with the regulations if Z_e is 0.03 Ω and Z_{pn} is 0.03 Ω.

Working

From Table MC1, 75 kW at 415 V gives f.l.c. of 136 A , HRC fuse size required is 200M250
Maximum overload setting is 160 A. This must be used since overloads can be turned up.
From CSG 14, two circuits with an overload setting of 160 A, so cable size required is 50 mm^2
From CR10, 50 mm^2 cable, has a current-carrying capacity of 197 A, and a Factor of 19080.

Formula 69   L (length for cable) $= \dfrac{19080}{160} \times \dfrac{(16.6 - 4)}{16.6} = 90.515\text{m}$

The cable length allowed is too short, so see if allowing for power factor will help.
From CR10, R = 0.86 and x = 0.135
From SPF 1 'Sine of Power Factor' table Part 4, for a power factor of 0.7 sine = 0.7141

Formula 75   Maximum length $= \dfrac{(16.6 - 4) \times 1000}{(0.7 \times 0.86 + 0.7141 \times 0.135)\, 160\, A} = 112.76\ m$

The length of the cable will be suitable for voltage drop.
Actual voltage drop in the cable if overloads are turned up to maximum.

Formula 74   Voltage drop $= \dfrac{(0.7 \times 0.86 + 0.7141 \times 0.135) \times 100\ m \times 160\ A}{1000} = 11.17\ V$

Add 4 V voltage drop up to distribution board, total voltage drop = 15.17 V.

Protection against indirect contact

Now check that protection against indirect contact is provided, i.e., Formula 33: $Z_s = Z_e + Z_{inst}$.
From Table ELI 3 Part 4, value of Z_{inst} per 1000 m = 3.468 Ω = 0.003468 Ω per m.
Z_s = 0.03 + (100 × 0.003468) = 0.3768
From ZS 1A Part 3, Z_s allowed for 250 A HRC fuse is 0.16, cable not therefore suitable.
Try using a 4 core cable with one core used as the protective conductor.
From R XLPE 1 Part 4, impedance of 50 mm² cable is 0.624 Ω per 1000 m.
Z_{inst} = 100 × 2 × 0.000624 = 0.1248, therefore Z_s = 0.03 + 0.1248 = 0.1548 Ω.
This arrangement would be suitable. The Z_s would be less than 0.1548 Ω since the armour would be in parallel with the protective conductor, additionally, there would be a reduction in the cable's resistance since it is not carrying the maximum permitted I_{tab} given by the table.

Protective conductor

Now check the thermal capacity of the protective conductor. From PCZ 1 Part 3, 250 A fuse 50 mm² CPC, Z_s allowed is 0.163 Ω, this is larger than Z_s so thermal protection is provided.

Short-circuit protection

Now check suitability of the live conductors for short-circuit protection. The Z_{pp} for the cable will be 100 × 2 × 0.000624 = 0.1248, This has to be divided by √3 to use the SCZ tables = 0.072. Z_{pn} up to the distribution board is 0.03 Ω, therefore, total Z_{pn} = 0.03 + 0.072 = 0.102 Ω.
From SCZ 7 Part 3, Maximum Z_{pn} allowed for 250 A fuse and 50 mm² cable is 0.232 Ω.
This is larger than the total Z_{pn} so live conductors are protected.

Example 3 (Fixed welding load)

A 415 V three-phase fixed 28 A welding load fed from an HRC fuse in a sub-distribution board. The PVC cables are to be installed in a conduit with a single-phase circuit using 1.5 mm² cable, and a three-phase circuit using 2.5 mm² cable. It is anticipated that there will be two bends between draw-in boxes, which will be spaced 3 m apart. The three-phase voltage drop up to the distribution board is 4 V, and the total voltage drop is to be limited to 4%. Z_s at the distribution board bus bars is 0.5 Ω, and Z_p at the bus bars is 0.3 Ω. What size cable will be required if the circuit length is 80 metres?

Working

Fuse size required for 28 A load is 32 A, assuming there is no inrush current.

Grouping - cable size

Number of circuits = 3. From CSG 1 Part 3, three-phase conductor size required = 10 mm²
From Table CR 1 Part 3, I_{tab} = 50 A, Factor = 4368.

Voltage drop

Formula 69   Cable length allowed $= \dfrac{4368}{28} \times \dfrac{(16.6 - 4)}{16.6} = 118.4$ m - satisfactory

Conduit size

From table CCC 1 Table 1 Part 3,   factor for 1.5 mm^2 cable $=$ $22 \times 2 =$ 44
                                factor for 2.5 mm^2 cable $=$ $30 \times 3 =$ 90
                                factor for 10 mm^2 cable $=$ $105 \times 3 =$ $\underline{315}$
                                                      Total $=$ 449

From CCC 1 Table 2 Part 3, two bends 3 m between boxes so conduit size required = 32 mm

Protection against indirect contact

From table RC 1 Part 4, resistance of 10 mm^2 cable at 115 °C $=$ 0.002525 Ω/m
From table ZCT 1 Part 4, impedance of 32 mm conduit $=$ $\underline{0.002}$ Ω/m
                                                  Total 0.004525 Ω/m

$Z_{inst} = 80 \times 0.004525$ $\Omega = 0.362$ Ω.
Now Z_S at the distribution board busbar is the Z_e for the following circuit.
Therefore Formula 33: $Z_S = 0.5 + 0.362$ $\Omega = 0.862$ Ω.
From ZS 1 A, for 5 s and 32 A fuse, Z_S allowed $= 1.92$ Ω - This is satisfactory.

Protective conductor

Checking with Table M54G, PCA 1 in Part 4.
10 mm^2 phase conductor, protective conductor area $= 10 \times 2.45 = 24.5$ mm^2.
From PCA 1 Part 4, area of 32 mm heavy gauge conduit $= 167$ mm^2.
So conduit is satisfactory as a circuit protective conductor.

Short-circuit protection

Z_p up to the distribution board is 0.3 Ω
Resistance of 10 mm^2 phase conductor $= 0.002525$ $\Omega \times 80 = 0.202$ Ω
Total $Z_{pp} = 2(0.3 + 0.2)$ $\Omega = 1.0$ Ω
To compare the value of 1 Ω with the SCZ 1 table in part 3 divide by $\sqrt{3}$, i.e., 1.732

$$Z_{pn} = \frac{1.0}{1.732} = 0.577 \ \Omega$$

From SCZ 7 Part 3, maximum impedance allowed $= 1.7$ Ω - This is satisfactory.

Example 4 (Lighting circuit voltage drop)

A lighting circuit comprising ten lighting points is wired in 1 mm^2 cable using the loop in system of wiring. If each lighting point is fitted with a 150 watt lamp, what will be the voltage drop at the sixth and last lighting point if the distance from the distribution board to the first lighting point is 10 m, with 5 m between lighting points and 11 m from the sixth and last lighting point to its switch? The electricity supply U_o is 240 V?

Working

The worst case would be with all the lamps on. INST 3 table Part 3, gives the voltage drop for 100 watt lamps, therefore the values in the table will require multiplying by 1.5.

    Table 1: voltage drop from distribution board to first lighting point 10 lights 10 m $=$ 1.83
    Table 2: voltage drop first to second light for 5 m $=$ 0.83
    Table 2: voltage drop second light to third light 5 m $=$ 0.73
    Table 2: voltage drop third light to fourth light 5 m $=$ 0.64

Table 2: voltage drop fourth light to fifth light 5 m	=	0.55
Table 2: voltage drop fifth light to sixth light 5 m	=	0.46
Total voltage drop to sixth light	=	5.04
Table 2: voltage drop sixth light to seventh light 5 m	=	0.37
Table 2: voltage drop seventh light to eighth light 5 m	=	0.28
Table 2: voltage drop eighth light to ninth light 5 m	=	0.18
Table 2: voltage drop ninth light to tenth light 5 m	=	0.09
Total		5.96
Table 3: add voltage drop from tenth light to switch 11 m	=	0.20
Total	=	6.16

This is the total for 100 watt lamps which must now be multiplied by 1.5.

$6.16 \times 1.5 = 9.24$ V, this is less than the 9.6 Volt drop allowed so is satisfactory.

Voltage drop at the sixth light $= (5.04 \times 1.5) + (0.2 \times 1.5)$ (Table 3 light to switch) $= 7.86$ V.

General

Short cuts could obviously be made in the examples given, but all the steps have been included to illustrate how the various tables are used. Additionally, the answers have been given to three or four decimal places; this has been done so that those who want to work through the examples can compare answers. In practice answers will probably be rounded up or down to whole numbers.

Conduit

As far as conduit is concerned, there will be no problem using it in a normal installation as a protective conductor, because its thermal capacity will always withstand the fault current flowing through it.

Installing a protective conductor within the conduit really is a waste of time and money. In the past, heavy gauge galvanised solid drawn conduit has been installed in hazardous areas without any additional protective conductors, and without any problems arising from the installation.

The point is often made that a protective conductor is installed in the conduit in case the conduit becomes broken, corroded, or otherwise damaged. The answer to this point is that the conduit has not been installed in accordance with the requirements in the first place, as can be seen if the requirements given under 'Conduit' in Part 1 are studied.

Conduit, like all other parts of the installation, relies very heavily on good workmanship (Regulation 13.01.01). Since conduit and trunking are often specified for the wrong environmental conditions, and rely heavily on good workmanship, the view could be taken that installations using conduit and trunking as a protective conductor should be subject to regular inspection and testing. The need for such inspection and testing is, however, the fault of the designer, and or installer, not of the conduit or trunking.

Welding supplies

Example 3 dealt with a fixed welding circuit and assumed no inrush current, in practice there would be an inrush current, but more important the load would contain current spikes. The true full load current would be the r.m.s. value of these current spikes, which can be higher than the full load current rating of the welder. Since this information may not be available when designing the installation the tendency should be to oversize the switchgear and cables.

Maximum lengths in metres of single-phase circuits with PVC cable with various loads in areas requiring a DISCONNECTION TIME OF 0.4 s

Maximum length in metres for voltage drop, protection against indirect contact, thermal protection of cpc and short-circuit current

Resistive load in KW	Full load current in amperes	Protective device type	Protective device rating In	1.0 mm² twin & cpc TN-C-S ZE=0.35	1.0 mm² twin & cpc TN-S ZE=0.8	1.5 mm² twin & cpc TN-C-S ZE=0.35	1.5 mm² twin & cpc TN-S ZE=0.8	2.5 mm² twin & cpc TN-C-S ZE=0.35	2.5 mm² twin & cpc TN-S ZE=0.8	4 mm² twin & cpc TN-C-S ZE=0.35	4 mm² twin & cpc TN-S ZE=0.8	6 mm² twin & cpc TN-C-S ZE=0.35	6 mm² twin & cpc TN-S ZE=0.8	10 mm² twin & cpc TN-C-S ZE=0.35	10 mm² twin & cpc TN-S ZE=0.8	16 mm² twin & cpc TN-C-S ZE=0.35	16 mm² twin & cpc TN-S ZE=0.8
1	4.2	BS 88 Fuse	6	52	52	79	79	128	128	209	209	315	315	523	523	822	822
3	12.5	"	16	17	17	26	26	42	42	69	69	105	105	174	174	274	274
6	25.0	"	25	–	–	–	–	21	21	34	30	52	48	87	78	137	119
7	29.2	"	32	–	–	–	–	–	–	29	12	45	20	74	32	117	49
8.3	34.6	"	40	–	–	–	–	–	–	19	–	35	4	57	6	87	10
10	41.7	"	50	–	–	–	–	–	–	–	–	19	–	31	–	47	–
1	4.2	BS 1361 Fuse	5	52	52	79	79	128	128	209	209	315	315	523	523	822	822
3	12.5	"	15	17	17	26	26	42	42	69	69	105	105	174	174	274	274
6	25.0	"	30	–	–	–	–	21	14	34	17	52	27	87	45	137	68
7	29.2	"	30	–	–	–	–	–	–	29	17	45	27	74	45	117	68
8.3	34.6	"	45	–	–	–	–	–	–	4	–	17	–	28	–	42	–
10	41.7	"	45	–	–	–	–	–	–	–	–	17	–	28	–	42	–
1	4.2	BS EN 60898 Type B MCB (Can be used for BS 3871 Type 1)	5	52	52	79	79	128	128	209	209	315	315	523	523	822	822
3	12.5		15	17	16	26	26	42	42	69	69	105	105	174	174	274	274
3	12.5		16	17	14	26	26	42	42	69	69	105	105	174	174	274	274
6	25.0		25	–	–	–	–	21	21	34	34	52	52	87	87	137	137
7	29.2		30	–	–	–	–	–	–	29	29	45	45	74	74	117	117
7	29.2		32	–	–	–	–	–	–	29	29	45	45	74	74	117	117
8.3	34.6		40	–	–	–	–	–	–	25	17	38	27	63	45	99	68
10	41.7		45	–	–	–	–	–	–	–	–	31	18	52	30	82	46
1	4.2	BS EN 60898 Type C MCB (Can be used for BS 3871 Type 2)	5	52	52	79	79	128	128	192	173	307	276	500	450	762	685
3	12.5		15	17	17	26	19	42	29	54	34	86	55	140	90	214	137
3	12.5		16	17	–	26	16	42	25	49	30	79	48	129	78	197	119
6	25.0		25	–	–	–	–	21	5	26	6	42	11	68	18	104	27
7	29.2		30	–	–	–	–	–	–	19	–	31	–	50	–	77	–
7	29.2		32	–	–	–	–	–	–	17	–	27	–	45	–	68	–
8.3	34.6		40	–	–	–	–	–	–	10	–	17	–	28	–	42	–
10	41.7		45	–	–	–	–	–	–	–	–	12	–	20	–	30	–
1	4.2	BS 3036 Fuse	5	52	52	79	79	128	128	209	209	315	315	523	523	822	822
3	12.5	"	15	17	17	26	26	42	42	69	69	105	105	174	174	274	274
6	25.0	"	30	–	–	–	–	21	12	34	14	52	23	87	38	135	58
7	29.2	"	30	–	–	–	–	–	–	29	14	45	23	74	38	117	58

Notes: Table based on 240 V and voltage drop of 4%. Cables to be installed open and clipped direct, derate circuits when grouped, in contact with thermal insulation, or when ambient temperature exceeds 30 °C, although the lengths in the table are still applicable. Where the load is not likely to cause overload, e.g. heating load, immersion heater etc., the cable's current-carrying capacity does not have to equal the rating of the protective device. The table gives lengths for different sizes of cable, for each load for TN-C-S and TN-S systems, and can be used in any area where disconnection time has to be 0.4 s, e.g. bathroom, or fixed equipment outdoors. The table can also be used for 70 °C sheathed MICC cable, or PVC cable single circuits in conduit and trunking. If installed with other circuits, the cable size would need increasing in accordance with the CSG Tables.

INST 2

DISCONNECTION TIME 5 s

MAXIMUM CABLE LENGTHS

Maximum lengths in metres of single-phase circuits with PVC cable with various loads in areas requiring a DISCONNECTION TIME OF 5 s

Maximum length in metres for voltage drop, protection against indirect contact, thermal protection of cpc and short-circuit current

Resistive load in KW	Full load current in amperes	Protective device type	Protective device rating I_n	1.0 mm² twin & cpc		1.5 mm² twin & cpc		2.5 mm² twin & cpc		4 mm² twin & cpc		6 mm² twin & cpc		10 mm² twin & cpc		16 mm² twin & cpc	
				TN-C-S $Z_E=0.35$	TN-S $Z_E=0.8$	TN-C-S $Z_E=0.35$	TN-S $Z_E=0.8$	TN-C-S $Z_E=0.35$	TN-S $Z_E=0.8$	TN-C-S $Z_E=0.35$	TN-S $Z_E=0.8$	TN-C-S $Z_E=0.35$	TN-S $Z_E=0.8$	TN-C-S $Z_E=0.35$	TN-S $Z_E=0.8$	TN-C-S $Z_E=0.35$	TN-S $Z_E=0.8$
1	4.2	BS 88 Fuse	6	52	52	79	79	128	128	209	209	315	315	523	523	822	822
3	12.5	"	16	17	17	26	26	42	42	69	69	105	105	174	174	274	274
6	25.0	"	25	-	-	-	-	21	21	34	34	52	52	87	87	137	137
7	29.2	"	32	-	-	-	-	-	-	29	21	45	45	74	74	117	117
8.3	34.6	"	40	-	-	-	-	-	-	19	-	38	24	63	63	99	99
10	41.7	"	50	-	-	-	-	-	-	-	-	26	-	52	32	82	49
1	4.2	BS 1361 Fuse	5	52	52	79	79	128	128	209	209	315	315	523	523	822	822
3	12.5	"	15	17	17	26	26	42	42	69	69	105	105	174	174	274	274
6	25.0	"	30	-	-	-	-	21	21	34	29	52	52	87	87	137	137
7	29.2	"	30	-	-	-	-	-	-	29	29	45	45	74	74	117	117
8.3	34.6	"	45	-	-	-	-	-	-	4	-	23	-	63	45	99	68
10	41.7	"	45	-	-	-	-	-	-	-	-	23	-	52	22	82	34
1	4.2	BS EN 60898 Type B MCB (Can be used for BS 3871 Type 1)	5	52	52	79	79	128	128	209	209	315	315	523	523	822	822
3	12.5		15	17	17	26	26	42	42	69	69	105	105	174	174	274	274
3	12.5		16	17	17	26	26	42	42	69	69	105	105	174	174	274	274
6	25.0		25	-	-	-	-	21	21	34	34	52	52	87	87	137	137
7	29.2		30	-	-	-	-	-	-	29	29	45	45	74	74	117	117
7	29.2		32	-	-	-	-	-	-	29	29	45	45	74	74	117	117
8.3	34.6		40	-	-	-	-	-	-	25	17	38	27	63	45	99	68
10	41.7		45	-	-	-	-	-	-	-	-	31	18	52	30	82	46
1	4.2	BS EN 60898 Type C MCB (can be used for BS 3871 Type 2)	5	52	52	79	79	128	128	192	173	307	276	500	450	762	685
3	12.5		15	17	16	26	19	29	29	54	34	86	55	140	90	214	137
3	12.5		16	17	14	26	16	25	25	49	30	79	48	129	78	197	119
6	25.0		25	-	-	-	-	5	5	26	6	42	11	68	18	104	27
7	29.2		30	-	-	-	-	-	-	19	-	31	-	50	-	77	-
7	29.2		32	-	-	-	-	-	-	17	-	27	-	45	-	68	-
8.3	34.6		40	-	-	-	-	-	-	10	-	17	-	28	-	42	-
10	41.7		45	-	-	-	-	-	-	-	-	12	-	20	-	30	-
1	4.2	BS 3036 Fuse	5	52	52	79	79	128	128	209	209	315	315	523	523	822	822
3	12.5	"	15	17	17	26	26	42	42	69	69	105	105	174	174	274	274
6	25.0	"	30	-	-	-	-	21	21	34	34	52	52	87	87	137	137
7	29.2	"	30	-	-	-	-	-	-	29	29	45	45	74	74	117	117

Notes: Table based on 240 V and voltage drop of 4%. Cables to be installed open and clipped direct, derate circuits when grouped, in contact with thermal insulation, or when ambient temperature exceeds 30 °C, although the lengths in the table are still applicable. Where the load is not likely to cause overload, e.g. heating load, the cable's current-carrying capacity does not have to equal the rating of the protective device. The table gives lengths for different sizes of cable, for each load for TN-C-S and TN-S systems, and can be used in any area where disconnection time has to be 5 seconds, e.g. fixed equipment, 110 V circuits etc. The table can also be used for 70°C sheath MICC cable, or PVC cable single circuits in conduit and trunking. If installed with other circuits, the cable size would need increasing in accordance with the CSC Tables.

VOLTAGE DROP

INST 3

LIGHTING

Maximum voltage drop in a lighting circuit with 240V single-phase supply, using the loop-in method of wiring (see notes).

TABLE 1 - Distance from distribution board to first lighting point in metres

Number of lighting points	3	4	5	6	7	8	9	10	11	12	13	14	15	16	17	18	19	20	21
									Voltage drop										
1	0.06	0.07	0.09	0.11	0.13	0.15	0.17	0.18	0.20	0.22	0.24	0.26	0.28	0.29	0.31	0.33	0.35	0.37	0.39
2	0.11	0.15	0.18	0.22	0.26	0.29	0.33	0.37	0.40	0.44	0.48	0.51	0.55	0.59	0.62	0.66	0.70	0.73	0.77
3	0.17	0.22	0.28	0.33	0.39	0.44	0.50	0.55	0.61	0.66	0.72	0.77	0.83	0.88	0.94	0.99	1.05	1.10	1.16
4	0.22	0.29	0.37	0.44	0.51	0.59	0.66	0.73	0.81	0.88	0.95	1.03	1.10	1.17	1.25	1.32	1.39	1.47	1.54
5	0.28	0.37	0.46	0.55	0.64	0.73	0.83	0.92	1.01	1.10	1.19	1.28	1.38	1.47	1.56	1.65	1.74	1.83	1.93
6	0.33	0.44	0.55	0.66	0.77	0.88	0.99	1.10	1.21	1.32	1.43	1.54	1.65	1.76	1.87	1.98	2.09	2.20	2.31
7	0.39	0.51	0.64	0.77	0.90	1.03	1.16	1.28	1.41	1.54	1.67	1.80	1.93	2.05	2.18	2.31	2.44	2.57	2.70
8	0.44	0.59	0.73	0.88	1.03	1.17	1.32	1.47	1.61	1.76	1.91	2.05	2.20	2.35	2.49	2.64	2.79	2.93	3.08
9	0.50	0.66	0.83	0.99	1.16	1.32	1.49	1.65	1.81	1.98	2.15	2.31	2.48	2.64	2.81	2.97	3.14	3.30	3.46
10	0.55	0.73	0.92	1.10	1.28	1.47	1.65	1.83	2.02	2.20	2.38	2.57	2.75	2.93	3.12	3.30	3.48	3.67	3.85

TABLE 2 - Voltage drop between lighting fittings

Length between lights in metres	1st & 2nd	2nd & 3rd	3rd & 4th	4th & 5th	5th & 6th	6th & 7th	7th & 8th	8th & 9th	9th & 10th
2	0.33	0.29	0.26	0.22	0.18	0.15	0.11	0.07	0.04
3	0.50	0.44	0.39	0.33	0.28	0.22	0.17	0.11	0.06
4	0.66	0.59	0.51	0.44	0.37	0.29	0.22	0.15	0.07
5	0.83	0.73	0.64	0.55	0.46	0.37	0.28	0.18	0.09
6	0.99	0.88	0.77	0.66	0.55	0.44	0.33	0.22	0.11
7	1.16	1.03	0.90	0.77	0.64	0.51	0.39	0.26	0.13
8	1.32	1.17	1.03	0.88	0.73	0.59	0.44	0.29	0.15
9	1.49	1.32	1.16	0.99	0.83	0.66	0.50	0.33	0.17
10	1.65	1.47	1.28	1.10	0.92	0.73	0.55	0.37	0.18
11	1.81	1.61	1.41	1.21	1.01	0.81	0.61	0.40	0.20
12	1.98	1.76	1.54	1.32	1.10	0.88	0.66	0.44	0.22
13	2.15	1.91	1.67	1.43	1.19	0.95	0.72	0.48	0.24
14	2.31	2.05	1.80	1.54	1.28	1.03	0.77	0.51	0.26
15	2.48	2.20	1.93	1.65	1.38	1.10	0.83	0.55	0.28

TABLE 3 - Voltage drop between lighting point and switch, where switch cables are connected at lighting point

Distance - lighting point to switch in metres							
	1	2	3	4	5	6	7
Volt drop	0.02	0.04	0.06	0.07	0.09	0.11	0.13

Distance - lighting point to switch in metres							
	8	9	10	11	12	13	14
Volt drop	0.15	0.17	0.18	0.20	0.22	0.24	0.26

Distance - lighting point to switch in metres							
	15	16	17	18	19	20	21
Volt drop	0.28	0.29	0.31	0.33	0.35	0.37	0.39

HOW TO USE: The table is based on 1 mm2 PVC cable, 240 V supply, with 100 watt load on each lighting point, and the loop-in system of wiring. Add the voltage drop from the distribution board to the first lighting point (Table 1), then add the voltage drop between each lighting point (Table 2) and then add the voltage drop for the length of cable from the lighting point to switch (Table 3). This will give the voltage drop at that lighting point. To obtain the voltage drop for a larger cable, multiply the total voltage drop obtained by 44, and divide by R volt drop from the CR Table for the larger cable. For larger, but equal loads on each lighting point, multiply the voltage drop by the increase in load, e.g. for 200 watt at each lighting point, multiply the voltage drop worked out by 2.

DISCONNECTION TIME 0.4 SECONDS   INST 4   MAXIMUM CABLE LENGTHS

240V SINGLE-PHASE BS 1363 SOCKET-OUTLET CIRCUITS (See notes)

Maximum length in metres for protection against indirect contact, thermal protection of cpc and short-circuit current

Circuit type	Protective device type	Protective device rating I_n	Maximum floor area allowed in sq. metres	2.5 mm² twin & CPC		4 mm² twin & CPC		2.5 mm² single core PVC in steel conduit		4 mm² single core PVC in steelconduit		2 core 2.5 mm² MICC cable		2 core 4 mm² MICC cable		2 core 1.5 mm² MICC cable	
				TN-C-S $Z_E=0.35$	TN-S $Z_E=0.8$	TN-C-S $Z_E=0.35$	TN-S $Z_E=0.8$	TN-C-S $Z_E=0.35$	TN-S $Z_E=0.8$	TN-C-S $Z_E=0.35$	TN-S $Z_E=0.8$	TN-C-S $Z_E=0.35$	TN-S $Z_E=0.8$	TN-C-S $Z_E=0.35$	TN-S $Z_E=0.8$	TN-C-S $Z_E=0.35$	TN-S $Z_E=0.8$
1	2	3	4	5	6	7	8	9	10	11	12	13	14	15	16	17	18
Ring	BS 88	32	100	95	43	128	50	95	72	150	94	100	87	166	134	17	18
Radial	"	32	50	-	-	27	12	-	-	27	23	-	-	48	48	-	-
Radial	"	20	20	26	26	43	43	26	26	27	27	28	28	48	48	17	17
Ring	BS 1361	30	100	100	59	147	69	100	99	160	130	100	100	178	178	-	-
Radial	"	30	50	-	-	29	17	-	-	29	29	18	18	32	32	-	-
Radial	"	20	20	26	26	43	42	26	26	43	43	28	28	48	48	17	17
Ring	BS EN 60898 Type B	32	100	100	100	160	138	100	100	160	160	100	100	178	178	-	-
Ring		30	100	95	95	150	121	95	95	150	150	-	-	166	166	-	-
Radial		32	50	-	-	29	29	-	-	29	29	18	18	32	32	-	-
Radial		30	50	-	-	27	27	-	-	27	27	-	-	30	30	-	-
Radial		20	20	26	26	43	43	26	26	43	43	28	28	48	48	17	17
Ring	BS EN 60898 Type C	32	100	66	-	78	-	100	-	147	-	100	-	178	-	-	-
Ring		30	100	59	-	69	-	95	-	130	-	100	-	166	-	-	-
Radial		32	50	-	-	19	-	-	-	29	-	-	-	32	-	-	-
Radial		30	50	-	-	17	-	-	-	27	-	-	-	30	-	-	-
Radial	"	20	20	26	14	36	17	26	24	43	32	28	28	48	48	17	17
Ring	BS 3036	30	100	100	50	137	58	100	84	160	111	100	100	178	157	17	-
Radial	"	20	20	26	26	43	43	26	26	43	43	28	28	48	48	17	17

Notes:

(1) Twin & cpc cable to be installed open and clipped direct. Where circuits are grouped direct, larger cables must be installed; the lengths given in the Table for each cable size will still be applicable, but see note 3. The table is equally applicable to single circuits in plastic conduit if a minimum size of 1.5mm² cpc is also installed.

(2) The Table gives the maximum length for protection against indirect contact, for the thermal suitability of the protective conductor and for live conductors with a fault. Voltage drop has been taken in no account by assuming that radial circuits are fully loaded. For ring circuits, the average of the minimum and maximum voltage drop that could occur with the ring fully loaded has been used (i.e. assume all load at mid point, and load evenly distributed round the ring). In practice, the voltage drop for socket circuits cannot be calculated, since the load can be changed at any time. Voltage drop is therefore limited by the floor area allowed for each type of circuit. The MICC cables have to be PVC sheathed or exposed to touch, i.e., maximum permitted operating temperature of sheath 70 °C.

(3) Grouping: Two circuits can be installed without derating for grouping. For ring circuits only, up to five circuits can be bunched together when 4mm² twin & cpc cable is used. For derating radial circuits, use I_n divided by derating factor G (Table GF1). For ring circuits use I_n x 0.67 divided by G, then select the cable size required to carry the calculated current. No further diversity is allowed. The formula for circuits not liable to simultaneous overload should not be used, because simultaneous overload can occur with socket-outlet circuits. Additionally, the conductors cannot be assumed to be lightly loaded.

Non-armoured single core PVC insulated copper cables with or without sheath to BS6004, BS 6231 or BS 6346

Full thermal ratings Ambient temperature 30 °C Conductor maximum operating temperature 70 °C

Conductor cross-sectional area in sq.mm.	Enclosed in conduit, trunking, underground conduit (with sheath). Not in contact with thermal insulation.								Open, clipped direct, or lying on a non-metallic surface, or enclosed in non-thermal insulating plaster. Sheaths touching.							
	Two cables single-phase a.c.				Three or four cables three-phase a.c.				Two cables single-phase a.c.				Three or four cables three-phase a.c.			
	Current rating amperes	Voltage drop mV/A/m		Factor length × amperes	Current rating amperes	Voltage drop mV/A/m		Factor length × amperes	Current rating amperes	Voltage drop mV/A/m		Factor length × amperes	Current rating amperes	Voltage drop mV/A/m		Factor length × amperes
		r	x			r	x			r	x			r	x	
1	2	3	4	5	6	7	8	9	10	11	12	13	14	15	16	17
1.0	13.5	44.0	-	218	12	38.0	-	437	15.5	44.0	-	218	14	38.0	-	437
1.5	17.5	29.0	-	331	15.5	25.0	-	664	20	29.0	-	331	18	25.0	-	664
2.5	24	18.0	-	533	21	15.0	-	1107	27	18.0	-	533	25	15.0	-	1107
4.0	32	11.0	-	873	28	9.50	-	1747	37	11.0	-	873	33	9.50	-	1747
6.0	41	7.30	-	1315	36	6.40	-	2594	47	7.30	-	1315	43	6.40	-	2594
10	57	4.40	-	2182	50	3.80	-	4368	65	4.40	-	2182	59	3.80	-	4368
16	76	2.80	-	3429	68	2.40	-	6917	87	2.80	-	3429	79	2.40	-	6917
25	101	1.80	0.33	5333	89	1.50	0.29	10710	114	1.75	0.200	5486	104	1.50	0.25	10710
35	125	1.30	0.31	7385	110	1.10	0.27	15091	141	1.25	0.195	7680	129	1.10	0.24	15091
50	151	0.95	0.30	9600	134	0.81	0.26	19529	182	0.93	0.190	10105	167	0.80	0.24	19762
70	192	0.65	0.29	13333	171	0.56	0.25	27213	234	0.63	0.185	14545	214	0.55	0.24	27667

Ambient temperature correction factors for PVC insulated cable

For rewirable fuses only

Ambient temperature °C	25	30	35	40	45	50	55	60	65
Correction factor	1.03	1.0	0.97	0.94	0.91	0.87	0.84	0.69	0.48

For all other protective devices

Ambient temperature °C	25	30	35	40	45	50	55	60	65
Correction factor	1.03	1.0	0.94	0.87	0.79	0.71	0.61	0.50	0.35

Note: Factor gives cable size for 4% voltage drop at 240 V and 415 V. Multiply load current by cable length, then select a factor equal to, or larger than the calculated value of load × length. For loads with a power factor where length is critical, use the values of r and x in the formulas given in the introduction, or in Part 2 Voltage drop. Cables need derating when BS 3036 fuses are used for overload protection.

(4D1)

I_{tab} SINGLE CORE CABLES   **CR 1(a)**   COPPER CONDUCTORS

Non-armoured single core p.v.c. insulated copper cables with or without sheath to BS6004, BS 6231 or BS 6346

Full thermal ratings   Ambient temperature 30 °C   Conductor maximum operating temperature 70 °C

	Enclosed in conduit, trunking, underground conduit (with sheath). Not in contact with thermal insulation								Open, clipped direct, or lying on a non-metallic surface, or enclosed in non-thermal insulating plaster. Sheaths touching							
Conductor cross-sectional area in sq.mm.	Two cables single-phase a.c.				Three or four cables three-phase a.c.				Two cables single-phase a.c.				Three or four cables three-phase a.c.			
	Current rating amperes	Voltage drop mV/A/m		Factor length × amperes	Current rating amperes	Voltage drop mV/A/m		Factor length × amperes	Current rating amperes	Voltage drop mV/A/m		Factor length × amperes	Current rating amperes	Voltage drop mV/A/m		Factor length × amperes
		r	x			r	x			r	x			r	x	
1	2	3	4	5	6	7	8	9	10	11	12	13	14	15	16	17
95	232	0.490	0.28	17143	207	0.42	0.24	34583	284	0.470	0.180	19200	261	0.410	0.23	35319
120	269	0.390	0.27	20426	239	0.33	0.23	40488	330	0.370	0.175	23415	303	0.320	0.23	41500
150	300	0.310	0.27	23415	262	0.27	0.23	46111	381	0.300	0.175	28235	349	0.260	0.23	48324
185	341	0.250	0.27	25946	296	0.22	0.23	51875	436	0.240	0.170	33103	400	0.210	0.22	53548
240	400	0.195	0.26	29091	346	0.17	0.23	57241	515	0.185	0.165	38400	472	0.160	0.22	61481
300	458	0.160	0.26	30968	394	0.14	0.23	61481	594	0.150	0.165	43636	545	0.130	0.22	66400
400	546	0.130	0.26	33103	467	0.12	0.22	66400	694	0.120	0.160	48000	634	0.105	0.21	69167
500	626	0.110	0.26	34286	533	0.10	0.22	66400	792	0.098	0.155	51892	723	0.086	0.21	72474
630	720	0.094	0.25	35556	611	0.08	0.22	69167	904	0.081	0.155	54857	826	0.072	0.21	75455
800	-	-	-	-	-	-	-	-	1030	0.068	0.150	58182	943	0.060	0.21	75455
1000	-	-	-	-	-	-	-	-	1154	0.059	0.150	60000	1058	0.052	0.20	79048

Ambient temperature correction factors for PVC insulated cable

For rewirable fuses only									For all other protective devices								
Ambient temperature °C	25	30	35	40	45	50	55	60	65	25	30	35	40	45	50	55	60	65
Correction factor	1.03	1.0	0.97	0.94	0.91	0.87	0.84	0.69	0.48	1.03	1.0	0.94	0.87	0.79	0.71	0.61	0.50	0.35

Note: Factor gives cable size for 4% voltage drop at 240 V and 415 V. Multiply load current by cable length, then select a factor equal to, or larger than the calculated value of load × length. For loads with a power factor where length is critical, use the values of r and x in the formulas given in the introduction, or in Part 2 Voltage drop. Cables need derating when BS 3036 fuses are used for overload protection.

(4D1)

I tab SINGLE CORE CABLES CR 2 COPPER CONDUCTORS

Non-armoured single core PVC insulated copper cables with or without sheath to BS6004, BS 6231 or BS 6346

Full thermal ratings | Ambient temperature 30 °C | Conductor maximum operating temperature 70 °C

Conductor cross-sectional area in sq.mm.	Enclosed in conduit in thermally insulating wall or ceiling, conduit in contact with thermally conductive surface								Conductor cross-sectional area in sq.mm.	Sheathed cables on perforated cable tray, bunched and unenclosed. Holes in tray occupy 30% of tray area							
	Two cables single-phase a.c.				Three or four cables three-phase a.c.					Two cables single-phase a.c.				Three or four cables three-phase a.c.			
	Current rating amperes	Voltage drop mV/A/m		Factor length × amperes	Current rating amperes	Voltage drop mV/A/m		Factor length × amperes		Current rating amperes	Voltage drop mV/A/m		Factor length × amperes	Current rating amperes	Voltage drop mV/A/m		Factor length × amperes
		r	x			r	x				r	x			r	x	
1	2	3	4	5	6	7	8	9	10	11	12	13	14	15	16	17	18
1.0	11.0	44	-	218	10.5	38.0	-	437	25	126	1.750	0.200	5486	112	1.5	0.25	10710
1.5	14.5	29	-	331	13.5	25.0	-	664	35	156	1.250	0.195	7680	141	1.1	0.24	15091
2.5	19.5	18	-	533	18	15.0	-	1107	50	191	0.930	0.190	10105	172	0.8	0.24	19762
4.0	26.0	11	-	873	24	9.50	-	1747	70	246	0.630	0.185	14545	223	0.55	0.24	27667
6.0	34.0	7.3	-	1315	31	6.40	-	2594	95	300	0.470	0.180	19200	273	0.41	0.23	35319
10	46.0	4.4	-	2182	42	3.80	-	4368	120	349	0.370	0.175	23415	318	0.32	0.23	41500
16	61.0	2.8	-	3429	56	2.40	-	6917	150	404	0.300	0.175	28235	369	0.26	0.23	48824
25	80.0	1.8	0.33	5333	73	1.50	0.29	10710	185	463	0.240	0.170	33103	424	0.21	0.22	53548
35	99.0	1.3	0.31	7385	89	1.10	0.27	15091	240	549	0.185	0.165	38400	504	0.16	0.22	61481
50	119	0.95	0.30	9600	108	0.81	0.26	19529	300	635	0.150	0.165	43636	584	0.13	0.22	66400
70	151	0.65	0.29	13333	136	0.56	0.25	27213	400	732	0.120	0.160	48000	679	0.105	0.21	69167
95	182	0.49	0.28	17143	164	0.42	0.24	34583	500	835	0.098	0.155	51892	778	0.086	0.21	72174

Ambient temperature correction factors for PVC insulated cable

For rewirable fuses only

Ambient temperature °C	25	30	35	40	45	50	55	60	65
Correction factor	1.03	1.0	0.97	0.94	0.91	0.87	0.84	0.69	0.48

For all other protective devices

Ambient temperature °C	25	30	35	40	45	50	55	60	65
Correction factor	1.03	1.0	0.94	0.87	0.79	0.71	0.61	0.50	0.35

Note: Factor gives cable size for 4% voltage drop at 240 V and 415 V. Multiply load current by cable length, then select a factor equal to, or larger than the calculated value of load × length. For loads with a power factor where length is critical, use the values of r and x in the formulas given in the introduction, or in Part 2 Voltage drop. Cables need derating when BS 3036 fuses are used for overload protection.

(4D1)

I_tab SINGLE CORE CABLES

CR 3   COPPER CONDUCTORS

Non-armoured single core thermosetting (XLPE) insulated copper cables with or without sheath to BS 7211 or BS 5467

Full thermal ratings   Ambient temperature 30 °C   Conductor maximum operating temperature 90 °C

Conductor cross-sectional area in sq.mm	Enclosed in conduit, trunking, underground conduit (with sheath). Not in contact with thermal insulation.								Open, clipped direct, or lying on a non-metallic surface, or enclosed in non-thermal insulating plaster.							
	Two cables single-phase a.c.				Three or four cables three-phase a.c.				Two cables single-phase a.c.				Three or four cables three-phase a.c.			
	Current rating amperes	Voltage drop mV/A/m r	x	Factor length × amperes	Current rating amperes	Voltage drop mV/A/m r	x	Factor length × amperes	Current rating amperes	Voltage drop mV/A/m r	x	Factor length × amperes	Current rating amperes	Voltage drop mV/A/m r	x	Factor length × amperes
1	2	3	4	5	6	7	8	9	10	11	12	13	14	15	16	17
1.0	17	46	-	209	15	40	-	415	19	46	-	209	17.5	40.0	-	415
1.5	22	31	-	310	19	27	-	615	25	31	-	310	23	27.0	-	615
2.5	30	19	-	505	26	16	-	1038	34	19	-	505	31	16.0	-	1038
4.0	40	12	-	800	35	10	-	1660	46	12	-	800	41	10.0	-	1660
6.0	51	7.90	-	1215	45	6.80	-	2441	59	7.90	-	1215	54	6.80	-	2441
10	71	4.70	-	2043	63	4.00	-	4150	81	4.70	-	2043	74	4.00	-	4150
16	95	2.90	-	3310	85	2.50	-	6640	109	2.90	-	3310	99	2.50	-	6640
25	126	1.85	0.31	5053	111	1.60	0.27	10061	143	1.85	0.19	5189	130	1.60	0.19	10375
35	156	1.35	0.29	7111	138	1.15	0.25	14435	176	1.35	0.18	7111	161	1.15	0.18	14435
50	189	1.00	0.29	9143	168	0.87	0.25	18444	228	0.99	0.18	9600	209	0.86	0.18	19030
70	240	0.70	0.28	12800	214	0.6	0.24	25538	293	0.68	0.175	13521	268	0.59	0.175	26774
95	290	0.51	0.27	16552	259	0.44	0.23	33200	355	0.49	0.17	18462	326	0.43	0.17	36087

Ambient temperature correction factors for thermosetting insulation

For rewirable fuses only

Ambient temperature °C	25	30	35	40	45	50	55	60	65
Correction factor	1.02	1.0	0.98	0.95	0.93	0.91	0.89	0.87	0.85

For all other protective devices

Ambient temperature °C	25	30	35	40	45	50	55	60	65
Correction factor	1.02	1.0	0.96	0.91	0.87	0.82	0.76	0.71	0.65

Note: Factor gives cable size for 4% voltage drop at 240 V and 415 V. Multiply load current by cable length, then select a factor equal to, or larger than the calculated value of load × length. For loads with a power factor where length is critical, use the values of r and x in the formulas given in the introduction, or in Part 2 Voltage drop. Cables need derating when BS 3036 fuses are used for overload protection.

(4E1)

Itab SINGLE CORE CABLES   **CR 3 (a)**   COPPER CONDUCTORS

Non-armoured single core thermosetting (XLPE) insulated copper cables with or without sheath to BS 7211 or BS 5467

Full thermal ratings   Ambient temperature 30 °C   Conductor maximum operating temperature 90 °C

	Enclosed in conduit, trunking, underground conduit (with sheath). Not in contact with thermal insulation.								Open, clipped direct, or lying on a non-metallic surface, or enclosed in non-thermal insulating plaster.							
	Two cables single-phase a.c.				Three or four cables three-phase a.c.				Two cables single-phase a.c.				Three or four cables three-phase a.c.			
Conductor cross-sectional area in sq.mm.	Current rating amperes	Voltage drop mV/A/m		Factor length × amperes	Current rating amperes	Voltage drop mV/A/m		Factor length × amperes	Current rating amperes	Voltage drop mV/A/m		Factor length × amperes	Current rating amperes	Voltage drop mV/A/m		Factor length × amperes
		r	x			r	x			r	x			r	x	
1	2	3	4	5	6	7	8	9	10	11	12	13	14	15	16	17
120	336	0.410	0.26	20000	299	0.350	0.23	39524	413	0.390	0.165	22326	379	0.340	0.165	43684
150	375	0.330	0.26	22326	328	0.290	0.23	44865	476	0.320	0.165	26667	436	0.280	0.165	51875
185	426	0.270	0.26	25946	370	0.230	0.23	51875	545	0.260	0.165	32000	500	0.220	0.165	59286
240	500	0.210	0.26	29091	433	0.185	0.22	57241	644	0.200	0.160	38400	590	0.170	0.165	69167
300	513	0.175	0.25	30968	493	0.150	0.22	61481	743	0.160	0.160	43636	681	0.135	0.160	79048
400	683	0.140	0.25	33103	584	0.125	0.22	66400	868	0.130	0.155	48000	793	0.110	0.160	85128
500	783	0.120	0.25	34286	666	0.100	0.22	69167	990	0.105	0.155	51892	904	0.088	0.160	92222
630	900	0.100	0.25	35556	764	0.088	0.21	72174	1130	0.086	0.155	54857	1033	0.071	0.160	97647
800	-	-	-	-	-	-	-	-	1288	0.072	0.150	56471	1179	0.059	0.155	100606
1000	-	-	-	-	-	-	-	-	1443	0.063	0.150	58182	1323	0.050	0.155	100606

Ambient temperature correction factors for thermosetting insulation

For rewirable fuses only

Ambient temperature °C	25	30	35	40	45	50	55		
Correction factor	1.02	1.0	0.98	0.95	0.93	0.91	0.89		

For all other protective devices

Ambient temperature °C	25	30	35	40	45	50	55	60	65
Correction factor	1.02	1.0	0.96	0.91	0.87	0.82	0.76	0.71	0.65

Note: Factor gives cable size for 4% voltage drop at 240 V and 415 V. Multiply load current by cable length, then select a factor equal to, or larger than the calculated value of load × length. For loads with a power factor where length is critical, use the values of r and x in the formulas given in the introduction, or in Part 2 Voltage drop.
Cables need derating when BS 3036 fuses are used for overload protection.

(4E1)

Itab SINGLE CORE CABLES   CR 4   COPPER CONDUCTORS

Non-armoured single core thermosetting (XLPE) insulated copper cables with or without sheath to BS 7211 or BS 5467

Full thermal ratings   Ambient temperature 30 °C   Conductor maximum operating temperature 90 °C

Left section: Enclosed in conduit in thermally insulating wall or ceiling, conduit in contact with thermally conductive surface.

Right section: Sheathed cables on perforated cable tray, bunched and unenclosed. Holes in tray occupy 30% of tray area.

Conductor cross-sectional area in sq.mm.	Two cables single-phase a.c. Current rating amperes	Voltage drop mV/A/m r	Voltage drop mV/A/m x	Factor length × amperes	Three or four cables three-phase a.c. Current rating amperes	Voltage drop mV/A/m r	Voltage drop mV/A/m x	Factor length × amperes	Conductor cross-sectional area in sq.mm	Two cables single-phase a.c. Current rating amperes	Voltage drop mV/A/m r	Voltage drop mV/A/m x	Factor length × amperes	Three or four cables three-phase a.c. Current rating amperes	Voltage drop mV/A/m r	Voltage drop mV/A/m x	Factor length × amperes
1	2	3	4	5	6	7	8	9	10	11	12	13	14	15	16	17	18
1.0	14	46	-	209	13	40	-	415	25	158	1.85	0.190	5189	140	1.6	0.19	10375
1.5	18	31	-	310	17	27	-	615	35	195	1.35	0.180	7111	176	1.15	0.18	14435
2.5	24	19	-	505	23	16	-	1038	50	293	0.99	0.180	9600	215	0.86	0.18	19080
4.0	33	12	-	800	30	10	-	1660	70	308	0.68	0.175	13521	279	0.59	0.175	26774
6.0	43	7.9	-	1215	39	6.8	-	2441	95	375	0.49	0.170	18462	341	0.43	0.170	36087
10	58	4.7	-	2043	53	4.0	-	4150	120	436	0.39	0.165	22326	398	0.34	0.165	43684
16	76	2.9	-	3310	70	2.5	-	6640	150	505	0.32	0.165	26667	461	0.28	0.165	51875
25	100	1.85	0.31	5053	91	1.6	0.27	10061	185	579	0.26	0.165	32000	530	0.22	0.165	59286
35	124	1.35	0.29	7111	111	1.15	0.25	14435	240	686	0.20	0.160	38400	630	0.17	0.160	69167
50	149	1.00	0.29	9143	135	0.87	0.25	18444	300	794	0.16	0.160	43636	730	0.135	0.160	79048
70	189	0.70	0.28	12800	170	0.60	0.24	25538	400	915	0.13	0.155	48000	849	0.110	0.160	85128
95	228	0.51	0.27	16552	205	0.44	0.23	33200	500	1044	0.105	0.155	51892	973	0.088	0.160	92222
120	263	0.41	0.26	20000	235	0.35	0.23	39524	630	1191	0.086	0.155	54857	1115	0.071	0.160	97647
150	300	0.33	0.26	22326	270	0.29	0.23	44865	800	1358	0.072	0.15	56471	1275	0.059	0.155	100606

Ambient temperature correction factors for thermosetting insulation

For rewirable fuses only

Ambient temperature °C	25	30	35	40	45	50	55	60	65
Correction factor	1.02	1.0	0.98	0.95	0.93	0.91	0.89	0.87	0.85

For all other protective devices

Ambient temperature °C	25	30	35	40	45	50	55	60	65
Correction factor	1.02	1.0	0.96	0.91	0.87	0.82	0.76	0.71	C.65

Note: Factor gives cable size for 4% voltage drop at 240 V and 415 V. Multiply load current by cable length, then select a factor equal to, or larger than the calculated value of load × length. For loads with a power factor where length is critical, use the values of r and x in the formulas given in the introduction, or in Part 2 Voltage drop. Cables need derating when BS 3036 fuses are used for overload protection.

(4E1)

Itab SINGLE CORE CABLES CR 5 COPPER CONDUCTORS

Non-armoured single core 85 °C rubber insulated copper cables with or without sheath to BS6007 or BS 6883

Full thermal ratings   Ambient temperature 30 °C   Conductor maximum operating temperature 85 °C

	Enclosed in conduit, trunking, underground conduit (with sheath). Not in contact with thermal insulation								Open, clipped direct, or lying on a non-metallic surface, or enclosed in non-thermal insulating plaster. Sheaths touching							
	Two cables single-phase a.c.				Three or four cables three-phase a.c.				Two cables single-phase a.c.				Three or four cables three-phase a.c.			
Conductor cross-sectional area in sq.mm.	Current rating	Voltage drop mV/A/m		Factor length ×	Current rating	Voltage drop mV/A/m		Factor length ×	Current rating	Voltage drop mV/A/m		Factor length ×	Current rating	Voltage drop mV/A/m		Factor length ×
	amperes	r	x	amperes	amperes	r	x	amperes	amperes	r	x	amperes	amperes	r	x	amperes
1	2	3	4	5	6	7	8	9	10	11	12	13	14	15	16	17
1.0	17	46	-	209	15	40	-	415	19	46	-	209	17.5	40	-	415
1.5	22	31	-	310	19.5	26	-	638	25	31	-	310	23	26	-	638
2.5	30	18	-	533	27	16	-	1038	34	18	-	533	31	16	-	1038
4.0	40	12	-	800	36	10	-	1660	45	12	-	800	42	10	-	1660
6.0	52	7.7	-	1247	46	6.7	-	2478	59	7.7	-	1247	54	6.7	-	2478
10	72	4.6	-	2087	63	4.0	-	4150	81	4.6	-	2087	75	4.0	-	4150
16	96	2.9	-	3310	85	2.5	-	6640	108	2.9	-	3310	100	2.5	-	6640
25	127	1.85	0.32	5053	112	1.60	0.28	10061	143	1.85	0.200	5189	133	1.60	0.25	10375
35	157	1.35	0.31	6857	138	1.15	0.27	13833	177	1.30	0.195	7111	164	1.15	0.24	14435
50	190	1.00	0.30	9143	167	0.87	0.26	18242	215	0.97	0.19	9697	199	0.84	0.24	18864
70	242	0.68	0.29	12973	213	0.6	0.25	25538	274	0.66	0.185	13913	254	0.57	0.24	26774

Ambient temperature correction factors for 85 °C rubber insulation

For rewirable fuses only									For all other protective devices								
Ambient temperature °C	25	30	35	40	45	50	55	60	65	25	30	35	40	45	50	55	60	65
Correction factor	1.02	1.0	0.97	0.95	0.93	0.91	0.88	0.86	0.83	1.02	1.0	0.95	0.90	0.85	0.80	0.74	0.67	0.60

Note: Factor gives cable size for 4% voltage drop at 240 V and 415 V. Multiply load current by cable length, then select a factor equal to, or larger than the calculated value of load × length. For loads with a power factor where length is critical, use the values of r and x in the formulas given in the introduction, or in Part 2 Voltage drop. Cables need derating when BS 3036 fuses are used for overload protection.

(4F1)

Itab SINGLE CORE CABLES   **CR 6**   **ALUMINIUM CONDUCTORS**

Non-armoured single core PVC insulated aluminium cables with or without sheath to BS6004 or BS 6346

Full thermal ratings   Ambient temperature 30 °C   Conductor operating temperature 70 °C

	Open, clipped direct, or lying on a non-metallic surface, or enclosed in non-thermal insulating plaster. Sheaths touching.								Installed in free air on perforated cable tray, where the perforations occupy 30% of the tray area. Laid flat and touching							
Conductor cross-sectional area in sq.mm.	Two cables single-phase a.c.				Three or four cables three-phase a.c.				Two cables single-phase a.c.				Three or four cables three-phase a.c.			
	Current rating amperes	Voltage drop mV/A/m		Factor length × amperes	Current rating amperes	Voltage drop mV/A/m		Factor length × amperes	Current rating amperes	Voltage drop mV/A/m		Factor length × amperes	Current rating amperes	Voltage drop mV/A/m		Factor length × amperes
		r	x			r	x			r	x			r	x	
1	2	3	4	5	6	7	8	9	10	11	12	13	14	15	16	17
50	134	1.55	0.190	6194	123	1.35	0.24	12296	144	1.55	0.190	6194	132	1.35	0.24	12296
70	172	1.05	0.185	9143	159	0.91	0.24	17660	185	1.05	0.185	9143	169	0.91	0.24	17660
95	210	0.77	0.185	12152	194	0.67	0.23	23380	225	0.77	0.185	12152	206	0.67	0.23	23380
120	245	0.61	0.180	15000	226	0.53	0.23	28621	261	0.61	0.180	15000	240	0.53	0.23	28621
150	283	0.49	0.175	18462	261	0.42	0.23	34583	301	0.49	0.175	18462	277	0.42	0.23	34583
185	324	0.40	0.175	22326	299	0.34	0.23	40488	344	0.40	0.175	22326	317	0.34	0.23	40488
240	384	0.30	0.170	27429	354	0.26	0.22	47429	407	0.30	0.170	27429	375	0.26	0.22	47429
300	444	0.24	0.170	32000	410	0.21	0.22	53548	469	0.24	0.170	32000	433	0.21	0.22	53548
380	511	0.195	0.165	36923	472	0.170	0.22	59286	543	0.195	0.165	36923	502	0.170	0.22	59286
480	591	0.155	0.165	41739	546	0.140	0.22	63846	629	0.155	0.165	41739	582	0.140	0.22	63846
600	679	0.130	0.160	45714	626	0.110	0.22	69167	722	0.130	0.160	45714	669	0.110	0.22	69167
740	771	0.105	0.160	50526	709	0.094	0.21	72174	820	0.105	0.160	50526	761	0.094	0.21	72174
960	900	0.086	0.155	53333	823	0.077	0.21	74107	953	0.086	0.155	53333	886	0.077	0.21	74107
1200	1022	0.074	0.155	56471	926	0.066	0.21	75455	1073	0.074	0.155	56471	999	0.066	0.21	75455

Ambient temperature correction factors for p.v.c insulated cable

For rewirable fuses only

Ambient temperature °C	25	30	35	40	45	50	55	60	65
Correction factor	1.03	1.0	0.97	0.94	0.91	0.87	0.84	0.69	0.48

For all other protective devices

Ambient temperature °C	25	30	35	40	45	50	55	60	65
Correction factor	1.03	1.0	0.94	0.87	0.79	0.71	0.61	0.50	0.35

Note: Factor gives cable size for 4% voltage drop at 240 V and 415 V. Multiply load current by cable length, then select a factor equal to, or larger than the calculated value of load × length. For loads with a power factor where length is critical, use the values of r and x in the formulas given in the introduction, or in Part 2 Voltage drop. Cables need derating when BS 3036 fuses are used for overload protection.

(4K1)

I_tab MULTICORE CABLES

CR 7 COPPER CONDUCTORS

Non-armoured multicore PVC insulated and sheathed copper cables to BS6004, BS 6346 or BS 7629

Full thermal ratings Ambient temperature 30 °C Conductor maximum operating temperature 70 °C

Conductor cross-sectional area in sq.mm.	Enclosed in conduit, trunking, underground conduit (with sheath). Not in contact with thermal insulation								Open, clipped direct, or lying on a non-metallic surface or enclosed in non-thermal insulating plaster							
	One twin cable or one twin and cpc cable single-phase a.c.				One 3 core, or 3 core & cpc cable or 4 core cable three-phase a/c.				One twin cable or one twin and cpc cable single-phase a.c.				One 3 core, or 3 core & cpc cable or 4 core cable three-phase a/c.			
	Current rating amperes	Voltage drop mV/A/m		Factor length × amperes	Current rating amperes	Voltage drop mV/A/m		Factor length × amperes	Current rating amperes	Voltage drop mV/A/m		Factor length × amperes	Current rating amperes	Voltage drop mV/A/m		Factor length × amperes
		r	x			r	x			r	x			r	x	
1	2	3	4	5	6	7	8	9	10	11	12	13	14	15	16	17
1.0	13	44	–	218	11.5	38	–	437	15	44	–	218	13.5	38	–	437
1.5	16.5	29	–	331	15	25	–	664	19.5	29	–	331	17.5	25	–	664
2.5	23	18	–	533	20	15	–	1107	27	18	–	533	24	15	–	1107
4.0	30	11	–	873	27	9.5	–	1747	36	11	–	873	32	9.5	–	1747
6.0	38	7.3	–	1315	34	6.4	–	2594	46	7.3	–	1315	41	6.4	–	2594
10	52	4.4	–	2182	46	3.8	–	4368	63	4.4	–	2182	57	3.8	–	4368
16	69	2.8	–	3429	62	2.4	–	6917	85	2.8	–	3429	76	2.4	–	6917
25	90	1.75	0.170	5486	80	1.5	0.145	11067	112	1.75	0.170	5486	96	1.5	0.145	11067
35	111	1.25	0.165	7680	99	1.1	0.145	15091	138	1.25	0.165	7680	119	1.1	0.145	15091

Ambient temperature correction factors for PVC insulated cable

For rewirable fuses only

Ambient temperature °C	25	30	35	40	45	50	55	60	65
Correction factor	1.03	1.0	0.97	0.94	0.91	0.87	0.84	0.69	0.48

For all other protective devices

Ambient temperature °C	25	30	35	40	45	50	55	60	65
Correction factor	1.03	1.0	0.94	0.87	0.79	0.71	0.61	0.50	0.35

Note: Factor gives cable size for 4% voltage drop at 240 V and 415 V. Multiply load current by cable length, then select a factor equal to, or larger than the calculated value of load × length. For loads with a power factor where length is critical, use the values of r and x in the formulas given in the introduction, or in Part 2 Voltage drop. Cables need derating when BS 3036 fuses are used for overload protection.

(4D2)

Itab MULTICORE CABLES

CR 7 (a)   COPPER CONDUCTORS

Non-armoured multicore PVC insulated and sheathed copper cables to BS6004, BS 6346 or BS 7629

Full thermal ratings   Ambient temperature 30 °C   Conductor maximum operating temperature 70 °C

Conductor cross-sectional area in sq.mm.	Enclosed in conduit, trunking, underground conduit (with sheath). Not in contact with thermal insulation								Open, clipped direct, or lying on a non-metallic surface or enclosed in non-thermal insulating plaster							
	One twin cable or one twin and cpc cable single-phase a.c.				One 3 core, or 3 core & cpc cable or 4 core cable three-phase a/c.				One twin cable or one twin and cpc cable single-phase a.c.				One 3 core, or 3 core & cpc cable or 4 core cable three-phase a/c.			
	Current rating amperes	Voltage drop mV/A/m		Factor length × amperes	Current rating amperes	Voltage drop mV/A/m		Factor length × amperes	Current rating amperes	Voltage drop mV/A/m		Factor length × amperes	Current rating amperes	Voltage drop mV/A/m		Factor length × amperes
		r	x			r	x			r	x			r	x	
1	2	3	4	5	6	7	8	9	10	11	12	13	14	15	16	17
50	133	0.93	0.165	10213	118	0.80	0.140	20494	168	0.93	0.165	10213	144	0.80	0.140	20494
70	168	0.63	0.160	14769	149	0.55	0.140	29123	213	0.63	0.160	14769	184	0.55	0.140	29123
95	201	0.47	0.155	19200	179	0.41	0.135	38605	258	0.47	0.155	19200	223	0.41	0.135	38605
120	232	0.38	0.155	23415	206	0.33	0.135	47429	299	0.38	0.155	23415	259	0.33	0.135	47429
150	258	0.30	0.155	28235	225	0.26	0.130	57241	344	0.30	0.155	28235	299	0.26	0.130	57241
185	294	0.25	0.150	33103	255	0.21	0.130	66400	392	0.25	0.150	33103	341	0.21	0.130	66400
240	344	0.190	0.150	40000	297	0.165	0.130	79048	461	0.190	0.150	40000	403	0.165	0.130	79048
300	394	0.155	0.145	45714	339	0.135	0.130	89730	530	0.155	0.145	45714	464	0.135	0.130	89730
400	470	0.115	0.145	51892	402	0.100	0.125	103750	634	0.115	0.145	51892	557	0.100	0.125	103750

Ambient temperature correction factors for PVC insulated cable

For rewirable fuses only

Ambient temperature °C	25	30	35	40	45	50	55	60	65
Correction factor	1.03	1.0	0.97	0.94	0.91	0.87	0.84	0.69	0.48

For all other protective devices

	25	30	35	40	45	50	55	60	65
	1.03	1.0	0.94	0.87	0.79	0.71	0.61	0.50	0.35

Note: Factor gives cable size for 4% voltage drop at 240 V and 415 V. Multiply load current by cable length, then select a factor equal to, or larger than the calculated value of load × length. For loads with a power factor where length is critical, use the values of r and x in the formulas given in the introduction, or in Part 2 Voltage drop. Cables need derating when BS 3036 fuses are used for overload protection.

(4D2)

Non-armoured multicore PVC insulated and sheathed copper cables to BS6004, BS 6346 or BS 7629

Full thermal ratings   Ambient temperature 30 °C   Conductor maximum operating temperature 70 °C

Conductor cross-sectional area in sq.mm.	Installed direct, or in conduit, in a thermally insulated wall or ceiling, in contact with a 10 W/m²K thermally conductive surface on one side only								Installed in free air on perforated cable tray, where the perforations occupy 30% of tray area, or on brackets with a space of 0.3 x cable diameter between wall and cable, or spaced 2 x cable diameter when installed on ladder racking.							
	One twin cable or one twin and cpc cable single-phase a.c.				One 3 core, or 3 core & cpc cable or 4 core cable three-phase a.c.				One twin cable or one twin and cpc cable single-phase a.c.				One 3 core, or 3 core & cpc cable or 4 core cable three-phase a.c.			
	Current rating amperes	Voltage drop mV/A/m		Factor length × amperes	Current rating amperes	Voltage drop mV/A/m		Factor length × amperes	Current rating amperes	Voltage drop mV/A/m		Factor length × amperes	Current rating amperes	Voltage drop mV/A/m		Factor length × amperes
		r	x			r	x			r	x			r	x	
1	2	3	4	5	6	7	8	9	10	11	12	13	14	15	16	17
1.0	11	44	-	218	10	38	-	437	17	44	-	218	14.5	38	-	437
1.5	14	29	-	331	13	25	-	664	22	29	-	331	18.5	25	-	664
2.5	18.5	18	-	533	17.5	15	-	1107	30	18	-	533	25	15	-	1107
4.0	25	11	-	873	23	9.5	-	1747	40	11	-	873	34	9.5	-	1747
6.0	32	7.3	-	1315	29	6.4	-	2594	51	7.3	-	1315	43	6.4	-	2594
10	43	4.4	-	2182	39	3.8	-	4368	70	4.4	-	2182	60	3.8	-	4368
16	57	2.8	-	3429	52	2.4	-	6917	94	2.8	-	3429	80	2.4	-	6917
25	75	1.75	0.170	5486	68	1.5	0.145	11067	119	1.75	0.170	5486	101	1.5	0.145	11067
35	92	1.25	0.165	7680	83	1.1	0.145	15091	148	1.25	0.165	7680	126	1.1	0.145	15091

Ambient temperature correction factors for PVC insulated cable

	For rewirable fuses only									For all other protective devices								
Ambient temperature °C	25	30	35	40	45	50	55	60	65	25	30	35	40	45	50	55	60	65
Correction factor	1.03	1.0	0.97	0.94	0.91	0.87	0.84	0.69	0.48	1.03	1.0	0.94	0.87	0.79	0.71	0.61	0.50	0.35

Note: Factor gives cable size for 4% voltage drop at 240 V and 415 V. Multiply load current by cable length, then select a factor equal to, or larger than the calculated value of load × length. For loads with a power factor where length is critical, use the values of r and x in the formulas given in the introduction, or in Part 2 Voltage drop. Cables need derating when BS 3036 fuses are used for overload protection.

(4D2)

I_tab MULTICORE CABLES   CR 8 (a)   COPPER CONDUCTORS

Non-armoured multicore PVC insulated and sheathed copper cables to BS6004, BS 6346 or BS 7629

Full thermal ratings   Ambient temperature 30 °C   Conductor maximum operating temperature 70 °C

Conductor cross-sectional area in sq.mm.	Installed direct, or in conduit, in a thermally insulated wall or ceiling, in contact with a 10 W/m²K thermally conductive surface on one side only								Installed in: free air on perforated cable tray, where the perforations occupy 30% of tray area, or on brackets with a space of 0.3 x cable diameter between wall and cable, or spaced 2 x cable diameter when installed on ladder racking.							
	One twin cable or one twin and cpc cable single-phase a.c.				One 3 core, or 3 core & cpc cable or 4 core cable three-phase a/c.				One twin cable or one twin and cpc cable single-phase a.c.				One 3 core, or 3 core & cpc cable or 4 core cable three-phase a/c.			
	Current rating amperes	Voltage drop mV/A/m r	x	Factor length × amperes	Current rating amperes	Voltage drop mV/A/m r	x	Factor length × amperes	Current rating amperes	Voltage drop mV/A/m r	x	Factor length × amperes	Current rating amperes	Voltage drop mV/A/m r	x	Factor length × amperes
1	2	3	4	5	6	7	8	9	10	11	12	13	14	15	16	17
50	110	0.93	0.165	10213	99	0.80	0.140	20494	180	0.93	0.165	10213	153	0.80	0.140	20494
70	139	0.63	0.160	14769	125	0.55	0.140	29123	232	0.63	0.160	14769	196	0.55	0.140	29123
95	167	0.47	0.155	19200	150	0.41	0.135	38605	282	0.47	0.155	19200	238	0.41	0.135	38605
120	192	0.38	0.155	23415	172	0.33	0.135	47429	328	0.38	0.155	23415	276	0.33	0.135	47429
150	219	0.30	0.155	28235	196	0.26	0.130	57241	379	0.30	0.155	28235	319	0.26	0.130	57241
185	248	0.25	0.150	33103	223	0.21	0.130	66400	434	0.25	0.150	33103	364	0.21	0.130	66400
240	291	0.190	0.150	40000	261	0.165	0.130	79048	514	0.190	0.150	40000	430	0.165	0.130	79048
300	334	0.155	0.145	45714	298	0.135	0.130	89730	593	0.155	0.145	45714	497	0.135	0.130	89730
400	-	-	-	-	-	-	-	-	715	0.115	0.145	51892	597	0.100	0.125	103750

Ambient temperature correction factors for PVC insulated cable

	For rewirable fuses only									For all other protective devices								
Ambient temperature °C	25	30	35	40	45	50	55	60	65	25	30	35	40	45	50	55	60	65
Correction factor	1.03	1.0	0.97	0.94	0.91	0.87	0.84	0.69	0.48	1.03	1.0	0.94	0.87	0.79	0.71	0.61	0.50	0.35

Note: Factor gives cable size for 4% voltage drop at 240 V and 415 V. Multiply load current by cable length, then select a cable size equal to, or larger than the calculated value of load × length. For loads with a power factor where length is critical, use the values of r and x in the formulas given in the introduction, or in Part 2 Voltage drop. Cables need derating when BS 3036 fuses are used for overload protection.

(4D2)

Itab ARMOURED CABLES CR 9 COPPER CONDUCTORS

Armoured multicore PVC insulated copper cables to BS 6346

Full thermal ratings Ambient temperature 30 °C Conductor maximum operating temperature 70 °C

	Open, clipped direct, or lying on a non-metallic surface, or enclosed in non-thermal insulating plaster.								Installed in free air on perforated cable tray, where the perforations occupy 30% of tray area, or on brackets with a space of 0.3 x cable diameter between wall and cable, or spaced 2 x cable diameter when installed on ladder racking.							
Conductor cross-sectional area in sq.mm.	One two core cables single-phase a.c.				One three or four core cable three-phase a.c.				One two core cables single-phase a.c.				One three or four core cable three-phase a.c.			
	Current rating amperes	Voltage drop mV/A/m r	x	Factor length × amperes	Current rating amperes	Voltage drop mV/A/m r	x	Factor length × amperes	Current rating amperes	Voltage drop mV/A/m r	x	Factor length × amperes	Current rating amperes	Voltage drop mV/A/m r	x	Factor length × amperes
1	2	3	4	5	6	7	8	9	10	11	12	13	14	15	16	17
1.5	21	29	-	331	18	25	-	664	22	29	-	331	19	25	-	664
2.5	28	18	-	533	25	15	-	1107	31	18	-	533	26	15	-	1107
4.0	38	11	-	873	33	9.5	-	1747	41	11	-	873	35	9.5	-	1747
6.0	49	7.3	-	1315	42	6.4	-	2594	53	7.3	-	1315	45	6.4	-	2594
10	67	4.4	-	2182	58	3.8	-	4368	72	4.4	-	2182	62	3.8	-	4368
16	89	2.8	-	3429	77	2.4	-	6917	97	2.8	-	3429	83	2.4	-	6917
25	118	1.75	0.170	5486	102	1.5	0.145	11067	128	1.75	0.170	5486	110	1.5	0.145	11067
35	145	1.25	0.165	7680	125	1.1	0.145	15091	157	1.25	0.165	7680	135	1.1	0.145	15091
50	175	0.93	0.165	10213	151	0.8	0.140	20494	190	0.93	0.165	10213	163	0.8	0.140	20494

Ambient temperature correction factors for PVC insulated cable

	For rewirable fuses only									For all other protective devices								
Ambient temperature °C	25	30	35	40	45	50	55	60	65	25	30	35	40	45	50	55	60	65
Correction factor	1.03	1.0	0.97	0.94	0.91	0.87	0.84	0.69	0.48	1.03	1.0	0.94	0.87	0.79	0.71	0.61	0.50	0.35

Note: Factor gives cable size for 4% voltage drop at 240 V and 415 V. Multiply load current by cable length, then select a factor equal to, or larger than the calculated value of load × length. For loads with a power factor where length is critical, use the values of r and x in the formulas given in the introduction, or in Part 2 Voltage drop. Cables need derating when BS 3036 fuses are used for overload protection.

(4D4)

Itab ARMOURED CABLES   **CR 9 (a)**   COPPER CONDUCTORS

Armoured multicore PVC insulated copper cables to BS 6346

Full thermal ratings   Ambient temperature 30 °C   Conductor maximum operating temperature 70 °C

	Open, clipped direct, or lying on a non-metallic surface, or enclosed in non-thermal insulating plaster.								Installed in free air on perforated cable tray, where the perforations occupy 30% of tray area, or on brackets with a space of 0.3 x cable diameter between wall and cable, or spaced 2 x cable diameter when installed on ladder racking.							
	One two core cables single-phase a.c.				One three or four core cable three-phase a.c.				One two core cables single-phase a.c.				One three or four core cable three-phase a.c.			
Conductor cross-sectional area in sq.mm.	Current rating amperes	Voltage drop mV/A/m		Factor length × amperes	Current rating amperes	Voltage drop mV/A/m		Factor length × amperes	Current rating amperes	Voltage drop mV/A/m		Factor length × amperes	Current rating amperes	Voltage drop mV/A/m		Factor length × amperes
		r	x			r	x			r	x			r	x	
1	2	3	4	5	6	7	8	9	10	11	12	13	14	15	16	17
70	222	0.63	0.160	14769	192	0.550	0.140	29123	241	0.63	0.160	14769	207	0.550	0.140	29123
95	269	0.47	0.155	19200	231	0.410	0.135	38605	291	0.47	0.155	19200	251	0.410	0.135	38605
120	310	0.38	0.155	23415	267	0.330	0.135	47429	336	0.38	0.155	23415	290	0.330	0.135	47429
150	356	0.30	0.155	28235	306	0.260	0.130	57241	386	0.30	0.155	28235	332	0.260	0.130	57241
185	405	0.25	0.150	33103	348	0.210	0.130	66400	439	0.25	0.150	33103	378	0.210	0.130	66400
240	476	0.19	0.150	40000	409	0.165	0.130	79048	516	0.19	0.150	40000	445	0.165	0.130	79048
300	547	0.155	0.145	45714	469	0.135	0.130	89730	592	0.155	0.145	45714	510	0.135	0.130	89730
400	621	0.115	0.145	51892	540	0.100	0.125	103750	683	0.115	0.145	51892	590	0.100	0.125	103750

Ambient temperature correction factors for PVC insulated cable

	For rewirable fuses only									For all other protective devices								
Ambient temperature °C	25	30	35	40	45	50	55	60	65	25	30	35	40	45	50	55	60	65
Correction factor	1.03	1.0	0.97	0.94	0.91	0.87	0.84	0.69	0.48	1.03	1.0	0.94	0.87	0.79	0.71	0.61	0.50	0.35

Note: Factor gives cable size for 4% voltage drop at 240 V and 415 V. Multiply load current by cable length, then select a factor equal to, or larger than the calculated value of load × length. For loads with a power factor where length is critical, use the values of r and x in the formulas given in the introduction, or in Part 2 Voltage drop.
Cables need derating when BS 3036 fuses are used for overload protection.

(4D4)

Itab ARMOURED CABLES   CR 10   COPPER CONDUCTORS

Armoured multicore thermosetting (XLPE) insulated copper cables to BS 5467 or BS 6724

Full thermal ratings   Ambient temperature 30 °C   Conductor maximum operating temperature 90 °C

	Open, clipped direct, or lying on a non-metallic surface, or enclosed in non-thermal insulating plaster.								Installed in free air on perforated cable tray, where the perforations occupy 30% of tray area, or on brackets with a space of 0.3 x cable diameter between wall and cable, or spaced 2 x cable diameter when installed on ladder racking.							
Conductor cross-sectional area in sq.mm.	One two core cables single-phase a.c.				One three or four core cable three-phase a.c.				One two core cables single-phase a.c.				One three or four core cable three-phase a.c.			
	Current rating amperes	Voltage drop mV/A/m		Factor length × amperes	Current rating amperes	Voltage drop mV/A/m		Factor length × amperes	Current rating amperes	Voltage drop mV/A/m		Factor length × amperes	Current rating amperes	Voltage drop mV/A/m		Factor length × amperes
		r	x			r	x			r	x			r	x	
1	2	3	4	5	6	7	8	9	10	11	12	13	14	15	16	17
1.5	27	31	-	310	23	27	-	615	29	31	-	310	25	27	-	615
2.5	36	19	-	505	31	16	-	1038	39	19	-	505	33	16	-	1038
4	49	12	-	800	42	10	-	1660	52	12	-	800	44	10	-	1660
6	62	7.9	-	1215	53	6.8	-	2441	66	7.9	-	1215	56	6.8	-	2441
10	85	4.7	-	2043	73	4.0	-	4150	90	4.7	-	2043	78	4.0	-	4150
16	110	2.9	-	3310	94	2.5	-	6640	115	2.9	-	3310	99	2.5	-	6640
25	146	1.85	0.160	5053	124	1.60	0.140	10061	152	1.85	0.160	5053	131	1.60	0.140	10061
35	180	1.35	0.155	7111	154	1.15	0.135	14435	188	1.35	0.155	7111	162	1.15	0.135	14435
50	219	0.99	0.155	9600	187	0.86	0.135	19080	228	0.99	0.155	9600	197	0.86	0.135	19080

Ambient temperature correction factors for thermosetting insulation

	For rewirable fuses only									For all other protective devices								
Ambient temperature °C	25	30	35	40	45	50	55	60	65	25	30	35	40	45	50	55	60	65
Correction factor	1.02	1.0	0.98	0.95	0.93	0.91	0.89	0.87	0.85	1.02	1.0	0.96	0.91	0.87	0.82	0.76	0.71	0.65

Note: Factor gives cable size for 4% voltage drop at 240 V and 415 V. Multiply load current by cable length, then select a factor equal to, or larger than the calculated value of load × length. For loads with a power factor where length is critical, use the values of r and x in the formulas given in the introduction, or in Part 2 Voltage drop. Cables need derating when BS 3036 fuses are used for overload protection.

(4E4)

Itab ARMOURED CABLES   CR 10 (a)   COPPER CONDUCTORS

Armoured multicore thermosetting (XLPE) insulated copper cables to BS 5467 or BS 6724

Full thermal ratings   Ambient temperature 30 °C   Conductor maximum operating temperature 90 °C

Open, clipped direct, or lying on a non-metallic surface, or enclosed in non-thermal insulating plaster.

Installed in free air on perforated cable tray, where the perforations occupy 30% of tray area, or on brackets with a space of 0.3 x cable diameter between wall and cable, or spaced 2 x cable diameter when installed on ladder racking.

Conductor cross-sectional area in sq.mm.	One two core cables single-phase a.c.				One three or four core cable three-phase a.c.				One two core cables single-phase a.c.				One three or four core cable three-phase a.c.			
	Current rating amperes	Voltage drop mV/A/m		Factor length × amperes	Current rating amperes	Voltage drop mV/A/m		Factor length × amperes	Current rating amperes	Voltage drop mV/A/m		Factor length × amperes	Current rating amperes	Voltage drop mV/A/m		Factor length × amperes
		r	x			r	x			r	x			r	x	
1	2	3	4	5	6	7	8	9	10	11	12	13	14	15	16	17
70	279	0.67	0.150	13913	238	0.59	0.130	27667	291	0.67	0.150	13913	251	0.59	0.130	27567
95	338	0.50	0.150	18462	289	0.43	0.130	36889	354	0.50	0.150	18462	304	0.43	0.130	36389
120	392	0.40	0.145	22857	335	0.34	0.130	44865	410	0.40	0.145	22857	353	0.34	0.130	44865
150	451	0.32	0.145	27429	386	0.28	0.125	55333	472	0.32	0.145	27429	406	0.28	0.125	55333
185	515	0.26	0.145	33103	441	0.22	0.125	63846	539	0.26	0.145	33103	463	0.22	0.125	63846
240	607	0.20	0.140	40000	520	0.175	0.125	79048	636	0.20	0.140	40000	546	0.175	0.125	79048
300	698	0.16	0.140	45714	599	0.140	0.120	89730	732	0.16	0.140	45714	628	0.140	0.120	89730
400	787	0.13	0.145	49231	673	0.115	0.125	97647	847	0.13	0.145	49231	728	0.115	0.125	97647

Ambient temperature correction factors for thermosetting insulation

For rewirable fuses only

Ambient temperature °C	25	30	35	40	45	50	55	60	65
Correction factor	1.02	1.0	0.98	0.95	0.93	0.91	0.89	0.87	0.85

For all other protective devices

Ambient temperature °C	25	30	35	40	45	50	55	60	65
Correction factor	1.02	1.0	0.96	0.91	0.87	0.82	0.76	0.71	0.65

Note: Factor gives cable size for 4% voltage drop at 240 V and 415 V. Multiply load current by cable length, then select a factor equal to, or larger than the calculated value of load × length. For loads with a power factor where length is critical, use the values of r and x in the formulas given in the introduction, or in Part 2 Voltage drop. Cables need derating when BS 3036 fuses are used for overload protection.

(4E4)

Itab ARMOURED CABLES   CR 11   ALUMINIUM CONDUCTORS

Armoured multicore PVC insulated aluminium cables to BS 6346

Full thermal ratings   Ambient temperature 30 °C   Conductor maximum operating temperature 70 °C

Open, clipped direct, or lying on a non-metallic surface, or enclosed in non-thermal insulating plaster.

Installed in free air on perforated cable tray, where the perforations occupy 30% of tray area, or on brackets with a space of 0.3 x cable diameter between wall and cable, or spaced 2 x cable diameter when installed on ladder racking.

Conductor cross-sectional area in sq.mm.	One two core cables single-phase a.c.				One three or four core cable three-phase a.c.				One two core cables single-phase a.c.				One three or four core cable three-phase a.c.			
	Current rating amperes	Voltage drop mV/A/m		Factor length × amperes	Current rating amperes	Voltage drop mV/A/m		Factor length × amperes	Current rating amperes	Voltage drop mV/A/m		Factor length × amperes	Current rating amperes	Voltage drop mV/A/m		Factor length × amperes
		r	x			r	x			r	x			r	x	
1	2	3	4	5	6	7	8	9	10	11	12	13	14	15	16	17
16	68	4.50	-	2133	58	3.90	-	4256	71	4.50	-	2133	61	3.90	-	4256
25	89	2.90	0.175	3310	76	2.50	0.150	6640	94	2.90	0.175	3310	80	2.50	0.150	6640
35	109	2.10	0.170	4571	94	1.80	0.150	9222	115	2.10	0.170	4571	99	1.80	0.150	9222
50	131	1.55	0.170	6194	113	1.35	0.145	12296	139	1.55	0.170	6194	119	1.35	0.145	12296
70	165	1.05	0.165	9143	143	0.90	0.140	18043	175	1.05	0.165	9143	151	0.90	0.140	18043
95	199	0.77	0.160	12152	174	0.67	0.140	24412	211	0.77	0.160	12152	186	0.67	0.140	24412
120	-	-	-	-	202	0.53	0.135	30182	-	-	-	-	216	0.53	0.135	30182
150	-	-	-	-	232	0.42	0.135	37727	-	-	-	-	250	0.42	0.135	37727
185	-	-	-	-	265	0.34	0.135	44865	-	-	-	-	287	0.34	0.135	44865
240	-	-	-	-	312	0.26	0.130	55333	-	-	-	-	342	0.26	0.130	55333
300	-	-	-	-	360	0.21	0.130	66400	-	-	-	-	399	0.21	0.130	66400

Ambient temperature correction factors for p.v.c insulated cable

For rewirable fuses only									For all other protective devices								
Ambient temperature °C	25	30	35	40	45	50	55	60	65	25	30	35	40	45	50	55	60	65
Correction factor	1.03	1.0	0.97	0.94	0.91	0.87	0.84	0.69	0.48	1.03	1.0	0.94	0.87	0.79	0.71	0.61	0.50	0.35

Note: Factor gives cable size for 4% voltage drop at 240 V and 415 V. Multiply load current by cable length, then select a factor equal to, or larger than the calculated value of load × length. For loads with a power factor where length is critical, use the values of r and x in the formulas given in the introduction, or in Part 2 Voltage drop. Cables need derating when BS 3036 fuses are used for overload protection.

(4K4)

Itab ARMOURED CABLES   **CR 12**   **ALUMINIUM CONDUCTORS**

Armoured multicore thermosetting (XLPE). insulated aluminium cables to BS 5467

Full thermal ratings   Ambient temperature 30 °C   Conductor operating temperature 90 °C

Conductor cross-sectional area in sq.mm.	Open, clipped direct, or lying on a non-metallic surface, or enclosed in non-thermal insulating plaster.								Installed in free air on perforated cable tray, where the perforations occupy 30% of tray area, or on brackets with a space of 0.3 x cable diameter between wall and cable, or spaced 2 x cable diameter when installed on ladder racking.							
	One two core cables single-phase a.c.				One three or four core cable three-phase a.c.				One two core cables single-phase a.c.				One three or four core cable three-phase a.c.			
	Current rating amperes	Voltage drop mV/A/m		Factor length × amperes	Current rating amperes	Voltage drop mV/A/m		Factor length × amperes	Current rating amperes	Voltage drop mV/A/m		Factor length × amperes	Current rating amperes	Voltage drop mV/A/m		Factor length × amperes
		r	x			r	x			r	x			r	x	
1	2	3	4	5	6	7	8	9	10	11	12	13	14	15	16	17
16	82	4.80	-	2000	71	4.20	-	3952	85	4.80	-	2000	74	4.20	-	3952
25	108	3.10	0.165	3097	92	2.70	0.140	6148	112	3.10	0.165	3097	98	2.70	0.140	6148
35	132	2.20	0.160	4364	113	1.90	0.140	8513	138	2.20	0.160	4364	120	1.90	0.135	8513
50	159	1.65	0.160	5818	137	1.40	0.135	11448	166	1.65	0.160	5818	145	1.40	0.135	11448
70	201	1.10	0.155	8348	174	0.96	0.135	17113	211	1.10	0.155	8348	185	0.96	0.135	17113
95	242	0.82	0.150	11429	214	0.71	0.130	23056	254	0.82	0.150	11429	224	0.71	0.130	23056
120	-	-	-	-	249	0.56	0.130	28621	-	-	-	-	264	0.56	0.130	28621
150	-	-	-	-	284	0.45	0.130	35319	-	-	-	-	305	0.45	0.130	35319
185	-	-	-	-	328	0.37	0.130	42564	-	-	-	-	350	0.37	0.130	42554
240	-	-	-	-	386	0.28	0.125	53548	-	-	-	-	418	0.28	0.125	53548
300	-	-	-	-	441	0.23	0.125	63846	-	-	-	-	488	0.23	0.125	63846

Ambient temperature correction factors for thermosetting insulation

	For rewirable fuses only									For all other protective devices								
Ambient temperature °C	25	30	35	40	45	50	55	60	65	25	30	35	40	45	50	55	60	65
Correction factor	1.02	1.0	0.98	0.95	0.93	0.91	0.89	0.87	0.85	1.02	1.0	0.96	0.91	0.87	0.82	0.76	0.71	0.65

Note: Factor gives cable size for 4% voltage drop at 240 V and 415 V. Multiply load current by cable length. then select a factor equal to, or larger than the calculated value of load × length. For loads with a power factor where length is critical. use the values of r and x in the formulas given in the introduction. or in Part 2 Voltage drop. Cables need derating when BS 3036 fuses are used for overload protection.

(4L4)

Itab MICC PVC CABLES

CR 13

COPPER SHEATH

Mineral insulated copper cables with PVC sheath to BS 6207
Open and clipped direct or lying on a non-metallic surface

Full thermal ratings   Ambient temperature 30 °C   Maximum sheath operating temperature 70 °C

Conductor cross sectional area in sq. mm.	Two single core, or one 2 core cable single-phase a.c.		Three single core in trefoil or 1-three core three-phase a.c.		Three single core in flat formation three-phase a.c.		One four core with three cores loaded three-phase a.c.		One four core all cores loaded three-phase a.c.		One seven core all cores loaded single-phase a.c.		One twelve core all cores loaded single-phase a.c.	
	Current rating amps	Factor length x amps	Current rating amps	Factor length x amps	Current rating amps	Factor length x amps	Current rating amps	Factor length x amps	Current rating amps	Factor length x amps	Current rating amps	Factor length x amps	Current rating amps	Factor length x amps
1	2	3	4	5	6	7	8	9	10	11	12	13	14	15
Light duty 500V														
1.0	18.5	229	15	461	17	461	15	461	13	461	10	229	–	–
1.5	23	343	19	692	21	692	19.5	692	16.5	692	13	343	–	–
2.5	31	565	26	1186	29	1186	26	1186	22	1186	17.5	565	–	–
4.0	40	960	35	1824	38	1824	–	–	–	–	–	–	–	–
Heavy duty 750V														
1.0	19.5	229	16	461	18	461	16.5	461	14.5	461	11.5	229	9.5	229
1.5	25	343	21	692	23	692	21	692	18	692	14.5	343	12	343
2.5	34	565	28	1186	31	1186	28	1186	25	1186	19.5	565	16	565
4.0	45	960	37	1824	41	1824	37	1824	32	1824	26	960	–	–
6.0	57	1371	48	2767	52	2767	47	2767	41	2767	–	–	–	–
10	77	2286	65	4611	70	4611	64	4611	55	4611	–	–	–	–
16	102	3692	86	7217	92	7217	85	7217	72	7217	–	–	–	–

Ambient temperature correction factors for MICC with 70 °C sheath temperature

For rewirable fuses only

Ambient temperature °C	25	30	35	40	45	50	55	60	65
Correction factor	1.03	1.0	0.96	0.93	0.89	0.86	0.79	0.62	0.42

For all other protective devices

Ambient temperature °C	25	30	35	40	45	50	55	60
Correction factor	1.03	1.0	0.93	0.85	0.77	0.67	0.57	0.45

Note: Factor gives cable size for 4% voltage drop at 240 V and 415 V. Multiply load current by cable length, then select a factor equal to, or larger than the calculated value of load × length. Single core ratings apply only where cables are bonded at both ends. For unsheathed cables exposed to touch, see Table CR14. Cables need derating when BS 3036 fuses are used for overload protection.

(411).

Itab MICC PVC CABLES

CR 14

COPPER SHEATH

Mineral insulated unsheathed copper cables exposed to touch to BS 6207
Open and clipped direct or lying on a non-metallic surface

Full thermal ratings Ambient temperature 30 °C Maximum sheath operating temperature 70 °C

Conductor cross sectional area in sq. mm.	Two single core, or one 2 core cable single-phase a.c.		Three single core in trefoil or 1-three core three-phase a.c.		Three single core in flat formation three-phase a.c.		One four core with three cores loaded three-phase a.c.		One four core all cores loaded three-phase a.c.		One seven core all cores loaded single-phase a.c.		One twelve core all cores loaded single-phase a.c.	
	Current rating amps	Factor length x amps	Current rating amps	Factor length x amps	Current rating amps	Factor length x amps	Current rating amps	Factor length x amps	Current rating amps	Factor length x amps	Current rating amps	Factor length x amps	Current rating amps	Factor length x amps
1	2	3	4	5	6	7	8	9	10	11	12	13	14	15
Light duty 500V														
1.0	16.65	229	13.5	461	15.3	461	13.5	461	11.7	461	9	229	-	-
1.5	20.70	343	17.1	692	18.9	692	17.6	692	14.9	692	11.7	343	-	-
2.5	27.90	565	23.4	1186	26.1	1186	23.4	1186	19.8	1186	15.8	565	-	-
4.0	36.00	960	31.5	1824	34.2	1824	-	-	-	-	-	-	-	-
Heavy duty 750V														
1.0	17.55	229	14.4	461	16.2	461	14.9	461	13.1	461	10.4	229	8.6	229
1.5	22.50	343	18.9	692	20.7	692	18.9	692	16.2	692	13.1	343	10.8	343
2.5	30.60	565	25.2	1186	27.9	1186	25.2	1186	22.5	1186	17.6	565	14.4	565
4.0	40.50	960	33.3	1824	36.9	1824	33.3	1824	28.8	1824	23.4	960	-	-
6.0	51.30	1371	43.2	2767	46.8	2767	42.3	2767	36.9	2767	-	-	-	-
10	69.30	2286	58.5	4611	63	4611	57.6	4611	49.5	4611	-	-	-	-
16	91.80	3692	77.4	7217	82.8	7217	76.5	7217	64.8	7217	-	-	-	-

Ambient temperature correction factors for MICC with 70 °C sheath

For rewirable fuses only

Ambient temperature °C	25	30	35	40	45	50	55	60	65
Correction factor	1.03	1.0	0.96	0.93	0.89	0.86	0.79	0.62	0.42

For all other protective devices

Ambient temperature °C	25	30	35	40	45	50	55	60
Correction factor	1.03	1.0	0.93	0.85	0.77	0.67	0.57	0.45

Note: Factor gives cable size for 4% voltage drop at 240 V and 415 V. Multiply load current by cable length, then select a factor equal to, or larger than the calculated value of load × length. Single core ratings apply only where cables are bonded at both ends. Cables need derating when BS 3036 fuses are used for overload protection.

(4J1)

Itab MICC PVC CABLES CR 15 COPPER SHEATH

Mineral insulated copper cables with PVC sheath to BS 6207. Installed on perforated cable tray, horizontal or vertical. (See notes for unsheathed cables.)

Full thermal ratings Ambient temperature 30 °C Maximum sheath operating temperature 70 °C

Conductor cross sectional area in sq. mm.	One two core cable single-phase a.c. or dc.		One three core cable three-phase a.c.		Three single core in flat formation Three-phase a.c.		One four core with three cores loaded three-phase a.c.		One four core all cores loaded three-phase a.c.		One seven core all cores loaded single-phase a.c.		One twelve core all cores loaded single-phase a.c.	
	Current rating amps	Factor length x amps	Current rating amps	Factor length x amps	Current rating amps	Factor length x amps	Current rating amps	Factor length x amps	Current rating amps	Factor length x amps	Current rating amps	Factor length x amps	Current rating amps	Factor length x amps
1	2	3	4	5	6	7	8	9	10	11	12	13	14	15
Light duty 500V														
1.0	19.5	229	16.5	461	17	461	16	461	14	461	11	229	-	-
1.5	25	343	21	692	22	692	21	692	18	692	14	343	-	-
2.5	33	565	28	1186	29	1186	28	1186	24	1186	19	565	-	-
4.0	44	960	37	1824	39	1824	-	-	-	-	-	-	-	-
Heavy duty 750V														
1.0	21	229	17.5	461	19	461	18	461	16	461	12	229	10	229
1.5	26	343	22	692	25	692	23	692	20	692	15.5	343	13	343
2.5	36	565	30	1186	32	1186	30	1186	27	1186	21	565	17	565
4.0	47	960	40	1824	43	1824	40	1824	35	1824	28	960	-	-
6.0	60	1371	51	2767	54	2767	51	2767	44	2767	-	-	-	-
10	82	2286	69	4611	73	4611	68	4611	59	4611	-	-	-	-
16	109	3692	92	7217	97	7217	89	7217	78	7217	-	-	-	-

Ambient temperature correction factors for MICC with 70 °C sheath

	For rewirable fuses only									For all other protective devices							
Ambient temperature °C	25	30	35	40	45	50	55	60	65	25	30	35	40	45	50	55	60
Correction factor	1.03	1.0	0.96	0.93	0.89	0.86	0.79	0.62	0.42	1.03	1.0	0.93	0.85	0.77	0.67	0.57	0.45

Note: Factor gives cable size for 4% voltage drop at 240 and 415 V. Multiply load current by cable length, then select a factor equal to, or larger than the calculated value of load × length. Single core cable ratings only apply where cables are bonded at both ends. Cable ratings to be derated by 0.9 for unsheathed cables exposed to touch. Cables need derating when BS 3036 fuses are used for overload protection.

(4J1).

Itab MICC CABLES   **CR 16**   COPPER SHEATH

Mineral insulated bare copper cables to BS 6207, not exposed to touch, or in contact with combustible materials. Open and clipped direct or lying on a non-metallic surface

Full thermal ratings   Ambient temperature 30 °C   Maximum sheath operating temperature 105 °C

Conductor cross sectional area in sq. mm.	Two single core, or one 2 core cable single-phase a.c.		Three single core in trefoil or 1-three core three-phase a.c.		Three single core in flat formation three-phase a.c.		One four core with three cores loaded three-phase a.c.		One four core all cores loaded three-phase a.c.		One seven core all cores loaded single-phase a.c.		One twelve core all cores loaded single-phase a.c.	
	Current rating amps	Factor length x amps	Current rating amps	Factor length x amps	Current rating amps	Factor length x amps	Current rating amps	Factor length x amps	Current rating amps	Factor length x amps	Current rating amps	Factor length x amps	Current rating amps	Factor length x amps
1	2	3	4	5	6	7	8	9	10	11	12	13	14	15
Light duty 500V														
1.0	22	204	19	415	21	415	18.5	415	16.5	415	13	204	-	-
1.5	28	310	24	615	27	615	24	615	21	615	16.5	310	-	-
2.5	38	505	33	1038	36	1038	33	1038	28	1038	22	505	-	-
4.0	51	800	44	1660	47	1660	-	-	-	-	-	-	-	-
Heavy duty 750V														
1.0	24	204	20	415	24	415	20	415	17.5	415	14	204	12	204
1.5	31	310	26	615	30	615	26	615	22	615	17.5	310	15.5	310
2.5	42	505	35	1038	41	1038	35	1038	30	1038	24	505	20	505
4.0	55	800	47	1660	53	1660	46	1660	40	1660	32	800	-	-
6.0	70	1231	59	2441	67	2441	58	2441	50	2441	-	-	-	-
10	96	2043	81	4049	91	4049	78	4049	68	4049	-	-	-	-
16	127	3200	107	6385	119	6385	103	6385	90	6385	-	-	-	-

Ambient temperature correction factors for 105 °C sheath MICC cable

For rewirable fuses only									For all other protective devices								
Ambient temperature °C	25	30	35	40	45	50	55	60	65	25	30	35	40	45	50	55	60	65
Correction factor	1.02	1.0	0.98	0.96	0.93	0.91	0.89	0.86	0.84	1.02	1.0	0.96	0.92	0.88	0.84	0.80	0.75	0.70

NO CORRECTION FOR GROUPING REQUIRED

Note: Factor gives cable size for 4% voltage drop at 240 and 415 V. Multiply load current by cable length, then select a factor equal to, or larger than the calculated value of load × length. Single core cable ratings only apply where cables are bonded at both ends. Cables need derating when BS 3036 fuses are used for over oad protection.

(4J2)

Itab FLEXIBLE CORDS

CR 17   COPPER CONDUCTORS

Flexible cords to BS 6500

Conductor cross-sectional area in sq. mm.	Current-carrying capacity		Maximum length in metres for 4% volt drop at rated current at 240 and 415 V.		Factor length × amperes		Maximum weight allowed (in kg) to be supported by a twin flexible cord
	Single-phase a.c.	Three-phase a.c.	Single-phase	Three-phase	Single-phase	Three-phase	
1	2	3	4	5	6	7	8
0.50	3	3	34.41	69.17	103	208	2
0.75	6	6	25.81	51.23	155	307	3
1.00	10	10	20.87	41.50	209	415	5
1.25	13	-	19.96	-	259	-	5
1.50	16	16	18.75	38.43	300	615	5
2.50	25	20	20.21	51.88	505	1038	5
4.00	32	25	25.00	66.40	800	1660	5

Ambient temperature correction factors

Ambient temperature °C	35	40	45	50	55	60	65	70	120	125	130	135	140	145	150	155	160	165	170
60 °C rubber and pvc cords	0.91	0.82	0.71	0.58	0.41														
85 °C rubber cords having HOFR or heat resisting pvc sheath	1.0	1.0	1.0	1.0	0.96	0.83	0.67	0.47											
150 °C rubber cords	1.0	1.0	1.0	1.0	1.0	1.0	1.0	1.0	0.96	0.85	0.74	0.60	0.42						
Glass fibre cords	1.0	1.0	1.0	1.0	1.0	1.0	1.0	1.0	1.0	1.0	1.0	1.0	1.0	1.0	1.0	0.92	0.82	0.71	0.57

Notes: To determine the voltage drop in mV/A/m for single phase, divide 9,600 by the factor for the cable size. For the mV/A/m for three phase, divide 16,600 by the factor for the cable size.

(4H3)

SHOCK PROTECTION 0.4 s

ZS 1

BS 88 HRC FUSES

Maximum design and testing values of earth loop impedance Z_S for BS 88 Parts 2 & 6 fuses when U_{oc} is 240V

DISCONNECTION TIME   0.4 s

Maximum values of Z_S in ohms for different HRC fuse ratings

Fuse rating in amperes for 0.4 second disconnection time

Temperature degrees C	2	4	6	10	16	20	25	32	35	40	50	63	80	100	125	160	200	250	315	400	500
									Maximum value of Z_S in ohms												
Design	36	17	8.89	5.33	2.82	1.85	1.5	1.09	0.97	0.86	0.63	0.48	0.32	0.24	0.18	0.14	0.11	0.086	0.063	0.048	0.035
Testing	For testing the following maximum values of Z_S are based on a design temperature of 115 °C																				
30	26.87	12.69	6.63	3.98	2.10	1.38	1.12	0.81	0.72	0.64	0.47	0.358	0.239	0.179	0.134	0.104	0.082	0.064	0.047	0.036	0.026
25	26.47	12.50	6.54	3.92	2.07	1.36	1.10	0.80	0.71	0.63	0.46	0.353	0.235	0.176	0.132	0.103	0.081	0.063	0.046	0.035	0.026
20	26.09	12.32	6.44	3.86	2.04	1.34	1.09	0.79	0.70	0.62	0.46	0.348	0.232	0.174	0.130	0.101	0.080	0.062	0.046	0.035	0.025
15	25.71	12.14	6.35	3.81	2.01	1.32	1.07	0.78	0.69	0.61	0.45	0.343	0.229	0.171	0.129	0.100	0.079	0.061	0.045	0.034	0.025
10	25.35	11.97	6.26	3.75	1.99	1.30	1.06	0.77	0.68	0.61	0.44	0.338	0.225	0.169	0.127	0.099	0.077	0.061	0.044	0.034	0.025
5	25.00	11.81	6.17	3.70	1.96	1.28	1.04	0.76	0.67	0.60	0.44	0.333	0.222	0.167	0.125	0.097	0.076	0.060	0.044	0.033	0.024
	24.66	11.64	6.09	3.65	1.93	1.27	1.03	0.75	0.66	0.59	0.43	0.329	0.219	0.164	0.123	0.096	0.075	0.059	0.043	0.033	0.024
-5	24.32	11.49	6.01	3.60	1.91	1.25	1.01	0.74	0.66	0.58	0.43	0.324	0.216	0.162	0.122	0.095	0.074	0.058	0.043	0.032	0.024

Notes: The above values for testing are based on the phase and CPC conductors, both being copper or aluminium. Calculations are based on the resistance-temperature coefficient of 0.004 per °C at 20 °C.

Where the open circuit voltage to earth is not 240 V, multiply the Z_S from the table by the actual open circuit voltage to earth, and then divide by 240 V, to give the revised Z_S value for the actual open circuit voltage to earth.

The above values can also be used for gM (motor) fuses.

When U_{oc} is 240 V the nominal voltage U_0 is 230 V.

SHOCK PROTECTION 5s

ZS 1A

BS 88 HRC FUSES

Maximum design and testing values of earth loop impedance Z_S for BS 88 Part 2.2 fuses when U_{oc} is 240V

DISCONNECTION TIME 5 s

Maximum values of Z_S in ohms for different HRC fuse ratings

Fuse rating in amperes for 5 second disconnection time

Temperature degrees C		2	4	6	10	16	20	25	32	35	40	50	63	80	100	125	160	200	250	315	400	500
									Maximum value of Z_S in ohms													
Design		47	23	14.1	7.74	4.36	3.04	2.4	1.92	1.48	1.41	1.09	0.86	0.6	0.44	0.35	0.27	0.2	0.16	0.12	0.09	0.065
Testing °C		For testing the following maximum values of Z_S are based on a design temperature of 115 °C																				
	30	35.1	17.2	10.52	5.78	3.25	2.27	1.79	1.43	1.10	1.05	0.81	0.64	0.448	0.328	0.261	0.201	0.149	0.122	0.087	0.067	0.049
	25	34.6	16.9	10.37	5.69	3.21	2.24	1.76	1.41	1.09	1.04	0.80	0.63	0.441	0.324	0.257	0.199	0.147	0.120	0.086	0.066	0.048
	20	34.1	16.7	10.22	5.61	3.16	2.20	1.74	1.39	1.07	1.02	0.79	0.62	0.435	0.319	0.254	0.196	0.145	0.118	0.085	0.065	0.047
	15	33.6	16.4	10.07	5.53	3.11	2.17	1.71	1.37	1.06	1.01	0.78	0.61	0.429	0.314	0.250	0.193	0.143	0.116	0.084	0.064	0.046
	10	33.1	16.2	9.93	5.45	3.07	2.14	1.69	1.35	1.04	0.99	0.77	0.61	0.423	0.310	0.246	0.190	0.141	0.115	0.082	0.063	0.046
	5	32.6	16.0	9.79	5.38	3.03	2.11	1.67	1.33	1.03	0.98	0.76	0.60	0.417	0.306	0.243	0.188	0.139	0.113	0.081	0.063	0.045
		32.2	15.8	9.66	5.30	2.99	2.08	1.64	1.32	1.01	0.97	0.75	0.59	0.411	0.301	0.240	0.185	0.137	0.112	0.080	0.062	0.045
	-5	31.8	15.5	9.53	5.23	2.95	2.05	1.62	1.30	1.00	0.95	0.74	0.58	0.405	0.297	0.236	0.182	0.135	0.110	0.079	0.061	0.044

Notes: The above values for testing are based on the phase and CPC conductors, both being copper or aluminium. Calculations are based on the resistance-temperature coefficient of 0.004 per °C at 20 °C.

Where the open circuit voltage to earth is not 240 V, multiply the Z_S from the table by the actual open circuit voltage to earth, and then divide by 240 V, to give the revised Z_S value for the actual open circuit voltage to earth.

The above values can also be used for gM (motor) fuses.

When U_{oc} is 240 V the nominal voltage U_0 is 230 V.

SHOCK PROTECTION 0.4 and 5 s   ZS 2   BS 1361 FUSES

Maximum design and testing values of earth loop impedance Z_S for BS 1361 fuses when U_{oc} is 240V

DISCONNECTION TIMES   0.4 s AND   5 s

Maximum values of Z_S in ohms for different fuse ratings

Temperature degrees C	Fuse rating in amperes for 0.4 second disconnection time								Fuse rating in amperes for 5 second disconnection time							
	Maximum value of Z_S in ohms								Maximum value of Z_S in ohms							
	5	15	20	30	45	60	80	100	5	15	20	30	45	60	80	100
Design	**10.91**	**3.43**	**1.78**	**1.2**	**0.6**	**0.4**	**0.3**	**0.2**	**17.14**	**5.22**	**2.93**	**1.92**	**1.0**	**0.73**	**0.52**	**0.38**
Testing	For testing the following maximum values of Z_S are based on a design temperature of 115 °C															
30	8.14	2.56	1.33	0.90	0.45	0.30	0.22	0.15	12.79	3.90	2.19	1.43	0.75	0.54	0.39	0.28
25	8.02	2.52	1.31	0.88	0.44	0.29	0.22	0.15	12.60	3.84	2.15	1.41	0.74	0.54	0.38	0.28
20	7.91	2.49	1.29	0.87	0.43	0.29	0.22	0.14	12.42	3.78	2.12	1.39	0.72	0.53	0.38	0.28
15	7.79	2.45	1.27	0.86	0.43	0.29	0.21	0.14	12.24	3.73	2.09	1.37	0.71	0.52	0.37	0.27
10	7.68	2.42	1.25	0.85	0.42	0.28	0.21	0.14	12.07	3.68	2.06	1.35	0.70	0.51	0.37	0.27
5	7.58	2.38	1.24	0.83	0.42	0.28	0.21	0.14	11.90	3.63	2.03	1.33	0.69	0.51	0.36	0.26
0	7.47	2.35	1.22	0.82	0.41	0.27	0.21	0.14	11.74	3.58	2.01	1.32	0.68	0.50	0.36	0.26
-5	7.37	2.32	1.20	0.81	0.41	0.27	0.20	0.14	11.58	3.53	1.98	1.30	0.68	0.49	0.35	0.26

Notes: The above values for testing are based on the phase and CPC conductors, both being copper or aluminium. Calculations are based on the resistance-temperature coefficient of 0.004 per °C at 20 °C.

Where the open circuit voltage to earth is not 240 V, multiply the Z_S from the table by the actual open circuit voltage to earth, and then divide by 240 V, to give the revised Z_S value for the actual open circuit voltage to earth.

When U_{oc} is 240 V the nominal voltage U_o is 230 V.

Maximum design and testing values of earth loop impedance ZS for BS 3036 fuses when U_{oc} is 240V

DISCONNECTION TIMES 0.4 s AND 5 s

Maximum values of Z_S in ohms for different fuse ratings

Temperature degrees C	Fuse rating in amperes for 0.4 second disconnection time							Fuse rating in amperes for 5 second disconnection time						
	5	15	20	30	45	60	100	5	15	20	30	45	60	100
	Maximum value of Z_S in ohms							Maximum value of Z_S in ohms						
Design	10	2.67	1.85	1.14	0.62	0.44	0.2	18.46	5.58	4.0	2.76	1.66	1.17	0.558
Testing	For testing the following maximum values of Z_S are based on a design temperature of 115 °C													
30	7.46	1.99	1.38	0.85	0.46	0.328	0.149	13.81	4.16	2.99	2.06	1.239	0.873	0.416
25	7.35	1.96	1.36	0.84	0.46	0.324	0.147	13.60	4.10	2.94	2.03	1.221	0.860	0.410
20	7.25	1.93	1.34	0.83	0.45	0.319	0.145	13.41	4.04	2.90	2.00	1.203	0.848	0.404
15	7.14	1.91	1.32	0.81	0.44	0.314	0.143	13.21	3.99	2.86	1.97	1.186	0.836	0.399
10	7.04	1.88	1.30	0.80	0.44	0.310	0.141	13.03	3.93	2.82	1.94	1.169	0.824	0.393
5	6.94	1.85	1.28	0.79	0.43	0.306	0.139	12.85	3.88	2.78	1.92	1.153	0.813	0.388
0	6.85	1.83	1.27	0.78	0.42	0.301	0.137	12.67	3.82	2.74	1.89	1.137	0.801	0.382
-5	6.76	1.80	1.25	0.77	0.42	0.297	0.135	12.50	3.77	2.70	1.86	1.122	0.791	0.377

Notes: The above values for testing are based on the phase and CPC conductors, both being copper or aluminium. Calculations are based on the resistance-temperature coefficient of 0.004 per °C at 20 °C.

Where the open circuit voltage to earth is not 240 V, multiply the Z_S from the table by the actual open circuit voltage to earth, and then divide by 240 V, to give the revised Z_S value for the actual open circuit voltage to earth.

When U_{oc} is 240 V the nominal voltage U_0 is 230 V.

SHOCK PROTECTION 0.4 and 5 s

ZS 4

BS 3871 TYPE 1 MCBs

Maximum design and testing values of earth loop impedance Z_S for BS 3871 Type 1 MCBs when U_{oc} is 240V

DISCONNECTION TIMES 0.4 s AND 5 s

Maximum values of Z_S in ohms for different Type 1 MCB ratings

Temperature degrees C	MCB rating for 0.1 second to 5 second disconnection time																	Other MCB ratings
	2	4	5	6	10	15	16	20	25	30	32	40	45	50	63	80	100	$60/I_n$ or $240V:4 I_n$
								Maximum value of Z_S in ohms										
Design	30	15	12	10	6	4	3.75	3	2.4	2.0	1.88	1.5	1.33	1.2	0.95	0.75	0.6	
Testing																		
30	22.39	11.19	8.96	7.46	4.48	2.99	2.80	2.24	1.79	1.49	1.40	1.12	1.00	0.90	0.71	0.56	0.45	60/1.34In
25	22.06	11.03	8.82	7.35	4.41	2.94	2.76	2.21	1.76	1.47	1.38	1.10	0.98	0.88	0.70	0.55	0.44	60/1.36In
20	21.74	10.87	8.70	7.25	4.35	2.90	2.72	2.17	1.74	1.45	1.36	1.09	0.97	0.87	0.69	0.54	0.43	60/1.38In
15	21.43	10.71	8.57	7.14	4.29	2.86	2.68	2.14	1.71	1.43	1.34	1.07	0.95	0.86	0.68	0.54	0.43	60/1.40In
10	21.13	10.56	8.45	7.04	4.23	2.82	2.64	2.11	1.69	1.41	1.32	1.06	0.94	0.85	0.67	0.53	0.42	60/1.42In
5	20.83	10.42	8.33	6.94	4.17	2.78	2.60	2.08	1.67	1.39	1.30	1.04	0.93	0.83	0.66	0.52	0.42	60/1.44In
0	20.55	10.27	8.22	6.85	4.11	2.74	2.57	2.05	1.64	1.37	1.28	1.03	0.91	0.82	0.65	0.51	0.41	60/1.46In
-5	20.27	10.14	8.11	6.76	4.05	2.70	2.53	2.03	1.62	1.35	1.27	1.01	0.90	0.81	0.64	0.51	0.41	60/1.48In

For testing the following maximum values of Z_S are based on a design temperature of 115 °C

Notes: The above values for testing are based on the phase and CPC conductors, both being copper or aluminium. Calculations are based on the resistance-temperature coefficient of 0.004 per °C at 20 °C.

Where the open circuit voltage to earth is not 240 V, multiply the Z_S from the table by the actual open circuit voltage to earth, and then divide by 240 V, to give the revised Z_S value for the actual open circuit voltage to earth.

When U_{oc} is 240 V the nominal voltage U_o is 230 V.

SHOCK PROTECTION 0.4 and 5 s ZS 5 BS 3871 TYPE 2 MCBs

Maximum design and testing values of earth loop impedance Z_S for BS 3871 Type 2 MCBs when U_{oc} is 240V

DISCONNECTION TIMES 0.4 s AND 5 s

Temperature degrees C	Maximum values of Z_S in ohms for different Type 2 MCB ratings																	
	MCB rating for 0.1 second to 5 second disconnection time																	Other MCB ratings
	2	4	5	6	10	15	16	20	25	30	32	40	45	50	63	80	100	34.3/I_n or 240V/7I_n
	Maximum value of Z_S in ohms																	
Design	17.1	8.6	6.86	5.71	3.43	2.29	2.14	1.71	1.37	1.14	1.07	0.86	0.76	0.69	0.54	0.43	0.34	
Testing	For testing the following maximum values of Z_S are based on a design temperature of 115 °C																	
30	12.79	6.40	5.12	4.26	2.56	1.71	1.60	1.28	1.02	0.85	0.80	0.64	0.57	0.51	0.41	0.32	0.26	34.3/1.34I_n
25	12.61	6.30	5.04	4.20	2.52	1.68	1.58	1.26	1.01	0.84	0.79	0.63	0.56	0.50	0.40	0.32	0.25	34.3/1.36I_n
20	12.42	6.21	4.97	4.14	2.48	1.66	1.55	1.24	0.99	0.83	0.78	0.62	0.55	0.50	0.39	0.31	0.25	34.3/1.38I_n
15	12.24	6.12	4.90	4.08	2.45	1.63	1.53	1.22	0.98	0.82	0.77	0.61	0.54	0.49	0.39	0.31	0.24	34.3/1.40I_n
10	12.07	6.04	4.83	4.02	2.41	1.61	1.51	1.21	0.97	0.80	0.75	0.60	0.54	0.48	0.38	0.30	0.24	34.3/1.42I_n
5	11.90	5.95	4.76	3.97	2.38	1.59	1.49	1.19	0.95	0.79	0.74	0.60	0.53	0.48	0.38	0.30	0.24	34.3/1.44I_n
0	11.74	5.87	4.70	3.91	2.35	1.57	1.47	1.17	0.94	0.78	0.73	0.59	0.52	0.47	0.37	0.29	0.23	34.3/1.46I_n
-5	11.58	5.79	4.63	3.86	2.32	1.54	1.45	1.16	0.93	0.77	0.72	0.58	0.51	0.46	0.37	0.29	0.23	34.3/1.48I_n

Notes: The above values for testing are based on the phase and CPC conductors, both being copper or aluminium. Calculations are based on the resistance-temperature coefficient of 0.004 per °C at 20 °C.

Where the open circuit voltage to earth is not 240 V, multiply the Z_S from the table by the actual open circuit voltage to earth, and then divide by 240 V, to give the revised Z_S value for the actual open circuit voltage to earth.

When U_{oc} is 240 V the nominal voltage U_0 is 230 V.

SHOCK PROTECTION 0.4 and 5 s

ZS 6

BS 3871 TYPE 3 MCBs

Maximum design and testing values of earth loop impedance Z_S for BS 3871 Type 3 MCBs when U_{oc} is 240V

DISCONNECTION TIMES 0.4 s AND 5 s

Maximum values of Z_S in ohms for different Type 3 MCB ratings

Temperature degrees C	MCB rating for 0.1 second to 5 second disconnection time																	Other MCB ratings
	2	4	5	6	10	15	16	20	25	30	32	40	45	50	63	80	100	$24/I_n$ or $240/10I_n$
	Maximum value of Z_S in ohms																	
Design	12	6.0	4.8	4	2.4	1.6	1.5	1.2	0.96	0.8	0.75	0.6	0.53	0.48	0.38	0.3	0.24	
Testing	For testing the following maximum values of Z_S are based on a design temperature of 115 °C																	
30	8.96	4.48	3.58	2.99	1.79	1.19	1.12	0.90	0.72	0.60	0.56	0.45	0.40	0.36	0.28	0.22	0.18	24/1.34In
25	8.82	4.41	3.53	2.94	1.76	1.18	1.10	0.88	0.71	0.59	0.55	0.44	0.39	0.35	0.28	0.22	0.18	24/1.36In
20	8.70	4.35	3.48	2.90	1.74	1.16	1.09	0.87	0.70	0.58	0.54	0.43	0.39	0.35	0.28	0.22	0.17	24/1.38In
15	8.57	4.29	3.43	2.86	1.71	1.14	1.07	0.86	0.69	0.57	0.54	0.43	0.38	0.34	0.27	0.21	0.17	24/1.40In
10	8.45	4.23	3.38	2.82	1.69	1.13	1.06	0.85	0.68	0.56	0.53	0.42	0.38	0.34	0.27	0.21	0.17	24/1.42In
5	8.33	4.17	3.33	2.78	1.67	1.11	1.04	0.83	0.67	0.56	0.52	0.42	0.37	0.33	0.26	0.21	0.17	24/1.44In
0	8.22	4.11	3.29	2.74	1.64	1.10	1.03	0.82	0.66	0.55	0.51	0.41	0.37	0.33	0.26	0.21	0.16	24/1.46In
-5	8.11	4.05	3.24	2.70	1.62	1.08	1.01	0.81	0.65	0.54	0.51	0.41	0.36	0.32	0.26	0.20	0.16	24/1.48In

Notes: The above values for testing are based on the phase and CPC conductors, both being copper or aluminium. Calculations are based on the resistance-temperature coefficient of 0.004 per °C at 20 °C.

Where the open circuit voltage to earth is not 240 V, multiply the Z_S from the table by the actual open circuit voltage to earth, and then divide by 240 V, to give the revised Z_S value for the actual open circuit voltage to earth.

When U_{oc} is 240 V the nominal voltage U_o is 230 V.

SHOCK PROTECTION 0.4 and 5 s　　　　　**ZS 7**　　　　　BS EN 60898 TYPE B MCBs

Maximum design and testing values of earth loop impedance Z_S for BS EN 60898 Type B MCBs when U_{oc} is 240V

DISCONNECTION TIMES 0.4 s AND 5 s

Maximum values of Z_S in ohms for different Type B MCB ratings

Temperature degrees C	MCB rating for 0.1 second to 5 second disconnection time																	Other MCB ratings
	2	4	5	6	10	15	16	20	25	30	32	40	45	50	63	80	100	$48/I_n$ or $240V/5I_n$
	Maximum value of Z_S in ohms																	
Design	24	12	9.6	8	4.8	3.2	3	2.4	1.92	1.6	1.5	1.2	1.07	0.96	0.76	0.6	0.48	
Testing	For testing the following maximum values of Z_S are based on a design temperature of 115 °C																	
30	17.91	8.96	7.16	5.97	3.58	2.39	2.24	1.79	1.43	1.19	1.12	0.90	0.80	0.72	0.57	0.45	0.36	$48/1.34I_n$
25	17.65	8.82	7.06	5.88	3.53	2.35	2.21	1.76	1.41	1.18	1.10	0.88	0.78	0.71	0.56	0.44	0.35	$48/1.36I_n$
20	17.39	8.70	6.96	5.80	3.48	2.32	2.17	1.74	1.39	1.16	1.09	0.87	0.77	0.70	0.55	0.43	0.35	$48/1.38I_n$
15	17.14	8.57	6.86	5.71	3.43	2.29	2.14	1.71	1.37	1.14	1.07	0.86	0.76	0.69	0.54	0.43	0.34	$48/1.40I_n$
10	16.90	8.45	6.76	5.63	3.38	2.25	2.11	1.69	1.35	1.13	1.06	0.85	0.75	0.68	0.54	0.42	0.34	$48/1.42I_n$
5	16.67	8.33	6.67	5.56	3.33	2.22	2.08	1.67	1.33	1.11	1.04	0.83	0.74	0.67	0.53	0.42	0.33	$48/1.44I_n$
0	16.44	8.22	6.58	5.48	3.29	2.19	2.05	1.64	1.32	1.10	1.03	0.82	0.73	0.66	0.52	0.41	0.33	$48/1.46I_n$
-5	16.22	8.11	6.49	5.41	3.24	2.16	2.03	1.62	1.30	1.08	1.01	0.81	0.72	0.65	0.51	0.41	0.32	$48/1.48I_n$

Notes:　The above values for testing are based on the phase and CPC conductors, both being copper or aluminium. Calculations are based on the resistance-temperature coefficient of 0.004 per °C at 20 °C.

Where the open circuit voltage to earth is not 240 V, multiply the Z_S from the table by the actual open circuit voltage to earth, and then divide by 240 V, to give the revised Z_S value for the actual open circuit voltage to earth.

When U_{oc} is 240 V the nominal voltage U_0 is 230 V.

SHOCK PROTECTION 0.4 and 5 s **ZS 8** BS EN 60898 TYPE C MCBs

Maximum design and testing values of earth loop impedance Z_S for BS EN 60898 Type C MCBs when U_{oc} is 240V

DISCONNECTION TIMES 0.4 s AND 5 s

Maximum values of Z_S in ohms for different Type C MCB ratings

Temperature degrees C	MCB rating for 0.1 second to 5 second disconnection time																	Other MCB ratings
	2	4	5	6	10	15	16	20	25	30	32	40	45	50	63	80	100	24/I_n or 240/10I_n
							Maximum value of Z_S in ohms										
Design	12	6	4.8	4	2.4	1.6	1.5	1.2	0.96	0.8	0.75	0.6	0.53	0.48	0.38	0.3	0.24	
Testing	For testing the following maximum values of Z_S are based on a design temperature of 115 °C																	
30	8.96	4.48	3.58	2.99	1.79	1.19	1.12	0.90	0.72	0.60	0.56	0.45	0.40	0.36	0.28	0.22	0.18	24/1.34In
25	8.82	4.41	3.53	2.94	1.76	1.18	1.10	0.88	0.71	0.59	0.55	0.44	0.39	0.35	0.28	0.22	0.18	24/1.36In
20	8.70	4.35	3.48	2.90	1.74	1.16	1.09	0.87	0.70	0.58	0.54	0.43	0.39	0.35	0.28	0.22	0.17	24/1.38In
15	8.57	4.29	3.43	2.86	1.71	1.14	1.07	0.86	0.69	0.57	0.54	0.43	0.38	0.34	0.27	0.21	0.17	24/1.40In
10	8.45	4.23	3.38	2.82	1.69	1.13	1.06	0.85	0.68	0.56	0.53	0.42	0.38	0.34	0.27	0.21	0.17	24/1.42In
5	8.33	4.17	3.33	2.78	1.67	1.11	1.04	0.83	0.67	0.56	0.52	0.42	0.37	0.33	0.26	0.21	0.17	24/1.44In
0	8.22	4.11	3.29	2.74	1.64	1.10	1.03	0.82	0.66	0.55	0.51	0.41	0.37	0.33	0.26	0.21	0.16	24/1.46In
-5	8.11	4.05	3.24	2.70	1.62	1.08	1.01	0.81	0.65	0.54	0.51	0.41	0.36	0.32	0.26	0.20	0.16	24/1.48In

Notes: The above values for testing are based on the phase and CPC conductors, both being copper or aluminium. Calculations are based on the resistance-temperature coefficient of 0.004 per °C at 20 °C.

Where the open circuit voltage to earth is not 240 V, multiply the Z_S from the table by the actual open circuit voltage to earth, and then divide by 240 V, to give the revised Z_S value for the actual open circuit voltage to earth.

When U_{oc} is 240 V the nominal voltage U_0 is 230 V.

Maximum design and testing values of earth loop impedance Z_S for BS EN 60898 Type D MCBs when U_{oc} is 240V

DISCONNECTION TIMES 0.4 s AND 5 s

Maximum values of Z_S in ohms for different Type D MCB ratings

Temperature degrees C	2	4	5	6	10	15	16	20	25	30	32	40	45	50	63	80	100	Other MCB ratings
	MCB rating for 0.1 second to 5 second disconnection time																	$12/I_n$ or $240V/20I_n$
	Maximum value of Z_S in ohms																	
Design	6.0	3.0	2.4	2.0	1.2	0.8	0.75	0.6	0.48	0.4	0.38	0.3	0.27	0.24	0.19	0.15	0.12	
Testing	For testing the following maximum values of Z_S are based on a design temperature of 115 °C																	
30	4.48	2.24	1.79	1.49	0.90	0.60	0.56	0.45	0.36	0.30	0.28	0.22	0.20	0.18	0.14	0.11	0.09	$12/1.34I_n$
25	4.41	2.21	1.76	1.47	0.88	0.59	0.55	0.44	0.35	0.29	0.28	0.22	0.20	0.18	0.14	0.11	0.09	$12/1.36I_n$
20	4.35	2.17	1.74	1.45	0.87	0.58	0.54	0.43	0.35	0.29	0.27	0.22	0.19	0.17	0.14	0.11	0.09	$12/1.38I_n$
15	4.29	2.14	1.71	1.43	0.86	0.57	0.54	0.43	0.34	0.29	0.27	0.21	0.19	0.17	0.14	0.11	0.09	$12/1.40I_n$
10	4.23	2.11	1.69	1.41	0.85	0.56	0.53	0.42	0.34	0.28	0.26	0.21	0.19	0.17	0.13	0.11	0.08	$12/1.42I_n$
5	4.17	2.08	1.67	1.39	0.83	0.56	0.52	0.42	0.33	0.28	0.26	0.21	0.19	0.17	0.13	0.10	0.08	$12/1.44I_n$
0	4.11	2.05	1.64	1.37	0.82	0.55	0.51	0.41	0.33	0.27	0.26	0.21	0.18	0.16	0.13	0.10	0.08	$12/1.46I_n$
-5	4.05	2.03	1.62	1.35	0.81	0.54	0.51	0.41	0.32	0.27	0.25	0.20	0.18	0.16	0.13	0.10	0.08	$12/1.48I_n$

Notes: The above values for testing are based on the phase and CPC conductors, both being copper or aluminium. Calculations are based on the resistance-temperature coefficient of 0.004 per °C at 20 °C.

Where the open circuit voltage to earth is not 240 V, multiply the Z_S from the table by the actual open circuit voltage to earth, and then divide by 240 V, to give the revised Z_S value for the actual open circuit voltage to earth.

When U_{oc} is 240 V and the nominal voltage U_0 is 230 V.

SHOCK PROTECTION 5 s   ZS 10   REDUCED VOLTAGE

Maximum design values of earth loop impedance for different types of protective device, when disconnection time required is 5 seconds and U_{oc} is 55 volts

MAXIMUM DISCONNECTION TIME 5 s

Protective device type	Protective device rating in amperes																				
	2A	4A	5A	6A	10A	15A	16A	20A	25A	30A	32A	40A	45A	50A	60A	63A	80A	100A	125A	160A	200A
	Maximum design value of earth loop impedance Z_S in ohms																				
BS 88 Fuse	10.77	5.27	-	-	1.77	-	1.00	0.70	0.55	-	0.44	0.323	-	0.250	-	0.197	0.138	0.101	0.080	0.062	0.046
BS1361 fuse	-	-	3.93	-	-	1.20	-	0.67	-	0.44	-	-	0.229	-	0.167	-	0.119	0.087	-	-	-
BS 3036 fuse	-	-	4.23	-	-	1.28	-	0.92	-	0.63	-	-	0.380	-	0.268	-	-	0.128	-	-	-
Type 1 MCB	6.90	3.45	2.76	2.30	1.38	0.92	0.86	0.69	0.55	0.46	0.43	0.345	0.307	0.276	0.230	0.219	0.173	0.138	0.110	0.086	0.069
Type 2 MCB	3.95	1.98	1.58	1.32	0.79	0.53	0.49	0.40	0.32	0.26	0.25	0.198	0.176	0.158	0.132	0.125	0.099	0.079	0.063	0.049	0.040
Type 3 MCB	5.50	2.75	2.20	1.83	1.10	0.73	0.69	0.55	0.44	0.37	0.34	0.275	0.244	0.220	0.183	0.175	0.138	0.110	0.088	0.069	0.055
Type B MCB	2.75	1.38	1.10	0.92	0.55	0.37	0.34	0.28	0.22	0.18	0.17	0.138	0.122	0.110	0.092	0.087	0.069	0.055	0.044	0.034	0.028
Type C MCB	2.75	1.38	1.10	0.92	0.55	0.37	0.34	0.28	0.22	0.18	0.17	0.138	0.122	0.110	0.092	0.087	0.069	0.055	0.044	0.034	0.028
Type D MCB	1.40	0.70	0.56	0.47	0.28	0.19	0.18	0.14	0.11	0.09	0.09	0.070	0.062	0.056	0.047	0.044	0.035	0.028	0.022	0.018	0.014

Notes: To obtain the value of earth loop impedance allowed for a U_{oc} voltage of 63.5 V, divide the values in the above table by 55 and multiply by 63.5.
471A

Maximum design values of earth loop impedance for different types of protective device, when disconnection time required is 0.2 s and U_{oc} is 240 volts

MAXIMUM DISCONNECTION TIME 0.2 s

Protective device type	Protective device rating in amperes																				
	2A	4A	5A	6A	10A	15A	16A	20A	25A	30A	32A	40A	45A	50A	60A	63A	80A	100A	125A	160A	200A
	Maximum design value of earth loop impedance Z_S in ohms																				
BS 88 Fuse	33.8	15.19	-	7.74	4.71	-	2.53	1.60	1.33	-	0.92	0.71	-	0.53	-	0.39	0.27	0.20	0.17	0.12	0.09
BS1361 fuse			9.6	-	-	3.0	-	1.55	-	1.0	-	-	0.51	-	0.33	-	0.28	0.17	-	-	-
BS 3036 fuse			7.5	-	-	1.92	-	1.33	-	0.80	-	-	0.40	-	0.30	-	-	0.13	-	-	-
Type 1 MCB	30	15	12	10	6.0	4.0	3.75	3.0	2.4	2.0	1.88	1.50	1.33	1.20	1.0	0.95	0.75	0.60	0.48	0.38	0.30
Type 2 MCB	17.1	8.57	6.86	5.7	3.43	2.29	2.14	1.71	1.37	1.14	1.07	0.86	0.76	0.69	0.57	0.54	0.43	0.34	0.27	0.21	0.17
Type 3 MCB	12	6.0	4.8	4.0	2.4	1.6	1.5	1.2	0.96	0.80	0.75	0.60	0.53	0.48	0.40	0.38	0.30	0.24	0.19	0.15	0.12
Type B MCB	24	12	9.6	8.0	4.8	3.2	3.0	2.4	1.92	1.6	1.5	1.2	1.1	0.96	0.80	0.76	0.60	0.48	0.38	0.30	0.24
Type C MCB	12	6	4.8	4	2.4	1.6	1.5	1.2	0.96	0.8	0.75	0.6	0.53	0.48	0.4	0.38	0.30	0.24	0.19	0.15	0.12
Type D MCB	6	3	2.4	2	1.2	0.8	0.75	0.6	0.48	0.4	0.38	0.3	0.27	0.24	0.2	0.19	0.15	0.12	0.10	0.08	0.06

Notes: To obtain the value of earth loop impedance allowed for any other open circuit voltage divide the values in the above table by 240, and multiply by the actual open circuit voltage. For motor fuses, check that the characteristic is the same as that for the standard fuse; if not, consult the manufacturer; the characteristics at the rear of this book are the same, so the Z_S figures given above can be used for motor fuses as well as for general purpose fuses, when using this type and make of fuse. When U_{oc} is 240 V U_0 is 230 V.

604/605B1/B2

THERMAL PROTECTION C.P.C.   PCZ 1   BS 88 HRC FUSES

Maximum values of Z_S for thermal protection of the protective conductor, when U_{oc} = 240V

Size of copper C.P.C in m m²	Rating of H.R.C. fuse in amperes																			
	2	4	6	10	16	20	25	32	40	50	63	80	100	125	160	200	250	315	400	500
	Maximum value of Z_S at design temperature																			
1	47	23	14	7.7	4.1	2.3	1.5	0.9	-	-	-	-	-	-	-	-	-	-	-	-
1.5	47	23	14	7.7	4.4	2.9	2.0	1.3	0.81	0.43	-	-	-	-	-	-	-	-	-	-
2.5	47	23	14	7.7	4.4	3.0	2.4	1.9	1.16	0.73	0.40	-	-	-	-	-	-	-	-	-
4	47	23	14	7.7	4.4	3.0	2.4	1.9	1.41	1.09	0.65	0.33	0.175	-	-	-	-	-	-	-
6	47	23	14	7.7	4.4	3.0	2.4	1.9	1.41	1.09	0.86	0.49	0.269	0.180	-	-	-	-	-	-
10	47	23	14	7.7	4.4	3.0	2.4	1.9	1.41	1.09	0.86	0.60	0.434	0.275	0.205	0.090	-	-	-	-
16	47	23	14	7.7	4.4	3.0	2.4	1.9	1.41	1.09	0.86	0.60	0.436	0.348	0.251	0.151	0.096	0.051	-	-
25	47	23	14	7.7	4.4	3.0	2.4	1.9	1.41	1.09	0.86	0.60	0.436	0.348	0.260	0.200	0.146	0.078	0.045	-
35	47	23	14	7.7	4.4	3.0	2.4	1.9	1.41	1.09	0.86	0.60	0.436	0.348	0.260	0.200	0.163	0.111	0.090	0.033
50	47	23	14	7.7	4.4	3.0	2.4	1.9	1.41	1.09	0.86	0.60	0.436	0.348	0.260	0.200	0.163	0.117	0.090	0.051
70	47	23	14	7.7	4.4	3.0	2.4	1.9	1.41	1.09	0.86	0.60	0.436	0.348	0.260	0.200	0.163	0.117	0.090	0.065
95 or larger	47	23	14	7.7	4.4	3.0	2.4	1.9	1.41	1.09	0.86	0.60	0.436	0.348	0.260	0.200	0.163	0.117	0.090	0.065

Notes: The impedances in the table are based on an initial conductor temperature of 70 °C and a k factor of 115 for a maximum disconnection time of 5 seconds, for thermal protection of the C.P.C. Where the open circuit voltage to earth is not 240V, multiply the Z_S from the table by the actual open circuit voltage to earth, and then divide by 240V, to give the revised Z_S value for the actual open circuit voltage to earth. When the open circuit voltage U_{oc} to earth is 240 V then the nominal voltage to earth U_o is 230 V.

How to use: Work out the Z_S for the circuit at the design temperature for the conductors, then compare this with the Z_S given in the table under the rating of the protective device for the size of circuit protective conductor being used. If the Z_S in the table is larger than the Z_S worked out for the circuit, the protective conductor is thermally protected. If the circuit Z_S is greater than the table Z_S, recalculate the circuit Z_S using a larger C.P.C. until the circuit Z_S is less than that given in the table. Now check the calculated Z_S with the appropriate ZS table, to ensure that the circuit gives protection against indirect contact.

THERMAL PROTECTION C.P.C. PCZ 2 BS 1361 FUSES

Maximum values of Z_S for thermal protection of the protective conductor, when U_{oc} = 240V

Size of copper C.P.C in mm²	Rating of BS 1361 fuse in amperes							
	5	15	20	30	45	60	80	100
	Maximum value of Z_S at design temperature							
1.0	17.14	5.22	2.38	1.00	0.286	-	-	-
1.5	17.14	5.22	2.79	1.48	0.462	-	-	-
2.5	17.14	5.22	2.93	1.92	0.686	0.286	-	-
4.0	17.14	5.22	2.93	1.92	1.000	0.453	0.267	-
6.0	17.14	5.22	2.93	1.92	1.000	0.686	0.393	0.218
10	17.14	5.22	2.93	1.92	1.000	0.727	0.522	0.324
16	17.14	5.22	2.93	1.92	1.000	0.727	0.522	0.381
25 or larger	17.14	5.22	2.93	1.92	1.000	0.727	0.522	0.381

Notes: The impedances in the table are based on an initial conductor temperature of 70 °C and a k factor of 115 for a maximum disconnection time of 5 seconds, for thermal protection of the C.P.C. Where the open circuit voltage to earth is not 240V, multiply the Z_S from the table by the actual open circuit voltage to earth, and then divide by 240V, to give the revised Z_S value for the actual open circuit voltage to earth. When the open circuit voltage U_{oc} to earth is 240 V then the nominal voltage to earth U_0 is 230 V.

How to use: Work out the Z_S for the circuit at the design temperature for the conductors, then compare this with the Z_S given in the table under the rating of the protective device for the size of circuit protective conductor being used. If the Z_S in the table is larger than the Z_S worked out for the circuit, the protective conductor is thermally protected. If the circuit Z_S is greater than the table Z_S, recalculate the circuit Z_S using a larger C.P.C. until the circuit Z_S is less than that given in the table. Now check the calculated Z_S with the appropriate ZS table, to ensure that the circuit gives protection against indirect contact.

THERMAL PROTECTION C.P.C.   **PCZ 3**   BS 3036 FUSES

Maximum values of Z_S for thermal protection of the protective conductor, when U_{oc} = 240V

Size of copper C.P.C. in mm².	Rating of BS 3036 fuse in amperes						
	5	15	20	30	45	60	100
	Maximum value of Z_S at design temperature						
1.0	18.46	5.58	3.44	-	-	-	-
1.5	18.46	5.58	4.00	2.39	-	-	-
2.5	18.46	5.58	4.00	2.76	1.38	1.17	-
4.0	18.46	5.58	4.00	2.76	1.66	1.17	-
6.0	18.46	5.58	4.00	2.76	1.66	1.17	-
10	18.46	5.58	4.00	2.76	1.66	1.17	0.558
16	18.46	5.58	4.00	2.76	1.66	1.17	0.558
25	18.46	5.58	4.00	2.76	1.66	1.17	0.558
35 or larger	18.46	5.58	4.00	2.76	1.66	1.17	0.558

Notes: The impedances in the table are based on an initial conductor temperature of 70 °C and a k factor of 115 for a maximum disconnection time of 5 seconds, for thermal protection of the C.P.C. Where the open circuit voltage to earth is not 240V, multiply the Z_S from the table by the actual open circuit voltage to earth, and then divide by 240V, to give the revised Z_S value for the actual open circuit voltage to earth. When the open circuit voltage U_{oc} to earth is 240 V then the nominal voltage to earth U_o is 230 V.

How to use: Work out the Z_S for the circuit at the design temperature for the conductors, then compare this with the Z_S given in the table under the rating of the protective device for the size of circuit protective conductor being used. If the Z_S in the table is larger than the Z_S worked out for the circuit, the protective conductor is thermally protected. If the circuit Z_S is greater than the table Z_S, recalculate the circuit Z_S using a larger C.P.C. until the circuit Z_S is less than that given in the table. Now check the calculated Z_S with the appropriate ZS table, to ensure that the circuit gives protection against indirect contact.

THERMAL PROTECTION C.P.C.   PCZ4   BS 3871 TYPE 1 MCB

Maximum values of Z_S for thermal protection of the protective conductor, when U_{oc} = 240V

Size of copper C.P.C. in mm².	Rating of BS 3871 TYPE 1 MCB in amperes																	Minimum value of impedance for all MCB ratings
	2	4	5	6	10	15	16	20	25	30	32	40	45	50	63	80	100	
	Maximum value of Z_S at design temperature																	
1.0	30	15	12.0	10.0	6.0	4.0	3.75	3.0	2.4	2.0	1.88	1.5	1.3	1.2	0.95	0.75	0.60	0.2200
1.5	30	15	12.0	10.0	6.0	4.0	3.75	3.0	2.4	2.0	1.88	1.5	1.3	1.2	0.95	0.75	0.60	0.1500
2.5	30	15	12.0	10.0	6.0	4.0	3.75	3.0	2.4	2.0	1.88	1.5	1.3	1.2	0.95	0.75	0.60	0.0880
4.0	30	15	12.0	10.0	6.0	4.0	3.75	3.0	2.4	2.0	1.88	1.5	1.3	1.2	0.95	0.75	0.60	0.0550
6.0	30	15	12.0	10.0	6.0	4.0	3.75	3.0	2.4	2.0	1.88	1.5	1.3	1.2	0.95	0.75	0.60	0.0365
10	30	15	12.0	10.0	6.0	4.0	3.75	3.0	2.4	2.0	1.88	1.5	1.3	1.2	0.95	0.75	0.60	0.0219
16	30	15	12.0	10.0	6.0	4.0	3.75	3.0	2.4	2.0	1.88	1.5	1.3	1.2	0.95	0.75	0.60	0.0142
25	30	15	12.0	10.0	6.0	4.0	3.75	3.0	2.4	2.0	1.88	1.5	1.3	1.2	0.95	0.75	0.60	0.0088
35	30	15	12.0	10.0	6.0	4.0	3.75	3.0	2.4	2.0	1.88	1.5	1.3	1.2	0.95	0.75	0.60	0.0063
50 or larger	30	15	12.0	10.0	6.0	4.0	3.75	3.0	2.4	2.0	1.88	1.5	1.3	1.2	0.95	0.75	0.60	0.0044

Notes: The impedances in the table are based on an initial conductor temperature of 70 °C and a k factor of 115 for a maximum disconnection time of 5 seconds, for thermal protection of the C.P.C. Where the open circuit voltage to earth is not 240V, multiply the Z_S from the table by the actual open circuit voltage to earth, and then divide by 240V, to give the revised Z_S value for the actual open circuit voltage to earth. When the open circuit voltage U_{oc} to earth is 240 V then the nominal voltage to earth U_o is 230 V.

How to use: Work out the Z_S for the circuit at the design temperature for the conductors, then compare this with the Z_S given in the table under the rating of the protective device for the size of circuit protective conductor being used. If the Z_S in the table is larger than the Z_S worked out for the circuit, the protective conductor is thermally protected. If the circuit Z_S is greater than the table Z_S, recalculate the circuit Z_S using a larger C.P.C. until the circuit Z_S is less than that given in the table. Now check the calculated Z_S with the appropriate ZS table, to ensure that the circuit gives protection against indirect contact.

THERMAL PROTECTION C.P.C. PCZ 5 BS 3871 TYPE 2 MCB

Maximum values of Z_S for thermal protection of the protective conductor, when U_{OC} = 240V

Size of copper C.P.C. in m m².	Rating of BS 3871 TYPE 2 MCB in amperes																	Minimum value of impedance for all MCB ratings
	2	4	5	6	10	15	16	20	25	30	32	40	45	50	63	80	100	
	Maximum value of Z_S at design temperature																	
1.0	17.1	8.6	6.86	5.71	3.43	2.29	2.14	1.71	1.37	1.14	1.07	0.86	0.76	0.69	0.54	0.43	0.34	0.2191
1.5	17.1	8.6	6.86	5.71	3.43	2.29	2.14	1.71	1.37	1.14	1.07	0.86	0.76	0.69	0.54	0.43	0.34	0.1462
2.5	17.1	8.6	6.86	5.71	3.43	2.29	2.14	1.71	1.37	1.14	1.07	0.86	0.76	0.69	0.54	0.43	0.34	0.0877
4.0	17.1	8.6	6.86	5.71	3.43	2.29	2.14	1.71	1.37	1.14	1.07	0.86	0.76	0.69	0.54	0.43	0.34	0.0548
6.0	17.1	8.6	6.86	5.71	3.43	2.29	2.14	1.71	1.37	1.14	1.07	0.86	0.76	0.69	0.54	0.43	0.34	0.0365
10	17.1	8.6	6.86	5.71	3.43	2.29	2.14	1.71	1.37	1.14	1.07	0.86	0.76	0.69	0.54	0.43	0.34	0.0219
16	17.1	8.6	6.86	5.71	3.43	2.29	2.14	1.71	1.37	1.14	1.07	0.86	0.76	0.69	0.54	0.43	0.34	0.0142
25	17.1	8.6	6.86	5.71	3.43	2.29	2.14	1.71	1.37	1.14	1.07	0.86	0.76	0.69	0.54	0.43	0.34	0.0088
35	17.1	8.6	6.86	5.71	3.43	2.29	2.14	1.71	1.37	1.14	1.07	0.86	0.76	0.69	0.54	0.43	0.34	0.0063
50 or larger	17.1	8.6	6.86	5.71	3.43	2.29	2.14	1.71	1.37	1.14	1.07	0.86	0.76	0.69	0.54	0.43	0.34	0.0044

Notes: The impedances in the table are based on an initial conductor temperature of 70 °C and a k factor of 115 for a maximum disconnection time of 5 seconds, for thermal protection of the C.P.C. Where the open circuit voltage to earth is not 240V, multiply the Z_S from the table by the actual open circuit voltage to earth, and then divide by 240V, to give the revised Z_S value for the actual open circuit voltage to earth. When the open circuit voltage U_{OC} to earth is 240 V then the nominal voltage to earth U_0 is 230 V.

How to use: Work out the Z_S for the circuit at the design temperature for the conductors, then compare this with the Z_S given in the table under the rating of the protective device for the size of circuit protective conductor being used. If the Z_S in the table is larger than the Z_S worked out for the circuit, the protective conductor is thermally protected. If the circuit Z_S is greater than the table Z_S, recalculate the circuit Z_S using a larger C.P.C. until the circuit Z_S is less than that given in the table. Now check the calculated Z_S with the appropriate ZS table, to ensure that the circuit gives protection against indirect contact.

THERMAL PROTECTION C.P.C.   PCZ 6   BS 3871 TYPE 3 MCB

Maximum values of Z_S for thermal protection of the protective conductor, when U_{oc} = 240V

Size of copper C.P.C. in m.m².	Rating of BS 3871 TYPE 3 MCB in amperes																	Minimum value of impedance for all MCB ratings
	2	4	5	6	10	15	16	20	25	30	32	40	45	50	63	80	100	
	Maximum value of Z_S at design temperature																	
1.0	12	6	4.8	4	2.4	1.6	1.5	1.2	0.96	0.8	0.75	0.6	0.53	0.48	0.38	0.3	0.24	0.2191
1.5	12	6	4.8	4	2.4	1.6	1.5	1.2	0.96	0.8	0.75	0.6	0.53	0.48	0.38	0.3	0.24	0.1462
2.5	12	6	4.8	4	2.4	1.6	1.5	1.2	0.96	0.8	0.75	0.6	0.53	0.48	0.38	0.3	0.24	0.0877
4.0	12	6	4.8	4	2.4	1.6	1.5	1.2	0.96	0.8	0.75	0.6	0.53	0.48	0.38	0.3	0.24	0.0548
6.0	12	6	4.8	4	2.4	1.6	1.5	1.2	0.96	0.8	0.75	0.6	0.53	0.48	0.38	0.3	0.24	0.0365
10	12	6	4.8	4	2.4	1.6	1.5	1.2	0.96	0.8	0.75	0.6	0.53	0.48	0.38	0.3	0.24	0.0219
16	12	6	4.8	4	2.4	1.6	1.5	1.2	0.96	0.8	0.75	0.6	0.53	0.48	0.38	0.3	0.24	0.0142
25	12	6	4.8	4	2.4	1.6	1.5	1.2	0.96	0.8	0.75	0.6	0.53	0.48	0.38	0.3	0.24	0.0088
35	12	6	4.8	4	2.4	1.6	1.5	1.2	0.96	0.8	0.75	0.6	0.53	0.48	0.38	0.3	0.24	0.0063
50 or larger	12	6	4.8	4	2.4	1.6	1.5	1.2	0.96	0.8	0.75	0.6	0.53	0.48	0.38	0.3	0.24	0.0044

Notes: The impedances in the table are based on an initial conductor temperature of 70 °C and a k factor of 115 for a maximum disconnection time of 5 seconds, for thermal protection of the C.P.C. Where the open circuit voltage to earth is not 240V, multiply the Z_S from the table by the actual open circuit voltage to earth, and then divide by 240V, to give the revised Z_S value for the actual open circuit voltage to earth. When the open circuit voltage U_{oc} to earth is 240 V then the nominal voltage to earth U_o is 230 V.

How to use: Work out the Z_S for the circuit at the design temperature for the conductors, then compare this with the Z_S given in the table under the rating of the protective device for the size of circuit protective conductor being used. If the Z_S in the table is larger than the Z_S worked out for the circuit, the protective conductor is thermally protected. If the circuit Z_S is greater than the table Z_S, recalculate the circuit Z_S using a larger C.P.C. until the circuit Z_S is less than that given in the table. Now check the calculated Z_S with the appropriate ZS table, to ensure that the circuit gives protection against indirect contact.

THERMAL PROTECTION C.P.C.   PCZ 7   BS EN 60898 TYPE B MCB

Maximum values of Z_S for thermal protection of the protective conductor, when $U_{oc} = 240V$

Size of copper C.P.C. in m m².	Rating of BS EN 60898 TYPE B MCB in amperes																	Minimum value of impedance for all MCB ratings
	2	4	5	6	10	15	16	20	25	30	32	40	45	50	63	80	100	
	Maximum value of Z_S at design temperature																	
1.0	24	12	9.6	8	4.8	3.2	3.0	2.4	1.92	1.6	1.5	1.2	1.07	0.96	0.76	0.6	0.48	0.2191
1.5	24	12	9.6	8	4.8	3.2	3.0	2.4	1.92	1.6	1.5	1.2	1.07	0.96	0.76	0.6	0.48	0.1462
2.5	24	12	9.6	8	4.8	3.2	3.0	2.4	1.92	1.6	1.5	1.2	1.07	0.96	0.76	0.6	0.48	0.0877
4.0	24	12	9.6	8	4.8	3.2	3.0	2.4	1.92	1.6	1.5	1.2	1.07	0.96	0.76	0.6	0.48	0.0548
6.0	24	12	9.6	8	4.8	3.2	3.0	2.4	1.92	1.6	1.5	1.2	1.07	0.96	0.76	0.6	0.48	0.0355
10	24	12	9.6	8	4.8	3.2	3.0	2.4	1.92	1.6	1.5	1.2	1.07	0.96	0.76	0.6	0.48	0.0219
16	24	12	9.6	8	4.8	3.2	3.0	2.4	1.92	1.6	1.5	1.2	1.07	0.96	0.76	0.6	0.48	0.0142
25	24	12	9.6	8	4.8	3.2	3.0	2.4	1.92	1.6	1.5	1.2	1.07	0.96	0.76	0.6	0.48	0.0088
35	24	12	9.6	8	4.8	3.2	3.0	2.4	1.92	1.6	1.5	1.2	1.07	0.96	0.76	0.6	0.48	0.0063
50 or larger	24	12	9.6	8	4.8	3.2	3.0	2.4	1.92	1.6	1.5	1.2	1.07	0.96	0.76	0.6	0.48	0.0044

Notes: The impedances in the table are based on an initial conductor temperature of 70 °C and a k factor of 115 for a maximum disconnection time of 5 seconds, for thermal protection of the C.P.C. Where the open circuit voltage to earth is not 240V, multiply the Z_S from the table by the actual open circuit voltage to earth, and then divide by 240V, to give the revised Z_S value for the actual open circuit voltage to earth. When the open circuit voltage U_{oc} to earth is 240 V then the nominal voltage to earth U_o is 230 V.

How to use: Work out the Z_S for the circuit at the design temperature for the conductors, then compare this with the Z_S given in the table under the rating of the protective device for the size of circuit protective conductor being used. If the Z_S in the table is larger than the Z_S worked out for the circuit, the protective conductor is thermally protected. If the circuit Z_S is greater than the table Z_S, recalculate the circuit Z_S using a larger C.P.C. until the circuit Z_S is less than that given in the table. Now check the calculated Z_S with the appropriate ZS table, to ensure that the circuit gives protection against indirect contact.

THERMAL PROTECTION C.P.C. PCZ 8 BS EN 60898 TYPE C MCB

Maximum values of Z_S for thermal protection of the protective conductor, when $U_{oc} = 240V$

Size of copper C.P.C. in mm².	Rating of BS EN 60898 TYPE C MCB in amperes																	Minimum value of impedance for all MCB ratings
	2	4	5	6	10	15	16	20	25	30	32	40	45	50	63	80	100	
	Maximum value of Z_S at design temperature																	
1.0	12	6	4.8	4.0	2.4	1.6	1.5	1.2	0.96	0.8	0.75	0.6	0.53	0.48	0.38	0.3	0.24	0.2191
1.5	12	6	4.8	4.0	2.4	1.6	1.5	1.2	0.96	0.8	0.75	0.6	0.53	0.48	0.38	0.3	0.24	0.1462
2.5	12	6	4.8	4.0	2.4	1.6	1.5	1.2	0.96	0.8	0.75	0.6	0.53	0.48	0.38	0.3	0.24	0.0877
4.0	12	6	4.8	4.0	2.4	1.6	1.5	1.2	0.96	0.8	0.75	0.6	0.53	0.48	0.38	0.3	0.24	0.0548
6.0	12	6	4.8	4.0	2.4	1.6	1.5	1.2	0.96	0.8	0.75	0.6	0.53	0.48	0.38	0.3	0.24	0.0365
10	12	6	4.8	4.0	2.4	1.6	1.5	1.2	0.96	0.8	0.75	0.6	0.53	0.48	0.38	0.3	0.24	0.0219
16	12	6	4.8	4.0	2.4	1.6	1.5	1.2	0.96	0.8	0.75	0.6	0.53	0.48	0.38	0.3	0.24	0.0142
25	12	6	4.8	4.0	2.4	1.6	1.5	1.2	0.96	0.8	0.75	0.6	0.53	0.48	0.38	0.3	0.24	0.0088
35	12	6	4.8	4.0	2.4	1.6	1.5	1.2	0.96	0.8	0.75	0.6	0.53	0.48	0.38	0.3	0.24	0.0063
50 or larger	12	6	4.8	4.0	2.4	1.6	1.5	1.2	0.96	0.8	0.75	0.6	0.53	0.48	0.38	0.3	0.24	0.0044

Notes: The impedances in the table are based on an initial conductor temperature of 70 °C and a k factor of 115 for a maximum disconnection time of 5 seconds, for thermal protection of the C.P.C. Where the open circuit voltage to earth is not 240V, multiply the Z_S from the table by the actual open circuit voltage to earth, and then divide by 240V, to give the revised Z_S value for the actual open circuit voltage to earth. When the open circuit voltage U_{oc} to earth is 240 V then the nominal voltage to earth U_o is 230 V.

How to use: Work out the Z_S for the circuit at the design temperature for the conductors, then compare this with the Z_S given in the table under the rating of the protective device for the size of circuit protective conductor being used. If the Z_S in the table is larger than the Z_S worked out for the circuit, the protective conductor is thermally protected. If the circuit Z_S is greater than the table Z_S, recalculate the circuit Z_S using a larger C.P.C. until the circuit Z_S is less than that given in the table. Now check the calculated Z_S with the appropriate ZS table, to ensure that the circuit gives protection against indirect contact.

THERMAL PROTECTION C.P.C.　　　PCZ 9　　　BS EN 60898 TYPE D MCB

Maximum values of Z_S for thermal protection of the protective conductor, when U_{oc} = 240V

Size of copper C.P.C. in mm².	Rating of BS EN 60898 TYPE D MCB in amperes																	Minimum value of impedance for all MCB ratings
	2	4	5	6	10	15	16	20	25	30	32	40	45	50	63	80	100	
	Maximum value of Z_S at design temperature																	
1.0	6	3	2.4	2.0	1.2	0.8	0.75	0.6	0.48	0.4	0.375	0.3	0.27	0.24	0.19	0.15	0.12	0.2191
1.5	6	3	2.4	2.0	1.2	0.8	0.75	0.6	0.48	0.4	0.375	0.3	0.27	0.24	0.19	0.15	0.12	0.1452
2.5	6	3	2.4	2.0	1.2	0.8	0.75	0.6	0.48	0.4	0.375	0.3	0.27	0.24	0.19	0.15	0.12	0.0877
4.0	6	3	2.4	2.0	1.2	0.8	0.75	0.6	0.48	0.4	0.375	0.3	0.27	0.24	0.19	0.15	0.12	0.0548
6.0	6	3	2.4	2.0	1.2	0.8	0.75	0.6	0.48	0.4	0.375	0.3	0.27	0.24	0.19	0.15	0.12	0.0365
10	6	3	2.4	2.0	1.2	0.8	0.75	0.6	0.48	0.4	0.375	0.3	0.27	0.24	0.19	0.15	0.12	0.0219
16	6	3	2.4	2.0	1.2	0.8	0.75	0.6	0.48	0.4	0.375	0.3	0.27	0.24	0.19	0.15	0.12	0.0142
25	6	3	2.4	2.0	1.2	0.8	0.75	0.6	0.48	0.4	0.375	0.3	0.27	0.24	0.19	0.15	0.12	0.0088
35	6	3	2.4	2.0	1.2	0.8	0.75	0.6	0.48	0.4	0.375	0.3	0.27	0.24	0.19	0.15	0.12	0.0065
50 or larger	6	3	2.4	2.0	1.2	0.8	0.75	0.6	0.48	0.4	0.375	0.3	0.27	0.24	0.19	0.15	0.12	0.0044

Notes:　The impedances in the table are based on an initial conductor temperature of 70 °C and a k factor of 115 for a maximum disconnection time of 5 seconds, for thermal protection of the C.P.C. Where the open circuit voltage to earth is not 240V, multiply the Z_S from the table by the actual open circuit voltage to earth, and then divide by 240V, to give the revised Z_S value for the actual open circuit voltage to earth. When the open circuit voltage U_{oc} to earth is 240 V fcvthen the nominal voltage to earth U_0 is 230 V.

How to use:　Work out the Z_S for the circuit at the design temperature for the conductors, then compare this with the Z_S given in the table under the rating of the protective device for the size of circuit protective conductor being used. If the Z_S in the table is larger than the Z_S worked out for the circuit, the protective conductor is thermally protected. If the circuit Z_S is greater than the table Z_S, recalculate the circuit Z_S using a larger C.P.C. until the circuit Z_S is less than that given in the table. Now check the calculated Z_S with the appropriate ZS table, to ensure that the circuit gives protection against indirect contact.

SHOCK PROTECTION 5 s

PCZ 10

SOCKET CIRCUITS

Maximum design values of circuit protective conductor impedance for different types of protective device, for use with socket-outlet circuits, Class I circuits, or circuits for portable equipment intended to be moved or for distribution boards with mixed 0.4 s and 5 s circuits.

MAXIMUM DISCONNECTION TIME 5 s

Protective device rating in amperes

Protective device type	2A	4A	5A	6A	10A	15A	16A	20A	25A	30A	32A	40A	45A	50A	60A	63A	80A	100A	125A	160A	200A
	Maximum design value of C.P.C. impedance in ohms																				
BS 88 Fuse	9.79	4.79	-	2.48	1.48	-	0.83	0.55	0.43	-	0.34	0.26	-	0.19	-	0.179	0.125	0.092	0.073	0.056	0.042
BS1361 fuse	-	-	3.25	-	-	0.96	-	0.55	-	0.36	-	-	0.18	-	0.15	-	0.108	0.079	-	-	-
BS 3036 fuse	-	-	3.25	-	-	0.96	-	0.63	-	0.43	-	-	0.24	-	0.24	-	-	0.116	-	-	-
Type 1 MCB	6.25	3.13	2.5	2.08	1.25	0.83	0.78	0.63	0.5	0.42	0.39	0.313	0.278	0.25	0.208	0.198	0.156	0.125	0.100	0.078	0.063
Type 2 MCB	3.57	1.79	1.43	1.19	0.71	0.48	0.45	0.36	0.29	0.24	0.22	0.179	0.159	0.143	0.119	0.113	0.089	0.071	0.057	0.045	0.036
Type 3 MCB	2.50	1.25	1.00	0.83	0.50	0.33	0.31	0.25	0.20	0.17	0.16	0.13	0.11	0.10	0.08	0.079	0.063	0.050	0.040	0.031	0.025
Type B MCB	5.00	2.50	2.00	1.67	1.00	0.67	0.63	0.50	0.40	0.33	0.31	0.25	0.22	0.20	0.17	0.159	0.125	0.100	0.080	0.063	0.050
Type C MCB	2.50	1.25	1.00	0.83	0.50	0.33	0.31	0.25	0.20	0.17	0.16	0.13	0.11	0.10	0.08	0.079	0.063	0.050	0.040	0.031	0.025
Type D MCB	1.25	0.63	0.5	0.42	0.25	0.17	0.16	0.13	0.1	0.08	0.08	0.063	0.056	0.05	0.042	0.040	0.031	0.025	0.020	0.016	0.013

Notes: The disconnection time for socket-outlet circuits, or for Class 1 hand-held equipment circuits, can be increased to 5 s, provided that the impedance of the protective conductor does not exceed the maximum value given in the above table. The impedance of the protective conductor is to be taken all the way back to the point where the protective conductor is connected to the main equipotential bonding conductors. The point of equipotential bonding can be at a local distribution board, providing it is bonded to the same types of extraneous conductive parts to which the main equipotential bonding conductors are connected with the same size of bonding conductors.

41C

SHORT-CIRCUIT PROTECTION

SCZ 1

BS 3871 TYPE 1 MCB

Maximum values of Z_{pn} for copper conductors, for thermal protection of the live conductors, when $U_{oc} = 240V$

Size of copper live conductors mm².	Rating of BS 3871 TYPE 1 MCB in amperes																	Minimum value of impedance for all MCB rating
	2	4	5	6	10	15	16	20	25	30	32	40	45	50	63	80	100	
	Maximum value of Z_{pn} at design temperature																	
1.0	30	15	12.0	10.0	6.0	4.0	3.75	3.0	2.4	2.0	1.88	1.5	1.3	1.2	0.95	0.75	0.60	0.2191
1.5	30	15	12.0	10.0	6.0	4.0	3.75	3.0	2.4	2.0	1.88	1.5	1.3	1.2	0.95	0.75	0.60	0.1462
2.5	30	15	12.0	10.0	6.0	4.0	3.75	3.0	2.4	2.0	1.88	1.5	1.3	1.2	0.95	0.75	0.60	0.1877
4.0	30	15	12.0	10.0	6.0	4.0	3.75	3.0	2.4	2.0	1.88	1.5	1.3	1.2	0.95	0.75	0.60	0.0548
6.0	30	15	12.0	10.0	6.0	4.0	3.75	3.0	2.4	2.0	1.88	1.5	1.3	1.2	0.95	0.75	0.60	0.0365
10	30	15	12.0	10.0	6.0	4.0	3.75	3.0	2.4	2.0	1.88	1.5	1.3	1.2	0.95	0.75	0.60	0.0219
16	30	15	12.0	10.0	6.0	4.0	3.75	3.0	2.4	2.0	1.88	1.5	1.3	1.2	0.95	0.75	0.60	0.0.42
25	30	15	12.0	10.0	6.0	4.0	3.75	3.0	2.4	2.0	1.88	1.5	1.3	1.2	0.95	0.75	0.60	0.0088
35	30	15	12.0	10.0	6.0	4.0	3.75	3.0	2.4	2.0	1.88	1.5	1.3	1.2	0.95	0.75	0.60	0.0063
50 or larger	30	15	12.0	10.0	6.0	4.0	3.75	3.0	2.4	2.0	1.88	1.5	1.3	1.2	0.95	0.75	0.60	0.0044

Notes: The impedances in the table are based on an initial conductor temperature of 70 °C and a k factor of 115. For protection of conductors, the maximum impedance of the conductors is required. With single phase circuits or three phase and neutral circuits, take the total impedance of the phase and neutral (Z_{pn}) at the design temperature, and compare it with the impedance given in the table. For safety reasons, the circuit impedance should be less than the maximum allowed by the table. For three-phase three-wire circuits, take the total impedance of two phases (Z_{pp}), and divide by 1.732; compare the resultant impedance with the value from the table. As before, the circuits impedance should be less than the figure in the table.

When the open circuit voltage U_{oc} is 240 V the nominal voltage U_o is 230 V. The single-phase U_{oc} of 240 V corresponds with a three-phase open circuit voltage U_{loc} of 415 V and a single-phase voltage U_o of 230 V corresponds with a three-phase nominal voltage U_{ol} of 400 V.

SHORT-CIRCUIT PROTECTION

SCZ 2

BS 3871 TYPE 2 MCB

Maximum values of Z_{pn} for copper conductors, for thermal protection
of the live conductors, when U_{oc} = 240V

Size of copper live conductors mm².	Rating of BS 3871 TYPE 2 MCB in amperes																	Minimum value of impedance for all MCB rating
	2	4	5	6	10	15	16	20	25	30	32	40	45	50	63	80	100	
	Maximum value of Z_{pn} at design temperature																	
1.0	17.1	8.6	6.86	5.71	3.43	2.29	2.14	1.71	1.37	1.14	1.07	0.86	0.76	0.69	0.54	0.43	0.34	0.2191
1.5	17.1	8.6	6.86	5.71	3.43	2.29	2.14	1.71	1.37	1.14	1.07	0.86	0.76	0.69	0.54	0.43	0.34	0.1462
2.5	17.1	8.6	6.86	5.71	3.43	2.29	2.14	1.71	1.37	1.14	1.07	0.86	0.76	0.69	0.54	0.43	0.34	0.0877
4.0	17.1	8.6	6.86	5.71	3.43	2.29	2.14	1.71	1.37	1.14	1.07	0.86	0.76	0.69	0.54	0.43	0.34	0.0548
6.0	17.1	8.6	6.86	5.71	3.43	2.29	2.14	1.71	1.37	1.14	1.07	0.86	0.76	0.69	0.54	0.43	0.34	0.0365
10	17.1	8.6	6.86	5.71	3.43	2.29	2.14	1.71	1.37	1.14	1.07	0.86	0.76	0.69	0.54	0.43	0.34	0.0219
16	17.1	8.6	6.86	5.71	3.43	2.29	2.14	1.71	1.37	1.14	1.07	0.86	0.76	0.69	0.54	0.43	0.34	0.0142
25	17.1	8.6	6.86	5.71	3.43	2.29	2.14	1.71	1.37	1.14	1.07	0.86	0.76	0.69	0.54	0.43	0.34	0.0088
35	17.1	8.6	6.86	5.71	3.43	2.29	2.14	1.71	1.37	1.14	1.07	0.86	0.76	0.69	0.54	0.43	0.34	0.0063
50 or larger	17.1	8.6	6.86	5.71	3.43	2.29	2.14	1.71	1.37	1.14	1.07	0.86	0.76	0.69	0.54	0.43	0.34	0.0044

Notes: The impedances in the table are based on an initial conductor temperature of 70 °C and a k factor of 115. For protection of conductors, the maximum impedance of the conductors is required. With single phase circuits or three phase and neutral circuits, take the total impedance of the phase and neutral (Z_{pn}) at the design temperature, and compare it with the impedance given in the table. For safety reasons, the circuit impedance should be less than the maximum allowed by the table. For three-phase three-wire circuits, take the total impedance of two phases (Z_{pp}), and divide by 1.732; compare the resultant impedance with the value from the table. As before, the circuits impedance should be less than the figure in the table.

When the open circuit voltage U_{oc} is 240 V the nominal voltage U_o is 230 V. The single-phase U_{oc} of 240 V corresponds with a three-phase open circuit voltage U_{loc} of 415 V and a single-phase voltage U_o of 230 V corresponds with a three-phase nominal voltage U_{ol} of 400 V.

SHORT-CIRCUIT PROTECTION   SCZ 3   BS 3871 TYPE 3 MCB

Maximum values of Z_{pn} for copper conductors, for thermal protection of the live conductors, when U_{oc} = 240V

Size of copper live conductors mm².	Rating of BS 3871 TYPE 3 MCB in amperes																	Minimum value of impedance for all MCB rating
	2	4	5	6	10	15	16	20	25	30	32	40	45	50	63	80	100	
	Maximum value of Z_{pn} at design temperature																	
1.0	12	6	4.8	4	2.4	1.6	1.5	1.2	0.96	0.8	0.75	0.6	0.53	0.48	0.38	0.3	0.24	0.2191
1.5	12	6	4.8	4	2.4	1.6	1.5	1.2	0.96	0.8	0.75	0.6	0.53	0.48	0.38	0.3	0.24	0.1462
2.5	12	6	4.8	4	2.4	1.6	1.5	1.2	0.96	0.8	0.75	0.6	0.53	0.48	0.38	0.3	0.24	0.0877
4.0	12	6	4.8	4	2.4	1.6	1.5	1.2	0.96	0.8	0.75	0.6	0.53	0.48	0.38	0.3	0.24	0.0548
6.0	12	6	4.8	4	2.4	1.6	1.5	1.2	0.96	0.8	0.75	0.6	0.53	0.48	0.38	0.3	0.24	0.0365
10	12	6	4.8	4	2.4	1.6	1.5	1.2	0.96	0.8	0.75	0.6	0.53	0.48	0.38	0.3	0.24	0.0219
16	12	6	4.8	4	2.4	1.6	1.5	1.2	0.96	0.8	0.75	0.6	0.53	0.48	0.38	0.3	0.24	0.0142
25	12	6	4.8	4	2.4	1.6	1.5	1.2	0.96	0.8	0.75	0.6	0.53	0.48	0.38	0.3	0.24	0.0088
35	12	6	4.8	4	2.4	1.6	1.5	1.2	0.96	0.8	0.75	0.6	0.53	0.48	0.38	0.3	0.24	0.0063
50 or larger	12	6	4.8	4	2.4	1.6	1.5	1.2	0.96	0.8	0.75	0.6	0.53	0.48	0.38	0.3	0.24	0.0044

Notes: The impedances in the table are based on an initial conductor temperature of 70 °C and a k factor of 115. For protection of conductors, the maximum impedance of the conductors is required. With single phase circuits or three phase and neutral circuits, take the total impedance of the phase and neutral (Z_{pn}) at the design temperature, and compare it with the impedance given in the table. For safety reasons, the circuit impedance should be less than the maximum allowed by the table. For three-phase three-wire circuits, take the total impedance of two phases (Z_{pp}), and divide by 1.732; compare the resultant impedance with the value from the table. As before, the circuits impedance should be less than the figure in the table. When the open circuit voltage U_{oc} is 240 V the nominal voltage U_{o} is 230 V. The single-phase U_{oc} of 240 V corresponds with a three-phase open circuit voltage U_{loc} of 415 V and a single-phase voltage U_{o} of 230 V corresponds with a three-phase nominal voltage U_{ol} of 400 V.

SHORT-CIRCUIT PROTECTION   SCZ 4   BS EN 60898 TYPE B MCB

Maximum values of Z_{pn} for copper conductors, for thermal protection of the live conductors, when U_{oc} = 240V

Size of copper live conductors mm².	Rating of BS EN 60898 TYPE B MCB in amperes																	Minimum value of impedance for all MCB rating
	2	4	5	6	10	15	16	20	25	30	32	40	45	50	63	80	100	
	Maximum value of Z_{pn} at design temperature																	
1.0	24	12	9.6	8	4.8	3.2	3.0	2.4	1.92	1.6	1.5	1.2	1.07	0.96	0.76	0.6	0.48	0.2191
1.5	24	12	9.6	8	4.8	3.2	3.0	2.4	1.92	1.6	1.5	1.2	1.07	0.96	0.76	0.6	0.48	0.1462
2.5	24	12	9.6	8	4.8	3.2	3.0	2.4	1.92	1.6	1.5	1.2	1.07	0.96	0.76	0.6	0.48	0.0877
4.0	24	12	9.6	8	4.8	3.2	3.0	2.4	1.92	1.6	1.5	1.2	1.07	0.96	0.76	0.6	0.48	0.0548
6.0	24	12	9.6	8	4.8	3.2	3.0	2.4	1.92	1.6	1.5	1.2	1.07	0.96	0.76	0.6	0.48	0.0365
10	24	12	9.6	8	4.8	3.2	3.0	2.4	1.92	1.6	1.5	1.2	1.07	0.96	0.76	0.6	0.48	0.0219
16	24	12	9.6	8	4.8	3.2	3.0	2.4	1.92	1.6	1.5	1.2	1.07	0.96	0.76	0.6	0.48	0.0142
25	24	12	9.6	8	4.8	3.2	3.0	2.4	1.92	1.6	1.5	1.2	1.07	0.96	0.76	0.6	0.48	0.0088
35	24	12	9.6	8	4.8	3.2	3.0	2.4	1.92	1.6	1.5	1.2	1.07	0.96	0.76	0.6	0.48	0.0063
50 or larger	24	12	9.6	8	4.8	3.2	3.0	2.4	1.92	1.6	1.5	1.2	1.07	0.96	0.76	0.6	0.48	0.0044

Notes: The impedances in the table are based on an initial conductor temperature of 70 °C and a k factor of 115. For protection of conductors, the maximum impedance of the conductors is required. With single phase circuits or three phase and neutral circuits, take the total impedance of the phase and neutral (Z_{pn}) at the design temperature, and compare it with the impedance given in the table. For safety reasons, the circuit impedance should be less than the maximum allowed by the table. For three-phase three-wire circuits, take the total impedance of two phases (Z_{pp}), and divide by 1.732; compare the resultant impedance with the value from the table. As before, the circuits impedance should be less than the figure in the table.

When the open circuit voltage U_{oc} is 240 V the nominal voltage U_0 is 230 V. The single-phase U_{oc} of 240 V corresponds with a three-phase open circuit voltage U_{loc} of 415 V and a single-phase voltage U_0 of 230 V corresponds with a three-phase nominal voltage U_{0l} of 400 V.

SHORT-CIRCUIT PROTECTION

SCZ 5

BS EN 60898 TYPE C MCB

Maximum values of Z_{pn} for copper conductors, for thermal protection of the live conductors, when U_{oc} = 240V

Size of copper live conductors mm².	Rating of BS EN 60898 TYPE C MCB in amperes																	Minimum value of impedance for all MCB rating
	2	4	5	6	10	15	16	20	25	30	32	40	45	50	63	80	100	
	Maximum value of Z_{pn} at design temperature																	
1.0	12	6	4.8	4.0	2.4	1.6	1.5	1.2	0.96	0.8	0.75	0.6	0.53	0.48	0.38	0.3	0.24	0.2191
1.5	12	6	4.8	4.0	2.4	1.6	1.5	1.2	0.96	0.8	0.75	0.6	0.53	0.48	0.38	0.3	0.24	0.1462
2.5	12	6	4.8	4.0	2.4	1.6	1.5	1.2	0.96	0.8	0.75	0.6	0.53	0.48	0.38	0.3	0.24	0.0877
4.0	12	6	4.8	4.0	2.4	1.6	1.5	1.2	0.96	0.8	0.75	0.6	0.53	0.48	0.38	0.3	0.24	0.0548
6.0	12	6	4.8	4.0	2.4	1.6	1.5	1.2	0.96	0.8	0.75	0.6	0.53	0.48	0.38	0.3	0.24	0.0365
10	12	6	4.8	4.0	2.4	1.6	1.5	1.2	0.96	0.8	0.75	0.6	0.53	0.48	0.38	0.3	0.24	0.0219
16	12	6	4.8	4.0	2.4	1.6	1.5	1.2	0.96	0.8	0.75	0.6	0.53	0.48	0.38	0.3	0.24	0.0142
25	12	6	4.8	4.0	2.4	1.6	1.5	1.2	0.96	0.8	0.75	0.6	0.53	0.48	0.38	0.3	0.24	0.0088
35	12	6	4.8	4.0	2.4	1.6	1.5	1.2	0.96	0.8	0.75	0.6	0.53	0.48	0.38	0.3	0.24	0.0063
50 or larger	12	6	4.8	4.0	2.4	1.6	1.5	1.2	0.96	0.8	0.75	0.6	0.53	0.48	0.38	0.3	0.24	0.0044

Notes: The impedances in the table are based on an initial conductor temperature of 70 °C and a k factor of 115. For protection of conductors, the maximum impedance of the conductors is required. With single phase circuits or three phase and neutral circuits, take the total impedance of the phase and neutral (Z_{pn}) at the design temperature, and compare it with the impedance given in the table. For safety reasons, the circuit impedance should be less than the maximum allowed by the table. For three-phase three-wire circuits, take the total impedance of two phases (Z_{pp}), and divide by 1.732; compare the resultant impedance with the value from the table. As before, the circuits impedance should be less than the figure in the table.

When the open circuit voltage U_{oc} is 240 V the nominal voltage U_o is 230 V. The single-phase U_{oc} of 240 V corresponds with a three-phase open circuit voltage U_{loc} of 415 V and a single-phase voltage U_o of 230 V corresponds with a three-phase nominal voltage U_{ol} of 400 V.

SHORT-CIRCUIT PROTECTION

SCZ 6

BS EN 60898 TYPE D MCB

Maximum values of Z_{pn} for copper conductors, for thermal protection of the live conductors, when U_{oc} = 240V

Size of copper live conductors mm².	Rating of BS EN 60898 TYPE D MCB in amperes																	Minimum value of impedance for all MCB rating
	2	4	5	6	10	15	16	20	25	30	32	40	45	50	63	80	100	
	Maximum value of Z_{pn} at design temperature																	
1.0	6	3	2.4	2.0	1.2	0.8	0.75	0.6	0.48	0.4	0.375	0.3	0.27	0.24	0.19	0.15	0.12	0.2191
1.5	6	3	2.4	2.0	1.2	0.8	0.75	0.6	0.48	0.4	0.375	0.3	0.27	0.24	0.19	0.15	0.12	0.1462
2.5	6	3	2.4	2.0	1.2	0.8	0.75	0.6	0.48	0.4	0.375	0.3	0.27	0.24	0.19	0.15	0.12	0.0877
4.0	6	3	2.4	2.0	1.2	0.8	0.75	0.6	0.48	0.4	0.375	0.3	0.27	0.24	0.19	0.15	0.12	0.0548
6.0	6	3	2.4	2.0	1.2	0.8	0.75	0.6	0.48	0.4	0.375	0.3	0.27	0.24	0.19	0.15	0.12	0.0365
10	6	3	2.4	2.0	1.2	0.8	0.75	0.6	0.48	0.4	0.375	0.3	0.27	0.24	0.19	0.15	0.12	0.0219
16	6	3	2.4	2.0	1.2	0.8	0.75	0.6	0.48	0.4	0.375	0.3	0.27	0.24	0.19	0.15	0.12	0.0142
25	6	3	2.4	2.0	1.2	0.8	0.75	0.6	0.48	0.4	0.375	0.3	0.27	0.24	0.19	0.15	0.12	0.0088
35	6	3	2.4	2.0	1.2	0.8	0.75	0.6	0.48	0.4	0.375	0.3	0.27	0.24	0.19	0.15	0.12	0.0063
50 or larger	6	3	2.4	2.0	1.2	0.8	0.75	0.6	0.48	0.4	0.375	0.3	0.27	0.24	0.19	0.15	0.12	0.0044

Notes: The impedances in the table are based on an initial conductor temperature of 70 °C and a k factor of 115. For protection of conductors, the maximum impedance of the conductors is required. With single phase circuits or three phase and neutral circuits, take the total impedance of the phase and neutral (Z_{pn}) at the design temperature, and compare it with the impedance given in the table. For safety reasons, the circuit impedance should be less than the maximum allowed by the table. For three-phase three-wire circuits, take the total impedance of two phases (Z_{pp}), and divide by 1.732; compare the resultant impedance with the value from the table. As before, the circuits impedance should be less than the figure in the table.

When the open circuit voltage U_{oc} is 240 V the nominal voltage U_o is 230 V. The single-phase U_{oc} of 240 V corresponds with a three-phase open circuit voltage U_{loc} of 415 V and a single-phase voltage U_o of 230 V corresponds with a three-phase nominal voltage U_{ol} of 400 V.

SHORT-CIRCUIT PROTECTION SCZ 7 BS 88 HRC FUSES

Maximum values of Z_{pn} for copper conductors, for thermal protection of the live conductors, when $U_{oc} = 240V$

Rating of BS 88 H.R.C. fuse in amperes

Maximum value of Z_{pn} in ohms with a 'k' factor of 115

Size of copper live conductors in mm²	2	4	6	10	16	20	25	32	40	50	63	80	100	125	160	200	250	315	400	500
1	60.5	27.8	16.6	8.0	3.6	2.1	1.3	0.8	-	-	-	-	-	-	-	-	-	-	-	-
1.5	64.2	29.0	18.6	9.2	4.3	2.6	1.7	1.2	0.72	0.38	-	-	-	-	-	-	-	-	-	-
2.5	64.2	30.9	20.9	9.9	5.2	3.2	2.3	1.7	1.03	0.65	0.36	-	-	-	-	-	-	-	-	-
4	64.2	30.9	21.4	11.3	6.0	4.0	3.0	1.7	1.45	0.95	0.58	0.30	0.156	-	-	-	-	-	-	-
6	64.2	30.9	21.4	12.3	6.5	4.6	3.5	1.7	1.74	1.21	0.80	0.43	0.240	0.161	0.145	-	-	-	-	-
10	64.2	30.9	21.4	12.3	7.5	5.5	4.3	1.7	2.20	1.58	1.10	0.63	0.386	0.246	0.240	0.080	-	-	-	-
16	64.2	30.9	21.4	12.3	7.5	6.5	4.9	1.7	2.61	1.90	1.35	0.85	0.557	0.360	0.334	0.135	0.085	0.045	-	-
25	64.2	30.9	21.4	12.3	7.5	6.5	4.9	1.7	2.98	2.26	1.74	1.04	0.730	0.497	0.417	0.193	0.130	0.070	0.041	-
35	64.2	30.9	21.4	12.3	7.5	6.5	4.9	1.7	2.98	2.46	1.81	1.30	0.835	0.580	0.522	0.255	0.180	0.099	0.060	0.030
50	64.2	30.9	21.4	12.3	7.5	6.5	4.9	1.7	2.98	2.46	2.05	1.36	0.949	0.696	0.614	0.326	0.232	0.137	0.087	0.046
70	64.2	30.9	21.4	12.3	7.5	6.5	4.9	1.7	2.98	2.46	2.05	1.49	1.098	0.803	0.673	0.398	0.286	0.174	0.116	0.061
95	64.2	30.9	21.4	12.3	7.5	6.5	4.9	1.7	2.98	2.46	2.05	1.49	1.193	0.907	0.727	0.474	0.334	0.209	0.137	0.085
120	64.2	30.9	21.4	12.3	7.5	6.5	4.9	1.7	2.98	2.46	2.05	1.49	1.193	0.994	0.773	0.522	0.379	0.246	0.159	0.104
150	64.2	30.9	21.4	12.3	7.5	6.5	4.9	1.7	2.98	2.46	2.05	1.49	1.193	0.994	0.773	0.549	0.401	0.268	0.174	0.123
185	64.2	30.9	21.4	12.3	7.5	6.5	4.9	1.7	2.98	2.46	2.05	1.49	1.193	0.994	0.773	0.580	0.435	0.298	0.199	0.139
240	64.2	30.9	21.4	12.3	7.5	6.5	4.9	1.7	2.98	2.46	2.05	1.49	1.193	0.994	0.773	0.614	0.474	0.334	0.232	0.168
300 or larger	64.2	30.9	21.4	12.3	7.5	6.5	4.9	1.7	2.98	2.46	2.05	1.49	1.193	0.994	0.773	0.614	0.497	0.363	0.246	0.186

Notes: The impedances in the table are based on an initial conductor temperature of 70 °C and a k factor of 115. For protection of conductors, the maximum impedance of the conductors is required. With single phase circuits or three phase and neutral circuits, take the total impedance of the phase and neutral circuits, take the total impedance given in the table. For safety reasons, the circuit impedance should be less than the maximum allowed by the table. For three-phase three-wire circuits, take the total impedance of two phases (Z_{pp}), and divide by 1.732; compare the resultant impedance with the value from the table. As before, the circuits impedance should be less than the figure in the table. When the open circuit voltage U_{oc} is 240 V the nominal voltage U_0 is 230 V. The single-phase U_{oc} of 240 V corresponds with a three-phase open circuit voltage U_{oc} of 415 V and a single-phase voltage U_0 of 230 V corresponds with a three-phase nominal voltage U_{ol} of 400 V.

Maximum values of Z_{pn} for copper conductors, for thermal protection of the live conductors, when U_{oc} = 240V

Size of copper live conductors in mm².	Rating of BS 1361 fuse in amperes							
	5	15	20	30	45	60	80	100
	Maximum value of Z_{pn} in ohms with a 'k' factor of 115							
1	20.87	4.54	2.07	0.87	0.25	–	–	–
1.5	20.87	5.22	2.43	1.28	0.40	–	–	–
2.5	23.19	5.64	3.02	1.67	0.60	0.248	–	–
4	23.19	6.14	3.48	2.24	0.77	0.394	0.232	–
6	24.27	6.73	4.17	2.43	0.97	0.596	0.342	0.190
10	24.27	7.32	4.85	2.90	1.35	0.835	0.469	0.282
16	24.27	7.59	5.42	3.26	1.61	1.128	0.652	0.409
25	24.27	8.03	5.80	3.60	2.04	1.391	0.835	0.522
35	24.27	8.03	6.14	3.94	2.27	1.670	0.994	0.673
50 or larger	24.27	8.03	6.14	4.17	2.55	1.988	1.128	0.773

Notes: The impedances in the table are based on an initial conductor temperature of 70 °C and a k factor of 115. For protection of conductors, the maximum impedance of the conductors is required. With single phase circuits or three phase and neutral circuits, take the total impedance of the phase and neutral (Z_{pn}) at the design temperature, and compare it with the impedance given in the table. For safety reasons, the circuit impedance should be less than the maximum allowed by the table. For three-phase three-wire circuits, take the total impedance of two phases (Z_{pp}), and divide by 1.732; compare the resultant impedance with the value from the table. As before, the circuits impedance should be less than the figure in the table. When the open circuit voltage U_{oc} is 240 V the nominal voltage U_0 is 230 V. The single-phase U_{oc} of 240 V corresponds with a three-phase open circuit voltage U_{loc} of 415 V and a single-phase voltage U_0 of 230 V corresponds with a three-phase nominal voltage U_{ol} of 400 V.

SHORT-CIRCUIT PROTECTION

SCZ 9

BS 3036 FUSES

Maximum values of Z_{pn} for copper conductors, for thermal protection of the live conductors, when U_{oc} = 240V

Size of copper live conductors in mm².	Rating of BS 3036 fuse in amperes						
	5	15	20	30	45	60	100
	Maximum value of Z_{pn} in ohms with a 'k' factor of 115						
1	20.87	5.22	3.07	-	-	-	-
1.5	21.74	5.96	3.86	2.13	-	-	-
2.5	21.74	6.42	4.35	2.90	1.23	-	-
4	21.74	6.73	4.64	3.31	1.74	1.043	-
6	21.74	6.73	4.85	3.54	1.90	1.391	-
10	21.74	7.07	5.09	3.73	2.13	1.739	0.835
16	21.74	7.20	5.22	3.73	2.13	1.815	0.773
25	21.74	7.26	5.22	3.73	2.13	1.897	0.870
35	21.74	7.26	5.22	3.73	2.13	1.988	0.928
50 or larger	21.74	7.26	5.22	3.73	2.13	2.036	0.971

Notes: The impedances in the table are based on an initial conductor temperature of 70 °C and a k factor of 115. For protection of conductors, the maximum impedance of the conductors is required. With single phase circuits or three phase and neutral circuits, take the total impedance of the phase and neutral (Z_{pn}) at the design temperature, and compare it with the impedance given in the table. For safety reasons, the circuit impedance should be less than the maximum allowed by the table. For three-phase three-wire circuits, take the total impedance of two phases (Z_{pp}), and divide by 1.732; compare the resultant impedance with the value from the table. As before, the circuits impedance should be less than the figure in the table. When the open circuit voltage U_{oc} is 240 V the nominal voltage U_o is 230 V. The single-phase U_{oc} of 240 V corresponds with a three-phase open circuit voltage U_{loc} of 415 V and a single-phase voltage U_o of 230 V corresponds with a three-phase nominal voltage U_{ol} of 400 V.

GROUPED SINGLE CORE CABLES   CSG 1   ENCLOSED

Cable sizes required for three-phase and single-phase circuits using PVC insulated copper single core cable with or without sheath, grouped and enclosed in conduit or trunking, not in contact with thermal insulation.

I_n or I_b in amperes, excluding BS 3036 fuses when they are used for overload protection

Single-phase circuits — Size of PVC single core cable required in mm².

Number of circuits grouped together	2 to 5	6	10	15	16	20	25	30	32	35	40
2	1.0	1.0	1.0	2.5	2.5	4	4	6	6	10	10
3	1.0	1.0	1.5	2.5	2.5	4	6	10	10	10	10
4	1.0	1.0	1.5	2.5	4	4	6	10	10	10	16
5	1.0	1.0	1.5	4	4	6	10	10	10	16	16
6	1.0	1.0	1.5	4	4	6	10	10	10	16	16
7	1.0	1.0	2.5	4	4	6	10	10	16	16	16
8	1.0	1.0	2.5	4	4	6	10	16	16	16	25
9	1.0	1.0	2.5	4	4	6	10	16	16	16	25
10	1.0	1.0	2.5	4	6	6	16	16	16	16	25
11	1.0	1.0	2.5	6	6	10	10	16	16	16	25
12	1.0	1.0	2.5	6	6	10	10	16	16	25	25
13	1.0	1.5	2.5	6	6	10	10	16	16	25	25
14	1.0	1.5	2.5	6	6	10	16	16	16	25	25
15	1.0	1.5	2.5	6	6	10	16	16	25	25	25
16	1.0	1.5	4	6	6	10	16	16	25	25	25
17	1.0	1.5	4	6	6	10	16	16	25	25	25
18	1.0	1.5	4	6	6	10	16	25	25	25	35
19	1.0	1.5	4	6	10	10	16	25	25	25	35
20	1.0	1.5	4	6	10	10	16	25	25	25	35

Three-phase circuits — Size of PVC single core cable required in mm².

Number of circuits grouped together	2 to 4	5	6	10	15	16	20	25	30	32	35	40
2	1.0	1.0	1.0	1.5	2.5	2.5	4	6	10	10	10	10
3	1.0	1.0	1.0	1.5	4	4	6	6	10	10	10	16
4	1.0	1.0	1.0	1.5	4	4	6	10	10	10	16	16
5	1.0	1.0	1.0	2.5	4	4	6	10	10	16	16	16
6	1.0	1.0	1.0	2.5	4	4	6	10	16	16	16	25
7	1.0	1.0	1.0	2.5	4	6	10	10	16	16	16	25
8	1.0	1.0	1.0	2.5	6	6	10	10	16	16	16	25
9	1.0	1.0	1.0	2.5	6	6	10	10	16	16	25	25
10	1.0	1.0	1.5	2.5	6	6	10	16	16	16	25	25
11	1.0	1.0	1.5	4	6	6	10	16	16	25	25	25
12	1.0	1.0	1.5	4	6	6	10	16	16	25	25	25
13	1.0	1.0	1.5	4	6	10	10	16	25	25	25	35
14	1.0	1.0	1.5	4	6	10	10	16	25	25	25	35
15	1.0	1.0	1.5	4	6	10	10	16	25	25	25	35
16	1.0	1.5	1.5	4	10	10	10	16	25	25	25	35
17	1.0	1.5	1.5	4	10	10	10	16	25	25	25	35
18	1.0	1.5	1.5	4	10	10	16	16	25	35	35	35
19	1.0	1.5	2.5	4	10	10	16	16	25	35	35	35
20	1.0	1.5	2.5	4	10	10	16	16	25	35	35	35

Note: Where a cable or circuit will not carry more than 30% of the tabulated current rating of its grouped cable size, it can be ignored when counting the number grouped together for the remaining circuits. Where the distance between adjacent conductors or cables is more than twice the diameter of the larger cable, no derating for grouping need be applied. Circuit groupings can be a mixture of three phase and single phase circuits. Where the circuit is not likely to carry overload current, use the full load current I_b of the circuit, and not the protective device size I_n, when sizing the tie cable from the above table.

GROUPED SINGLE CORE CABLES CSG 2 BUNCHED

Cable sizes required for three-phase and single-phase circuits using PVC insulated copper single core cable with or without sheath, bunched and clipped direct to a non-metallic surface, not in contact with thermal insulation.

I_n or I_b in amperes, excluding BS 3036 fuses when they are used for overload protection

Single-phase circuits — Size of PVC single core cable required in mm².

Number of circuits grouped together	2 to 5	6	10	15	16	20	25	30	32	35	40
2	1.0	1.0	1.0	1.5	1.5	2.5	4	6	6	6	10
3	1.0	1.0	1.0	2.5	2.5	2.5	4	6	6	10	10
4	1.0	1.0	1.0	2.5	2.5	4	6	6	10	10	10
5	1.0	1.0	1.5	2.5	2.5	4	6	10	10	10	16
6	1.0	1.0	1.5	2.5	4	4	6	10	10	10	16
7	1.0	1.0	1.5	4	4	4	6	10	10	10	16
8	1.0	1.0	1.5	4	4	6	10	10	10	16	16
9	1.0	1.0	1.5	4	4	6	10	10	16	16	16
10	1.0	1.0	2.5	4	4	6	10	10	16	16	16
11	1.0	1.0	2.5	4	4	6	10	10	16	16	16
12	1.0	1.0	2.5	4	4	6	10	16	16	16	25
13	1.0	1.0	2.5	4	4	6	10	16	16	16	25
14	1.0	1.0	2.5	4	6	6	10	16	16	16	25
15	1.0	1.0	2.5	4	6	10	16	16	16	16	25
16	1.0	1.0	2.5	4	6	10	10	16	16	16	25
17	1.0	1.0	2.5	6	6	10	10	16	25	25	25
18	1.0	1.0	2.5	6	6	10	10	16	16	25	25
19	1.0	1.5	2.5	6	6	10	10	16	16	25	25
20	1.0	1.5	2.5	6	6	10	16	16	16	25	25

Three-phase circuits — Size of PVC single core cable required in mm².

Number of circuits grouped together	2 to 5	6	10	15	16	20	25	30	32	35	40	45
2	1.0	1.0	1.0	2.5	2.5	2.5	4	6	6	10	10	10
3	1.0	1.0	1.5	2.5	2.5	4	6	6	10	10	10	16
4	1.0	1.0	1.5	2.5	2.5	4	6	10	10	10	16	16
5	1.0	1.0	1.5	2.5	4	6	6	10	10	10	16	16
6	1.0	1.0	1.5	4	4	6	10	10	10	16	16	16
7	1.0	1.0	2.5	4	4	6	10	10	16	16	16	25
8	1.0	1.0	2.5	4	4	6	10	10	16	16	16	25
9	1.0	1.0	2.5	4	6	6	10	16	16	16	25	25
10	1.0	1.0	2.5	4	6	6	10	16	16	16	25	25
11	1.0	1.0	2.5	4	6	6	10	16	16	16	25	25
12	1.0	1.0	2.5	6	6	10	10	16	16	16	25	25
13	1.0	1.0	2.5	6	6	10	10	16	16	25	25	25
14	1.0	1.0	2.5	6	6	10	16	16	16	25	25	35
15	1.0	1.5	2.5	6	6	10	16	16	16	25	25	35
16	1.0	1.5	2.5	6	6	10	16	16	16	25	25	35
17	1.0	1.5	2.5	6	6	10	16	16	25	25	25	35
18	1.0	1.5	4	6	6	10	16	16	25	25	25	35
19	1.0	1.5	4	6	6	10	16	16	25	25	25	35
20	1.0	1.5	4	6	6	10	16	16	25	25	35	35

Note: Where a cable or circuit will not carry more than 30% of the tabulated current rating of its grouped cable size, it can be ignored when counting the number grouped together for the remaining circuits. Where the distance between adjacent conductors or cables is more than twice the diameter of the larger cable, no derating for grouping need be applied. Circuit groupings can be a mixture of three phase and single phase circuits. Where the circuit is not likely to carry overload current, use the full load current I_b of the circuit, and not the protective device size I_n, when sizing the cable from the above table.

GROUPED SINGLE CORE CABLES   CSG 3   ENCLOSED/THERMAL

Cable sizes required for three-phase and single-phase circuits using PVC insulated copper single core cable with or without sheath, grouped and enclosed, in conduit or trunking in contact with thermal insulation, one side being in contact with a thermally conductive surface.

I_n or I_b in amperes, excluding BS 3036 fuses when they are used for overload protection

Number of circuits grouped together	Single-phase circuits											Three-phase circuits											
	2 TO 4	5	6	10	15	16	20	25	30	32	35	2 to 4	5	6	10	15	16	20	25	30	32	35	40
	Size of PVC single core cable required in mm².											Size of PVC single core cable required in mm².											
2	1.0	1.0	1.0	1.5	2.5	4	4	6	10	10	10	1.0	1.0	1.0	1.5	4	4	6	10	10	10	16	16
3	1.0	1.0	1.0	1.5	4	4	6	10	10	10	16	1.0	1.0	1.0	2.5	4	4	6	10	16	16	16	25
4	1.0	1.0	1.0	2.5	4	4	6	10	16	16	16	1.0	1.0	1.0	2.5	4	6	6	10	16	16	16	25
5	1.0	1.0	1.0	2.5	4	6	6	10	16	16	16	1.0	1.0	1.0	2.5	6	6	10	10	16	16	25	25
6	1.0	1.0	1.0	2.5	6	6	10	10	16	16	25	1.0	1.0	1.0	2.5	6	6	10	16	16	16	25	25
7	1.0	1.0	1.0	2.5	6	6	10	16	16	16	25	1.0	1.0	1.5	4	6	6	10	16	16	25	25	35
8	1.0	1.0	1.5	2.5	6	6	10	16	16	25	25	1.0	1.0	1.5	4	6	6	10	16	25	25	25	35
9	1.0	1.0	1.5	4	6	6	10	16	16	25	25	1.0	1.0	1.5	4	6	10	10	16	25	25	25	35
10	1.0	1.0	1.5	4	6	6	10	16	25	25	25	1.0	1.0	1.5	4	10	10	16	16	25	25	25	35
11	1.0	1.0	1.5	4	6	10	10	16	25	25	25	1.0	1.5	1.5	4	10	10	16	16	25	25	35	35
12	1.0	1.0	1.5	4	6	10	10	16	25	25	25	1.0	1.5	1.5	4	10	10	16	16	25	25	35	35
13	1.0	1.5	1.5	4	6	10	10	16	25	25	25	1.0	1.5	2.5	4	10	10	16	25	25	35	35	50
14	1.0	1.5	1.5	4	10	10	16	16	25	25	35	1.0	1.5	2.5	4	10	10	16	25	25	35	35	50
15	1.0	1.5	1.5	4	10	10	16	16	25	25	35	1.0	1.5	2.5	4	10	10	16	25	25	35	35	50
16	1.0	1.5	2.5	4	10	10	16	16	25	25	35	1.0	1.5	2.5	6	10	10	16	25	35	35	35	50
17	1.0	1.5	2.5	4	10	10	16	25	25	25	35	1.0	1.5	2.5	6	10	10	16	25	35	35	35	50
18	1.0	1.5	2.5	4	10	10	16	25	25	35	35	1.0	1.5	2.5	6	10	10	16	25	35	35	50	50
19	1.0	1.5	2.5	6	10	10	16	25	25	35	35	1.0	1.5	2.5	6	10	10	16	25	35	35	50	50
20	1.0	1.5	2.5	6	10	10	16	25	25	35	35	1.0	1.5	2.5	6	10	10	16	25	35	35	50	50

Note: Where a cable or circuit will not carry more than 30% of the tabulated current rating of its grouped cable size, it can be ignored when counting the number grouped together for the remaining circuits. Where the distance between adjacent conductors or cables is more than twice the diameter of the larger cable, no derating for grouping need be applied. Circuit groupings can be a mixture of three phase and single phase circuits. Where the circuit is not likely to carry overload current, use the full load current I_b of the circuit, and not the protective device size I_n, when sizing the cable from the above table.

GROUPED MULTICORE CABLES CSG 4 ENCLOSED

Cable sizes required for three-phase and single-phase circuits using PVC insulated multicore copper cable or twin and CPC copper cable, grouped and enclosed in conduit or trunking, not in contact with thermal insulation.

I_n or I_b in amperes, excluding BS 3036 fuses when they are used for overload protection

Number of grouped multicore cables	Single-phase circuits											Three-phase circuits											
	Size of PVC single core cable required in mm²											Size of PVC single core cable required in mm²											
	2 TO 4	5	6	10	15	16	20	25	30	32	35	2 to 4	5	6	10	15	16	20	25	30	32	35	40
2	1.0	1.0	1.0	1.0	2.5	2.5	4	6	6	10	10	1.0	1.0	1.0	1.5	2.5	2.5	4	6	10	10	10	16
3	1.0	1.0	1.0	1.5	2.5	2.5	4	6	10	10	10	1.0	1.0	1.0	1.5	4	4	6	10	16	16	16	16
4	1.0	1.0	1.0	1.5	2.5	2.5	6	10	10	10	16	1.0	1.0	1.0	2.5	4	4	6	16	16	16	16	16
5	1.0	1.0	1.0	2.5	4	4	6	10	10	16	16	1.0	1.0	1.0	2.5	4	4	6	10	16	16	16	25
6	1.0	1.0	1.0	2.5	4	4	6	10	16	16	16	1.0	1.0	1.0	2.5	4	6	10	10	16	16	16	25
7	1.0	1.0	1.0	2.5	4	4	6	10	16	16	16	1.0	1.0	1.0	2.5	6	6	10	16	16	16	16	25
8	1.0	1.0	1.0	2.5	4	6	10	10	16	16	16	1.0	1.0	1.0	2.5	6	6	10	16	16	16	25	25
9	1.0	1.0	1.0	2.5	4	6	10	10	16	16	25	1.0	1.0	1.5	2.5	6	6	10	16	16	25	25	25
10	1.0	1.0	1.0	2.5	6	6	10	10	16	16	25	1.0	1.0	1.5	4	6	6	10	16	25	25	25	35
11	1.0	1.0	1.0	2.5	6	6	10	16	16	16	25	1.0	1.0	1.5	4	6	10	10	16	25	25	25	35
12	1.0	1.0	1.5	2.5	6	6	10	16	16	25	25	1.0	1.0	1.5	4	6	10	10	16	25	25	25	35
13	1.0	1.0	1.5	2.5	6	6	10	16	16	25	25	1.0	1.0	1.5	4	6	10	10	16	25	25	25	35
14	1.0	1.0	1.5	4	6	6	10	16	25	25	25	1.0	1.0	1.5	4	10	10	16	16	25	25	35	35
15	1.0	1.0	1.5	4	6	6	10	16	25	25	25	1.0	1.0	1.5	4	10	10	16	16	25	25	35	35
16	1.0	1.0	1.5	4	6	10	10	16	25	25	25	1.0	1.0	1.5	4	10	10	16	16	25	25	35	35
17	1.0	1.0	1.5	4	6	10	10	16	25	25	25	1.0	1.0	1.5	4	10	10	16	25	25	25	35	50
18	1.0	1.0	1.5	4	10	10	10	16	25	25	25	1.0	1.0	2.5	4	10	10	16	25	25	35	35	50
19	1.0	1.0	1.5	4	10	10	16	16	25	25	35	1.0	1.0	2.5	4	10	10	16	25	25	35	35	50
20	1.0	1.5	1.5	4	10	10	16	16	25	25	35	1.0	1.5	2.5	4	10	10	16	25	25	35	35	50

Note: Where a cable or circuit will not carry more than 30% of the tabulated current rating of its grouped cable size, it can be ignored when counting the number grouped together for the remaining circuits. Where the distance between adjacent conductors or cables is more than twice the diameter of the larger cable, no derating for grouping need be applied. Circuit groupings can be a mixture of three phase and single phase circuits. Where the circuit is not likely to carry overload current, use the full load current I_b of the circuit, and not the protective device size I_n, when sizing the cable from the above table.

GROUPED MULTICORE CABLES   CSG 5   BUNCHED

Cable sizes required for three-phase and single-phase circuits using PVC insulated multicore copper cable or twin and CPC copper cable, bunched and clipped direct to a non-metallic surface, not in contact with thermal insulation.

I_n or I_b in amperes, excluding BS 3036 fuses when they are used for overload protection

Size of PVC multicore cable required in mm².

Number of grouped multicore cables	Single-phase circuits											Three-phase circuits											
	2 TO 5	6	10	15	16	20	25	30	32	35	40	2 to 5	6	10	15	16	20	25	30	32	35	40	45
2	1.0	1.0	1.0	1.5	2.5	2.5	4	6	6	6	10	1.0	1.0	1.0	2.5	2.5	4	4	6	6	10	10	10
3	1.0	1.0	1.0	2.5	2.5	4	4	6	6	10	10	1.0	1.0	1.5	2.5	2.5	4	6	10	10	10	10	16
4	1.0	1.0	1.5	2.5	2.5	4	6	10	10	10	10	1.0	1.0	1.5	2.5	4	4	6	10	10	16	16	16
5	1.0	1.0	1.5	2.5	2.5	4	6	10	10	10	16	1.0	1.0	1.5	4	4	6	10	10	16	16	16	16
6	1.0	1.0	1.5	2.5	4	4	6	10	10	10	16	1.0	1.0	1.5	4	4	6	10	10	16	16	16	25
7	1.0	1.0	1.5	4	4	6	10	10	10	16	16	1.0	1.0	2.5	4	4	6	10	16	16	16	16	25
8	1.0	1.0	1.5	4	4	6	10	10	16	16	16	1.0	1.0	2.5	4	4	6	10	16	16	16	25	25
9	1.0	1.0	2.5	4	4	6	10	10	16	16	16	1.0	1.0	2.5	4	4	6	10	16	16	16	25	25
10	1.0	1.0	2.5	4	4	6	10	10	16	16	16	1.0	1.0	2.5	4	6	10	10	16	16	16	25	25
11	1.0	1.0	2.5	4	4	6	10	16	16	16	25	1.0	1.0	2.5	6	6	10	10	16	16	16	25	35
12	1.0	1.0	2.5	4	4	6	10	16	16	16	25	1.0	1.0	2.5	6	6	10	10	16	16	25	25	35
13	1.0	1.0	2.5	4	6	6	10	16	16	16	25	1.0	1.5	2.5	6	6	10	16	16	16	25	25	35
14	1.0	1.0	2.5	4	6	10	10	16	16	16	25	1.0	1.5	2.5	6	6	10	16	16	16	25	25	35
15	1.0	1.0	2.5	4	6	10	10	16	16	16	25	1.0	1.5	2.5	6	6	10	16	16	25	25	25	35
16	1.0	1.0	2.5	6	6	6	10	16	16	25	25	1.0	1.5	4	6	6	10	16	16	25	25	35	35
17	1.0	1.0	2.5	6	6	6	10	16	16	25	25	1.0	1.5	4	6	6	10	16	16	25	25	35	35
18	1.0	1.5	2.5	6	6	6	16	16	16	25	25	1.0	1.5	4	6	6	10	16	25	25	25	35	35
19	1.0	1.5	2.5	6	6	10	16	16	16	25	25	1.0	1.5	4	6	10	10	16	25	25	25	35	35
20	1.0	1.5	2.5	6	6	10	16	16	16	25	25	1.0	1.5	4	6	10	10	16	25	25	25	35	35

Note: Where a cable or circuit will not carry more than 30% of the tabulated current rating of its grouped cable size, it can be ignored when counting the number grouped together for the remaining circuits. Where the distance between adjacent conductors or cables is more than twice the diameter of the larger cable, no derating for grouping need be applied. Circuit groupings can be a mixture of three phase and single phase circuits. Where the circuit is not likely to carry overload current, use the full load current I_b of the circuit, and not the protective device size I_n, when sizing the cable from the above table.

GROUPED MULTICORE CABLES   CSG 6   BUNCHED/THERMAL

Cable sizes required for three-phase & single-phase circuits using PVC insulated multicore copper cable, or twin and CPC copper cable, bunched and clipped direct to a non-metallic surface, in contact with thermal insulation, one side being in contact with a thermally conductive surface.

I_n or I_b in amperes, excluding BS 3036 fuses when they are used for overload protection

Size of PVC multicore cable required in mm².

Number of grouped multicore cables	Single-phase circuits											Three-phase circuits											
	2 TO 4	5	6	10	15	16	20	25	30	32	35	2 to 4	5	6	10	15	16	20	25	30	32	35	40
2	1.0	1.0	1.0	1.5	4	4	4	6	10	10	16	1.0	1.0	1.0	1.5	4	4	6	10	10	16	16	16
3	1.0	1.0	1.0	2.5	4	4	6	10	10	16	16	1.0	1.0	1.0	2.5	4	4	6	10	16	16	16	25
4	1.0	1.0	1.0	2.5	4	4	6	10	16	16	16	1.0	1.0	1.0	2.5	4	6	10	10	16	16	25	25
5	1.0	1.0	1.0	2.5	4	6	10	10	16	16	25	1.0	1.0	1.0	2.5	6	6	10	16	16	25	25	25
6	1.0	1.0	1.0	2.5	6	6	10	16	16	16	25	1.0	1.0	1.5	2.5	6	6	10	16	16	25	25	35
7	1.0	1.0	1.0	2.5	6	6	10	16	16	25	25	1.0	1.0	1.5	4	6	10	10	16	25	25	25	35
8	1.0	1.0	1.5	4	6	6	10	16	25	25	25	1.0	1.0	1.5	4	6	10	10	16	25	25	25	35
9	1.0	1.0	1.5	4	6	6	10	16	25	25	25	1.0	1.0	1.5	4	10	10	16	16	25	25	35	35
10	1.0	1.0	1.5	4	6	10	10	16	25	25	25	1.0	1.5	1.5	4	10	10	16	16	25	25	35	50
11	1.0	1.0	1.5	4	10	10	10	16	25	25	35	1.0	1.5	1.5	4	10	10	16	25	25	35	35	50
12	1.0	1.0	1.5	4	10	10	16	16	25	25	35	1.0	1.5	2.5	4	10	10	16	25	25	35	35	50
13	1.0	1.5	1.5	4	10	10	16	16	25	25	35	1.0	1.5	2.5	4	10	10	16	25	35	35	35	50
14	1.0	1.5	1.5	4	10	10	16	25	25	35	35	1.0	1.5	2.5	6	10	10	16	25	35	35	35	50
15	1.0	1.5	2.5	4	10	10	16	25	25	35	35	1.0	1.5	2.5	6	10	10	16	25	35	35	35	50
16	1.0	1.5	2.5	4	10	10	16	25	25	35	35	1.0	1.5	2.5	6	10	10	16	25	35	35	35	50
17	1.0	1.5	2.5	4	10	10	16	25	25	35	35	1.0	1.5	2.5	6	10	16	16	25	35	35	50	70
18	1.0	1.5	2.5	6	10	10	16	25	35	35	35	1.0	1.5	2.5	6	10	16	16	25	35	35	50	70
19	1.0	1.5	2.5	6	10	10	16	25	35	35	35	1.5	1.5	2.5	6	10	16	16	25	35	35	50	70
20	1.0	1.5	2.5	6	16	16	16	25	35	35	35	1.5	2.5	2.5	6	16	16	25	25	35	50	50	70

Note: Where a cable or circuit will not carry more than 30% of the tabulated current rating of its grouped cable size, it can be ignored when counting the number grouped together for the remaining circuits. Where the distance between adjacent conductors or cables is more than twice the diameter of the larger cable, no derating for grouping need be applied. Circuit groupings can be a mixture of three phase and single phase circuits. Where the circuit is not likely to carry overload current, use the full load current I_b of the circuit, and not the protective device size I_n, when sizing the cable from the above table.

GROUPED MULTICORE CABLES

CSG 7

CLIPPED DIRECT

Cable sizes required for three-phase and single-phase circuits using PVC insulated multicore copper cable or twin and CPC copper cable, clipped direct in a single layer to a non-metallic surface with cable sheaths touching, not in contact with thermal insulation.

I_n or I_b in amperes, excluding BS 3036 fuses when they are used for overload protection

Single-phase circuits

Size of PVC multicore cable required in mm^2.

Number of grouped multicore cables	2 TO 10	15	16	20	25	30	32	35	40	45	50
2	1.0	1.5	1.5	2.5	4	4	6	6	10	10	10
3	1.0	1.5	2.5	2.5	4	6	6	6	10	10	16
4	1.0	2.5	2.5	2.5	4	6	6	10	10	10	16
5	1.0	2.5	2.5	4	4	6	6	10	10	10	16
6	1.0	2.5	2.5	4	4	6	6	10	10	10	16
7	1.0	2.5	2.5	4	4	6	6	10	10	10	16
8	1.0	2.5	2.5	4	4	6	6	10	10	16	16
9	1.0	2.5	2.5	4	4	6	6	10	10	16	16

Three-phase circuits

Size of PVC multicore cable required in mm^2.

Number of grouped multicore cables	2 to 6	10	15	16	20	25	30	32	35	40	45	50
2	1.0	1.0	2.5	2.5	2.5	4	6	6	10	10	10	16
3	1.0	1.0	2.5	2.5	4	4	6	6	10	10	10	16
4	1.0	1.0	2.5	2.5	4	6	6	10	10	10	16	16
5	1.0	1.5	2.5	2.5	4	6	6	10	10	10	16	16
6	1.0	1.5	2.5	2.5	4	6	10	10	10	10	16	16
7	1.0	1.5	2.5	2.5	4	6	10	10	10	10	16	16
8	1.0	1.5	2.5	2.5	4	6	10	10	10	10	16	16
9	1.0	1.5	2.5	2.5	4	6	10	10	10	10	16	16

Note: Where a cable or circuit will not carry more than 30% of the tabulated current rating of its grouped cable size, it can be ignored when counting the number grouped together for the remaining circuits. Where the distance between adjacent conductors or cables is more than twice the diameter of the larger cable, no derating for grouping need be applied. Circuit groupings can be a mixture of three phase and single phase circuits. Where the circuit is not likely to carry overload current, use the full load current I_b of the circuit, and not the protective device size I_n, when sizing the cable from the above table.

GROUPED MULTICORE CABLES CSG 8 CABLE TRAY

Cable sizes required for three-phase and single-phase circuits using PVC insulated multicore copper cable or twin and CPC copper cable, installed in a single layer on perforated metal cable tray.

I_n or I_b in amperes, excluding BS 3036 fuses when they are used for overload protection

Single-phase circuits — Size of PVC multicore cable required in mm²

Number of grouped multicore cables	2 TO 10	15	16	20	25	30	32	35	40	45	50
2	1.0	1.5	1.5	2.5	2.5	4	4	6	6	10	10
3	1.0	1.5	1.5	2.5	4	4	4	6	6	10	10
4	1.0	1.5	1.5	2.5	4	4	6	6	10	10	10
5	1.0	1.5	1.5	2.5	4	4	6	6	10	10	10
6	1.0	1.5	1.5	2.5	4	6	6	6	10	10	10
7	1.0	1.5	1.5	2.5	4	6	6	6	10	10	10
8	1.0	1.5	1.5	2.5	4	6	6	6	10	10	10
9	1.0	1.5	2.5	2.5	4	6	6	6	10	10	10
10	1.0	1.5	2.5	2.5	4	6	6	6	10	16	16
11	1.0	1.5	2.5	2.5	4	6	6	6	10	10	16
12	1.0	1.5	2.5	2.5	4	6	6	6	10	10	16

Three-phase circuits — Size of PVC multicore cable required in mm²

Number of grouped multicore cables	2 to 10	15	16	20	25	30	32	35	40	45	50	60
2	1.0	1.5	2.5	2.5	4	6	6	6	10	10	10	16
3	1.0	1.5	2.5	2.5	4	6	6	10	10	10	16	16
4	1.0	1.5	2.5	4.0	4	6	6	10	10	10	16	16
5	1.0	2.5	2.5	4.0	4	6	6	10	10	10	16	16
6	1.0	2.5	2.5	4.0	4	6	10	10	10	16	16	25
7	1.0	2.5	2.5	4.0	6	6	10	10	10	16	16	25
8	1.0	2.5	2.5	4.0	6	6	10	10	10	16	16	25
9	1.0	2.5	2.5	4.0	6	6	10	10	10	16	16	25
10	1.0	2.5	2.5	4.0	6	6	10	10	10	16	16	25
11	1.0	2.5	2.5	4.0	6	6	10	10	10	16	16	25
12	1.0	2.5	2.5	4.0	6	6	10	10	10	16	16	25

Note: Where a cable or circuit will not carry more than 30% of the tabulated current rating of its grouped cable size, it can be ignored when counting the number grouped together for the remaining circuits. Where the distance between adjacent conductors or cables is more than twice the diameter of the larger cable, no derating for grouping need be applied. Circuit groupings can be a mixture of three phase and single phase circuits. Where the circuit is not likely to carry overload current, use the full load current I_b of the circuit, and not the protective device size I_n, when sizing the cable from the above table.

GROUPED ARMOURED CABLES   **CSG 9**   BUNCHED

Cable sizes required for three phase & single phase circuits using p.v.c insulated multicore armoured copper cable, bunched and clipped direct to a non-metallic surface, not in contact with thermal insulation.

I_n or I_b in amperes, excluding BS 3036 fuses when they are used for overload protection

Number of grouped armoured cables	Single-phase circuits											Three-phase circuits											
	Size of armoured PVC cable required in mm².											Size of armoured PVC cable required in mm².											
	2 TO 6	10	15	16	20	25	30	32	35	40	45	2 to 6	10	15	16	20	25	30	32	35	40	45	50
2	1.5	1.5	1.5	1.5	2.5	4	4	6	6	10	10	1.5	1.5	2.5	2.5	2.5	4	6	6	10	10	10	16
3	1.5	1.5	2.5	2.5	4	4	6	6	10	10	10	1.5	1.5	2.5	2.5	4	6	10	10	10	10	16	16
4	1.5	1.5	2.5	2.5	4	6	6	10	10	10	16	1.5	1.5	2.5	2.5	4	6	10	10	16	16	16	16
5	1.5	1.5	2.5	2.5	4	6	10	10	10	10	16	1.5	1.5	2.5	4	6	6	10	10	16	16	16	25
6	1.5	1.5	2.5	2.5	4	6	10	10	10	16	16	1.5	1.5	4	4	6	10	10	10	16	16	25	25
7	1.5	1.5	2.5	4	4	6	10	10	10	16	16	1.5	2.5	4	4	6	10	10	16	16	16	25	25
8	1.5	1.5	4	4	6	6	10	10	16	16	16	1.5	2.5	4	4	6	10	10	16	16	16	25	25
9	1.5	1.5	4	4	6	10	10	10	16	16	25	1.5	2.5	4	4	6	10	16	16	16	25	25	25
10	1.5	1.5	4	4	6	10	16	10	16	16	25	1.5	2.5	4	6	6	10	16	16	16	25	25	35
11	1.5	2.5	4	4	6	10	10	16	16	16	25	1.5	2.5	4	6	10	10	16	16	16	25	25	35
12	1.5	2.5	4	4	6	10	10	16	16	16	25	1.5	2.5	6	6	10	10	16	16	25	25	35	35
13	1.5	2.5	4	4	6	16	16	16	16	25	25	1.5	2.5	6	6	10	10	16	16	25	25	35	35
14	1.5	2.5	4	4	6	10	16	16	16	25	25	1.5	2.5	6	6	10	10	16	16	25	25	35	35
15	1.5	2.5	4	4	6	10	16	16	16	25	25	1.5	2.5	6	6	10	16	16	16	25	25	35	35
16	1.5	2.5	4	6	6	10	16	16	16	25	25	1.5	2.5	6	6	10	16	16	16	25	25	25	35
17	1.5	2.5	4	6	10	10	16	16	16	25	25	1.5	2.5	6	6	10	16	16	25	25	25	35	35
18	1.5	2.5	6	6	10	10	16	16	25	25	25	1.5	4	6	6	10	16	16	25	25	35	35	50
19	1.5	2.5	6	6	10	10	16	16	25	25	25	1.5	4	6	6	10	16	25	25	25	35	35	50
20	1.5	2.5	6	6	10	10	16	16	25	35	35	1.5	4	6	6	10	16	25	25	35	35	35	50

Note: Where a cable or circuit will not carry more than 30% of the tabulated current rating of its grouped cable size, it can be ignored when counting the number grouped together for the remaining circuits. Where the distance between adjacent conductors or cables is more than twice the diameter of the larger cable, no derating for grouping need be applied. Circuit groupings can be a mixture of three phase and single phase circuits. Where the circuit is not likely to carry overload current, use the full load current I_b of the circuit, and not the protective device size I_n, when sizing the cable from the above table.

GROUPED ARMOURED CABLES   CSG 10   CLIPPED DIRECT

Cable sizes required for three-phase and single-phase circuits using armoured multicore PVC insulated copper cable, clipped direct in a single layer to a non-metallic surface with cable sheaths touching, not in contact with thermal insulation.

I_n or I_b in amperes, excluding BS 3036 fuses when they are used for overload protection

Single-phase circuits
Size of armoured PVC cable required in mm^2.

Number of grouped armoured cables	2 TO 10	15	16	20	25	30	32	35	40	45	50
2	1.5	1.5	1.5	2.5	4	4	4	6	6	10	10
3	1.5	1.5	1.5	2.5	4	4	6	6	10	10	10
4	1.5	1.5	2.5	2.5	4	6	6	6	10	10	10
5	1.5	1.5	2.5	2.5	4	6	6	6	10	10	16
6	1.5	1.5	2.5	2.5	4	6	6	6	10	10	16
7	1.5	1.5	2.5	2.5	4	6	6	6	10	10	16
8	1.5	1.5	2.5	4	4	6	6	10	10	10	16
9	1.5	2.5	2.5	4	4	6	6	10	10	10	16

Three-phase circuits
Size of armoured PVC cable required in mm^2.

Number of grouped armoured cables	2 to 10	15	16	20	25	30	32	35	40	45	50	60
2	1.5	1.5	2.5	2.5	4	6	6	6	10	10	16	16
3	1.5	2.5	2.5	4	4	6	6	10	10	10	16	16
4	1.5	2.5	2.5	4	6	6	10	10	10	16	16	25
5	1.5	2.5	2.5	4	6	6	10	10	10	16	16	25
6	1.5	2.5	2.5	4	6	6	10	10	10	16	16	25
7	1.5	2.5	2.5	4	6	6	10	10	10	16	16	25
8	1.5	2.5	2.5	4	6	10	10	10	10	16	16	25
9	1.5	2.5	2.5	4	6	10	10	10	10	16	16	25

Note: Where a cable or circuit will not carry more than 30% of the tabulated current rating of its grouped cable size, it can be ignored when counting the number grouped together for the remaining circuits. Where the distance between adjacent conductors or cables is more than twice the diameter of the larger cable, no derating for grouping need be applied. Circuit groupings can be a mixture of three phase and single phase circuits. Where the circuit is not likely to carry overload current, use the full load current I_b of the circuit, and not the protective device size I_n, when sizing the cable from the above table.

GROUPED ARMOURED CABLES   CSG 11   CABLE TRAY

Cable sizes required for three-phase and single-phase circuits using armoured multicore PVC insulated copper cable, installed in a single layer on perforated metal cable tray with cable sheaths touching, not in contact with thermal insulation.

I_n or I_b in amperes, excluding BS 3036 fuses when they are used for overload protection

Number of grouped armoured cables	Single-phase circuits											Three-phase circuits											
	Size of armoured PVC cable required in mm².											Size of armoured PVC cable required in mm².											
	2 TO 15	16	20	25	30	32	35	40	45	50	60	2 to 10	15	16	20	25	30	32	35	40	45	50	60
2	1.5	1.5	2.5	2.5	4	4	4	6	6	10	10	1.5	1.5	1.5	2.5	4	4	4	6	10	10	10	16
3	1.5	1.5	2.5	2.5	4	4	6	6	10	10	16	1.5	1.5	2.5	2.5	4	6	6	6	10	10	10	16
4	1.5	1.5	2.5	4	4	6	6	6	10	10	16	1.5	2.5	2.5	2.5	4	6	6	10	10	10	16	16
5	1.5	1.5	2.5	4	4	6	6	10	10	10	16	1.5	2.5	2.5	4	4	6	6	10	10	16	16	16
6	1.5	1.5	2.5	4	4	6	6	10	10	10	16	1.5	2.5	2.5	4	4	6	6	10	10	16	16	16
7	1.5	1.5	2.5	4	4	6	6	10	10	10	16	1.5	2.5	2.5	4	4	6	6	10	10	16	16	16
8	1.5	1.5	2.5	4	4	6	6	10	10	10	16	1.5	2.5	2.5	4	4	6	6	10	10	16	16	16
9	1.5	2.5	2.5	4	6	6	6	10	10	10	16	1.5	2.5	2.5	4	4	6	6	10	10	16	16	25
10	1.5	2.5	2.5	4	6	6	6	10	10	10	16	1.5	2.5	2.5	4	6	6	6	10	10	16	16	25
11	1.5	2.5	2.5	4	6	6	6	10	10	10	16	1.5	2.5	2.5	4	6	6	10	10	16	16	16	25
12	1.5	2.5	2.5	4	6	6	6	10	10	10	16	1.5	2.5	2.5	4	6	6	10	10	16	16	16	25

Note: Where a cable or circuit will not carry more than 30% of the tabulated current rating of its grouped cable size, it can be ignored when counting the number grouped together for the remaining circuits. Where the distance between adjacent conductors or cables is more than twice the diameter of the larger cable, no derating for grouping need be applied. Circuit groupings can be a mixture of three phase and single phase circuits. Where the circuit is not likely to carry overload current, use the full load current I_b of the circuit, and not the protective device size I_n, when sizing the cable from the above table.

GROUPED ARMOURED CABLES CSG 12 BUNCHED XLPE CABLES

Cable sizes required for three-phase and single-phase circuits using multicore armoured copper cable with thermosetting (XLPE) insulation, bunched and clipped direct to a non-metallic surface, not in contact with thermal insulation.

I_n or I_b in amperes, excluding BS 3036 fuses when they are used for overload protection

Single-phase circuits

Size of armoured XLPE cable required in mm².

Number of grouped armoured cables	2 to 10	15	16	20	25	30	32	35	40	45	50	60	63	80
2	1.5	1.5	1.5	1.5	2.5	4	4	4	6	6	10	10	10	16
3	1.5	1.5	1.5	2.5	2.5	4	4	6	6	10	10	16	16	25
4	1.5	1.5	1.5	2.5	4	4	6	6	6	10	10	16	16	25
5	1.5	1.5	1.5	2.5	4	6	6	6	10	10	10	16	16	25
6	1.5	1.5	2.5	2.5	4	6	6	6	10	10	16	16	16	25
7	2.5	2.5	2.5	4	4	6	6	10	10	10	16	25	25	35
8	2.5	2.5	2.5	4	4	6	6	10	10	16	16	25	25	35
9	2.5	2.5	2.5	4	6	6	6	10	10	16	16	25	25	35
10	2.5	2.5	2.5	4	6	6	10	10	16	16	16	25	25	35
11	2.5	2.5	2.5	4	6	10	10	10	16	16	16	25	25	35
12	2.5	2.5	4	4	6	10	10	10	16	16	25	25	25	35
13	2.5	4	4	4	6	10	10	10	16	16	25	25	25	50
14	2.5	4	4	4	6	10	10	10	16	16	25	25	35	50
15	2.5	2.5	4	4	6	10	10	16	16	16	25	25	35	50
16	2.5	4	4	4	6	10	10	16	16	16	25	25	35	50
17	4	4	4	6	10	10	10	16	16	25	25	35	35	50
18	4	4	4	6	10	10	10	16	16	25	25	35	35	50
19	4	4	4	6	10	10	10	16	25	25	25	35	35	50
20	4	4	4	6	10	10	10	16	25	25	25	35	35	50

Three-phase circuits

Size of armoured XLPE cable required in mm².

Number of grouped armoured cables	2 to 6	10	15	16	20	25	30	32	35	40	45	50	60	63	80
2	1.5	1.5	1.5	1.5	2.5	4	4	4	6	6	10	10	16	16	25
3	1.5	1.5	1.5	1.5	2.5	4	6	6	6	10	10	16	16	16	25
4	1.5	1.5	1.5	2.5	2.5	4	6	6	10	10	16	16	16	25	25
5	1.5	1.5	1.5	2.5	4	6	6	10	10	10	16	16	25	25	35
6	1.5	1.5	2.5	2.5	4	6	6	10	10	16	16	16	25	25	35
7	1.5	1.5	2.5	2.5	4	6	10	10	10	16	16	16	25	25	35
8	1.5	1.5	2.5	2.5	4	6	10	10	16	16	25	25	25	25	50
9	1.5	1.5	2.5	4	4	6	10	10	16	16	25	25	25	35	50
10	1.5	1.5	2.5	4	4	6	10	10	16	16	16	25	35	35	50
11	1.5	1.5	2.5	4	6	10	10	16	16	25	25	25	35	35	50
12	1.5	2.5	4	4	6	10	10	16	16	16	25	25	35	35	50
13	1.5	2.5	4	4	6	10	10	16	16	16	25	25	35	35	50
14	1.5	2.5	4	4	6	10	10	16	16	25	25	25	35	35	50
15	1.5	2.5	4	4	6	10	16	16	16	25	25	25	35	35	70
16	1.5	2.5	4	4	6	10	16	16	16	25	25	35	35	35	70
17	1.5	2.5	4	4	6	10	16	16	16	25	25	35	35	50	70
18	1.5	2.5	4	6	10	16	16	16	16	25	25	35	35	50	70
19	1.5	2.5	4	6	10	16	16	16	25	25	35	35	50	50	70
20	1.5	2.5	4	6	10	16	16	16	25	25	35	35	50	50	70

Note:
Where a cable or circuit will not carry more than 30% of the tabulated current rating of its grouped cable size, it can be ignored when counting the number grouped together for the remaining circuits. Where the distance between adjacent conductors or cables is more than twice the diameter of the larger cable, no derating for grouping need be applied. Circuit groupings can be a mixture of three phase and single phase circuits. Where the circuit is not likely to carry overload current, use the full load current I_b of the circuit, and not the protective device size I_n, when sizing the cable from the above table.

GROUPED ARMOURED CABLES CSG 13 XLPE CLIPPED DIRECT

Cable sizes required for three-phase and single-phase circuits using multicore armoured copper cable with thermosetting (XLPE) insulation, clipped direct in a single layer to a non-metallic surface with cable sheaths touching, not in contact with thermal insulation.

I_n or I_b in amperes, excluding BS 3036 fuses when they are used for overload protection

Single-phase circuits

Size of armoured XLPE cable required in mm².

Number of grouped armoured cables	2 to 16	20	25	30	32	35	40	45	50	60	63	80	100	125
2	1.5	1.5	2.5	2.5	4	4	4	6	6	10	10	16	25	35
3	1.5	1.5	2.5	4	4	4	6	6	10	10	16	25	35	35
4	1.5	1.5	2.5	4	4	4	6	6	10	10	16	25	35	35
5	1.5	2.5	2.5	4	4	4	6	6	10	16	16	25	35	35
6	1.5	2.5	2.5	4	4	4	6	10	10	16	16	25	35	35
7	1.5	2.5	2.5	4	4	4	6	10	10	16	16	25	35	35
8	1.5	2.5	2.5	4	4	6	6	10	10	16	16	25	35	35
9	1.5	2.5	2.5	4	4	6	6	10	16	16	16	25	35	35

Three-phase circuits

Size of armoured XLPE cable required in mm².

Number of grouped armoured cables	2 to 16	20	25	30	32	35	40	45	50	60	63	80	100	125	160
2	1.5	2.5	2.5	4	4	4	6	6	10	10	16	16	25	35	70
3	1.5	2.5	4	4	6	6	10	10	10	16	16	25	35	50	70
4	1.5	2.5	4	4	6	6	10	10	10	16	16	25	35	50	70
5	1.5	2.5	4	4	6	6	10	10	10	16	16	25	35	50	70
6	1.5	2.5	4	4	6	6	10	10	10	16	16	25	35	50	70
7	1.5	2.5	4	4	6	6	10	10	10	16	16	25	35	50	70
8	1.5	2.5	4	4	6	6	10	10	10	16	16	25	35	50	70
9	1.5	2.5	4	4	6	6	10	10	10	16	16	25	35	50	70

Note: Where a cable or circuit will not carry more than 30% of the tabulated current rating of its grouped cable size, it can be ignored when counting the number grouped together for the remaining circuits. Where the distance between adjacent conductors or cables is more than twice the diameter of the larger cable, no derating for grouping need be applied. Circuit groupings can be a mixture of three phase and single phase circuits. Where the circuit is not likely to carry overload current, use the full load current I_b of the circuit, and not the protective device size I_n, when sizing the cable from the above table.

GROUPED ARMOURED CABLES   CSG 14   XLPE ON TRAY

Cable sizes required for three-phase and single-phase circuits using multicore armoured copper cable with thermosetting (XLPE) insulation, installed in a single layer on perforated metal cable tray with cable sheaths touching, not in contact with thermal insulation.

I_t or I_b in amperes, excluding BS 3036 fuses when they are used for overload protection

Number of grouped armoured cables	Single-phase circuits														Three-phase circuits														
	Size of armoured XLPE cable required in mm².														Size of armoured XLPE cable required in mm².														
	2 to 20	25	30	32	35	40	45	50	60	63	80	100	125	160	2 to 16	20	25	30	32	35	40	45	50	60	63	80	100	125	160
2	1.5	2.5	2.5	2.5	4	4	6	6	10	10	16	25	25	35	1.5	1.5	2.5	4	4	4	6	6	10	10	10	16	25	35	50
3	1.5	2.5	2.5	4	4	4	6	6	10	10	16	25	35	50	1.5	1.5	2.5	4	4	4	6	6	10	10	10	16	25	35	70
4	1.5	2.5	2.5	4	4	4	6	6	10	10	16	25	35	50	1.5	2.5	2.5	4	4	6	6	10	10	10	16	25	25	50	70
5	1.5	2.5	4	4	4	6	6	10	10	10	16	25	35	50	1.5	2.5	4	4	4	6	6	10	10	16	16	25	35	50	70
6	1.5	2.5	4	4	4	6	6	10	10	10	16	25	35	50	1.5	2.5	4	4	4	6	6	10	10	16	16	25	35	50	70
7	1.5	2.5	4	4	4	6	6	10	10	10	16	25	35	50	1.5	2.5	4	4	4	6	6	10	10	16	16	25	35	50	70
8	1.5	2.5	4	4	4	6	6	10	10	10	16	25	35	50	1.5	2.5	4	4	4	6	6	10	10	16	16	25	35	50	70
9	1.5	2.5	4	4	4	6	6	10	10	10	16	25	35	50	1.5	2.5	4	4	4	6	6	10	10	16	16	25	35	50	70
10	1.5	2.5	4	4	4	6	6	10	10	10	16	25	35	50	1.5	2.5	4	4	6	6	10	10	10	16	16	25	35	50	70
11	1.5	2.5	4	4	4	6	6	10	10	10	16	25	35	50	1.5	2.5	4	4	6	6	10	10	10	16	16	25	35	50	70
12	1.5	2.5	4	4	4	6	6	10	10	10	16	25	35	70	1.5	2.5	4	4	6	6	10	10	10	16	16	25	35	50	70

Note: Where a cable or circuit will not carry more than 30% of the tabulated current rating of its grouped cable size, it can be ignored when counting the number grouped together for the remaining circuits. Where the distance between adjacent conductors or cables is more than twice the diameter of the larger cable, no derating for grouping need be applied.   Circuit groupings can be a mixture of three phase and single phase circuits. Where the circuit is not likely to carry overload current, use the full load current I_b of the circuit, and not the protective device size I_n, when sizing the cable from the above table.

GROUPED MICC CABLES   CSG 15   CLIPPED DIRECT

Cable sizes required for three-phase and single-phase circuits using MICC insulated copper cable PVC sheathed, clipped direct in a single layer to a non-metallic surface with cable sheaths touching.

I_n or I_b in amperes, excluding BS 3036 fuses when they are used for overload protection

Light duty 500V

Number of cables grouped together	One two core cable					One three core cable					One 4 core cable 3 cores loaded					One 4 core cable 4 cores loaded					
	2 to 10	15	16	20	25	2 to 10	15	16	20	25	2 to 10	15	16	20	25	2 to 6	10	15	16	20	25
2	1.0	1.0	1.5	2.5	2.5	1.0	1.5	1.5	2.5	4	1.0	1.5	1.5	2.5	-	1.0	1.0	2.5	2.5	-	-
3	1.0	1.5	1.5	2.5	4	1.0	1.5	2.5	2.5	4	1.0	1.5	2.5	2.5	-	1.0	1.0	2.5	2.5	-	-
4	1.0	1.5	1.5	2.5	4	1.0	2.5	2.5	4	4	1.0	2.5	2.5	-	-	1.0	1.5	2.5	2.5	-	-
5	1.0	1.5	1.5	2.5	4	1.0	2.5	2.5	4	4	1.0	2.5	2.5	-	-	1.0	1.5	2.5	2.5	-	-
6	1.0	1.5	1.5	2.5	4	1.0	2.5	2.5	4	4	1.0	2.5	2.5	-	-	1.0	1.5	2.5	-	-	-
7	1.0	1.5	1.5	2.5	4	1.0	2.5	2.5	4	4	1.0	2.5	2.5	-	-	1.0	1.5	2.5	-	-	-
8	1.0	1.5	1.5	2.5	4	1.0	2.5	2.5	4	-	1.0	2.5	2.5	-	-	1.0	1.5	2.5	-	-	-
9	1.0	1.5	1.5	2.5	4	1.0	2.5	2.5	4	-	1.0	2.5	2.5	-	-	1.0	1.5	2.5	-	-	-

Heavy duty 750V

Number of cables grouped together	One two core cable					One three core cable					One 4 core cable 3 cores loaded					One 4 core cable 4 cores loaded					
	2 to 10	15	16	20	25	2 to 10	15	16	20	25	2 to 10	15	16	20	25	2 to 6	10	15	16	20	25
2	1.0	1.0	1.0	1.5	2.5	1.0	1.5	1.5	2.5	4	1.0	1.5	1.5	2.5	4	1.0	1.0	1.5	2.5	2.5	4
3	1.0	1.5	1.5	2.5	2.5	1.0	1.5	1.5	2.5	4	1.0	1.5	1.5	2.5	4	1.0	1.0	2.5	2.5	4	4
4	1.0	1.5	1.5	2.5	2.5	1.0	1.5	2.5	2.5	4	1.0	1.5	2.5	2.5	4	1.0	1.0	2.5	2.5	4	6
5	1.0	1.5	1.5	2.5	4	1.0	1.5	2.5	2.5	4	1.0	1.5	2.5	2.5	4	1.0	1.0	2.5	2.5	4	6
6	1.0	1.5	1.5	2.5	4	1.0	1.5	2.5	2.5	4	1.0	1.5	2.5	2.5	4	1.0	1.0	2.5	2.5	4	6
7	1.0	1.5	1.5	2.5	4	1.0	1.5	2.5	2.5	4	1.0	1.5	2.5	2.5	4	1.0	1.0	2.5	2.5	4	6
8	1.0	1.5	1.5	2.5	4	1.0	1.5	2.5	4	4	1.0	1.5	2.5	4.0	4	1.0	1.0	2.5	2.5	4	6
9	1.0	1.5	1.5	2.5	4	1.0	2.5	2.5	4	4	1.0	2.5	2.5	4.0	4	1.0	1.0	2.5	2.5	4	6

Note: Where a cable or circuit will not carry more than 30% of the tabulated current rating of its grouped cable size, it can be ignored when counting the number grouped together for the remaining circuits. Where the distance between adjacent conductors or cables is more than twice the diameter of the larger cable, no derating for grouping need be applied. Circuit groupings can be a mixture of three phase and single phase circuits. Where the circuit is not likely to carry overload current, use the full load current I_b of the circuit, and not the protective device size I_n, when sizing the cable from the above table.

GROUPED MICC CABLES

CSG 16

ON TRAY

Cable sizes required for three-phase and single-phase circuits using MICC PVC sheathed cable installed on horizontal or vertical (see notes) perforated metal cable tray with cable sheaths touching.

I_t or I_b in amperes, excluding BS 3036 fuses when they are used for overload protection

Number of cables grouped together	One two core cable					One three core cable					One 4 core cable 3 cores loaded					One 4 core cable 4 cores loaded				
	2 to 10	15	16	20	25	2 to 10	15	16	20	25	2 to 10	15	16	20	25	2 to 10	15	16	20	25
Light duty 500V																				
2	1.0	1.0	1.0	1.5	2.5	1.0	1.5	1.5	2.5	2.5	1.0	1.5	1.5	2.5	2.5	1.0	1.5	1.5	2.5	-
3	1.0	1.0	1.5	1.5	2.5	1.0	1.5	1.5	2.5	4	1.0	1.5	1.5	2.5	-	1.0	2.5	2.5	-	-
4	1.0	1.0*	1.5	1.5*	2.5*	1.0	1.5	1.5*	2.5	4	1.0	1.5	1.5*	2.5	-	1.0	2.5	2.5	-	-
5	1.0	1.0*	1.5	2.5	2.5*	1.0	1.5	1.5*	2.5	4	1.0	1.5	1.5*	2.5	-	1.0	2.5	2.5	-	-
6	1.0	1.5	1.5	2.5	4	1.0	1.5	2.5	2.5	4	1.0	1.5	2.5	2.5	-	1.0	2.5	2.5	-	-
7	1.0	1.5	1.5	2.5	4	1.0	1.5	2.5	2.5	4	1.0	1.5	2.5	2.5	-	1.0	2.5	2.5	-	-
8	1.0	1.5	1.5	2.5	4	1.0	1.5	2.5	2.5	4	1.0	1.5	2.5	2.5	-	1.0	2.5	2.5	-	-
9	1.0	1.5	1.5	2.5	4	1.0	1.5*	2.5	2.5*	4	1.0	1.5*	2.5	2.5†	-	1.0*	2.5	2.5	-	-
Heavy duty 750V																				
2	1.0	1.0	1.0	1.5	2.5	1.0	1	1.5	2.5	2.5	1.0	1.0	1.0	1.5	2.5	1.0	1.5	1.5	2.5	4
3	1.0	1.0	1.0	1.5	2.5	1.0	1.5	1.5	2.5	4	1.0	1.5	1.5	2.5	4	1.0	1.5	2.5	2.5	4
4	1.0	1.0*	1.5*	1.5*	2.5	1.0	1.5	1.5	2.5	4	1.0	1.5	1.5	2.5	4	1.0	1.5	1.5*	2.5	4
5	1.0	1.0*	1.5*	2.5	2.5	1.0	1.5	1.5	2.5	4	1.0	1.5	1.5	2.5	4	1.0	1.5	2.5	2.5	4
6	1.0	1.5	1.5	2.5	2.5	1.0	1.5	1.5	2.5	4	1.0	1.5	1.5	2.5	4	1.0	1.5	2.5	2.5	4
7	1.0	1.5	1.5	2.5	2.5	1.0	1.5	1.5	2.5	4	1.0	1.5	1.5	2.5	4	1.0	1.5	2.5	2.5	4
8	1.0	1.5	1.5	2.5	2.5	1.0	1.5	1.5	2.5	4	1.0	1.5	1.5	2.5	4	1.0	2.5	2.5	2.5	4
9	1.0	1.0*	1.5	2.5	2.5	1.0	1.5	1.5*	2.5	4	1.0	1.5	1.5	2.5	4	1.0	1.5*	2.5	2.5*	4*

Note: Cable sizes marked with * should be increased to the next larger size cable when installed on a vertical cable tray.

Where a cable or circuit will not carry more than 30% of the tabulated current rating of its grouped cable size, it can be ignored when counting the number of cables grouped together for the remaining circuits. Where the distance between adjacent conductors or cables is more than twice the diameter of the larger cable, no derating for grouping need be applied. Circuit groupings can be a mixture of three phase and single phase circuits. Where the circuit is not likely to carry overload current, use the full load current I_b of the circuit, and not the protective device size I_n, when sizing the cable from the above table.

GROUPED ARMOURED CABLES

CSG 17
ALUMINIUM

CLIPPED DIRECT

Cable sizes required for three-phase and single-phase circuits using armoured multicore aluminium PVC insulated cable, clipped direct in a single layer to a non-metallic surface with cable sheaths touching, not in contact with thermal insulation.

I_n or I_b in amperes, excluding BS 3036 fuses when they are used for overload protection

Single-phase circuits

Size of armoured PVC cable required in mm².

Number of grouped armoured cables	2 TO 45	50	60	63	80	100	125	160
2	16	16	25	25	35	50	70	95
3	16	16	25	25	35	50	70	-
4	16	16	25	25	35	70	95	-
5	16	25	25	25	50	70	95	-
6	16	25	25	25	50	70	95	-
7	16	25	25	25	50	70	95	-
8	16	25	25	35	50	70	95	-
9	16	25	25	35	50	70	95	-

Three-phase circuits

Size of armoured PVC cable required in mm².

Number of grouped armoured cables	2 to 40	45	50	60	63	80	100	125	160	200	250
2	16	16	25	25	25	35	70	95	120	185	240
3	16	16	25	25	35	50	70	95	150	185	300
4	16	25	25	35	35	50	70	95	150	240	300
5	16	25	25	35	35	50	70	95	150	240	300
6	16	25	25	35	35	50	70	95	150	240	300
7	16	25	25	35	35	50	70	95	150	240	300
8	16	25	25	35	35	50	70	120	150	240	300
9	16	25	25	35	35	70	70	120	150	240	300

Note: Where a cable or circuit will not carry more than 30% of the tabulated current rating of its grouped cable size, it can be ignored when counting the number of cables grouped together for the remaining circuits. Where the distance between adjacent conductors or cables is more than twice the diameter of the larger cable, no derating for grouping need be applied. Circuit groupings can be a mixture of three phase and single phase circuits. Where the circuit is not likely to carry overload current, use the full load current I_b of the circuit, and not the protective device size I_n, when sizing the cable from the above table.

GROUPED ARMOURED CABLES

CSG 18
ALUMINIUM

Cable sizes required for three-phase and single-phase circuits using armoured multicore aluminium PVC insulated cable, installed in a single layer on perforated metal cable tray with cable sheaths touching, not in contact with thermal insulation.

I_n or I_b in amperes, excluding BS 3036 fuses when they are used for overload protection

Number of grouped armoured cables	Single-phase circuits								Three-phase circuits										
	Size of armoured PVC cable required in mm².								Size of armoured PVC cable required in mm².										
	2 TO 45	50	60	63	80	100	125	160	2 to 40	45	50	60	63	80	100	125	160	200	250
2	16	16	16	25	25	50	70	95	16	16	16	25	25	35	50	70	95	150	240
3	16	16	25	25	35	50	70	95	16	16	25	25	25	35	70	95	120	150	240
4	16	16	25	25	35	50	70	95	16	16	25	25	35	50	70	95	120	185	240
5	16	16	25	25	35	50	70	70	16	16	25	25	35	50	70	95	120	185	240
6	16	16	25	25	35	50	70	-	16	16	25	25	35	50	70	95	150	185	240
7	16	16	25	25	35	50	70	-	16	25	25	35	35	50	70	95	150	185	300
8	16	16	25	25	35	50	70	-	16	25	25	35	35	50	70	95	150	185	300
9	16	16	25	25	35	50	70	-	16	25	25	35	35	50	70	95	150	185	300
10	16	16	25	25	35	70	95	-	16	25	25	35	35	50	70	95	150	185	300
11	16	16	25	25	35	70	95	-	16	25	25	35	35	50	70	95	150	185	300
12	16	25	25	25	35	70	95	-	16	25	25	35	35	50	70	95	150	185	300

Note: Where a cable or circuit will not carry more than 30% of the tabulated current rating of its grouped cable size, it can be ignored when counting the number of cables grouped together for the remaining circuits. Where the distance between adjacent conductors or cables is more than twice the diameter of the larger cable, no derating for grouping need be applied. Circuit groupings can be a mixture of three phase and single phase circuits. Where the circuit is not likely to carry overload current, use the full load current I_b of the circuit, and not the protective device size I_n, when sizing the cable from the above table.

GROUPED ARMOURED CABLES

CSG 19
ALUMINIUM

XLPE CLIPPED DIRECT

Cable sizes required for three-phase and single-phase circuits using armoured multicore aluminium cable with thermosetting (XLPE) insulation, clipped direct in a single layer to a non-metallic surface with cable sheaths touching, not in contact with thermal insulation.

I_n or I_b in amperes, excluding BS 3036 fuses when they are used for overload protection

Number of grouped armoured cables	Single-phase circuits								Three-phase circuits										
	Size of armoured XLPE cable required in mm².								Size of armoured XLPE cable required in mm².										
	2 TO 50	60	63	80	100	125	160	200	2 to 45	50	60	63	80	100	125	160	200	250	315
2	16	16	16	25	35	50	70	95	16	16	16	25	35	50	70	95	120	185	240
3	16	16	16	25	35	50	95	-	16	16	25	25	35	50	70	95	150	185	300
4	16	16	25	25	50	70	95	-	16	16	25	25	35	50	70	95	150	240	300
5	16	25	25	35	50	70	95	-	16	16	25	25	35	50	70	120	150	240	300
6	16	25	25	35	50	70	95	-	16	16	25	25	35	70	70	120	150	240	300
7	16	25	25	35	50	70	95	-	16	16	25	25	35	70	70	120	150	240	300
8	16	25	25	35	50	70	95	-	16	16	25	25	35	70	95	120	150	240	-
9	16	25	25	35	50	70	95	-	16	25	25	25	50	70	95	120	185	240	-

Note: Where a cable or circuit will not carry more than 30% of the tabulated current rating of its grouped cable size, it can be ignored when counting the number of cables grouped together for the remaining circuits. Where the distance between adjacent conductors or cables is more than twice the diameter of the larger cable, no derating for grouping need be applied. Circuit groupings can be a mixture of three phase and single phase circuits. Where the circuit is not likely to carry overload current, use the full load current I_b of the circuit, and not the protective device size I_n, when sizing the cable from the above table.

GROUPED ARMOURED CABLES

CSG 20
ALUMINIUM

XLPE ON TRAY

Cable sizes required for three-phase and single-phase circuits using multicore armoured aluminium cable with thermosetting (XLPE) insulation, installed in a single layer on perforated metal cable tray with cable sheaths touching, not in contact with thermal insulation.

I_n or I_b in amperes, excluding BS 3036 fuses when they are used for overload protection

Number of grouped armoured cables	Single-phase circuits								Three-phase circuits										
	Size of armoured XLPE cable required in mm².								Size of armoured XLPE cable required in mm².										
	2 TO 50	60	63	80	100	125	160	200	2 to 50	60	63	80	100	125	160	200	250	315	355
2	16	16	16	25	35	50	70	95	16	16	16	25	35	70	95	120	150	240	240
3	16	16	16	25	35	50	70	95	16	16	25	35	50	70	95	120	185	240	300
4	16	16	16	25	35	50	70	-	16	25	25	35	50	70	95	120	185	240	300
5	16	16	16	25	35	70	95	-	16	25	25	35	50	70	95	150	185	300	300
6	16	16	16	25	35	70	95	-	16	25	25	35	50	70	95	150	185	300	300
7	16	16	25	25	35	70	95	-	16	25	25	35	50	70	95	150	185	300	300
8	16	16	25	25	35	70	95	-	16	25	25	35	50	70	95	150	185	300	300
9	16	16	25	25	35	70	95	-	16	25	25	35	50	70	95	150	185	300	-
10	16	16	25	35	50	70	95	-	16	25	25	35	50	70	120	150	240	300	-
11	16	16	25	35	50	70	95	-	16	25	25	35	50	70	120	150	240	300	-
12	16	25	25	35	50	70	95	-	16	25	25	35	50	70	120	150	240	300	-

Note: Where a cable or circuit will not carry more than 30% of the tabulated current rating of its grouped cable size, it can be ignored when counting the number of cables grouped together for the remaining circuits. Where the distance between adjacent conductors or cables is more than twice the diameter of the larger cable, no derating for grouping need be applied. Circuit groupings can be a mixture of three phase and single phase circuits. Where the circuit is not likely to carry overload current, use the full load current I_b of the circuit, and not the protective device size I_n, when sizing the cable from the above table..

GROUPED MULTICORE CABLES CSG 21 BUNCHED

OVERLOAD PROTECTION BY BS 3036 FUSES

Cable sizes required for three-phase and single-phase circuits using PVC insulated multicore copper cable or twin and CPC copper cable, bunched and clipped direct to a non-metallic surface, not in contact with thermal insulation.

Rating of BS 3036 protective device in amperes when BS 3036 fuse is used for overload protection

Number of grouped multicore cables	Single-phase circuits									Three-phase circuits								
	Size of PVC multicore cable required in mm².									Size of PVC multicore cable required in mm².								
	5	10	15	20	30	45	60	80	100	5	10	15	20	30	45	60	80	100
2	1.0	1.5	2.5	4	10	16	25	35	70	1.0	1.5	4	6	10	25	35	50	70
3	1.0	2.5	4	6	10	25	35	50	70	1.0	2.5	4	6	16	25	35	70	95
4	1.0	2.5	4	6	16	25	35	70	70	1.0	2.5	4	10	16	25	50	70	95
5	1.0	2.5	4	6	16	25	35	70	95	1.0	2.5	6	10	16	35	50	70	120
6	1.0	2.5	6	10	16	25	50	70	95	1.0	4	6	10	16	35	70	95	120
7	1.0	2.5	6	10	16	35	50	70	95	1.0	4	6	10	25	35	70	95	120
8	1.0	2.5	6	10	16	35	50	70	120	1.0	4	6	10	25	50	70	95	150
9	1.0	4	6	10	16	35	50	95	120	1.5	4	10	10	25	50	70	95	150
10	1.0	4	6	10	25	35	70	95	120	1.5	4	10	16	25	50	70	120	150
11	1.0	4	6	10	25	35	70	95	120	1.5	4	10	16	25	50	70	120	150
12	1.5	4	6	10	25	35	70	95	150	1.5	4	10	16	25	50	70	120	185
13	1.5	4	10	10	25	50	70	95	150	1.5	4	10	16	25	50	95	120	185
14	1.5	4	10	16	25	50	70	95	150	1.5	4	10	16	35	70	95	120	185
15	1.5	4	10	16	25	50	70	120	150	1.5	6	10	16	35	70	95	150	185
16	1.5	4	10	16	25	50	70	120	150	1.5	6	10	16	35	70	95	150	185
17	1.5	4	10	16	25	50	70	120	185	1.5	6	10	16	35	70	95	150	240
18	1.5	4	10	16	25	50	70	120	185	2.5	6	10	16	35	70	95	150	240
19	1.5	6	10	16	25	50	95	120	185	2.5	6	10	16	35	70	95	150	240
20	1.5	6	10	16	25	50	95	120	185	2.5	6	10	16	35	70	95	150	240

Note: Where a cable or circuit will not carry more than 30% of the tabulated current rating of its grouped cable size, it can be ignored when counting the number grouped together for the remaining circuits. Where the distance between adjacent conductors or cables is more than twice the diameter of the larger cable, no derating for grouping need be applied. Circuit groupings can be a mixture of three phase and single phase circuits.

GROUPED MULTICORE CABLES   CSG 22   CLIPPED DIRECT

OVERLOAD PROTECTION BY BS 3036 FUSES

Cable sizes required for three-phase and single-phase circuits using PVC insulated multicore copper cable or twin and CPC copper cable, clipped direct in a single layer to a non-metallic surface with cable sheaths touching, not in contact with thermal insulation.

Rating of BS 3036 protective device in amperes, when BS 3036 fuse is used for overload protection

Number of grouped multicore cables	Single-phase circuits									Three-phase circuits								
	5	10	15	20	30	45	60	80	100	5	10	15	20	30	45	60	80	100
	Size of PVC multicore cable required in mm².									Size of PVC multicore cable required in mm².								
2	1.0	1.5	2.5	4	10	16	25	35	50	1.0	1.5	4.0	6	10	16	35	50	70
3	1.0	1.5	2.5	4	10	16	25	50	70	1.0	1.5	4.0	6	10	25	35	50	70
4	1.0	1.5	4.0	6	10	16	25	50	70	1.0	2.5	4.0	6	10	25	35	70	70
5	1.0	1.5	4.0	6	10	16	35	50	70	1.0	2.5	4.0	6	10	25	35	70	95
6	1.0	1.5	4.0	6	10	25	35	50	70	1.0	2.5	4.0	6	16	25	35	70	95
7	1.0	1.5	4.0	6	10	25	35	50	70	1.0	2.5	4.0	6	16	25	35	70	95
8	1.0	1.5	4.0	6	10	25	35	50	70	1.0	2.5	4.0	6	16	25	35	70	95
9	1.0	2.5	4.0	6	10	25	35	50	70	1.0	2.5	4.0	6	16	25	35	70	95

Note: Where a cable or circuit will not carry more than 30% of the tabulated current rating of its grouped cable size, it can be ignored when counting the number grouped together for the remaining circuits. Where the distance between adjacent conductors or cables is more than twice the diameter of the larger cable, no derating for grouping need be applied. Circuit groupings can be a mixture of three phase and single phase circuits.

GROUPED MULTICORE CABLES   CSG 23   CABLE TRAY

OVERLOAD PROTECTION BY BS 3036 FUSES

Cable sizes required for three-phase and single-phase circuits using PVC insulated multicore copper cable or twin and CPC copper cable, installed in a single layer on perforated metal cable tray with cable sheaths touching, not in contact with thermal insulation.

Rating of BS 3036 protective device in amperes, when BS 3036 fuse is used for overload protection

Number of grouped multicore cables	Single-phase circuits									Three-phase circuits								
	Size of PVC multicore cable required in mm².									Size of PVC multicore cable required in mm².								
	5	10	15	20	30	45	60	80	100	5	10	15	20	30	45	60	80	100
2	1.0	1.0	2.5	4	6	16	25	35	50	1.0	1.5	2.5	4	10	16	25	50	70
3	1.0	1.0	2.5	4	6	16	25	35	50	1.0	1.5	4.0	4	10	16	35	50	70
4	1.0	1.5	2.5	4	10	16	25	35	50	1.0	1.5	4.0	6	10	25	35	50	70
5	1.0	1.5	2.5	4	10	16	25	35	70	1.0	1.5	4.0	6	10	25	35	50	70
6	1.0	1.5	2.5	4	10	16	25	50	70	1.0	2.5	4.0	6	10	25	35	50	70
7	1.0	1.5	2.5	4	10	16	25	50	70	1.0	2.5	4.0	6	10	25	35	50	70
8	1.0	1.5	2.5	4	10	16	25	50	70	1.0	2.5	4.0	6	10	25	35	50	70
9	1.0	1.5	2.5	4	10	16	25	50	70	1.0	2.5	4.0	6	10	25	35	50	70
10	1.0	1.5	2.5	4	10	16	25	50	70	1.0	2.5	4.0	6	10	25	35	70	70
11	1.0	1.5	2.5	4	10	16	25	50	70	1.0	2.5	4.0	6	10	25	35	70	70
12	1.0	1.5	2.5	4	10	16	25	50	70	1.0	2.5	4.0	6	10	25	35	70	95

Note: Where a cable or circuit will not carry more than 30% of the tabulated current rating of its grouped cable size, it can be ignored when counting the number grouped together for the remaining circuits. Where the distance between adjacent conductors or cables is more than twice the diameter of the larger cable, no derating for grouping need be applied. Circuit groupings can be a mixture of three phase and single phase circuits.

GROUPED MULTICORE CABLES   CSG 24   ENCLOSED

OVERLOAD PROTECTION BY BS 3036 FUSES

Cable sizes required for three-phase and single-phase circuits using PVC insulated multicore copper cable or twin and CPC copper cable, grouped and enclosed in conduit or trunking, not in contact with thermal insulation.

Rating of BS 3036 protective device in amperes, when BS 3036 fuse is used for overload protection

Single-phase circuits

Size of PVC multicore cable required in mm².

Number of grouped multicore cables	5	10	15	20	30	45	60	80	100
2	1.0	2.5	4	6	10	25	35	70	95
3	1.0	2.5	4	10	16	25	50	70	95
4	1.0	2.5	6	10	16	35	50	95	120
5	1.0	2.5	6	10	16	35	70	95	120
6	1.0	4	6	10	25	35	70	95	150
7	1.0	4	10	10	25	50	70	120	150
8	1.5	4	10	16	25	50	70	120	185
9	1.5	4	10	16	25	50	70	120	185
10	1.5	4	10	16	25	50	95	120	185
11	1.5	4	10	16	25	70	95	150	240
12	1.5	6	10	16	35	70	95	150	240
13	1.5	6	10	16	35	70	95	150	240
14	1.5	6	10	16	35	70	95	150	240
15	1.5	6	10	16	35	70	95	185	240
16	2.5	6	10	16	35	70	120	185	240
17	2.5	6	10	16	35	70	120	185	300
18	2.5	6	16	25	35	70	120	185	300
19	2.5	6	16	25	35	70	120	185	300
20	2.5	6	16	25	35	70	120	185	300

Three-phase circuits

Size of PVC multicore cable required in mm².

Number of grouped multicore cables	5	10	15	20	30	45	60	80	100
2	1.0	2.5	4	10	16	25	50	70	95
3	1.0	2.5	6	10	16	35	70	95	120
4	1.0	4	6	10	25	35	70	95	150
5	1.0	4	10	10	25	50	70	120	185
6	1.5	4	10	16	25	50	70	120	185
7	1.5	4	10	16	25	50	95	120	240
8	1.5	4	10	16	25	70	95	150	240
9	1.5	6	10	16	35	70	95	150	240
10	1.5	6	10	16	35	70	95	185	240
11	1.5	6	10	16	35	70	95	185	240
12	2.5	6	10	16	35	70	120	185	300
13	2.5	6	16	25	35	70	120	185	300
14	2.5	6	16	25	35	70	120	240	300
15	2.5	6	16	25	35	70	120	240	300
16	2.5	6	16	25	50	95	120	240	300
17	2.5	10	16	25	50	95	150	240	400
18	2.5	10	16	25	50	95	150	240	400
19	2.5	10	16	25	50	95	150	240	400
20	2.5	10	16	25	50	95	150	240	400

Note: Where a cable or circuit will not carry more than 30% of the tabulated current rating of its grouped cable size, it can be ignored when counting the number grouped together for the remaining circuits. Where the distance between adjacent conductors or cables is more than twice the diameter of the larger cable, no derating for grouping need be applied.   Circuit groupings can be a mixture of three phase and single phase circuits.

CABLE CAPACITY OF CONDUIT

Multiply the quantity of each size of cable by the cable factor from Table 1; add the total factors obtained for each cable size together, then select a factor from the appropriate column of Table 2 which factor is equal to, or greater than, the total of the cable factors. The conduit size required is given at the top of each column of factors.

Table 1

Cable size mm²	Cable factor
1.0	16
1.5	22
2.5	30
4.0	43
6.0	58
10	105
16	145
25	218
35	275
50	383

Table 2

Conduit length in metres between draw-in boxes	Select column for straight or number of bends between draw in boxes																		
	Straight runs					One bend					Two bends				Three bends			
	Conduit size mm					Conduit size mm					Conduit size mm				Conduit size mm			
	16	20	25	32	38	16	20	25	32	38	16	20	25	32	16	20	25	32
1	207	329	571	1000	1234	188	303	543	947	1128	177	286	514	900	158	256	463	818
2	207	329	571	1000	1234	177	286	514	900	1072	158	256	463	818	130	213	388	692
3	207	329	571	1000	1234	167	270	487	857	1020	143	233	422	750	111	182	333	600
4	177	286	514	900	1110	158	256	463	818	973	130	213	388	692	97	159	292	529
5	171	278	500	878	1082	150	244	442	783	931	120	196	356	643	86	141	260	474
6	167	270	487	857	1056	143	233	422	750	891	111	182	333	600				
7	162	263	475	837	1030	136	222	404	720	855	103	169	311	563				
8	158	256	463	818	1005	130	213	388	692	822	97	159	292	529				
9	154	250	452	800	980	125	204	373	667	792	91	149	275	500				
10	150	244	442	783	956	120	196	358	643	763	86	141	260	474				

Note: The 38mm column is for 1½ inch conduit.

CCT 1

CABLE CAPACITY OF TRUNKING

Multiply each size of cable to be used by the factor for the cable size from table 1; add together the total obtained for each size of cable, then compare the total with the factors for trunking from table 2, selecting a trunking factor equal to, or greater than, the total factor obtained for the cables.

Table 1

Cable factors

Cable Type	Cable size mm^2	Factor
Solid	1.5	7.1
"	2.5	10.2
Stranded	1.5	8.1
"	2.5	11.4
"	4	15.2
"	6	22.9
"	10	36.3
"	16	50.3
"	25	75.4
"	35	95.0
"	50	132.7
"	70	176.7
"	95	227.0
"	120	284.0
"	150	346.0

Table 2

Trunking factors

Trunking size mm	Factor
75 x 25	738
50 x 37.5	767
100 x 25	993
50 x 50	1037
75 x 37.5	1146
100 x 37.5	1542
75 x 50	1555
100 x 50	2091
75 x 75	2371
150 x 50	3161
100 x 75	3189
100 x 100	4252
150 x 75	4787
150 x 100	6414
150 x 150	9575

PART FOUR

ELECTRICAL DESIGN TABLES

SPACE FOR NOTES OR AMENDMENTS:

Contents Part 4

K factors

Circuit breaker characteristics

Motor currents and comparative fuse sizes

Miscellaneous formula and tables

Resistance and impedance

Resistor colour code

Rewirable fuses

Temperature conversion

Thermal insulation

Thermal capacity of cable armour

Transformer impedance

DERATING FACTORS GF 1 GROUPED CABLES

Derating factors for two or more circuits using single core cable, or two or more multicore cables, single-phase or three-phase circuits grouped separately or mixed together.

Method of installing circuits or cables		Number of circuits or multicore cables grouped together																		
		2	3	4	5	6	7	8	9	10	11	12	13	14	15	16	17	18	19	20
Installed in trunking or conduit		0.80	0.70	0.65	0.60	0.57	0.54	0.52	0.50	0.48	0.465	0.45	0.44	0.43	0.42	0.41	0.40	0.39	0.385	0.38
Bunched together & clipped direct to a non-metallic surface		0.80	0.70	0.65	0.60	0.57	0.54	0.52	0.50	0.48	0.465	0.45	0.44	0.43	0.42	0.41	0.40	0.39	0.385	0.38
Single layer clipped direct to, or lying on a non-metallic surface, cable sheaths touching		0.85	0.79	0.75	0.73	0.72	0.72	0.71	0.70	0.70	0.70	0.70	0.70	0.70	0.70	0.70	0.70	0.70	0.70	0.70
Single layer clipped direct to, or lying on a non-metallic surface, with one cable diameter between cables		0.94	0.9	0.9	0.9	0.9	0.9	0.9	0.9	0.9	0.9	0.9	0.9	0.9	0.9	0.9	0.9	0.9	0.9	0.9
Single layer multicore cable on perforated metal cable tray, horizontal or vertical with sheaths touching *		0.86	0.81	0.77	0.75	0.74	0.73	0.73	0.72	0.71	0.705	0.70	0.70	0.70	0.70	0.70	0.70	0.70	0.70	0.70
Single layer multicore cable on perforated metal cable tray, horizontal or vertical with one cable diameter between cables *		0.91	0.89	0.88	0.87	0.87	0.87	0.87	0.87	0.87	0.87	0.87	0.87	0.87	0.87	0.87	0.87	0.87	0.87	0.87
Single layer multicore cable on ladder support with cable sheaths touching		0.86	0.82	0.80	0.79	0.78	0.78	0.78	0.77	0.77	0.77	0.77	0.77	0.77	0.77	0.77	0.77	0.77	0.77	0.77
Single layer, single core cable on perforated metal cable tray cable sheaths touching.	Horizontal	0.90	0.85	0.85	0.85	0.85	0.85	0.85	0.85	0.85	0.85	0.85	0.85	0.85	0.85	0.85	0.85	0.85	0.85	0.85
	Vertical	0.85	0.85	0.85	0.85	0.85	0.85	0.85	0.85	0.85	0.85	0.85	0.85	0.85	0.85	0.85	0.85	0.85	0.85	0.85
Micc cables on perforated cable tray cables sheaths touching	Horizontal	0.9	0.8	0.8	0.775	0.75	0.75	0.75	0.75	0.75	0.75	0.75	0.75	0.75	0.75	0.75	0.75	0.75	0.75	0.75
	Vertical	0.9	0.8	0.75	0.75	0.75	0.75	0.75	0.70	0.70	0.70	0.70	0.70	0.70	0.70	0.70	0.70	0.70	0.70	0.70
Multicore cable on ladder sheaths touching		0.86	0.82	0.8	0.79	0.78	0.78	0.78	0.77	0.77	0.77	0.77	0.77	0.77	0.77	0.77	0.77	0.77	0.77	0.77

* Excluding micc cable

Note: Where a cable or circuit will not carry more than 30% of the tabulated current rating of its grouped cable size, it can be ignored when counting the number of cables grouped together for the remaining circuits. Where the distance between adjacent conductors or cables is more than twice the diameter of the larger cable, no derating for grouping need be applied. Circuit groupings can be a mixture of three-phase and single-phase circuits.

GDF 1   GROUPED CABLES

NON-SIMULTANEOUS OVERLOAD

Graphical determination of formula to use when grouped circuits are not subject to simultaneous overload

For all other types of protective device

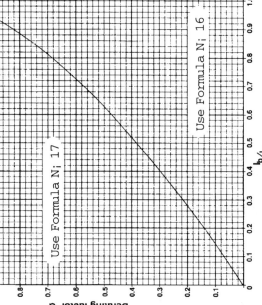

Use Formula Nᵢ 17

Use Formula Nᵢ 16

I_b/I_n

Derating factor 'G'

Notes: Compare grouping factor G against I_b/I_n to see which side of the line the point will be, then use the appropriate formula. If the point falls very close to, or on, the line use both formula to determine which has the higher I_t.

For BS 3036 fuses only

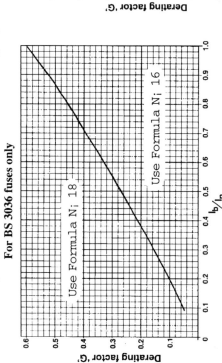

Use Formula Nᵢ 18

Use Formula Nᵢ 16

I_b/I_n

Derating factor 'G'

Formula 16 $I_{t1} \leq \dfrac{I_b}{G \times A \times T}$

Formula 17 $I_{t2} \geq \dfrac{1}{A \times T} \sqrt{I_n^2 + 0.48 I_b^2 \left(\dfrac{1 - G^2}{G^2}\right)}$

Formula 18 $I_{t3} \geq \dfrac{1}{A \times T} \sqrt{1.9 I_n^2 + 0.48 I_b^2 \left(\dfrac{1 - G^2}{G^2}\right)}$

FULL LOAD CURRENT (FLC.) MC1 HRC FUSE RATINGS

Full load currents for three-phase induction motors at average efficiencies, power factor, with recommended fuse ratings

Motor full load current (FLC) in amperes								Direct on line start (DOL) (See notes)				Assisted Start (See notes)		
								Motor FLC amperes		HRC fuse	Motor fuse	Motor FLC ampere		HRC fuse
KW	H.P.	220V	240V	380V	415V	440V	550V	From	To	Type 'gG'	Type 'gM'	From	To	Type 'gG'
0.37	0.50	2.00	1.75	1.15	1.05	1.00	0.80	0	0.70	2	-	0	1.40	2
0.55	0.75	2.70	2.60	1.60	1.50	1.40	1.10	0.80	1.40	4	-	1.50	2.10	4
0.75	1.0	3.90	3.30	2.30	2.00	1.90	1.50	1.50	2.00	6	-	2.20	3.10	6
1.10	1.5	4.70	4.35	2.80	2.50	2.40	1.90	2.10	3.00	10	-	3.20	5.50	10
1.50	2.0	6.50	5.90	3.80	3.50	3.30	2.60	3.10	6.10	16	-	5.60	10.0	16
2.20	3.0	9.30	8.30	5.40	5.00	4.70	3.80	6.20	9.00	20	-	10.1	14.0	20
3.00	4.0	12	11.0	7.10	6.50	6.10	4.90	9.10	11.00	25	20M25	14.1	18.0	25
4.00	5.5	15.4	14.0	9.00	8.40	7.90	6.40	11.1	14.4	32	20M32	18.1	22.0	32
5.50	7.5	20.7	20.0	11.9	11	10.3	8.2	14.5	15.4	35	32M35	22.1	28.0	35
7.50	10	28	25.0	16.1	14.4	14	11.2	15.5	18	40	32M40	28.1	32.0	40
11.0	15	39.1	36.5	23	21	19.8	15.8	18.1	22	50	32M50	32.1	40.0	50
15.0	20	52.8	48	30.5	28	26.4	21.1	22.1	28	63	32M63	40.1	51.0	63
18.5	25	66	60	38	35	33	26.4	28.1	45	80	63M80	51.1	80.0	80
22	30	77	72	45	41	39	31	45.1	58	100	63M100	80.1	100	100
30	40	103	96	60	55	52	42	58.1	80	125	100M125	100.1	125	125
37	50	128	120	75	69	65	52	80.1	99	160	100M160	125.1	160	160
45	60	151	144	87	80	75	60	99.1	128	200	-	160.1	200	200
55	75	185	171	107	98	92	74	128.1	180	250	200M250	200.1	250	250
75	100	257	235	148	136	128	102	180.1	216	315	200M315	250.1	315	315
90	120	308	278	180	164	154	123	216.1	270	355	315M355	315.1	355	355
110	150	370	345	214	196	185	148	270.1	328	400	-	355.1	400	400
132	175	426	403	247	226	213	170	328.1	385	450	400M450	400.1	450	450
150	200	500	454	292	268	252	202	385.1	430	500	-	450.1	500	500

Assumed starting conditions D.O.L up to 1 kW = 5 x FLC for 5 secs: 1.1 kW to 7.5 kW = 6 x FLC for 10 secs: 7.6 kW to 75 kW = 7 x FLC for 10 secs: over 75 kW = 6 x FLC for 15 secs Assisted start up to 1 kW = 2.5 x FLC for 20 secs: over 1 kW = 3.5 x FLC for 20 secs.

H.R.C. FUSES

HRC 1

COMPARATIVE TYPES

Industrial HRC fuse comparative types

BS 88 size reference N°	Hawker		GEC	MEM	Lawson	Fluvent	Reyrolle Belmos	Old GEC
-	D19L	STD	LST	LS	LST	-	RLST	-
A1	F21	NITD	NIT	SA2	NIT	FOBX	RIT	Y-PA/Y-QA
A2	H07	AAO	TIA	SB3	TIA	FOMX	GPE-6E/TIA	Y-RA
A3	K07	BAO	TIS	SB4	TIS	FOSX	GPE-6/RIS	Y-SA
A4	L14	CEO	TCP	SD5	TCP	FO1X	GPE-6F/RCP	Y-TA
-	M14	DEO	TFP	SD6	TFP	F763/2X	GPG-6F/RFP	Y-VAD
-	K08	AC/BC	TB	SE4	TB	754/M-754/S	GPE-5E/RC	Y-SB
-	K09	AD-BD	TBC	SF4	TBC	755/M-755/S	GPE-5F/RC	Y-TB
B1	L09	CD	TC	SF5	TC	FB1	GPF-5F/RC	Y-TB
B2	M09	DD	TF	SF6	TF	FB2	GPG-G/RF	Y-VB
B3	N09	ED	TKF	SF7	TFK	FB3	GPH-5G/RKF	Y-WB
B4	P09	ED	TMF	SF8	TMF	FB4	GPH-5G/RMF	Y-XBD
-	N11	EFS	TKM	SG7	TKM	F772/3	RKM	Y-WBL
C1	P11	EF	TM	SH8	TM	F773/4	GPH-5H/RM	Y-XB
-	P20	ES	85TM	SX8	85TM	-	-	Y-XA
-	R20	FS	86TT	SX9	86TT	-	-	Y-YA
C2	R11	FF	TTM	SH9	TTM	TB6	GPJ-5H/RTM	Y-YB
-	R12	FG	TT	SY9	TT	F781-6	GPH5-5J	Y-YBL
C3	S11	GF	TLM	SH10	TLM	FB6-F851/8	GPK-5H/RLM	Y-ZBD
-	S12	GG	TLT	SY10	TLT	F781-6/F781-8	GPH-SJ	Y-Z8
D1	U44	GH	TXU	SJ11	-	-	-	-
E1	DO4	SSD	SS	SS	SS	-	RSS	-
E1	FO6	NSD	NS	SN2/SN21	NS	-	RS	-
F2	ESD	ESD	ES	-	MES	F986B	-	Y-PF/Y-QF
-	K07R	OSD	OS	-	-	-	-	-

Note: This list is given for guidance only, since identical performance for the comparative types shown cannot be guaranteed. Under normal circumstances satisfactory interchangeability should be achieved.

RESISTANCE & IMPEDANCE   RA 1   ALUMINIUM CABLES

Resistance, reactance and impedance of solid aluminium conductors for two, three and four core armoured cables, ohms per 1000 metres

Aluminium conductors PVC insulated to BS 6346

Cable Size mm²	Temperature 20 °C			Temperature 70 °C			Temperature 95 °C			Temperature 115 °C		
	R	X	Z	R	X	Z	R	X	Z	R	X	Z
16	1.910	-	1.910	2.292	-	2.292	2.483	-	2.483	2.636	-	2.636
25	1.200	-	1.200	1.440	-	1.440	1.560	-	1.560	1.656	-	1.656
35	0.868	0.082	0.872	1.042	0.082	1.045	1.128	0.082	1.131	1.198	0.082	1.201
50	0.641	0.082	0.646	0.769	0.082	0.774	0.833	0.082	0.837	0.885	0.082	0.888
70	0.443	0.079	0.450	0.532	0.079	0.537	0.576	0.079	0.581	0.611	0.079	0.616
95	0.320	0.078	0.329	0.384	0.078	0.392	0.416	0.078	0.423	0.442	0.078	0.448
120	0.253	0.077	0.264	0.304	0.077	0.313	0.329	0.077	0.338	0.349	0.077	0.358
150	0.206	0.077	0.220	0.247	0.077	0.259	0.268	0.077	0.279	0.284	0.077	0.295
185	0.164	0.077	0.181	0.197	0.077	0.211	0.213	0.077	0.227	0.226	0.077	0.239
240	0.125	0.076	0.146	0.150	0.076	0.168	0.163	0.076	0.179	0.173	0.076	0.189
300	0.100	0.076	0.126	0.120	0.076	0.142	0.130	0.076	0.151	0.138	0.076	0.158

Aluminium conductors XLPE insulated to BS 5467

Cable Size mm²	Temperature 20 °C			Temperature 90 °C			Temperature 140 °C			Temperature 170 °C		
	R	X	Z	R	X	Z	R	X	Z	R	X	Z
16	1.910	-	1.910	2.445	-	2.445	2.827	-	2.827	3.056	-	3.056
25	1.200	-	1.200	1.536	-	1.536	1.776	-	1.776	1.920	-	1.920
35	0.868	0.077	0.871	1.111	0.077	1.114	1.285	0.077	1.287	1.389	0.077	1.391
50	0.641	0.077	0.646	0.820	0.077	0.824	0.949	0.077	0.952	1.026	0.077	1.028
70	0.443	0.075	0.449	0.567	0.075	0.572	0.656	0.075	0.660	0.709	0.075	0.713
95	0.320	0.073	0.328	0.410	0.073	0.416	0.474	0.073	0.479	0.512	0.073	0.517
120	0.253	0.073	0.263	0.324	0.073	0.332	0.374	0.073	0.381	0.405	0.073	0.411
150	0.206	0.074	0.219	0.264	0.074	0.274	0.305	0.074	0.314	0.330	0.074	0.338
185	0.164	0.074	0.180	0.210	0.074	0.223	0.243	0.074	0.254	0.262	0.074	0.273
240	0.125	0.073	0.145	0.160	0.073	0.176	0.185	0.073	0.199	0.200	0.073	0.213
300	0.100	0.072	0.123	0.128	0.072	0.147	0.148	0.072	0.165	0.160	0.072	0.175

Reactance can be ignored for cables larger than 35 mm². The temperatures of 95 °C and 140 °C are for CPCs or earthing conductors installed in an ambient temperature of 30 °C separate from live conductors. The temperature of 20 °C is used for calculating the maximum fault current for the withstand capacity of equipment. The temperatures of 115 °C and 170 °C are for calculating the minimum fault current when determining the withstand capacity of cables.

RESISTANCE & IMPEDANCE   **RA 2**   ALUMINIUM SINGLE CORE CABLE

Resistance, reactance and impedance of single core cables installed in trefoil formation ohms per 1000 metres

Three single core solid aluminium conductors, PVC insulated, aluminium wire armoured and PVC oversheath to BS 6346

Cable Size mm²	Temperature 20 °C			Temperature 70 °C			Temperature 95 °C			Temperature 115 °C		
	R	X	Z	R	X	Z	R	X	Z	R	X	Z
50	0.641	0.112	0.6507	0.769	0.112	0.7773	0.833	0.112	0.8408	0.885	0.112	0.8916
70	0.443	0.107	0.4557	0.532	0.107	0.5423	0.576	0.107	0.5858	0.611	0.107	0.6206
95	0.320	0.103	0.3362	0.384	0.103	0.3976	0.416	0.103	0.4286	0.442	0.103	0.4535
120	0.253	0.103	0.2732	0.304	0.103	0.3206	0.329	0.103	0.3447	0.349	0.103	0.3640
150	0.206	0.101	0.2294	0.247	0.101	0.2670	0.268	0.101	0.2862	0.284	0.101	0.3017
185	0.164	0.099	0.1916	0.197	0.099	0.2203	0.213	0.099	0.2351	0.226	0.099	0.2470
240	0.125	0.096	0.1576	0.150	0.096	0.1781	0.163	0.096	0.1887	0.173	0.096	0.1974
300	0.100	0.094	0.1372	0.120	0.094	0.1524	0.130	0.094	0.1604	0.138	0.094	0.1670
380	0.080	0.094	0.1234	0.096	0.094	0.1344	0.104	0.094	0.1402	0.110	0.094	0.1450
480	0.0633	0.092	0.1117	0.076	0.092	0.1193	0.082	0.092	0.1234	0.087	0.092	0.1269
600	0.0515	0.089	0.1028	0.062	0.089	0.1084	0.067	0.089	0.1114	0.071	0.089	0.1139
740	0.0410	0.089	0.0980	0.049	0.089	0.1017	0.053	0.089	0.1037	0.057	0.089	0.1055
960	0.0313	0.087	0.0925	0.038	0.087	0.0948	0.041	0.087	0.0960	0.043	0.087	0.0971
1200	0.0250	0.085	0.0886	0.030	0.085	0.0901	0.033	0.085	0.0910	0.035	0.085	0.0917

Three single core solid aluminium conductors XLPE insulation, aluminium wire armoured, PVC oversheath to BS 5467

Cable Size mm²	Temperature 20 °C			Temperature 90 °C			Temperature 140 °C			Temperature 170 °C		
	R	X	Z	R	X	Z	R	X	Z	R	X	Z
50	0.641	0.107	0.6499	0.820	0.107	0.8274	0.949	0.107	0.9547	1.026	0.107	1.0312
70	0.443	0.104	0.4550	0.567	0.104	0.5765	0.656	0.104	0.6638	0.709	0.104	0.7164
120	0.253	0.097	0.2710	0.324	0.097	0.3381	0.374	0.097	0.3868	0.405	0.097	0.4163
150	0.206	0.097	0.2277	0.264	0.097	0.2810	0.305	0.097	0.3199	0.330	0.097	0.3436
185	0.164	0.097	0.1908	0.210	0.097	0.2316	0.243	0.097	0.2618	0.263	0.097	0.2802
240	0.125	0.093	0.1558	0.160	0.093	0.1851	0.185	0.093	0.2071	0.200	0.093	0.2206
300	0.100	0.091	0.1352	0.128	0.091	0.1571	0.148	0.091	0.1737	0.160	0.091	0.1841
380	0.080	0.091	0.1212	0.102	0.091	0.1370	0.118	0.091	0.1493	0.128	0.091	0.1571
480	0.063	0.089	0.1092	0.081	0.089	0.1204	0.094	0.089	0.1292	0.101	0.089	0.1348
480	0.0633	0.086	0.1067	0.081	0.086	0.1181	0.094	0.086	0.1271	0.101	0.086	0.1328
600	0.0515	0.087	0.1011	0.066	0.087	0.1092	0.076	0.087	0.1157	0.082	0.087	0.1198
740	0.0410	0.087	0.0962	0.052	0.087	0.1016	0.061	0.087	0.1061	0.066	0.087	0.1090
960	0.0313	0.085	0.0906	0.040	0.085	0.0940	0.046	0.085	0.0968	0.050	0.085	0.0987
1200	0.0250	0.083	0.0867	0.032	0.083	0.0890	0.037	0.083	0.0909	0.040	0.083	0.0921

Notes: 1)   For practical purposes the above values can be used for unarmoured single core cables although there will be a difference in the values of reactance X.
         2)   Similarly, where single core cables are spaced one cable diameter apart the reactance values given above should be increased by about 0.06 Ω.

Resistance and impedance of PVC insulated copper conductors for single core cables or two, three and four core steel wire armoured cables

Resistance, reactance and impedance per 1000 metres for copper cable at different temperatures

Cable Size mm²	Temperature 20 °C			Temperature 70 °C			Temperature 95 °C			Temperature 115 °C		
	R	X	Z	R	X	Z	R	X	Z	R	X	Z
1.0	18.10	-	18.100	21.720	-	21.720	23.530	-	23.530	24.978	-	24.978
1.5	12.10	-	12.100	14.520	-	14.520	15.730	-	15.730	16.698	-	16.698
2.5	7.410	-	7.410	8.892	-	8.892	9.633	-	9.633	10.226	-	10.226
4.0	4.610	-	4.610	5.532	-	5.532	5.993	-	5.993	6.362	-	6.362
6.0	3.080	-	3.080	3.696	-	3.696	4.004	-	4.004	4.250	-	4.250
10	1.830	-	1.830	2.196	-	2.196	2.379	-	2.379	2.525	-	2.525
16	1.150	-	1.150	1.380	-	1.380	1.495	-	1.495	1.587	-	1.587
25	0.727	-	0.727	0.872	-	0.872	0.945	-	0.945	1.003	-	1.003
35	0.524	0.081	0.530	0.629	0.081	0.634	0.681	0.081	0.686	0.723	0.081	0.728
50	0.387	0.081	0.395	0.464	0.081	0.471	0.503	0.081	0.510	0.534	0.081	0.540
70	0.268	0.079	0.279	0.322	0.079	0.331	0.348	0.079	0.357	0.370	0.079	0.378
95	0.193	0.077	0.208	0.232	0.077	0.244	0.251	0.077	0.262	0.266	0.077	0.277
120	0.153	0.076	0.171	0.184	0.076	0.199	0.199	0.076	0.213	0.211	0.076	0.224
150	0.124	0.076	0.145	0.149	0.076	0.167	0.161	0.076	0.178	0.171	0.076	0.187
185	0.0991	0.076	0.1249	0.119	0.076	0.1411	0.129	0.076	0.1496	0.1368	0.076	0.1565
240	0.0754	0.075	0.1063	0.090	0.075	0.1175	0.098	0.075	0.1234	0.1041	0.075	0.1283
300	0.0601	0.075	0.0961	0.072	0.075	0.1040	0.078	0.075	0.1083	0.0829	0.075	0.1118
400	0.0470	0.074	0.0877	0.056	0.074	0.0930	0.061	0.074	0.0960	0.0649	0.074	0.0984

Notes: Reactance only has to be taken into account on cables larger than 35 mm². The temperature of 95 °C = (30 °C + 160 °C) ÷ 2 and is used when the conductor is used as a CPC, or earthing conductor, at an ambient temperature of 30 °C and not installed with other live conductors. The temperature 115 °C = (70 °C + 160 °C) ÷ 2 is used when the conductor is carrying its full rated current, or the conductor's temperature is 70 °C.

RESISTANCE & IMPEDANCE   RC 2   COPPER SINGLE CORE CABLE

Resistance, reactance and impedance of single core cables installed in trefoil formation ohms per 1000 metres

Three single core stranded copper conductors, PVC insulated, aluminium wire armoured and PVC oversheath to BS 6346

Cable Size mm²	Temperature 20 °C			Temperature 70 °C			Temperature 95 °C			Temperature 115 °C		
	R	X	Z	R	X	Z	R	X	Z	R	X	Z
50	0.387	0.109	0.4021	0.464	0.109	0.4770	0.503	0.109	0.5148	0.534	0.109	0.5451
70	0.268	0.104	0.2875	0.322	0.104	0.3380	0.348	0.104	0.3636	0.370	0.104	0.3842
95	0.193	0.100	0.2174	0.232	0.100	0.2523	0.251	0.100	0.2701	0.266	0.100	0.28<5
120	0.153	0.099	0.1822	0.184	0.099	0.2086	0.199	0.099	0.2222	0.211	0.099	0.2332
150	0.124	0.097	0.1574	0.149	0.097	0.1776	0.161	0.097	0.1881	0.171	0.097	0.1967
185	0.099	0.095	0.1373	0.119	0.095	0.1522	0.129	0.095	0.1601	0.137	0.095	0.1665
240	0.0754	0.093	0.1197	0.090	0.093	0.1298	0.098	0.093	0.1351	0.104	0.093	0.1396
300	0.0601	0.091	0.1091	0.072	0.091	0.1161	0.078	0.091	0.1199	0.083	0.091	0.1231
400	0.0470	0.091	0.1024	0.056	0.091	0.1071	0.061	0.091	0.1096	0.065	0.091	0.1117
500	0.0366	0.089	0.0962	0.044	0.089	0.0992	0.048	0.089	0.1009	0.051	0.089	0.1023
630	0.0283	0.086	0.0905	0.034	0.086	0.0925	0.037	0.086	0.0935	0.039	0.086	0.0945
800	0.0221	0.086	0.0888	0.027	0.086	0.0900	0.029	0.086	0.0907	0.030	0.086	0.0912
1000	0.0176	0.089	0.0907	0.021	0.089	0.0915	0.023	0.089	0.0919	0.024	0.089	0.092?

Three single core stranded copper conductors XLPE insulation, aluminium wire armoured, PVC oversheath to BS 5467

Cable Size mm²	Temperature 20 °C			Temperature 90 °C			Temperature 140 °C			Temperature 170 °C		
	R	X	Z	R	X	Z	R	X	Z	R	X	Z
50	0.387	0.106	0.4013	0.495	0.106	0.5066	0.573	0.106	0.5825	0.619	0.106	0.6282
70	0.268	0.103	0.2871	0.343	0.103	0.3582	0.397	0.103	0.4098	0.429	0.103	0.4410
95	0.193	0.098	0.2165	0.247	0.098	0.2658	0.286	0.098	0.3020	0.309	0.098	0.3240
120	0.153	0.096	0.1806	0.196	0.096	0.2181	0.226	0.096	0.2459	0.245	0.096	0.2630
150	0.124	0.096	0.1568	0.159	0.096	0.1855	0.184	0.096	0.2071	0.198	0.096	0.2204
185	0.099	0.095	0.1373	0.127	0.095	0.1585	0.147	0.095	0.1747	0.159	0.095	0.1848
240	0.0754	0.092	0.1190	0.097	0.092	0.1333	0.112	0.092	0.1446	0.121	0.092	0.1517
300	0.0601	0.090	0.1082	0.077	0.090	0.1184	0.089	0.090	0.1265	0.096	0.090	0.1317
400	0.0470	0.089	0.1006	0.060	0.089	0.1074	0.070	0.089	0.1130	0.075	0.089	0.1165
500	0.0366	0.087	0.0944	0.047	0.087	0.0988	0.054	0.087	0.1025	0.059	0.087	0.1049
630	0.0283	0.085	0.0896	0.036	0.085	0.0924	0.042	0.085	0.0948	0.045	0.085	0.0963
800	0.0221	0.085	0.0878	0.028	0.085	0.0896	0.033	0.085	0.0911	0.035	0.085	0.0921
1000	0.0176	0.084	0.0858	0.023	0.084	0.0870	0.026	0.084	0.0879	0.028	0.084	0.0886

Notes:
1) For practical purposes the above values can be used for unarmoured single core cables although there will be a difference in the values of reactance X.
2) Similarly, where single core cables are spaced one cable diameter apart the reactance values given above should be increased by about 0.06 Ω.

RESISTANCE & IMPEDANCE   **R XLPE 1**   XLPE COPPER CABLES

Resistance and impedance of XLPE insulated copper conductors for single core cable or two, three, and four core armoured cables

Resistance, reactance and impedance per 1000 metres at different temperatures

Cable size in sq. mm.	Temperature 20 °C			Temperature 90 °C			Temperature 140 °C			Temperature 170 °C		
	R	X	Z	R	X	Z	R	X	Z	R	X	Z
1	18.1	-	18.1	23.168	-	23.168	26.788	-	26.788	28.960	-	28.960
1.5	12.1	-	12.1	15.488	-	15.488	17.908	-	17.908	19.360	-	19.360
2.5	7.41	-	7.41	9.485	-	9.485	10.967	-	10.967	11.856	-	11.856
4	4.61	-	4.61	5.901	-	5.901	6.823	-	6.823	7.376	-	7.376
6	3.08	-	3.08	3.942	-	3.942	4.558	-	4.558	4.928	-	4.928
10	1.83	-	1.83	2.342	-	2.342	2.708	-	2.708	2.928	-	2.928
16	1.15	-	1.15	1.472	-	1.472	1.702	-	1.702	1.840	-	1.840
25	0.727	-	0.727	0.931	-	0.931	1.076	-	1.076	1.163	-	1.163
35	0.524	0.077	0.530	0.671	0.077	0.675	0.776	0.077	0.776	0.838	0.077	0.842
50	0.387	0.076	0.394	0.495	0.076	0.501	0.573	0.076	0.578	0.619	0.076	0.624
70	0.268	0.075	0.278	0.343	0.075	0.351	0.397	0.075	0.404	0.429	0.075	0.435
95	0.193	0.073	0.206	0.247	0.073	0.258	0.286	0.073	0.295	0.309	0.073	0.317
120	0.153	0.072	0.169	0.196	0.072	0.209	0.226	0.072	0.238	0.245	0.072	0.255
150	0.124	0.073	0.144	0.159	0.073	0.175	0.184	0.073	0.198	0.198	0.073	0.211
185	0.0991	0.073	0.1231	0.127	0.073	0.1464	0.147	0.073	0.1638	0.1586	0.073	0.1746
240	0.0754	0.072	0.1043	0.097	0.072	0.1204	0.112	0.072	0.1328	0.1206	0.072	0.1405
300	0.0601	0.072	0.0938	0.077	0.072	0.1054	0.089	0.072	0.1144	0.0962	0.072	0.1201

Notes: Reactance only needs to be taken into consideration with cables larger than 35 mm^2. The temperature 140 °C = (30 °C + 250 °C) ÷ 2 is used when the conductor is used as a CPC, or earthing conductor, and the ambient temperature is 30 °C and the conductor is not installed with other live conductors. The temperature 170 °C = (90 °C + 250 °C) ÷ 2 is used when the conductor is carrying its full rated current, or the conductor's temperature is 90 °C.

CONDUIT - TRUNKING ZCT 1

Values of impedance to be used when calculating phase earth loop impedance Z_S when conduit or trunking is used as the circuit protective conductor

IMPEDANCE OF CONDUIT in Ω per metre

Conduit size in mm	For fault currents up to 100A				For fault currents over 100A			
	Light gauge		Heavy gauge		Light gauge		Heavy gauge	
	Conduit	With joints	Conduit	With joints	Conduit	With joints	Conduit	With joints
16	0.0078	0.00975	0.0076	0.0095	0.005	0.00625	0.0038	0.00475
20	0.0054	0.00675	0.0047	0.005875	0.004	0.005	0.0025	0.003125
25	0.0035	0.004375	0.0032	0.004	0.0022	0.00275	0.0017	0.002125
32	0.0024	0.003	0.002	0.0025	0.0015	0.001875	0.0011	0.001375

IMPEDANCE OF HEAVY GAUGE TRUNKING WITHOUT LID in Ω per metre

Trunking size			
50 mm × 50 mm	0.00345 Ω	100 mm × 100 mm	0.00123 Ω
75 mm × 75 mm	0.00194 Ω	150 mm × 150 mm	0.00074 Ω

Notes: The impedance of trunking behaves in a similar manner to conduit with a fault current flowing, except the reduction in impedance does not occur until the fault current exceeds 200A; for practical purposes only the worst conditions are required for calculations, so it was decided that the lower impedances were not necessary. As far as conduit is concerned, joints should not contribute any additional impedance to the circuit; the above values being given in case the designer wishes to make an allowance for deterioration.

Values of 'k' for live and protective conductors with various types of insulation and with various initial and final conductor temperatures, for conductors up to 300 mm²

Type of conductor insulation	Initial temperature at start of fault °C	Limit temperature of conductor insulation °C	Live conductors, or protective conductors in a cable or bunched with live conductors		Protective conductors installed separately and not grouped or bunched with other conductors		Protective conductor as the sheath or armour of a cable			Protective conductor is conduit, trunking, or steel ducting
			Copper	Aluminium	Copper	Aluminium	Aluminium	Steel	Lead	Steel
70 °C P.V.C.	30	160	-	-	143	95	-	-	-	-
	50	160	-	-	-	-	-	-	-	47
	60	200	-	-	-	-	93	51	26	-
	70	160	115	76	-	-	-	-	-	-
85 °C P.V.C.	30	160	-	-	143	95	-	-	-	-
	58	160	-	-	-	-	-	-	-	45
	75	200	-	-	-	-	87	48	24	-
	85	160	104	69	-	-	-	-	-	-
90 °C P.V.C.	30	160	-	-	143	95	-	-	-	-
	60	160	-	-	-	-	-	-	-	44
	80	200	-	-	-	-	85	46	23	-
	90	160	100	66	-	-	-	-	-	-
60 °C Rubber	30	200	-	-	159	105	-	-	-	-
	60	200	141	93	-	-	-	-	-	-

Notes: The 'k' factor should be chosen by selecting the initial temperature of the conductor at the start of the fault and the limit temperature of the conductor's insulation.
Guide to initial temperatures:

Assuming that live conductors are carrying their tabulated current-carrying capacity from the CR tables, the initial temperature will be the maximum permitted operating temperature of the conductor as given in the CR tables. Where protective conductors are installed with live conductors the initial temperature of the protective conductor will be the same as the live conductors. Where a protective conductor is not installed in contact with live conductors, its initial temperature will be ambient temperature (30 °C). The initial temperature of the metal cable sheath or armour of a cable will be approximately 10 °C lower than the live conductors when they are carrying their tabulated current : i.e., for pvc cable, the initial temperature of the armour will be 70 - 10 = 60 °C

'K' FACTORS

K 1B

DIFFERENT MATERIALS

Values of 'k' for live and protective conductors with various types of insulation and with various initial and final conductor temperatures, for conductors up to 300 mm²

Type of conductor insulation	Initial temperature at start of fault °C	Limit temperature of conductor insulation °C	Live conductors, or protective conductors in a cable or bunched with live conductors		Protective conductors installed separately and not grouped or bunched with other conductors		Protective conductor as the sheath or armour of a cable			Protective conductor conduit, trunking, or steel ducting
			Copper	Aluminium	Copper	Aluminium	Aluminium	Steel	Lead	Steel
85 °C Rubber	30	220	-	-	166	110	-	-	-	-
	58	220	-	-	-	-	-	-	-	54
	75	220	-	-	-	-	93	51	26	-
	85	220	134	89	-	-	-	-	-	-
90 °C Thermosetting XLPE	30	250	-	-	176	116	-	-	-	-
	60	250	-	-	-	-	-	-	-	58
	80	200	-	-	-	-	85	46	23	-
	90	250	143	94	-	-	-	-	-	-
Paper cable	70	160	-	-	-	-	76	42	21	-
	80	160	108	71	-	-	-	-	-	-
MICC PVC	70	160	115	-	-	-	-	-	-	-
MICC bare	105	250	135	-	-	-	-	-	-	-

Notes: The 'k' factor should be chosen by selecting the initial temperature of the conductor at the start of the fault and the limit temperature of the conductor's insulation. Guide to initial temperatures:

Assuming that live conductors are carrying their tabulated current-carrying capacity from the CR tables, the initial temperature will be the maximum permitted operating temperature of the conductor as given in the CR tables. Where protective conductors are installed with live conductors the initial temperature of the protective conductor will be the same as the live conductors. Where a protective conductor is not installed in contact with live conductors, its initial temperature will be ambient temperature (30 °C). The initial temperature of the metal cable sheath or armour of a cable will be approximately 10 °C lower than the live conductors when they are carrying their tabulated current : i.e., for pvc cable, the initial temperature of the armour will be 70 - 10 = 60 °C

CABLES TO TRANSFORMERS   TC 1   COPPER CABLES

Cable sizes to be installed between switchgear and transformers when using closed and open cable trenches

Cables enclosed in a concrete trench minimum internal depth 300 mm (including 100 mm thick trench cover); minimum internal width 450 mm.

Open concrete trench; no cover — Cable diameter D_e.

Transformer rating	Maximum demand	50 mm between cables Non-armoured single core.		50 mm between cables Steel armoured multicore.	50 mm between cables. Non-armoured multicore	Cable diameter space between phases. Non-armoured single core	
kVA	A	Phase mm²	Neutral mm²	Cable size mm²	Cable size mm²	Phase mm²	Neutral mm²
PVC CABLES COPPER CONDUCTORS BS 6346							
200	279	3 x 120	1 x 120	One 4 core x 150	One 4 core x 185	3 x 120	1 x 120
315	438	3 x 240	1 x 240	One 4 core x 400	One 4 core x 400	3 x 240	1 x 240
500	696	3 x 500	1 x 185	Two 4 core x 300	Two 4 core x 400	3 x 500	1 x 500
800	1113	6 x 500	1 x 500			6 x 500	1 x 500
1000	1391	6 x 800	1 x 800			6 x 630	1 x 630
1500	2087	9 x 800	2 x 800			9 x 630	2 x 1000
2000	2783	12 x 1000	2 x 1000			12 x 1000	2 x 1000
XLPE CABLES COPPER CONDUCTORS BS 5467							
200	279	3 x 70	1 x 70	One 4 core x 120	One 4 core x 120	3 x 95	1 x 95
315	438	3 x 150	1 x 150	One 4 core x 240	One 4 core x 240	3 x 185	1 x 185
500	696	3 x 400	1 x 400	Two 4 core x 240	Two 4 core x 240	3 x 400	1 x 400
800	1113	3 x 1000	1 x 1000	Three 4 core x 300	Three 4 core x 240	3 x 800	1 x 800
1000	1391	6 x 500	1 x 500			6 x 400	1 x 400
1500	2087	9 x 630	2 x 800			6 x 800	1 x 800
2000	2783	9 x 1000	2 x 1000			9 x 630	2 x 630

Note: Cable sizes based on the three-phase load being balanced so that the current in the neutral can be ignored. Cable spacing must be strictly observed. Where XLPE cables are used the manufacturer of the transformer and switchgear must be consulted since the conductors operate at a temperature exceeding 70 °C. Due to the short cable length required between switchgear and transformer it is more economic to use the phase conductor size for the neutral.

456

CABLES TO TRANSFORMERS   TC 2   ALUMINIUM CABLES

Cable sizes to be installed to transformers when using closed and open cable trenches

Concrete trench: Cables enclosed in a concrete trench minimum internal depth 300 mm (including 100 mm thick trench cover); minimum internal width 450 mm.

Open concrete trench; no cover. Cable diameter D_e.

		Non-armoured single core (50 mm between cables)		Steel armoured multicore (50 mm between cables)	Non-armoured multicore (50 mm between cables)	Non-armoured single core (open trench)	
Transformer rating kVA	Maximum demand A	Phase mm²	Neutral mm²	Cable size mm²	Cable size mm²	Phase mm²	Neutral mm²

PVC CABLES ALUMINIUM CONDUCTORS BS 6346

kVA	A	Phase mm²	Neutral mm²	Steel armoured (mm²)	Non-armoured (mm²)	Phase mm²	Neutral mm²
200	279	3 x 185	1 x 185	One 4 core x 240	One 4 core x 300	3 x 185	1 x 185
315	438	3 x 380	1 x 380	Two 4 core x 240	Two 4 core x 240	3 x 380	1 x 380
500	696	3 x 960	1 x 960	Three 4 core x 300	Three 4 core x 300	3 x 740	1 x 740
800	1113	6 x 960	1 x 960			6 x 600	1 x 630
1000	1391	6 x 1200	1 x 1200			6 x 960	1 x 950
1500	2087	12 x 960	2 x 960			9 x 960	2 x 960
2000	2783						

XLPE CABLES ALUMINIUM CONDUCTORS BS 5467

kVA	A	Phase mm²	Neutral mm²	Steel armoured (mm²)	Non-armoured (mm²)	Phase mm²	Neutral mm²
200	279	3 x 120	1 x 120	One 4 core x 185	One 4 core x 185	3 x 150	1 x 150
315	438	3 x 240	1 x 240	Two 4 core x 150	Two 4 core x 150	3 x 240	1 x 240
500	696	3 x 740	1 x 740	Three 4 core x 240	Three 4 core x 240	3 x 600	1 x 600
800	1113	3 x 1200	1 x 1200			3 x 1200	1 x 1200
1000	1391	6 x 960	1 x 960			6 x 600	1 x 600
1500	2087	9 x 960	2 x 960			6 x 1200	1 x 1200
2000	2783	12 x 960	2 x 960			9 x 960	2 x 960

Note: Cable sizes based on the three-phase load being balanced so that the current in the neutral can be ignored. Cable spacing must be strictly observed. Where XLPE cables are used the manufacturer of the transformer and switchgear must be consulted since the conductors operate at a temperature exceeding 70 °C. Due to the short cable length required between switchgear and transformer it is more economic to use the phase conductor size for the neutral.

THERMAL CAPACITY   TCCA 1   CABLE ARMOUR
Steel wire armoured cables whose armour complies with the thermal requirements of table 54G

70 °C PVC copper conductors to BS 6346

Phase conductor	Area required	Cross-sectional area of steel armour					
		2 core	3 core	4 core	5 core	7 core	10 core
1.5	3.38	15	16	17	19	20	36
2.5	5.64	17	19	20	22	24	44
4	9.02	21	23	35	39	42	72
6	13.53	24	36	40			
10	22.55	41	44	49			
16	36.08	46	50	72			
25	36.08	60	66	76			
35	36.08	66	74	84			
50	56.37	74	84	122			
70	78.92	84	119	138			
95	107.11	122	138	160			
120	135.29	131	150	220			
150	169.12	144	211	240			
185	208.58	201	230	265			
240	270.59	225	260	299			
300	338.24	250	289	333			

Shaded area ☐

Cables not suitable

70 °C PVC aluminium conductors to BS 6346

Phase conductor	Area required	2 core	3 core	4 core
25	23.84	54	62	70
35	23.84	58	68	78
50	37.25	66	78	113
70	52.16	74	113	128
95	70.78	109	128	147
120	89.41	-	138	201
150	111.76	-	191	220
185	137.84	-	215	245
240	178.82	-	240	274
300	223.53	-	265	304

90 °C XLPE copper conductors to BS 5467

Phase conductor	Area required	Cross-sectional area of steel armour		
		2 core	3 core	4 core
1.5	4.66	16	17	18
2.5	7.77	17	19	20
4	12.43	19	21	23
6	18.65	22	23	36
10	31.09	26	39	43
16	49.74	41	44	49
25	49.74	42	62	70
35	49.74	62	70	80
50	77.72	68	78	90
70	108.80	80	90	131
95	147.66	113	128	147
120	186.52	125	141	206
150	233.15	138	201	230
185	287.55	191	220	255
240	373.04	215	250	289
300	466.30	235	269	319

90 °C XLPE aluminium conductors to BS 5467

Phase conductor	Area required	2 core	3 core	4 core
25	32.70	38	58	66
35	32.70	54	64	72
50	51.09	60	72	82
70	71.52	70	84	122
95	97.07	100	119	135
120	122.61		131	191
150	153.26		181	211
185	189.02		206	235
240	245.22		230	265
300	306.52		250	289

Shaded area ☐

Cables not suitable

ELI 1

Earth loop impedance for PVC and XLPE insulated copper conductor steel wire armoured cables to BS 6346, BS 5467 & BS 6742

Cable type		PVC insulated cables to BS 6346 copper conductors			XLPE insulated cables to BS 5467 and BS6742 copper conductors		
		Maximum resistance of armour Ω / 1000 m	Earth fault loop impedance per 1000 metres	Gross area of armour	Maximum resistance of armour Ω / 1000 m	Earth fault loop impedance per 1000 metres	Gross area of armour
Cores	Size mm^2.	Design Ω	Design Ω	mm^2	Design Ω	Design Ω	mm^2
2	1.5	14.26	30.96	15	14.35	33.71	16.0
3	1.5	12.91	29.61	16	12.87	32.23	17.0
4	1.5	11.72	28.42	17	11.60	30.96	18.0
5	1.5	10.37	27.06	19	-	-	-
7	1.5	9.75	26.45	20	-	-	-
10	1.5	5.43	22.13	36	-	-	-
2	2.5	12.13	22.36	17	13.43	25.29	17.0
3	2.5	11.14	21.36	19	11.59	23.45	19.0
4	2.5	9.75	19.97	20	10.51	22.36	20.0
5	2.5	8.88	19.11	22	-	-	-
7	2.5	8.39	18.62	24	-	-	-
10	2.5	4.57	14.79	44	-	-	-
2	4	10.00	16.36	21	12.06	19.44	19.0
3	4	8.86	15.22	23	10.61	17.98	21.0
4	4	5.68	12.04	35	9.28	16.65	23.0
5	4	5.06	11.42	39	-	-	-
7	4	4.81	11.17	42	-	-	-
10	4	2.71	9.08	72	-	-	-
2	6	9.06	13.31	24	10.69	15.61	22.0
3	6	5.82	10.07	36	9.33	14.26	23.0
4	6	5.06	9.31	40	5.87	10.80	36.0
2	10	5.20	7.72	41	9.16	12.09	26.0
3	10	4.68	7.21	44	5.66	8.58	39.0
4	10	4.20	6.72	49	5.05	7.98	43.0
2	16	4.67	6.25	46	5.80	7.64	41.0
3	16	4.05	5.64	50	5.09	6.93	44.0
4	16	2.71	4.30	72	4.37	6.21	49.0

Note: Impedance of armour is based on the final temperature which the armour can be expected to reach with fault current flowing. Earth fault loop impedance includes the impedance of one phase conductor and the armour impedance.

ELI 2

Earth loop impedance for BS 6346 : 1969 steel wire armoured p.v.c. insulated cables with copper and aluminium conductors

Cable		Steel wire armouring stranded copper conductors			Steel wire armouring solid aluminium conductors		
Number of cores	Cable size in mm².	Design value armour impedance Ω / 1000 m	Gross area of armour mm².	Design value earth loop impedance Ω / 1000 m.	Design value armour impedance Ω / 1000 m	Gross area of armour mm².	Design value earth loop impedance Ω / 1000 m.
1	2	3	4	5	6	7	8
2	16	4.666	46	6.25	4.932	42	7.57
3	16	4.050	50	5.64	4.303	46	6.94
4	16	2.715	72	4.30	2.962	66	5.60
2	25	3.466	60	4.47	3.866	54	5.52
3	25	3.037	66	4.04	3.164	62	4.82
4	25	2.591	76	3.59	2.838	70	4.49
2	35	3.213	66	3.94	3.612	58	4.81
3	35	2.674	74	3.40	2.926	68	4.12
4	35	2.364	84	3.09	2.486	78	3.68
2	50	2.815	74	3.36	3.213	66	4.10
3	50	2.423	84	2.96	2.549	78	3.44
4	50	1.632	122	2.17	1.753	113	2.64
2	70	2.550	84	2.93	2.815	74	3.43
3	70	1.797	119	2.17	1.797	113	2.41
4	70	1.511	138	1.89	1.632	128	2.25
2	95	1.759	122	2.04	2.022	109	2.47
3	95	1.548	138	1.82	1.672	128	2.12
4	95	1.246	160	1.52	1.390	147	1.84
2	120	1.627	131	1.85	-	-	-
3	120	1.424	150	1.65	1.548	138	1.91
4	120	0.926	220	1.15	1.008	201	1.37
2	150	1.497	144	1.68	-	-	-
3	150	0.983	211	1.17	1.080	191	1.38
4	150	0.856	240	1.04	0.926	220	1.22
2	185	1.082	201	1.24	-	-	-
3	185	0.911	230	1.07	0.971	215	1.21
4	185	0.787	265	0.94	0.845	245	1.08
2	240	0.967	225	1.10	-	-	-
3	240	0.816	260	0.94	0.876	240	1.06
4	240	0.708	299	0.84	0.765	274	0.95
2	300	0.892	250	1.00	-	-	-
3	300	0.746	289	0.86	0.805	265	0.96
4	300	0.653	333	0.76	0.708	304	0.87

Note: Impedance of armour is based on the final temperature which the armour can be expected to reach with fault current flowing. Earth fault loop impedance includes the impedance of one phase conductor and the armour impedance.

ELI 3

Earth loop impedance for BS 5467 steel wire armoured XLPE insulated cables with copper and aluminium conductors

Cable		Steel wire armouring stranded copper conductors			Steel wire armouring solid aluminium conductors		
Number of cores	Cable size in mm².	Design value armour impedance Ω / 1000 m.	Gross area of armour in mm²	Design value earth loop impedance Ω / 1000 m.	Design value armour impedance Ω / 1000 m.	Gross area of armour in mm².	Design value earth loop impedance Ω / 1000 m.
1	2	3	4	5	6	7	8
2	16	5.801	41	7.641	5.953	40	9.009
3	16	5.090	44	6.930	5.232	42	8.288
4	16	4.366	49	6.206	4.639	46	7.695
2	25	5.648	42	6.811	6.259	38	8.179
3	25	3.535	62	4.698	3.818	58	5.738
4	25	3.138	70	4.302	3.275	66	5.195
2	35	3.874	62	4.712	4.437	54	5.826
3	35	3.266	70	4.104	3.548	64	4.937
4	35	2.745	80	3.584	3.017	72	4.406
2	50	3.524	68	4.148	3.980	60	5.008
3	50	2.844	78	3.468	3.125	72	4.153
4	50	2.474	90	3.098	2.610	82	3.638
2	70	3.068	80	3.503	3.524	70	4.237
3	70	2.563	90	2.998	2.703	84	3.416
4	70	1.665	131	2.100	1.799	122	2.512
2	95	2.158	113	2.475	2.461	100	2.978
3	95	1.863	128	2.180	2.002	119	2.519
4	95	1.531	147	1.848	1.665	135	2.182
2	120	2.007	125	2.262	-	-	-
3	120	1.723	141	1.978	1.723	131	2.134
4	120	1.080	206	1.335	1.158	191	1.569
2	150	1.856	138	2.067	-	-	-
3	150	1.143	201	1.354	1.252	181	1.590
4	150	0.975	230	1.186	1.053	211	1.391
2	185	1.287	191	1.462	-	-	-
3	185	1.048	220	1.222	1.116	206	1.389
4	185	0.885	255	1.059	0.962	235	1.235
2	240	1.154	215	1.295	-	-	-
3	240	0.940	250	1.080	1.007	230	1.220
4	240	0.796	289	0.936	0.859	265	1.072
2	300	1.066	235	1.186	-	-	-
3	300	0.873	269	0.993	0.940	250	1.115
4	300	0.733	319	0.853	0.796	289	0.971

Note: Impedance of armour is based on the final temperature which the armour can be expected to reach with fault current flowing. Earth fault loop impedance includes the impedance of one phase conductor and the armour impedance.

ELI 4

Earth loop impedance for aluminium strip armoured solidal aluminium PVC insulated cables to BS 6346 and XLPE insulated cables to BS 5467

Number of cores	Cable size in mm².	PVC insulated solidal aluminium strip aluminium armoured cables			XLPE insulated solidal aluminium strip aluminium armoured cables		
		Design value armour impedance Ω / 1000 m.	Gross area of armour in mm².	Design value earth loop impedance Ω / 1000 m.	Design value armour impedance Ω / 1000 m.	Gross area of armour in mm².	Design value earth loop impedance Ω / 1000 m.
1	2	3	4	5	6	7	8
2	16	1.944	23	9.800	2.349	22	5.405
3	16	1.730	26	4.366	2.052	23	5.108
4	16	1.450	29	4.086	1.854	26	4.910
2	25	2.074	22	3.730	2.496	20	4.416
3	25	1.607	27	3.263	1.915	24	3.835
4	25	1.450	30	3.106	1.589	29	3.509
2	35	1.816	24	3.014	2.203	23	3.592
3	35	1.486	30	2.684	1.780	27	3.169
4	35	1.332	34	2.530	1.459	31	2.848
2	50	1.557	29	2.446	2.056	26	3.084
3	50	0.709	60	1.598	1.507	31	2.535
4	50	0.623	67	1.512	0.721	63	1.749
2	70	0.781	56	1.397	0.942	53	1.655
3	70	0.635	67	1.251	0.743	63	1.456
4	70	0.539	78	1.155	0.615	74	1.328
2	95	0.703	63	1.152	0.839	60	1.356
3	95	0.415	104	0.864	0.661	70	1.178
4	95	0.349	123	0.798	0.394	117	0.911
3	120	0.366	117	0.723	0.431	110	0.842
4	120	0.314	136	0.671	0.367	130	0.778
3	150	0.342	130	0.636	0.391	123	0.729
4	150	0.290	149	0.584	0.329	143	0.667
3	185	0.306	143	0.545	0.350	136	0.623
4	185	0.211	214	0.450	0.290	162	0.563
3	240	0.211	214	0.400	0.310	156	0.523
4	240	0.188	248	0.377	0.215	237	0.428
3	300	0.200	237	0.358	0.218	226	0.393
4	300	0.177	271	0.335	0.191	260	0.366

Note: Impedance of armour is based on the final temperature which the armour can be expected to reach with fault current flowing. Earth fault loop impedance includes the impedance of one phase conductor and the armour impedance.

ELI 5

Characteristics of mineral insulated copper sheathed cables giving effective sheath area and earth loop impedance when carrying fault current

Cable		Resistance in ohms per 1000 metres					
		Light duty 500 V cable			Heavy duty 750 V cable		
		Earth loop impedance		Effective sheath area in sq.mm.	Earth loop impedance		Effective sheath area in sq.mm.
Number of cores	Size sq. mm.	Exposed to touch, bare or covered	NOT exposed to touch and bare		Exposed to touch, bare or covered	NOT exposed to touch and bare	
1	2	3	4	5	6	7	8
4 x 1	1H6	-	-	-	5.1	5.33	31
4 x 1	1H10	-	-	-	3.2	3.34	38
4 x 1	1H16	-	-	-	2.11	2.21	46
4 x 1	1H25	-	-	-	1.4	1.46	60
4 x 1	1H35	-	-	-	1.05	1.10	71
4 x 1	1H50	-	-	-	0.77	0.807	88
4 x 1	1H70	-	-	-	0.57	0.598	109
4 x 1	1H95	-	-	-	0.44	0.465	130
4 x 1	1H120	-	-	-	0.36	0.382	150
4 x 1	1H150	-	-	-	0.30	0.316	175
2	1.0	30.7	32.0	5.4	-	-	-
2	1.5	21.2	22.1	6.3	19.8	20.7	11
2	2.5	13.3	13.9	8.2	12.4	12.8	13
2	4.0	8.64	9.03	10.7	8.04	8.39	16
2	6.0	-	-	-	5.59	5.86	18
2	10	-	-	-	3.56	3.73	24
2	16	-	-	-	2.36	2.48	30
2	25	-	-	-	1.62	1.70	38
3	1.0	30.0	31.3	6.7	-	-	-
3	1.5	20.6	21.4	7.8	19.6	20.4	12
3	2.5	12.9	13.5	9.5	12.2	12.7	14
3	4.0	-	-	-	7.92	8.25	17
3	6.0	-	-	-	5.49	5.74	20
3	10	-	-	-	3.46	3.63	27
3	16	-	-	-	2.29	2.40	34
3	25	-	-	-	1.57	1.64	42
4	1.0	29.6	30.7	7.7	-	-	-
4	1.5	20.2	21.1	9.1	19.4	20.2	14
4	2.5	12.6	13.1	11.3	12.0	12.6	16
4	4.0	-	-	-	7.74	8.08	20
4	6.0	-	-	-	5.35	5.59	24
4	10	-	-	-	3.36	3.53	30
4	16	-	-	-	2.20	2.31	39
4	25	-	-	-	1.50	1.57	49
7	1.0	28.9	30.1	10.2	-	-	-
7	1.5	19.7	20.5	11.8	19.0	19.8	18
7	2.5	12.1	12.6	15.4	11.7	12.2	22
12	2.5	-	-	-	11.4	11.8	34
19	1.5	-	-	-	18.5	18.9	37

ELI 6

Phase earth loop impedance of twin and CPC cables and single core cables enclosed in steel conduit

Twin and CPC copper cables using 70 °C PVC insulation

Cable size (R_1)	CPC size (R_2)	Ohms per 1000 metres for $R_1 + R_2$ at		
mm^2	mm^2	20 °C	70 °C	115 °C
1.0	1.0	36.2	43.44	49.96
1.5	1.0	30.2	36.24	41.68
2.5	1.5	19.51	23.41	26.92
4.0	1.5	16.71	20.05	23.06
6.0	2.5	10.49	12.59	14.48
10.0	4.0	6.44	7.73	8.89
16.0	6.0	4.23	5.08	5.84

Twin and CPC copper cables using 90 °C PVC or XLPE insulation

Cable size (R_1)	CPC size (R_2)	Ohms per 1000 metres for $R_1 + R_2$ at		
mm^2	mm^2	20 °C	90 °C	170 °C
1.0	1.0	36.2	46.34	57.92
1.5	1.0	30.2	38.66	48.32
2.5	1.5	19.51	24.97	31.22
4.0	1.5	16.71	21.39	26.74
6.0	2.5	10.49	13.43	16.78
10.0	4.0	6.44	8.24	10.30
16.0	6.0	4.23	5.41	6.77

Single core 70 °C PVC insulated copper cables in heavy gauge steel conduit

Cable size	Ohms per 1000 metres $R_1 + R_2$ in heavy gauge conduit at								
	20 °C Conduit size mm			70 °C Conduit size mm			115 °C Conduit size mm		
mm^2	20	25	32	20	25	32	20	25	32
1.0	22.80	21.30	20.10	26.42	24.92	23.72	29.68	28.18	26.98
1.5	16.80	15.30	14.10	19.22	17.72	16.52	21.40	19.90	18.70
2.5	12.11	10.61	9.41	13.59	12.09	10.89	14.93	13.43	12.23
4.0	9.31	7.81	6.61	10.23	8.73	7.53	11.06	9.56	8.36
6.0	7.78	6.28	5.08	8.40	6.90	5.70	8.95	7.45	6.25
10.0	6.53	5.03	3.83	6.90	5.40	4.20	7.23	5.73	4.53
16.0	5.85	4.35	3.15	6.08	4.58	3.38	6.29	4.79	3.59

CD 1

CABLE DIAMETERS

PVC insulated and PVC sheathed flat cables - BS 6004

Conductor area mm²	Single core mm	Two core mm	Three core mm	Twin and CPC mm	Three core and CPC mm
1*	4.5	4.7 × 7.4	4.7 × 9.8	4.7 × 8.6	4.7 × 11.0
1.5*	4.9	5.4 × 8.4	5.4 × 11.5	5.4 × 9.6	5.4 × 12.5
2.5*	5.8	6.2 × 9.8	6.2 × 13.5	6.2 × 11.5	6.2 × 14.5
4	6.8	7.2 × 11.5	7.4 × 16.5	7.2 × 13	7.4 × 18
6	7.4	8.0 × 13	8.0 × 18	8.0 × 15	8.0 × 20
10	8.8	9.6 × 16	9.6 × 22.5	9.6 × 19	9.6 × 25.5
16	10.5	11.0 × 18.5	11.0 × 26.6	11.0 × 22.5	11.0 × 29.5
25	12.5				
35	13.5	Notes: Dimensions approximate; *denotes solid conductors			

Multicore steel wire armoured cables up to 16 mm²

Number of cores	Cross-sectional area of conductor in mm²							
	1.5*	1.5	2.5*	2.5	4.0	6.0	10	16
	Diameter in mm PVC insulated PVC bedded SWA and PVC sheathed							
2	11.7	12.3	13.1	13.6	15.1	16.5	20.1	21.9
3	12.3	12.8	13.6	14.1	15.8	18.0	21.2	23.1
4	13.0	13.5	14.5	15.0	17.8	19.2	22.8	26.3
5	13.8	-	15.4	-	19.0	-	-	-
7	14.5	15.2	16.6	18.0	20.5	-	-	-
10	18.1	-	20.9	-	26.1	-	-	-
12	18.6	19.4	21.4	22.4	26.8	-	-	-
19	21.1	22.2	25.4	26.6	30.5	-	-	-
	Diameter in mm XLPE insulated LSF bedded SWA and LSF sheathed							
2	-	12.5	-	13.6	14.7	15.9	18.0	20.0
3	-	13.0	-	14.1	15.3	16.6	19.5	21.2
4	-	14.0	-	15.0	16.4	18.7	21.1	22.9
7	-	15.9	-	17.1	19.7	-	-	-
12	-	20.2	-	22.4	25.7	-	-	-
19	-	23.2	-	26.6	29.3	-	-	-
	Diameter in mm XLPE insulated PVC bedded SWA and PVC sheathed							
2	-	12.5	-	13.6	14.7	15.9	18.0	20.0
3	-	13.0	-	14.1	15.3	16.6	19.5	21.2
4	-	14.0	-	15.0	16.4	18.7	21.1	22.9
7	-	15.9	-	17.1	19.7	-	-	-
12	-	20.2	-	22.4	25.7	-	-	-
19	-	23.2	-	26.6	29.3	-	-	-

Notes: Dimensions approximate; *denotes solid conductors; PVC to BS 6346; LSF to BS 6724, XLPE to BS 5467

CABLE DIAMETERS

Single core copper power cables

Cable	Cross-sectional area of conductor in mm^2												
type	50	70	95	120	150	185	240	300	400	500	630	800	1000
	Approximate diameter in mm of PVC insulated copper cables - BS 6346												
Unarmoured	14.3	16.0	18.4	19.9	21.9	24.3	27.4	30.4	34.0	37.9	41.9	46.5	51.7
Aluminium armoured	18.3	20.2	22.4	25.2	27.0	29.4	32.5	35.2	40.2	44.9	47.9	54.1	59.0
	Approximate diameter in mm of XLPE insulated copper cables - BS 5467												
Unarmoured	13.6	15.6	17.5	19.2	21.3	23.4	26.3	29.0	32.7	36.6	41.3	46.3	51.3
Aluminium armoured	17.7	19.6	21.5	23.2	26.3	28.7	31.4	34.1	38.9	42.8	47.3	53.9	58.8

Single core aluminium power cables

Cable	Cross-sectional area of conductor in mm^2												
type	50	70	95	120	150	185	240	300	380	480	600	740	960	1200
	Approximate diameter in mm of PVC insulated aluminium cables - BS 6346												
Unarmoured	13.8	15.4	17.7	19.0	21.0	23.3	26.1	28.9	32.4	35.7	38.7	42.2	47.4	52.0
Aluminium armoured	17.8	19.6	21.7	24.3	26.1	28.3	31.2	33.7	38.4	41.7	44.7	49.6	54.9	59.7
	Approximate diameter in mm of XLPE insulated aluminium cables PVC sheathed - BS 5467												
Unarmoured	13.1	14.9	16.8	18.3	20.3	22.4	25	27.5	30.8	34.2	37.6	41.7	46.9	52.0
Aluminium armoured	17.2	18.9	20.8	22.3	25.4	27.6	30.1	32.6	37.1	40.4	43.8	49.1	54.4	59.7

Multicore armoured power cables

Cable	Cross-sectional area in mm^2											
type	16	25	35	50	70	95	120	150	185	240	300	400
	Approximate diameter in mm of PVC insulated copper conductor SWA cables - BS 6346											
Two core	21.9	23.0	24.8	27.2	29.5	34.4	37.1	40.2	45.1	50.5	55.4	60.8
Three core	23.1	25.1	27.3	30.5	34.8	39.1	41.9	47.2	51.4	57.3	62.6	68.8
Four core	26.3	27.5	30.0	34.8	38.4	43.3	48.1	52.3	57.5	63.9	69.9	78.8
	Approximate diameter in mm of XLPE insulated copper conductor SWA cables - BS 5467											
Two core	20.6	24.3	27.9	25.3	28.1	31.9	35.1	38.2	43.3	48.3	52.5	58.0
Three core	21.8	26.9	29.6	28.4	32.0	36.7	40.0	45.1	49.3	54.5	59.5	65.8
Four core	23.6	29.1	32.1	31.3	36.8	40.6	45.8	49.9	54.9	61.0	66.5	75.4
	Approximate diameter in mm of PVC insulated aluminium conductor SWA cables - BS 6346											
Two core	20.9	21.6	23.2	25.8	28.1	32.9	-	-	-	-	-	-
Three core	22.0	24.1	26.1	29.1	33.2	37.2	39.8	44.9	48.8	54.3	59.3	-
Four core	25.0	26.7	28.9	33.6	36.9	41.5	46.0	50.0	54.9	60.9	66.6	-

PCA 1

Protective conductor area

Table M 54 G

Minimum size of CPC when phase conductor is copper and CPC is a dissimilar material				
Type of insulation CPC conductor in contact with	Type of conductor material used for cpc	Cross-sectional area 'S' of **copper** phase conductor		
		Up to 16 mm^2	16 to 35 mm^2	Over 35 mm^2
		Cross-sectional area required for CPC		
70 °C P.V.C.	Aluminium sheath or armour	1.24 S	19.78	0.62 S
	Steel wire armour	2.25 S	36.08	1.13 S
	Steel conduit or trunking	2.45 S	39.15	1.22 S
	lead sheath	4.42 S	70.77	2.21 S
90 °C P.V.C.	Aluminium sheath or armour	1.18 S	18.82	0.59 S
	Steel wire armour	2.17 S	34.78	1.09 S
	Steel conduit or trunking	2.27 S	36.36	1.14 S
	lead sheath	4.35 S	69.57	2.17 S
85 °C Rubber	Aluminium sheath or armour	1.44 S	23.05	0.72 S
	Steel wire armour	2.63 S	42.04	1.31 S
	Steel conduit or trunking	2.48 S	39.70	1.24 S
	lead sheath	5.15 S	82.46	2.58 S
90 ° C Thermosetting XLPE	Aluminium sheath or armour	1.68 S	26.92	0.84 S
	Steel wire armour	3.11 S	49.74	1.55 S
	Steel conduit or trunking	2.47 S	39.45	1.23 S
	lead sheath	6.22 S	99.48	3.11 S

Area of C.P.Cs, conduit, cable armouring and trunking

Area of C. P. C. in twin and CPC cable		Cross-sectional area of steel conduit			Cross-sectional area steel trunking BS 4768: Part 1	
Cable size mm^2	C. P. C. size mm^2	Conduit size mm	Area mm^2		Size mm	Area mm^2
			Light gauge	Heavy gauge		
1.0	1.0	16	47	75	50 × 37.5	123
					50 × 50	148
1.5	1.0	20	59	107	75 × 50	207.12
					75 × 75	267.12
2.5	1.5	25	89	132		
4.0	1.5				100 × 50	237.12
		32	116	167	100 × 75	297.12
6.0	2.5	38	-	186	100 × 100	416.08
10	4.0				150 × 50	346.08
					150 × 75	416.08
16	6.0				150 × 100	554.88
					150 × 150	714.88

Area of cable armour or MICC sheath
See tables ELI 1, ELI 2, ELI 3, ELI 4 and ELI 5

SPF 1

SINE OF POWER FACTOR

Conversion table - Power Factor to Sines

Power factor	Sine	Radians	Degrees	Power factor	Sine	Radians	Degrees
0.10	0.9950	1.4706	84.26	0.55	0.8352	0.9884	56.63
0.11	0.9939	1.4606	83.68	0.56	0.8285	0.9764	55.94
0.12	0.9928	1.4505	83.11	0.57	0.8216	0.9643	55.25
0.13	0.9915	1.4404	82.53	0.58	0.8146	0.9521	54.55
0.14	0.9902	1.4303	81.95	0.59	0.8074	0.9397	53.84
0.15	0.9887	1.4202	81.37	0.60	0.8000	0.9273	53.13
0.16	0.9871	1.4101	80.79	0.61	0.7924	0.9147	52.41
0.17	0.9854	1.4000	80.21	0.62	0.7846	0.9021	51.68
0.18	0.9837	1.3898	79.63	0.63	0.7766	0.8892	50.95
0.19	0.9818	1.3796	79.05	0.64	0.7684	0.8763	50.21
0.20	0.9798	1.3694	78.46	0.65	0.7599	0.8632	49.46
0.21	0.9777	1.3592	77.88	0.66	0.7513	0.8500	48.70
0.22	0.9755	1.3490	77.29	0.67	0.7424	0.8366	47.93
0.23	0.9732	1.3387	76.70	0.68	0.7332	0.8230	47.16
0.24	0.9708	1.3284	76.11	0.69	0.7238	0.8093	46.37
0.25	0.9682	1.3181	75.52	0.70	0.7141	0.7954	45.57
0.26	0.9656	1.3078	74.93	0.71	0.7042	0.7813	44.77
0.27	0.9629	1.2974	74.34	0.72	0.6940	0.7670	43.95
0.28	0.9600	1.2870	73.74	0.73	0.6834	0.7525	43.11
0.29	0.9570	1.2766	73.14	0.74	0.6726	0.7377	42.27
0.30	0.9539	1.2661	72.54	0.75	0.6614	0.7227	41.41
0.31	0.9507	1.2556	71.94	0.76	0.6499	0.7075	40.54
0.32	0.9474	1.2451	71.34	0.77	0.6380	0.6920	39.65
0.33	0.9440	1.2345	70.73	0.78	0.6258	0.6761	38.74
0.34	0.9404	1.2239	70.12	0.79	0.6131	0.6600	37.81
0.35	0.9367	1.2132	69.51	0.80	0.6000	0.6435	36.87
0.36	0.9330	1.2025	68.90	0.81	0.5864	0.6266	35.90
0.37	0.9290	1.1918	68.28	0.82	0.5724	0.6094	34.92
0.38	0.9250	1.1810	67.67	0.83	0.5578	0.5917	33.90
0.39	0.9208	1.1702	67.05	0.84	0.5426	0.5735	32.86
0.40	0.9165	1.1593	66.42	0.85	0.5268	0.5548	31.79
0.41	0.9121	1.1483	65.80	0.86	0.5103	0.5355	30.68
0.42	0.9075	1.1374	65.17	0.87	0.4931	0.5156	29.54
0.43	0.9028	1.1263	64.53	0.88	0.4750	0.4949	28.36
0.44	0.8980	1.1152	63.90	0.89	0.4560	0.4735	27.13
0.45	0.8930	1.1040	63.26	0.90	0.4359	0.4510	25.84
0.46	0.8879	1.0928	62.61	0.91	0.4146	0.4275	24.49
0.47	0.8827	1.0815	61.97	0.92	0.3919	0.4027	23.07
0.48	0.8773	1.0701	61.31	0.93	0.3676	0.3764	21.57
0.49	0.8717	1.0587	60.66	0.94	0.3412	0.3482	19.95
0.50	0.8660	1.0472	60.00	0.95	0.3122	0.3176	18.19
0.51	0.8602	1.0356	59.34	0.96	0.2800	0.2838	16.26
0.52	0.8542	1.0239	58.67	0.97	0.2431	0.2456	14.07
0.53	0.8480	1.0122	57.99	0.98	0.1990	0.2003	11.48
0.54	0.8417	1.0004	57.32	0.99	0.1411	0.1415	8.11

TCF 1

Temperature Conversion Factors

Factors for converting the resistance of copper or aluminium from 20 °C
to the average of: actual conductor operating temperature + limit temperature of the
conductor's insulation

PVC insulated copper or aluminium conductor				XLPE insulated copper or aluminium conductors			
Conductor operating temperature in °C	Multiply resistance at 20 °C by	Conductor operating temperature in °C	Multiply resistance at 20 °C by	Conductor operating temperature in °C	Multiply resistance at 20 °C by	Conductor operating temperature in °C	Multiply resistance at 20 °C by
30	1.300	61	1.362	30	1.480	61	1.542
31	1.302	62	1.364	31	1.482	62	1.544
32	1.304	63	1.366	32	1.484	63	1.546
33	1.306	64	1.368	33	1.486	64	1.548
34	1.308	65	1.370	34	1.488	65	1.550
35	1.310	66	1.372	35	1.490	66	1.552
36	1.312	67	1.374	36	1.492	67	1.554
37	1.314	68	1.376	37	1.494	68	1.556
38	1.316	69	1.378	38	1.496	69	1.558
39	1.318	70	1.380	39	1.498	70	1.560
40	1.320	71	1.382	40	1.500	71	1.562
41	1.322	72	1.384	41	1.502	72	1.564
42	1.324	73	1.386	42	1.504	73	1.566
43	1.326	74	1.388	43	1.506	74	1.568
44	1.328	75	1.390	44	1.508	75	1.570
45	1.330	76	1.392	45	1.510	76	1.572
46	1.332	77	1.394	46	1.512	77	1.574
47	1.334	78	1.396	47	1.514	78	1.576
48	1.336	79	1.398	48	1.516	79	1.578
49	1.338	80	1.400	49	1.518	80	1.580
50	1.340	81	1.402	50	1.520	81	1.582
51	1.342	82	1.404	51	1.522	82	1.584
52	1.344	83	1.406	52	1.524	83	1.586
53	1.346	84	1.408	53	1.526	84	1.588
54	1.348	85	1.410	54	1.528	85	1.590
55	1.350	86	1.412	55	1.530	86	1.592
56	1.352	87	1.414	56	1.532	87	1.594
57	1.354	88	1.416	57	1.534	88	1.596
58	1.356	89	1.418	58	1.536	89	1.598
59	1.358	90	1.420	59	1.538	90	1.600
60	1.360	91	1.422	60	1.540	91	1.602

Note: Factors will give the resistance at the average of: conductor operating temperature plus the limit temperature for the conductor's insulation.; e.g., PVC cable operating temperature 70 °C: factor from table gives resistance at 115 °C.

Transformer impedance data

kVA rating	Resistance per phase when cold 15 / 20 °C Ω per phase	Resistance per phase fully loaded 75 °C Ω per phase	Reactance Ω per phase
Resistance & reactance in Ω of 11 kV/415V Class T1 transformers referred to the 415V system			
200	0.012	0.014	0.0423
315	0.0071	0.0085	0.027
500	0.0038	0.0045	0.0173
800	0.0022	0.0027	0.011
1000	0.0016	0.0019	0.0087
1500	0.00099	0.0012	0.0068
2000	0.00079	0.001	0.0056

My thanks to John Boulton GEC Transformers for the above typical values of transformer data

Fuse data

Breaking capacity of fuses

Type	Voltage	Breaking capacity kA
BS 3036 S1 duty	240V	1.0
S2 duty	240V	2.0
S3 duty	240V	4.0
BS 1361 Type I	240V	16.5
Type II	415V	33.0
BS1362	250V	6.0
BS 88 Type 2	250V	40.0
Type 2	415V	80.0

BS 3871 MCB data

Breaking capacity of BS 3871 Part 1 MCBs

Category of duty	Rated breaking capacity kA	Test power factor
M1	1.0	0.85 - 0.9
M1.5	1.5	0.80 - 0.85
M2	2.0	0.75 - 0.8
M3	3.0	0.75 - 0.8
M4	4.0	0.75 - 0.8
M6	6.0	0.75 - 0.8
M9	9.0	0.55 - 0.6

Thermal Insulation factors

Length in mm conductor enclosed in insulation	De-rating factor
50	0.89
100	0.81
200	0.68
400	0.55
500 plus	0.50

Correction factor for BS 3036 fuses when used for overload protection (S)

$$I_t = \text{Fuse rating} \div 0.725$$

Temperature conversion

°C	°F	°C	°F	°C	°F	°C	°F
0	32	35	95	60	140	85	185
10	50	40	104	60.8	141.44	90	194
15	59	47.66	117.79	65.2	149.36	105	221
20	68	50	122	68.4	155.12	115	239
25	77	55.6	132.08	70	158	160	320
30	86	57.53	135.55	80	176	170	338

MISC 2

Resistor colour code

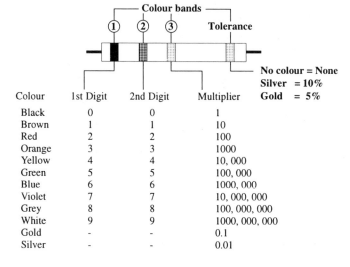

Colour	1st Digit	2nd Digit	Multiplier
Black	0	0	1
Brown	1	1	10
Red	2	2	100
Orange	3	3	1000
Yellow	4	4	10, 000
Green	5	5	100, 000
Blue	6	6	1000, 000
Violet	7	7	10, 000, 000
Grey	8	8	100, 000, 000
White	9	9	1000, 000, 000
Gold	-	-	0.1
Silver	-	-	0.01

No colour = None
Silver = 10%
Gold = 5%

Conversion Factors

Length

1 inch	=	1000 mils	1mm	=	39.37 mils
1 inch	=	25.4 mm	1mm	=	0.03937 inches
1 foot	=	304.8 mm	1 centimetre	=	0.3937 inches
1 yard	=	914.4 mm	1 metre	=	1.094 yards
1 yard	=	0.9144 m	1 metre	=	3.282 ft
1 mile	=	1 609.34m	1 kilometre	=	0.621373 miles

Area

1 sq.inch	=	645.16 sq.mm	1 sq.mm	=	0.00155 sq.inches
1 sq.ft	=	0.092903 sq.m	1 sq.m	=	10.764 sq. ft
1 sq.yard	=	0.8361 sq.m	1 sq.m	=	1.196 sq.yards
1 acre	=	4 840 sq.yards	1 acre	=	4046.86 sq.m
1 acre	=	2.471 Hectares	1 Hectares	=	0.4046.86 acres

Liquid

1 lb water	=	0.454 litres	1 ltres	=	2.202 lbs
1 pint	=	0.568 litres	1 litre	=	1.76 pints
1 gallon	=	4.546 litres	1 litre	=	0.22 gallons

Horse-power

1 h.p.	=	746 watts	1 watt	=	0.00134 h.p.
1 h.p	=	0.746 kW	1 kW	=	1.34 h.p.
1 h.p.	=	33 000 ft.lb./min	1 h.p.	=	0.759 kg.m/s

Weight

1 lb	=	0.4536 kg	1 kg	=	2.205 lbs
1 Cwt	=	112 lbs	20 Cwt	=	1 Ton (English)
1 Cwt	=	50.8 kg	1 kg	=	0.01968 Cwt
1 Ton	=	1 016.1 kg	1 kg	=	0.000984 Ton

MISC 3

Conversion factors continued

1 kWh	= 3415 BthU	= 860 kilogramme calories	= 3.6 Mega Joules
1 BthU	= 252 calories	= 778 ft lbf	= 1055 joules
1 joule	= 0.239 calories	= 1 watt second	= 0.737 ft lbf
1 gallon	= 4.546 Litres	= 276.9 cu inches	= 10 lb pure water
1 cu ft	= 28.3 litres	= 6.23 gallons	= 62.23 lb pure water
1 cu metre	= 1.308 cu yards	1 kWh	= 2.6552 x 10^6 ft.lbf
1 Newton	= 9.807 kgf	1 litre pure water	= 1 kg

Mathematical

Laws of indices

$$a^m \times a^n = a^{m+m} \ : \ a^m \div a^n = a^{m-n} \ : \ (a^m)^n = a^{mn} \ : \ a^{-n} = \ ^1/a^n \ : \ a^{p/q} = \ ^q\sqrt{a^p}$$

Note: $1^n = 1$, and $0^n = 0$: $a^0 = 1$ except when $a = 0$

$\pi = 3.142$: Circumference of circle = Diameter $\times \pi$: Area of circle = πr^2 or $\pi D^2 \div 4$

Water

Capacity of containers

$$\text{Rectangular container capacity in gallons} = \frac{L \times B \times H \, (cm)}{4546}$$

$$\text{Rectangular container capacity in litres} = \frac{L \times B \times H \, (cm)}{1000}$$

$$\text{Cylrindical container capacity in gallons} = \frac{D^2 \times H \, (cm)}{5788}$$

$$\text{Cylrindical container capacity in litres} = \frac{D^2 \times H \, (cm)}{1273}$$

Power required and time taken

$$\text{Time taken in minutes to heat water} = \frac{\text{Litres} \times \text{Temperature rise } ^\circ C}{14.33 \times kW \times \text{Efficiency}}$$

$$\text{Loading required in kW to heat water} = \frac{\text{Litres} \times \text{Temperature rise } ^\circ C}{14.33 \times \text{minutes} \times \text{Efficiency}}$$

Efficiency is approximately 0.9 for a lagged tank, and 0.8 for an unlagged tank.

Three phase formula

$$\text{h.p.} = \frac{kW \times \text{Efficiency}}{0.746} = \frac{kVA \times \text{Efficiency} \times p.f.}{0.746} = \frac{I_L \times V_L \times \text{Efficiency} \times 1.732 \times p.f.}{746}$$

$$I_L = \frac{kW \times 1000}{V_L \times 1.732 \times p.f.} = \frac{kVA \times 1000}{V_L \times 1.732} = \frac{\text{h.p.} \times 746}{V_L \times 1.732 \times \text{Efficiency} \times p.f.}$$

$$kW = \frac{\text{h.p.} \times 746}{1000 \times \text{Efficiency}} = \frac{I_L \times V_L \times 1.732 \times p.f.}{1000} = kVA \times p.f.$$

where V_L = Line volts, I_L = Line amperes, p.f. = Power factor and h.p. = Horse power.

MISC 4

Lighting

The lumen method is usually used for determining the amount of illumination required for a given area.

$$\text{Total lumens required [F]} = \frac{A \times E_{av}}{CU \times M}$$

Where A   is the Area to be illuminated
E_{av} is the average illumination required on the working plane
CU   is the coefficient of utilisation
M   is the maintenance factor, usually taken as 0.8, but reduced to 0.6 in dirty areas

The average illumination required on the working plane is taken from a table giving the recommended values of illumination for different conditions. Some typical values are :-

Area	lux	Approximate lm/ft²
General office	500	50
Drawing office boards	750	75
Store rooms, corridors	300	30
Shop counters	500	50
Business machine operation	750	75
Watch repair, and fine soldering	3000	300
Proof reading- print works	750	75
Living rooms, halls and landings	100	10
Bedrooms	50	5
Reading	150/300	15/30

To select the coefficient of utilisation from the manufacturer's data, the room index is required:

$$\text{Room index} = \frac{L \times W}{H_m (L \times W)}$$

where L is the length of the room, W the width and H_m is the mounting height of the lighting fitting above the working plane.Where the working plane is a desk top or work bench, the working plane is taken to be 0.85m above the floor. The coefficient of utilisation is then selected from the manufacturer's design data. Having worked out the total illumination required, the design lumens of the lamp is chosen, and the number of lighting points calculated.

$$\text{Total number of lamps} = \frac{\text{Illumination required [F]}}{\text{Design lumens per lamp}}$$

To ensure even illumination, fittings are arranged so that the space to mounting height ratio does not exceed 1.5:1; this means that the spacing between fittings shall not exceed $1.5H_m$. Glare index calculations may also be required.

Conversion factors

1 lumen/ sq.ft (Foot-lambert) = 10.76 Lux or 10.76 apostilb, or 3.426 cd/m^2, or 0.00221 cd/in^2

Heating calculations

Heat lost from a building is the product of the area of the surface, its thermal transmission coefficient, and the temperature difference between the inside and the outside temperatures.

$$Q_f = AU(t_1 - t_2)$$

where Q_f is the heat lost through the fabric of the building in Watts, A is the area of the surface in square metres, U is the thermal transmission coefficient W/m^2 °C, t_1 is the inside temperature required and t_2 is the outside temperature. Allowance must also be made for the heat lost through floors and ceilings. The power required can be determined approximately by allowing 15 watts per square foot to give a temperature of 20 °C (70 °F), with an outside temperature of 0 °C (32 F).

BS 3036 Fuse Characteristics
(Courtesy of WYLEX Ltd)

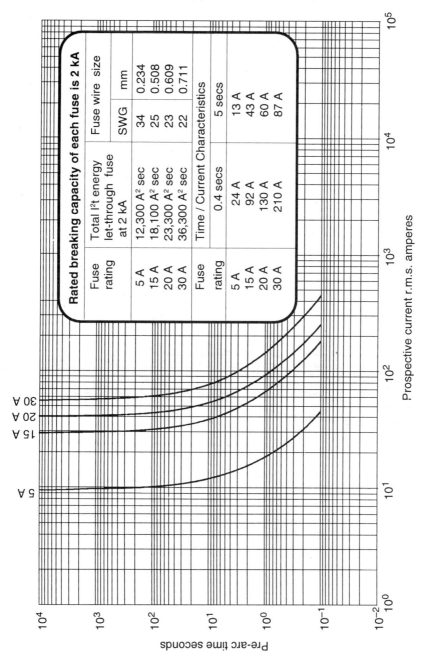

Rated breaking capacity of each fuse is 2 kA

Fuse wire size		
	SWG	mm
	34	0.234
	25	0.508
	23	0.609
	22	0.711

Fuse rating	Total I²t energy let-through fuse at 2 kA
5 A	12,300 A² sec
15 A	18,100 A² sec
20 A	23,300 A² sec
30 A	36,300 A² sec

Time / Current Characteristics

Fuse rating	0.4 secs	5 secs
5 A	24 A	13 A
15 A	92 A	43 A
20 A	130 A	60 A
30 A	210 A	87 A

Prospective current r.m.s. amperes

Pre-arc time seconds

BS 1361 Type 1 Cartridge Fuses - Domestic type

(Courtesy of Wylex Ltd)

BS 1361 characteristics		
Fuse Rating	Current for time	
	0.4 s	5s
5A	21.7A	14.2A
10A	44.9A	30.7A
15A	71.3A	48.5A
20A	133 A	83.7A
30A	205A	125A
35A	262A	158A
40A	336A	201A
45A	395A	237A

Note: Protective device ratings 10A, 35A, and 40A are not in BS 1361. They are manufactured to the same requirements but then tested and certified to BS 88.

Pre-arc time seconds

Prospective current r.m.s. amperes

Type B MCBs to BS EN 60898
(Courtesy of Crabtree Electrical Industries Ltd)

Typical values of I²t Energy Let-through for Types B, C and D Polestar MCBs

MCB rating	Total I²t Let-through in A²sec for breaking capacity		
	6 kA	10 kA	16 kA
6 A	8000	12000	16000
10 A	15000	25000	34000
16 A	20000	35000	53000
20 A	25000	40000	64000
32 A	28000	50000	70000
40 A	30000	65000	85000
50 A	36000	75000	110000
63 A	36000	75000	110000

See page 461 for I²t characteristics

Type B MCB typical tripping times

MCB Rating	Current required in amperes to operate MCB in				
	0.02 S	0.1 S	0.4 S	5 S	60 S
6 A	24	24	24	24	12
10 A	40	40	40	40	19
16 A	64	64	64	64	30
20 A	80	80	80	80	40
32 A	128	128	128	128	67
40 A	160	160	160	160	80
50 A	200	200	200	200	100
63 A	252	252	252	252	138

Crabtree Polestar Type B - Typical time current characteristics

476

Type C MCBs to BS EN 60898
(Courtesy of Crabtree Electrical Industries Ltd)

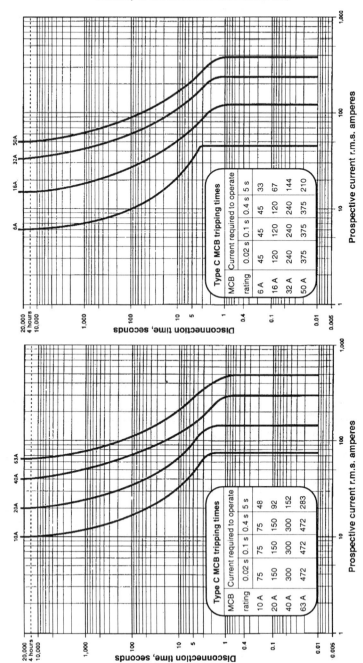

Crabtree Polestar Type C - Typical time current characteristics

Type C MCB tripping times

MCB rating	Current required to operate			
	0.02 s	0.1 s	0.4 s	5 s
6 A	45	45	45	33
16 A	120	120	120	67
32 A	240	240	240	144
50 A	375	375	375	210

Type C MCB tripping times

MCB rating	Current required to operate			
	0.02 s	0.1 s	0.4 s	5 s
10 A	75	75	75	48
20 A	150	150	150	92
40 A	300	300	300	152
63 A	472	472	472	283

Notes: For the typical values of I^2t Energy Let-through of Type C MCBs, see page 459 and for the characteristics see page 461.

Type D MCB to BS EN 60898
(Courtesy of Crabtree Electrical Industries Ltd)

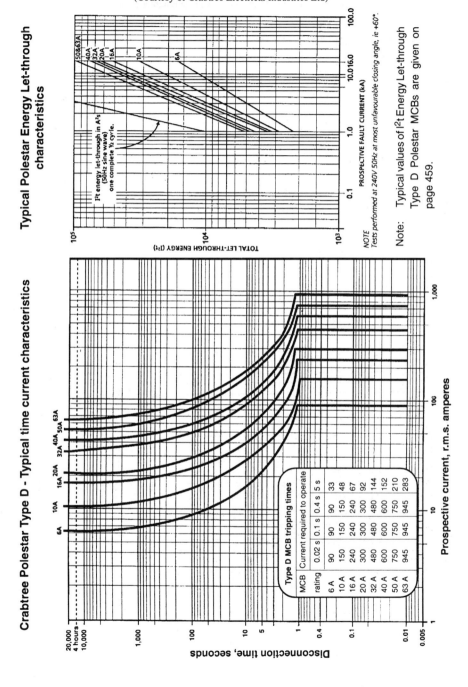

Crabtree Polestar Type D - Typical time current characteristics

Typical Polestar Energy Let-through characteristics

Type D MCB tripping times				
MCB	Current required to operate			
rating	0.02 s	0.1 s	0.4 s	5 s
6 A	90	90	90	33
10 A	150	150	150	48
16 A	240	240	240	67
20 A	300	300	300	92
32 A	480	480	480	144
40 A	600	600	600	152
50 A	750	750	750	210
63 A	945	945	945	283

NOTE
Tests performed at 240V 50Hz at most unfavourable closing angle, ie +60°.

Note: Typical values of I²t Energy Let-through Type D Polestar MCBs are given on page 459.

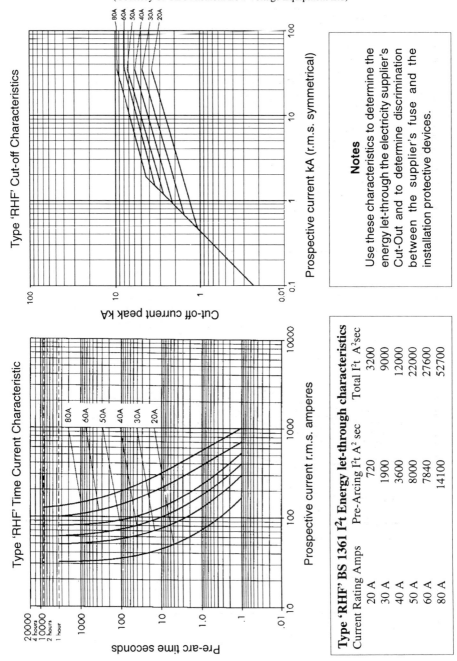

BS 1361 Type 1 fuses used in RECs Service Cut-outs for Domestic Premises

(Courtesy of GEC Alsthom Low Voltage Equipment Ltd)

Notes

Use these characteristics to determine the energy let-through the electricity supplier's Cut-Out and to determine discrimination between the supplier's fuse and the installation protective devices.

Type 'RHF' BS 1361 I²t Energy let-through characteristics

Current Rating Amps	Pre-Arcing I²t A² sec	Total I²t A²sec
20 A	720	3200
30 A	1900	9000
40 A	3600	12000
50 A	8000	22000
60 A	7840	27600
80 A	14100	52700

BS 1361 Type 2 fuses used in RECs Service Cut-Outs for Industrial Premises
(Courtesy of GEC Alsthom Low Voltage Equipment Ltd)

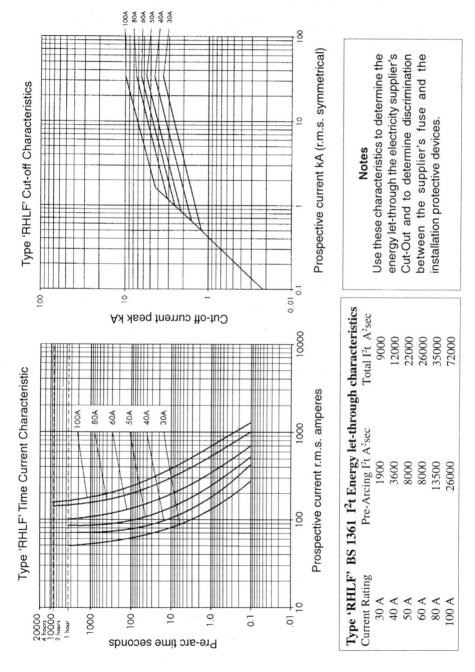

Notes

Use these characteristics to determine the energy let-through of the electricity supplier's Cut-Out and to determine discrimination between the supplier's fuse and the installation protective devices.

Type 'RHLF' BS 1361 I²t Energy let-through characteristics

Current Rating	Pre-Arcing I²t A²sec	Total I²t A²sec
30 A	1900	9000
40 A	3600	12000
50 A	8000	22000
60 A	8000	26000
80 A	13500	35000
100 A	26000	72000

2 - 20 Amp BS 88 Type NIT Time/ Current Characteristics
(Courtesy GEC Alsthom Low Voltage Equipment Ltd)

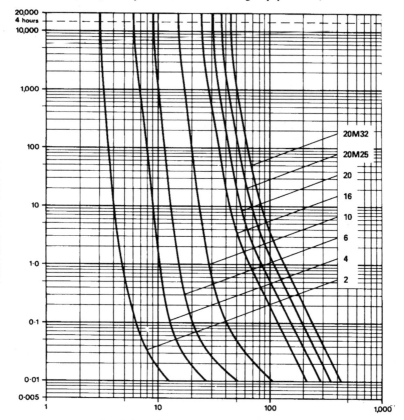

R.M.S. SYMMETRICAL PROSPECTIVE CURRENT IN AMPERES

Type NIT Cut-off Current Characteristic

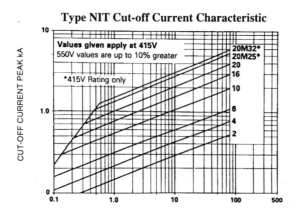

PROSPECTIVE CURRENT kA (R.M.S. SYMMETRICAL)

2 - 63 Amp BS 88 Part 2   Time/Current Characteristics
(Courtesy GEC Alsthom Low Voltage Equipment Ltd)

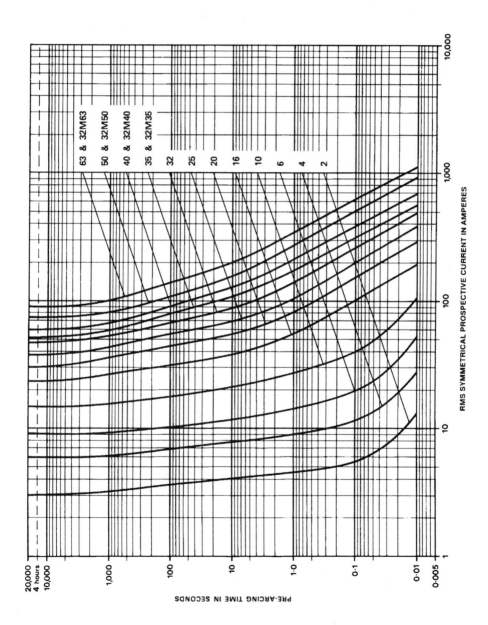

80 - 1250 Amp BS 88 Part 2 Time/Current Characteristics
(Courtesy GEC Alsthom Low Voltage Equipment Ltd)

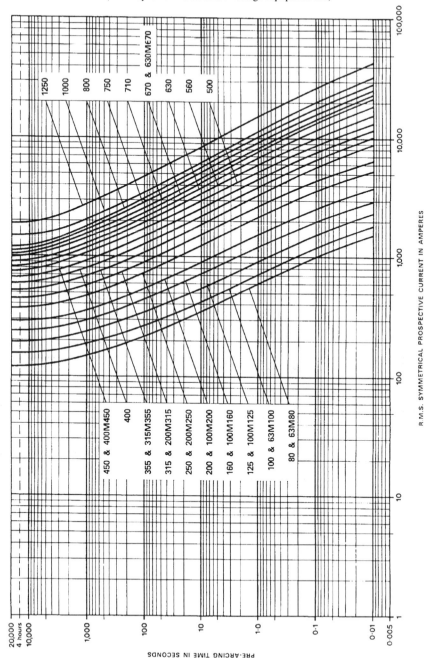

THIS IS THE LAW

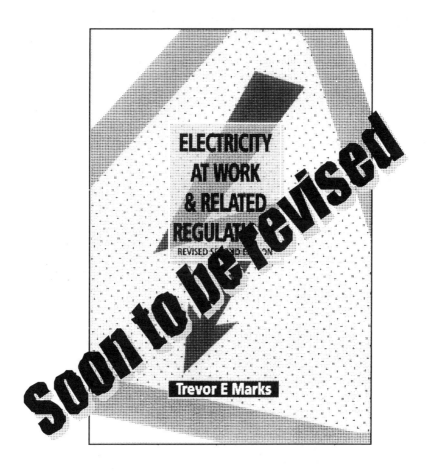

ELECTRICITY AT WORK & RELATED REGULATIONS

REVISED SECOND EDITION

Trevor E Marks

Soon to be revised

A Handbook for Compliance

William Ernest Publishing Limited

P O Box 206, Liverpool L69 4PZ

484

2 - 1250 Amp BS 88 Part 2 Cut-off Characteristics
(Courtesy GEC Alsthom Low Voltage Equipment Ltd)

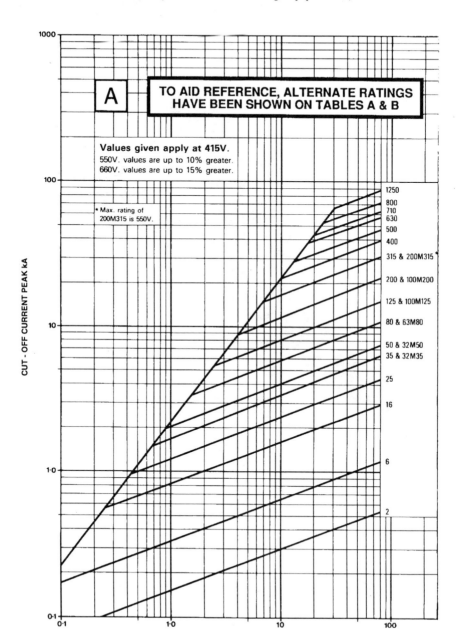

PROSPECTIVE CURRENT kA (R.M.S. SYMMETRICAL)

2 - 1250 Amp BS 88 Part 2 Cut-off Characteristics
(Courtesy GEC Alsthom Low Voltage Equipment Ltd)

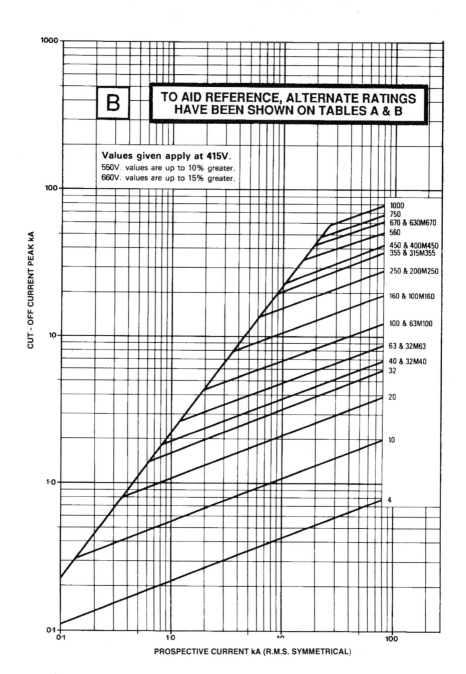

ENERGY LET-THROUGH HRC FUSE

2 to 1250 Amp BS 88 I^2t Fuse Characteristics (Courtesy GEC Alsthom)

Type of fuse	Current rating of fuse in amps	Pre-Arcing I^2t (A^2sec × 1,000)	Total I^2t (A^2 sec × 1,000)		
			415V	550V	660V
		A^2s	A^2s	A^2s	A^2s
NIT	2	0.0022	0.0054	0.031	-
NIT	4	0.0072	0.018	0.07	-
NIT	6	0.021	0.06	0.4	-
NIT	10	0.10	0.28	1.00	-
NIT	16	0.30	0.85	2.00	-
NIT	20	0.54	1.00	2.50	-
NIT	20M25	0.90	3.00	-	-
NIT	20M32	1.10	4.00	-	-
T	2	0.0022	0.0055	0.0074	0.015
T	4	0.007	0.0185	0.023	0.05
T	6	0.021	0.06	0.08	0.15
T	10	0.1	0.28	0.37	0.7
T	16	0.25	0.55	0.74	1.8
T	20	0.54	1.1	1.4	2.5
T	25	0.85	1.85	2.3	3.7
T	32	1.6	3.4	5.4	8.7
T	35 & 32M35	2.7	5.3	8	15
T	40 & 32M40	4	8.5	11	20.5
T	50 & 32M50	6.3	13.5	18.5	28
T	63 & 32M63	11	24	36	50
T	80 & 63M80	14	40	52	66
T	100 & 63M100	17	60	80	100
T	125 & 100M125	25	85	110	140
T	160 & 100M160	62	160	210	270
T	200 & 100M200	105	260	330	430
T	250 & 200M250	200	550	700	870
T	315 & 200M315*	300	800	1050	1350
T	400	640	1800	2500	3000
T	450 & 400M450	800	2200	3000	3800
T	500	1050	3000	3800	4500
T	560	1400	3800	4250	5400
T	630	2000	5200	6000	7500
T	670 & 630M670	2400	6400	7400	9000
T	710	2800	7000	8000	9700
T	750	3700	7500	10000	12000
T	800	4400	9600	12500	15000
T	1000	5300	12000	14500	17500
T	1250	10000	20000	24000	29000

*Maximum rating of 200M315 fuse is 550V.